THE BALTIMORE CASE

ALSO BY DANIEL J. KEVLES

The Physicists: The History of a Scientific
Community in Modern America

In the Name of Eugenics: Genetics and the
Uses of Human Heredity

The Code of Codes: Scientific and Social Issues in
the Human Genome Project (co-editor)

THE

BALTIMORE

CASE

A Trial of Politics, Science, and Character

DANIEL J. KEVLES

W. W. NORTON & COMPANY

NEW YORK · LONDON

In Memory of My Father, David Kevles,
and for
Michael and Joel,
the Next Generation

Portions of this work originally appeared in *The New Yorker.*

For information about permission to reproduce selections from this book, write to Permissions,
W. W. Norton & Company, Inc., 500 Fifth Avenue, New York, NY 10110.
The text of this book is composed in Electra
with the display set in Centaur
Composition by Binghamton Valley Composition
Manufacturing by Maple-Vail Book Manufacturing Group
Book design by Jo Anne Metsch

Library of Congress Cataloging-in-Publication Data

Kevles, Daniel J.
The Baltimore case : a trial of politics, science, and character /
by Daniel J. Kevles.
p. cm.
Includes bibliographical references and index.
ISBN 0-393-04103-4
1. Cellular immunity—Research—Moral and ethical aspects.
2. Fraud in science. 3. Baltimore, David. 4. O'Toole, Margot.
5. Imanishi-Kari, Thereza. I. Title.
QR185.5.K48 1998
364.16'3—dc21   97-51774
                                    CIP

ISBN 0-393-31970-9 pbk.

W. W. Norton & Company, Inc., 500 Fifth Avenue, New York, N.Y. 10110
www.wwnorton.com

W. W. Norton & Company Ltd., 10 Coptic Street, London WC1A 1PU

1 2 3 4 5 6 7 8 9 0

Contents

Preface

DAVID BALTIMORE won the Nobel Prize for physiology or medicine in 1975, when he was a thirty-seven-year-old professor of biology at the Massachusetts Institute of Technology (M.I.T.). Baltimore had been known among biologists as a wunderkind for some time. The work for which he shared his prize, a study of how a special class of viruses reproduce themselves (the AIDS virus was later shown to be one of them), ran contrary to most contemporary beliefs on the subject. After receiving the Nobel Prize, he continued doing research, but he also began to take a leading role in public debate about genetic engineering, the AIDS epidemic, and other issues over which science and public policy meet. He brought to whatever he did a degree of self-confidence that some of his colleagues called arrogance but that was integral to his achievements. In 1990, when he was fifty-two, he became president of Rockefeller University, one of the world's distinguished centers of teaching and research.

Eighteen months after he went to Rockefeller, David Baltimore fell from grace. He resigned, citing pressure from his colleagues and the personal toll of fighting a long battle over what was alleged to be a fraudulent research paper that he had collaborated on when he was at M.I.T. A front-page article in the *New York Times* noted that the "spectacle" of Baltimore's downfall made it seem "larger than life, with an effect greater than any case of scientific fraud in memory."*

David Baltimore was never suspected of faking anything himself, but he had stubbornly defended the work of someone who was—a biomedical scientist at M.I.T. named Thereza Imanishi-Kari. She was one of six coauthors of the disputed paper, which reported on an experiment in immunology and was published in the journal *Cell* in 1986. Baltimore's support of her work was perceived to be unprofessional and unwise, if not irresponsible. He was the senior author of the paper, and because of the notoriety of his involvement, the affair became popularly known as "the Baltimore case." David Baltimore went back to M.I.T., resuming his professorship of biology. He continued to do brilliant work, but he was dishonored as a public figure.

From its inception, the Baltimore case piqued my interest as a student of the affairs of science in American society. At first, I had no intention of writing about it. I knew none of the principals until 1991, when I first met David Baltimore in another connection, and so far as I could tell from the press coverage, Imanishi-Kari seemed guilty and Baltimore foolhardy in defending her so vigorously. The Baltimore case seemed to touch deep-seated doubts about the scientific enterprise. Many people thought it high time that scientists answered to the public that in large part pays their bills, and I thought they had a point in demanding an enlargement of accountability.

However, others—a minority, to be sure—considered Baltimore and Imanishi-Kari victims, unfairly pursued by witch-hunting zealots ignorant of the way science works. The case dragged on for a decade, leaving wrecked careers in its wake, pitting congressmen against scientists, and producing both martyrs and tormentors. I had been wondering how and why scientific fraud and misconduct had emerged as an issue in the United States during the 1980s, when it was of little consequence in any other scientifically vital nation. The sustained ferocity of the case in and out of the media prompted me to suspect that an analysis of it might throw some light on science in late-twentieth-century American society and would be revealing in and of itself. I began looking into the Baltimore case, suspending judgment on questions of guilt or innocence as well as foolhardiness or courage until I had mastered the facts on my own.

There was plenty to look into. The case had been covered in numerous newspaper and magazine articles, probed in several congressional hearings,

---

*Philip J. Hilts, "Nobelist Caught Up in Fraud Case Resigns as Head of Rockefeller U.," *New York Times*, Dec. 3, 1991, p. 1

and exposed in the reports of more than one investigative agency of government. Most of the people involved in the case granted me interviews and some gave me access to their files of memoranda and correspondence. Contests over the charges against Imanishi-Kari generated extensive testimony and opened many previously confidential documents. Ultimately, the case proved to be a rich site for contemporary history, providing both abundant public and private documents and access to the recollections of living participants.

I am a historian by training and practice, and I have approached this vast body of material with a strong sense of the historian's respect for evidence, duty to weigh contradictory forms of it, and obligation to achieve a balanced understanding of the story. I have also felt it imperative to deal with the science to the extent necessary to appreciate what came to be contested. The case started as a small dispute in a laboratory over an experiment and then exploded into the larger sphere of politics and the media, but it remained fraught with technical issues throughout its life.

I have written the book to make its scientific content accessible to nonbiologists as well as to biologists, keeping discussions of the technical issues as concise as possible and relegating elaborative material to the notes. I have provided a brief account of the disputed experiment in the latter part of Chapter One. It was technically intricate, and lay readers should not be discouraged if they have trouble grasping all of it. I know biologists who find it difficult to comprehend. For assistance on the main scientific points, I have provided illustrations and a glossary of technical terms and concepts.

Following the case itself, the large majority of this book reaches far beyond technical matters. It is about individual character and behavior in science and the interactions of scientists with each other as human beings and professionals. It is about the relationship of science to the investigative powers of Congress and the executive branch; about the media's treatment of scientific ethics and practices; about the material dependency of science on the federal government; about tensions emergent in the late twentieth century between the biomedical sciences and American political culture.

But this book is also about the civil rights of scientists, particularly Thereza Imanishi-Kari. Once I started studying the record of the case, several points became quickly evident:

- Imanishi-Kari had not had a fair trial.
- She had been convicted in the court of public opinion and nowhere else.

• Those who condemned Baltimore for defending his collaborator overlooked or were indifferent to those crucial aspects of the case, among
others.

Eventually, I became persuaded that Imanishi-Kari was innocent of the
charges against her and said so, explaining why, in an article that appeared
in *The New Yorker* magazine in May 1996. In subsequently writing this
book, I found no reason to modify the fundamental judgments expressed
there—except to have been reinforced in them by the outcome of the
case. In June 1996, Thereza Imanishi-Kari was officially exonerated on all
the counts that had been brought against her. David Baltimore began to
re-enter public life, and in 1997 he was appointed president of the California Institute of Technology (where I have been a member of the faculty
for more than thirty years). At its core, this book is the story of how a
great injustice was perpetrated in the name of scientific integrity and the
public trust and how it then came to be remedied, or remedied as much
it could be after its weight had been endured for a decade.

Daniel J. Kevles
Pasadena, California
March 1998

Headlines, 1991

Science as something already in existence, already completed, is the most objective, impersonal thing that we humans know. Science as something coming into being, as a goal, is just as subjectively, psychologically conditioned as are all other human endeavors.

—Albert Einstein
Address, 1932

And the significance of this great organization, gentlemen? It consists in this, that innocent persons are accused of guilt, and senseless proceedings are put in motion against them. . . .

—Franz Kafka,
The Trial

True gold fears no fire.

—Chinese Proverb

THE BALTIMORE CASE

ONE

■

"A Beautiful Paper"

THE BALTIMORE case originated with Margot O'Toole, a postdoctoral fellow then in her early thirties whom Thereza Imanishi-Kari had hired to work in her laboratory at the Massachusetts Institute of Technology (M.I.T.) in the summer of 1985 and who eventually blew the whistle on her boss. O'Toole was asked to do experiments that would extend the work described in the contested *Cell* paper, and her unhappiness at not being able to get the results she sought led to the first complaint about Imanishi-Kari's data. O'Toole's dogged insistence that she was right and her supervisor was wrong lay at the heart of the affair. She became a symbol of the heroic young scientist who takes a stand against the system and prevails over powerful figures like David Baltimore.

Margot O'Toole now does research at the Genetics Institute, a biotechnology company in Cambridge, Massachusetts, but her reputation as a scientist rests almost entirely on her conflict with Baltimore and Imanishi-Kari. She has received several awards emanating from her actions, among them the Humanist of the Year Award from the Ethical Society of Boston, and the Ethics Award of the American Institute of Chemists. O'Toole has an open Irish face and a manner that prompted a congressional investigator to say, "The first time you meet her she just reeks with integrity and credibility."[1] I first met her one day in Cambridge, in 1992, when I picked her up for lunch. She is a compelling storyteller, and she held me in thrall for hours with her tales of the Baltimore case.

Margot O'Toole, her mother once remarked, was virtually bred to confront trouble. She was raised in a strong Catholic household in Dublin, Ireland, and spent two years in a convent school. Family lore told of one grandfather, a miller, who lost his house to flames during the famine for surreptitiously diverting food from the landlords to the people; and of the other who was turned out on the street for involvement in the Land League Movement. O'Toole says that her mother relished battles, as did her father. He was an engineer for the Electricity Supply Board, and also a radio commentator and playwright. He wrote of "speaking out in the workplace, not going along," O'Toole says. One of his plays, *Man Alive*, which satirizes the bureaucratic complacency of a giant utility company, uncannily foreshadows key elements in the Baltimore case. O'Toole says that the Electricity Supply Board tried to block the production when it was in dress rehearsal and that, though she was merely a child when the play was produced, the fight about it was a staple of her growing up. The central character is an outspoken engineer named Tim O'Malley, who is told to keep his dissident thoughts to himself and is declared an incompetent troublemaker. He nonetheless refuses to quit, pledging at the end of the play, "As long as I stay I'll be a thorn in their backside, and every time they sit on anyone again they'll think of me."[2]

In 1966, when Margot was 14, the family moved to Boston, where her father eventually directed a program in science writing at Boston University. She was multiply talented, adept at swimming and French, and shared her parents' attachment to poetry.[3] O'Toole graduated with honors in biology from Brandeis University, in 1973: Civil rights protests and demonstrations against the Vietnam War had flourished during her undergraduate years, likely encouraging her familial propensity for dissent. O'Toole spent the next academic year at Harvard, working as an assistant in the laboratory of a young immunologist named Thomas G. Wegmann. He later said that he found her very naive and her qualities of mind characteristic more of her religious upbringing than of her scientific training. He also considered her very good and productive at her technical work. Well recommended by Wegmann, she went for graduate work to Tufts University Medical School, entering in 1974 and specializing in immunology under the auspices of Henry Wortis.[4]

Wortis, then approaching forty and recently promoted to associate professor, was a respected immunologist who was also known among biologists of his generation for his political activism. A tall rangy man with an easygoing manner and a broad, rugged face, he wears jeans, a plaid workshirt, and athletic shoes in the lab. He describes himself as a "red diaper baby." He is the son of politically left psychiatrists who middle-named him Havelock, after their friend, the famed sexologist Havelock Ellis; his parents spent time in the Soviet Union between the wars and were eventually called before the House Un-American Activities Committee. Wortis says that during the 1950s he was thrown out of the University of Wisconsin for his political associations, notably his chairmanship of the Marxist Labor Youth League. He militantly protested the draft and the war in Vietnam during the mid-1960s, when he was a postdoctoral fellow in genetics at Stanford University. O'Toole was his first doctoral student— her husband, Peter Brodeur, whom she married in 1978, was his third— and he admired her social conscience. He also found her bright, insightful, creative, and engaging. Even now, he keeps a photograph of her tacked to his bulletin board, a snapshot of an attractive young woman, her hair wet from the rain and her eyes alight with mischief.[5]

In 1979, O'Toole received her Ph.D., and early in 1980, she and Brodeur, both funded by fellowships from the National Institutes of Health (N.I.H.), took up postdoctoral appointments at what is now called the Fox Chase Cancer Research Center in Philadelphia. Postdoctoral fellowships, which had long been desirable apprenticeships in the biomedical sciences, were especially choice in the 1980s. American universities were then more than doubling their annual output of biomedical doctorates. The expansion both fueled the rapidly burgeoning biotechnology industry and exacerbated competition for grants and positions in academic biomedicine. Life was tense low down on the professional ladder, where fledgling scientists ambitious to stay in the game had to prove that they could fly on their own in the high-pressure atmosphere of creative research. Postdoctoral fellows normally pursue the general line of investigation under way in the supervisor's laboratory. But labs can be launching pads, too, providing the freedom and facilities to open a line of independent research—one promising enough to win grant money and perhaps a university post.

O'Toole struck scientists at Fox Chase as well versed in her field, and Donald Mosier, in whose lab she came to work, thought that she had a good conceptual grasp of scientific problems. Mosier says, however, that she made "no progress" at the bench, largely because she wanted "to be a research manager without having acquired sufficient bench skills." He expected O'Toole, as he did all his postdocs, to spend much of her first year learning the experimental techniques necessary for the project she was supposed to pursue. Mosier says that she had trouble accomplishing the relatively simple task of purifying an organic chemical needed for her research, even though she had help from others in the lab. He continues that she wanted to rely on technicians to perform the requisite tests and procedures before she had mastered them herself and that she devised "grand, complicated experiments," tried them periodically, and failed to make them work. Mosier remembers urging that she consider a career in teaching or science writing because she commanded the concepts of science, if not manipulations at the bench.[6]

O'Toole, on her part, was increasingly unhappy with the situation. She told another young scientist at Fox Chase that she was uncomfortable in Mosier's laboratory because his wife, a postdoctoral fellow whom he had recently married and who remained working in the lab, received unfairly favorable treatment. Mosier calls the charge "absolutely false," explaining that he leaned over backward not to favor his wife and that, in any case, her project was remote from O'Toole's. He says that finally, after many months without significant experimental accomplishment, he asked O'Toole to leave his laboratory. In May 1982, she moved to the Fox Chase lab of the husband-and-wife team of Melvin and Gayle Bosma. By then, with Mosier's help, she had obtained new fellowship support from the National Arthritis Foundation and she settled down to a productive line of work.[7]

Mel Bosma found her lively and captivating, fun to be with, though he recalls that she seemed at least as absorbed with the play of personality in the laboratory as with the work at the bench. (Wortis had the impression that O'Toole was "more sensitive to the social—what I would call scientific-political—interactions around her to the point where her concern about those issues might keep her from focusing on her work.") Mosier declares that she "had an instinct for polarizing laboratory members over minor issues." By way of example, he points to her pitting his technicians against each other because one of them, who he says was better qualified than the rest, was given more authority in the lab. He says that in some respects O'Toole reminded him of "a labor organizer": She was "intense, political," ready to leap on an issue."[8]

People then at Fox Chase still talk about how O'Toole grappled with the issue of day care. Fox Chase was renovating several rooms in an old nearby school for the care of staff infants. Completion of the renovation was scheduled for June of 1981, but it was delayed until September. During that summer, O'Toole gave birth to a son. Other mothers made temporary alternative arrangements for their children. O'Toole brought the baby to her laboratory, in defiance of the institution's rules. The rules were intended to protect children against exposure to radioactive substances and the institution against liability. O'Toole, told to stop keeping the baby in the lab, insisted to Patricia Harsche, the administrator in charge of the renovation, that Fox Chase owed her assistance, declaring in the hall one day, "If it hadn't been for you Pat, I wouldn't be pregnant." Harsche laughed, noting, "Margot, I've been accused of many things in my life but never of having made another woman pregnant." O'Toole, Harsche remembers, did not laugh in return. Harsche, who sympathized with O'Toole as she did with other mothers with day-care problems, quickly arranged to have a small room fitted out as a nursery where O'Toole and her husband cared for the baby until their son entered the day-care facility in October.[9]

A year or so later, Fox Chase enlarged its child-care accommodations by renovating another old school. After this facility was occupied, the heating system gave evidence of needing replacement and inspection revealed that it was lined with asbestos. Fox Chase had the asbestos removed with the special care that the law required and engaged an independent consulting firm to evaluate whether the job had rendered the building free of asbestos. The firm said that it had, but O'Toole joined several other mothers in declaring that the consultant's report was untrustworthy because Fox Chase had paid for it. The mothers insisted that the cancer research center obtain a second evaluation by a firm acceptable to them, which it did, and the new assessment reached the same conclusion as the first.[10]

Late in 1984, Brodeur was offered an assistant professorship at Tufts University. Not wanting to split the family by staying in Philadelphia, O'Toole followed her husband back to Boston. The N.I.H., to which O'Toole had applied for support, told her that she might obtain funding for the project she had started with Bosma if she could devise certain experimental materials that would bolster its promise. Her thesis adviser, Henry Wortis, who says that he thought of her as a "bright, insightful scientist," helped O'Toole get temporary space in the lab of his colleague, Sidney Leskowitz. She was appointed to an assistant research professorship, a position that the department chairman could create at will for people who had their own research funds.[11] She could work on her project until her postdoctoral money from the Arthritis Foundation ran out and continue with it if she managed to get new grant support. "That seemed great for me," she says, "because I wanted to continue it and I wanted to be in Boston. It was risky, because I had no backup. If I didn't get funded, I couldn't do it. So, I took the plunge and applied for the grants and went to Boston. But I didn't get them."[12]

Some time that spring, Wortis invited O'Toole to a party at his home, and there she met Thereza Imanishi-Kari. Imanishi-Kari was then forty-one, almost nine years older than O'Toole, but still a junior faculty member at M.I.T. O'Toole had certain skills that would be useful in a project that Imanishi-Kari wanted to pursue, an outgrowth of her collaboration with David Baltimore. Wortis had alerted Imanishi-Kari to O'Toole's need for grant support. On the spot, she offered O'Toole a one-year postdoctoral training fellowship supported by the N.I.H. O'Toole was to have time and facilities to strengthen her own project's eligibility for funding while she collaborated with Imanishi-Kari on extending the research that she was doing with Baltimore. O'Toole's research money was about to run out, and the offer was a godsend.

Thereza Imanishi-Kari is now a member of the pathology department at the Tufts University Medical School in Boston, where she went after leaving M.I.T. Her laboratory is a bright open room with several working credenzas laden with glassware, chemicals, and cultures, on the top floor of an old brick building with an elevator that might momentarily strand passengers between floors. Imanishi-Kari was born into an immigrant Japanese family in Brazil. She is a kind of cultural hybrid, giving the appearance of Japanese reserve but regularly shattering it with Latin expressiveness. ("I say things and face up to person," she told me.) She speaks seven languages, but her English, which is even now sometimes difficult to understand, was especially poor in the mid-1980s, when the disputed paper was published.

Thereza Imanishi-Kari grew up in Indaiatuba, a small town near São Paulo, where her parents were tenant farmers, growing cotton, vegetables, and coffee. Eventually they got a mule, started transporting the neighborhood's produce to market, and soon became the owners of a small trucking business. They wanted their five children to attend school and do well, but they expected their three daughters to devote their lives to marriage and family. Thereza and her sisters fought to get an education; after her older sister left home over the battle, her parents permitted Thereza to go to high school and then a university in São Paolo. Her grandfather wanted her to learn about her family's culture, and in 1968 she went to Kyoto University to do graduate work in biology. Hardly any women were studying science there, so she hung out with the men, perfecting the Japanese she learned as a child into the male rather than the female version of the language.[13]

Kyoto University was in a state of upheaval in 1968, like most universities at the time, with fights constantly breaking out on the campus. Imanishi-Kari repaired to cafes with other students, where they studied immunology and talked about the imitative tendencies of Japanese scientists, particularly their reliance on experimental systems developed in other countries. The students considered the dependency self-defeating.

MARGOT O'TOOLE

Photo Credit: Seth Resnick © 1997

THEREZA IMANISHI-KARI

Photo Credit: Mary Ellen Mark

DAVID BALTIMORE

Photo Credit: Warren Roos

Japanese students would work late hours applying a borrowed research system to a scientific problem only to find themselves preempted by the foreigners who had devised the system originally. She resolved to invent and rely on her own research system, which is what she did while she was completing her graduate work at the University of Helsinki, in Finland.[14] She had gone there in 1971 because she felt that the continuing disruptions at Kyoto made it impossible for her to do serious work. Using a chemical called NP ("nip," she pronounces it) that was well known to immunologists, she hit upon a method of tracking the behavior of certain immune genes in mice, the common laboratory surrogate for human beings.

In 1974, she obtained her doctorate and married a Finnish architect, Markku Tapani Kari. She spent several postdoctoral years in the laboratory of Klaus Rajewsky in Cologne, Germany, and became known for her work on the NP system. When M.I.T. was looking for a cellular immunologist to add to its faculty in 1979, she was encouraged to apply for the position by an M.I.T. biologist named Malcolm Gefter. Imanishi-Kari was the biology department's first choice in an international search that produced about thirty candidates. When she got the offer from M.I.T., she asked Rajewsky for advice. He counseled against acceptance, observing, "M.I.T. is a very competitive place. It's like a sea full of sharks and they eat the little ones very fast." She went anyway. "That was the beginning of my nightmare," she says.[15]

Imanishi-Kari arrived at M.I.T. in March 1981 as an assistant professor and moved into a laboratory on the first floor at the M.I.T. Center for Cancer Research on Ames Street. She had brought with her from Germany a freezerful of valuable NP research materials. She was vivacious, competent, quick on her feet, and formidably smart. Even O'Toole says that initially she found her "very unusual and quite charming."[16] Imanishi-Kari's group comprised several students and junior scientists on temporary billets and a technician named Chris Albanese, who had recently obtained a master's degree in biology from the State University of New York. Some—though not all—of the laboratory regulars, like Moema Reis, a visiting scientist from the Instituto Biologico in São Paolo, Brazil, were devoted to her. Reis, who was in her late thirties and had first gotten to know Imanishi-Kari in 1983, when she spent three months in her laboratory. She returned for a year beginning in February 1985 and became a contributor to the *Cell* paper. She says that the laboratory was "extremely pleasant," that Imanishi-Kari was welcoming and open, regularly sharing data and discussing experiments, and reviewing individual projects at lunch on Fridays. Reis considered her year in the laboratory "very helpful . . . because I had the opportunity to work closely with somebody who considers science a vital activity," somebody of "high intelligence who has a very hard drive for working," somebody who "carries out science and research as it should be carried out."[17]

Imanishi-Kari broke the laboratory rules against smoking and neglected to meet M.I.T.'s requirements for getting ahead. By M.I.T.'s standards, she published too few scientific papers, and part of what she published struck others as being a narrow extension of her earlier work with the NP system.[18] However, she was enlarging her repertoire of expertise by learning to use the techniques of molecular biology. And in 1984, during her collaboration with David Baltimore on a study of the production of antibodies in a special breed of mice, her tracking system helped expose some surprising and peculiar results. The findings made no difference for her future at M.I.T. She had already been looking around for another job when in July 1985 she was told that she would not be put up for tenure. She had an offer at a biotechnology research institute in La Jolla, California, another at Mt. Sinai Medical Center in New York City, and a strong prospect at Tufts in the department where O'Toole's husband now worked and where she had several friends, including Henry Wortis, with whom she was collaborating on a research project. Her daughter wanted to remain in Boston, so she preferred Tufts.[19] In mid-August 1985, the head of the Tufts Department of Pathology talked with Imanishi-Kari about a position, and she said that she would accept an offer when it became firm and final.[20]

In May 1986, when Tufts was moving to make the offer final, Baltimore declared in a letter of recommendation that, although Imanishi-Kari had been "slow to get an innovative research program going," in the last few years her research had begun "to take an interesting shape." He explained, referring to the work that they had just reported in *Cell*, that "it was the expertise in her laboratory that allowed us to understand" the odd immune response in the mice they had used in the experiment.[21]

THE EXPERIMENT

Background: The Immune System and Rearrangement

The immune system had long interested David Baltimore and he had turned to studying it in the mid-1970s, after he won the Nobel Prize for his work in virology. In mammals, the system's principal component comprises white blood cells called *lymphocytes*. In 1974, at a symposium in

Paris, the distinguished Danish immunologist Niels K. Jerne remarked that just twenty years earlier scientists "hardly even suspected" that "lymphocytes had anything to do with the immune system," adding that "now we know they *are* the immune system, or at least 98% of it." Billions of white blood cells—in fact, roughly a thousand billion of them—are present in the body, enough, taken together, to add up to a mass comparable to that of the liver or brain. They are found in the body's specialized immune organs—the thymus, bone marrow, spleen, and lymph nodes—and they circulate in the bloodstream.[22]

Two types of lymphocytes are central to the immune response in mammals: T cells, which are manufactured in the thymus, and B cells, which are generated in the bone marrow. Both eventually migrate to the lymph nodes and the spleen, reacting there with invading agents such as a virus or a bacterium. T cells perform several functions, one of which is to assist B cells in doing their job. B cells produce antibodies, which latch to and inactivate what the body takes to be hostile invaders. Antibodies are exquisitely specific, fitting to the invader the way a key fits to a lock.

The variety of infectious agents that invade the body is enormous, and the most striking feature of the immune system is that it generates a comparably enormous range of antibodies. Antibodies are constructed, like an erector set, from discretely identifiable elements, but, when they are completed, most antibodies structurally resemble the letter Y (*Figure 1*). Each tip of the Y comprises two independent variable regions that run up opposite sides of each of the arms. Together, they provide the antibody's specificity—its ability to fit the invading agent's lock.

Each variable region is the product of an independent set of genes which are formed from among many thousands of segments of DNA in the nucleus of the B cell. In the process of making antibodies, a small number of segments from different elements in the nucleus rearrange themselves and combine to produce genes for the two variable regions (*Figure 2*). The possible combinations of segments is huge, which is to say that the process generates a giant number of possibilities for each variable region. (If, say, one variable region derives from any one of a thousand genes and the other from any one of another thousand, then the number of combinations would be the product of the two, or one million from just two thousand genes.) It is this heterogeneity that allows the immune system to manufacture its mightily resourceful arsenal of antibodies. Antibody production occurs continuously throughout an animal's life, producing numerous incipient B cells. Each is committed to generating only one particular antibody, and all the B cells together account for the diversity of the animal's immune response.

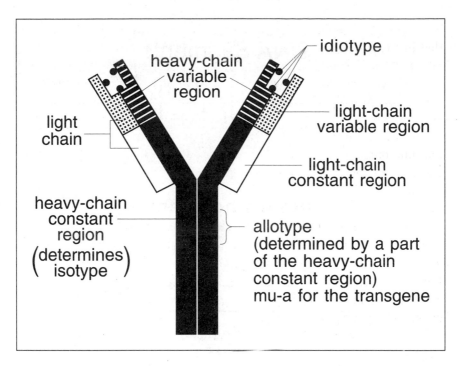

FIGURE 1
An Antibody and Its Elements

Each side of the antibody Y is constructed of a light chain of amino acids and a heavy chain of amino acids. Each chain consists of two elements—the variable region and the constant region. The two variable regions—the one with the stripes for the heavy chain and the one with the dots for the light chain—together give the antibody its ability to attack a specific invading agent. The idiotype—the pimple-like protrusions in the notch at the top of the Y—comprises particular surface features on the variable region. The constant region of the heavy chain—the dark element that starts at mid-arm then turns to run down to the bottom of the Y—defines the antibody's isotype. Its allotype is determined by a small chemical variation in the heavy-chain constant region.

When Baltimore began immunological research, one of the great puzzles in the field was how exactly the process of genetic rearrangement is stimulated and controlled. He took up that conundrum, among others, as he increasingly oriented his laboratory to apply the powerful and rapidly developing techniques of molecular biology to problems in immunology. Perhaps the most powerful of these techniques, an invention of the mid-1970s, was recombinant DNA, which permitted scientists to snip out a gene from one organism and insert it into another—to put a human gene into a mouse, for example. Recombinant DNA provided scientists a powerful tool for studying the nature and mechanisms of gene action, and

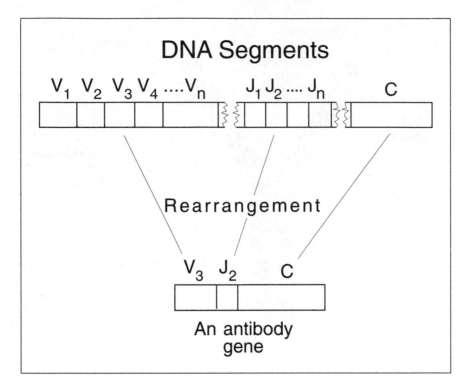

FIGURE 2
Rearrangement

Segments of DNA—sections V_1, V_2, V_3, V_4, . . . V_n plus J_1, J_2, . . . J_n plus C—are arranged on a chromosome in the nucleus of a B cell. Each of the V segments comprises the DNA for a specific variable region, one or more of which may carry DNA that encodes a particular idiotype. The J segments join the variable region to the DNA segment, C that encodes a constant region. As the cell matures, some of the sections— say, V_3, J_2, and C—rearrange themselves, joining to form an antibody gene, in this case for a light chain. The heavy chain is similarly assembled from V, J, and C segments of DNA plus a D segment.

Baltimore thought he could exploit it to study the process of genetic rearrangement.[23]

Imanishi-Kari's Method: Idiotype Tracking

Imanishi-Kari's NP tracking system provided another way of getting at the genetics of antibody formation. NP is one of a class of small organic chemicals that, when combined with a protein, will stimulate the generation of an antibody against itself.[24] In certain strains of mice, the antibodies display a distinctive chemical feature called an *idiotype*. Although located in the variable regions at the tips of the antibody's arms (see *Figure 1*), idiotypes have nothing to do with giving antibodies the specificity of their response to a foreign agent like NP. It is more like a birthmark—an identifying signature.

Imanishi-Kari obtained antibodies from fast-growing cell cultures called *hybridomas*. She formed them by joining cells from the different immune organs of her mice with rapidly multiplying myeloma cancer cells. Hybridomas are nourished in a bath of nutrients, and the antibodies they generate make their way into the fluid—a *supernatant*. Imanishi-Kari relied on serological methods—that is, methods used to identify the properties or contents of organic fluids such as blood sera—to characterize the antibodies, testing them with chemical or biological substances. Such substances are called *reagents* when they are deployed as tools in experiments, and Imanishi-Kari managed to devise reagents indicating the presence of idiotypes on antibodies to NP.

While in Europe during the 1970s, she demonstrated that the ability of certain mice to mark such antibodies with an idiotypic birthmark is passed down from one generation to the next in accord with Mendel's laws of inheritance.[25] Such inheritability meant that the idiotype is a product of a particular segment of DNA. Thus, the antibody's birthmark provided a serologically detectable feature that permitted tracking the behavior of the segment and its surrounding DNA in the operation of the immune system (see *Figure 2*). "It was a purely accidental finding," Imanishi-Kari says, "but I think I was one of the first to show that some antibodies have inheritable specificity" and to reveal "a serological marker for the variable regions."[26]

Imanishi-Kari's NP system exploited the marker to study antibody responses.[27] In 1976, while working in Rajewsky's laboratory in Cologne, she and Rajewsky attended a meeting at the N.I.H. in Bethesda, Maryland, and encountered David Baltimore. Baltimore knew Klaus Rajewsky and thought highly of him, and he was intrigued to learn about Imanishi-Kari's NP system. "So, through my association with Klaus," Baltimore says, "I became associated with Thereza."[28]

Initiating the Experiment: Mice and Molecules

In 1982, Baltimore had an idea that led eventually to the *Cell* paper. Imanishi-Kari had recently obtained antibodies against NP, with a distinctive idiotype, from hybridomas that she had developed from an inbred strain of mice called BALB/c. Scientists in Baltimore's laboratory, using hybridomas from Imanishi-Kari's lab, isolated and characterized the DNA responsible for the part of the variable region of these antibodies that included its idiotypic birthmark.[2] Baltimore's idea was to use a gene engineered to contain this DNA in an experiment with a new genre of mice that had recently been introduced into the laboratory scene. These animals were just like any other laboratory mice except for one feature: A gene from another animal had been inserted into them when they were just newly fertilized eggs. When the mice were born, they contained a copy of the gene in every cell in their bodies. Scientists called the animals *transgenic mice*. They were highly promising instruments for biomedical research because they enabled observation and analysis of the inserted gene's impact on a living mammalian system. Baltimore expected that a suitably constructed transgenic mouse might reveal something about the action of immune genes.

He thus proposed to insert the gene engineered from BALB/c DNA into an inbred strain of mice designated C57BL/6—black mice sometimes called "Black/6" for short. Baltimore wanted to see whether the gene would express itself—that is, contribute to the formation of antibodies— in the recipient mice as though it was native to them. Beyond that, he hoped that the presence of the foreign gene in the Black/6 mice would reveal something about how the process of rearrangement is controlled. Biologists thought that once the DNA segments in a cell were rearranged to produce a specific antibody, a kind of negative feedback process prevented any rearrangement from occurring in parallel parts of the cell's DNA. They believed that this process prevented any one B cell from producing more than one kind of antibody. The engineered gene would comprise DNA segments that had already been rearranged. The question was whether it would, as theory suggested, inhibit rearrangement of the truly native DNA segments in the mice so that they would not produce antibodies of their own.[30]

In 1983, Baltimore sent a suitably engineered gene to a biologist at Columbia University named Frank Constantini, who knew how to make transgenic mice. The gene was designated "17.2.25," after the hybridoma from which it had been obtained, and would serve as what scientists call the *transgene*—that is, the foreign gene that would be inserted into the mouse. Constantini inserted 17.2.25 into the newly fertilized eggs of normal Black/6 mice. The eggs were introduced into the wombs of surrogate mother mice, and some of the altered eggs developed into transgenic mice—that is, mice that carried the foreign gene for the antibody to NP[31] (*Figure 3*).

Baltimore next established a small colony of the transgenic mice from

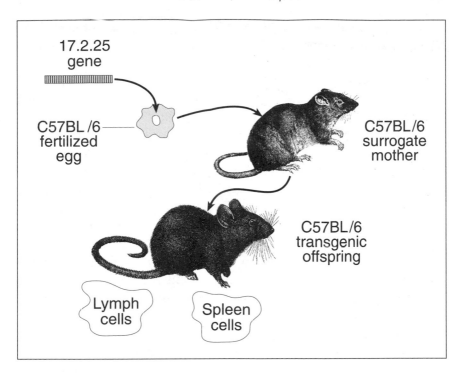

FIGURE 3
Making a Transgenic C57BL/6 (Black/6) Mouse

The 17.2.25 gene, engineered from the DNA of BALB/c mice, is inserted into the nucleus of a newly fertilized egg of a C57BL/6 mouse. The transgenic egg is then surgically placed into the uterus of a Black/6 surrogate mother, where it develops to term. The mouse born of the egg is transgenic, having the 17.2.25 gene in every cell of its body. Lymph cells and spleen cells taken from the transgenic mouse are then used to form transgenic hybridomas.

Constantini in his laboratory on the fifth floor of the M.I.T. cancer center. He enlisted two of his postdoctoral fellows—David Weaver and Rudolf Grosschedl—to work with them, but he remembers remarking one day, "You know, we really ought to look at the immune response in these animals in a more serious way, and I don't understand how to do that." Baltimore wanted to learn about the kind of antibodies that were circulating in the blood of the mice, and such research required serology. He had neither the skills nor the tools of serology in his laboratory, but Imanishi-Kari commanded both.[32] Since her arrival at M.I.T., Imanishi-Kari had permitted herself little involvement with Baltimore's research. Baltimore says that the fact that they had previously published together had "sort of chilled our ability to collaborate because, as a young person,

she was very afraid of being seen as the adjunct of the senior person. . . .
I tried to keep a distance." Once he had the transgenic mice, Baltimore
recalls, "I went to Thereza and I said, 'Look. I know you've been worried
about the relationship, but here's an opportunity where your expertise will
make a big difference. And I think you can only benefit both of us and
science if you'll put some effort into it.' And she agreed."[33]

In late 1983, Grosschedl and Weaver went downstairs to Imanishi-Kari's
laboratory for instruction in the rudiments of serology and hybridomas.[34]
Their job was ultimately to analyze the antibodies produced by the hybridomas using the techniques of molecular biology. They first checked the
antibodies that were circulating in the blood of the living transgenic mice.
Baltimore's group found elevated levels of NP-sensitive antibodies that
had been produced by the 17.2.25 gene. However, they also detected normal levels of antibodies native to the mice, which suggested that the
introduction of the transgene had not inhibited rearrangement of the
DNA that generated them.[35] In mid-1984, intrigued by the result, Baltimore's group started to investigate the phenomenon in the region of the
mice where the antibodies were actually produced—that is, at the level
of individual B cells.

About this time, Grosschedl left for a post on the West Coast, so Weaver became the linchpin between the laboratories of Baltimore and Imanishi-Kari. With a Ph.D. in biochemistry from Harvard Medical School,
he was particularly interested in DNA rearrangement, which occurs in
cancer as well as in immune responses.[36] In a sense, the experiment that
led to the disputed *Cell* paper was his; he would be the senior author on
the publication. It was commonly referred to as "Weaver *et al.*," and
Weaver himself, a leanly built man who is precise in speech and quietly
resolute in manner, says that it gave him an early but uncomfortable kind
of notoriety.[37]

Imanishi-Kari instructed Weaver in how to grow hybridomas. In late
1984, after a number of failed attempts, he succeeded with her help in
obtaining growth in four batches of hybridomas that she had given him—
two of lymph cells and two of spleen cells from the transgenic Black/6
mice.[38] The experiment now proceeded along two complementary lines:
serological analysis by Imanishi-Kari of the antibodies produced in the
hybridomas and molecular analysis of them by Weaver. (The fact that
the experiment depended on these two types of research, which are technically quite different from each other, helped to make the effort difficult
to understand for non-specialists, including some federal investigators.)
David Baltimore recalls that he and his collaborators had expected the

DAVID WEAVER

Photo Credit: Courtesy David Weaver

experiment "to be fairly straightforward" but that it wasn't. In fact, the results that came from both Imanishi-Kari's and Weaver's investigations were unexpected, right from the beginning.[39]

Imanishi-Kari's Contribution: Serology

To characterize the antibodies, Imanishi-Kari decanted the supernatants into small cuplike wells—they measure about a quarter inch across the top—that were indented in parallel rows on a rectangular plate of clear plastic measuring several inches on a side. (Most of the plates used in the experiment contained 96 wells, in eight rows of twelve.) The sides of the wells were coated with either a reagent that would grab antibodies against NP or a reagent that would capture antibodies with the telltale idiotype. Imanishi-Kari then washed the supernatant out of the wells and determined the characteristics of whatever antibodies had stuck to the sides (*Figure 4*). She was first interested in identifying their isotypes—a characteristic of antibodies that would help reveal whether they had been produced by the transgene or genes native to the mice.

Antibodies are also called *immunoglobulins*—"Ig" (scientists say "eye-gee"), for short. In vertebrates—mice, for example—they are divided into

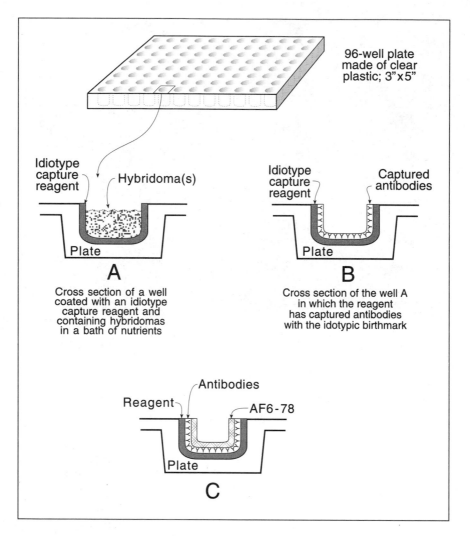

FIGURE 4-A
Serological Characterization of Antibodies
Figures 4-A and 4-B

A. The wells on a 96-well plate are coated with a reagent—say, one that will capture antibodies with the idiotypic birthmark (see cross section A of a well). Hybridomas formed from the lymph or spleen cells of a transgenic mouse are then placed in wells together with a bath of nutrients that permits them to grow and produce antibodies. After a suitable period of such growth, the plates are washed. If antibodies with the idiotypic birthmark have been produced, they will have stuck to the reagent coat (see cross section B of a well). Both to test whether any antibodies have stuck and to characterize those that have, a second reagent—for example, radioiodinated AF6-78.25—is introduced into the wells. If the antibodies have the allotype *mu-b*, the second reagent will attach itself to them (see cross section C). The plates are then washed again, leaving the following sandwich: idiotype reagent/antibody/radioiodinated AF6-78.25.

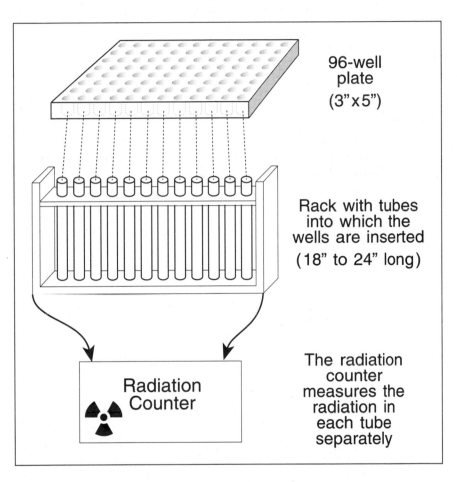

**96-well
plate
(3"x5")**

**Rack with tubes
into which the
wells are inserted
(18" to 24" long)**

**Radiation
Counter**

**The radiation
counter
measures the
radiation in
each tube
separately**

FIGURE 4-B

B. The wells are then cut out of the plate. Each well is inserted into a tube on a rack that holds a large number of them. The rack is then placed into a radiation counter, where each tube is separately checked for whether radiation is emanating from it and, if so, the intensity of the radiation is measured in counts per minute. Counts registered for a particular well that are above a certain level—Imanishi-Kari's cutoff was 1,000—indicate that the radioiodinated AF6-78.25 has found antibodies in that well to attach itself to, which means that the antibodies carry the idiotypic birthmark and have a *mu-b* allotype.

five general classes. One of them, labeled *IgM*, is the first class of antibody produced by a developing B cell. Another, labeled *IgG*, is the most common immunoglobulin in the blood.[40] All antibodies that belong to one class of immunoglobulins are said to have the same "isotype" (not to be confused with "idiotype"). Isotype is determined by the composition of a long chain of amino acids—called the *heavy chain*—that runs from the upper tip of each arm of the Y down to the bottom of its trunk. The *variable region* of the chain forms its upper stretch. The rest of the chain is called the *constant region*. The chemical composition of the heavy-chain constant region defines the antibody's isotype; it is essentially the same in all the antibodies that comprise any one class. In IgM antibodies, the isotype is termed *mu*. In the IgG variety, it is termed *gamma*.

When Imanishi-Kari checked the isotypes of the antibodies generated by the first transgenic hybridomas analyzed in the experiment, she discovered that many of them were *gamma*. Here was one of the first surprises. The transgene could only produce antibodies that were *mu*. Since the antibodies were *gamma* rather than *mu*, Imanishi-Kari was forced to conclude that they could not have come from the transgene.[41] She says she wondered about the result, noting, "If you have something very unusual, you think you maybe did something wrong. . . . Because with serology, you don't know whether what you're seeing is real or whether something's wrong with the reagent you're using. It's always ambiguous."[42]

To learn more about what was happening, Imanishi-Kari sought to characterize more specifically the antibodies against NP that the hybridomas were generating. Like the transgene, native genes in the Black/6 mice could also produce *mu* antibodies. But *mu* constant regions can vary slightly from each other by a tiny bit of chemistry, producing differences in what scientists call their *allotypes*. The allotype of the transgenic antibody was designated *mu-a*; that of the native antibody was designated *mu-b*. To determine the origin of the antibodies that the hybridomas were producing, Imanishi-Kari needed reagents that would distinguish between the two. Some time after August 1984, she obtained a newly available reagent from Henry Wortis that would react only with *mu-b* antibodies—that is, only with antibodies emanating from a native gene. It was called AF6-78.25. Early in 1985, she got hold of a reagent for detecting transgenic antibodies. It had been devised in the laboratory of a scientist at the N.I.H. named William Paul, and it was called *Bet-1*. Under suitable conditions, it was far more likely to latch on to *mu-a*, the transgenic antibody, than to *mu-b*, the native one.[43]

Imanishi-Kari tested for *mu-a* antibodies using a radioimmune assay.

For this procedure, the wells in the plastic plates were coated with a reagent sensitive to the idiotype and then filled with supernatant from the hybridomas. Antibodies with the idiotypic birthmark would stick to the coat. She then washed the wells. She made Bet-1 radioactive by combining it chemically with radioactive iodine, which gives off gamma rays (a form of high-energy radiation; not to be confused with *gamma* isotypes; which the rays have nothing to do with). Bet-1 was then applied to the wells and would attach itself to any *mu-a* antibodies that might have adhered to the coated sides. The wells were then cut out of the plate by using a hot wire to slice horizontally through the plastic, a process that generated a lot of smoke. Each well was tweezed into its own small test tube. A rack held the tubes side by side in the order in which the wells in them had occurred on the plastic plate (see *Figure 4*).

The rack was designed for insertion in a gamma radiation counter that was available in a small room near Imanishi-Kari's laboratory. She shared the counter, along with additional equipment, with other scientists on the first floor of the M.I.T. cancer center. The machine processed each tube successively, measuring the gamma radiation emanating from each well for up to a minute. The detection of gamma radiation meant that Bet-1 and, hence, transgenic antibody was present (because it was to that antibody that radioactive Bet-1 preferentially attached itself). The intensity of the radiation, quantified as gamma counts per minute, indicated how much of the antibody was there. The radiation counter was hooked up to a teletypewriter that printed a record of the counts from each well in a column on a roll of paper[44] (*Figure 5*). A similar procedure carried out with the AF6-78.25 reagent tested whether the supernatants contained *mu-b* antibodies, that is, antibodies originating from a native gene.

The serological analysis demanded huge amounts of care and labor. It meant preparing and checking reagents like Bet-1. It required the management of hundreds of wells at any one time—keeping track of the reagents used on the hybridomas, of the concentrations of antibodies in the supernatants, and of which wells went into which radiation-counter tubes. "Sometimes I made mistakes," Imanishi-Kari says. Her collaborator Moema Reis recalls that she would start an experiment, run the tubes with their cutout wells into a radiation counter, and find the data emerging from the counter when she was already absorbed with a second experiment. She would set the counter printouts aside for later collection and collation. Imanishi-Kari might leave hers in file folders or drawers or on window sills for months.[45]

By the summer of 1985, Imanishi-Kari and Reis had publishable data

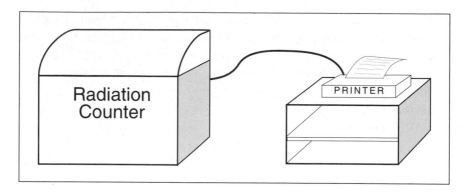

FIGURE 5-A
Radiation Counter, Printer, and Data Tape
Figures 5-A, 5-B, and 5-C

A. The radiation counter prints out on a tape the number of radiation counts per minute that it detects in each tube on the rack that has been placed into the machine. Each number in the data column on the tape is a measure of the counts per minute for a single tube, which corresponds to the particular well on the original plate that was inserted into the tube. Imanishi-Kari would usually cut out the data column from the tape and mount it on a loose page, often taken from a spiral notebook. Sometimes, however, she would simply write down in a column on the page the numbers from the data column of the tape. The pages that follow, which are taken from Moema Reis's notebooks, show both types of data entry.

on transgenic hybridomas in 340 wells. The antibodies produced in 172 of them displayed idiotypic birthmarks that were not the same as those from the transgene but were closely related to them. However, only 42, fewer than a quarter, of these wells contained *mu-a* antibodies, which meant that the antibodies in fewer than a quarter of the wells derived from the transgene. Another 11 wells contained antibodies that were *mu-b*, which meant that they had been produced by native genes. The remaining 119 wells generated antibodies that were neither *mu-a* nor *mu-b*; they all had to come from native genes, too. Thus, the hybridomas in the vast majority of the wells—119 plus 11, making a total of 130 in all—produced antibodies to NP with idiotypic birthmarks similar to the transgene's, but these antibodies had been generated by genes native to the normal Black/6 mice.[46]

Weaver and Baltimore's Contribution: Molecular Biology

All the while, Weaver had investigated the behavior of the hybridomas at a molecular level. He first looked for expression of the transgene's DNA. The machinery of gene expression produces a molecule called *RNA*

Test supernatants with AF6 and BET1

3/22/85
3/22/85
119

α.17.2.25
BET1
18547.36
18381.82
9833.88
8543.88
6798.88
BET1 P9.37.9
591.88
332.88
242.88
227.88
224.88
AF6-78
3896.88
1788.88
1218.88
812.88
648.88
AF6-78
435.88
275.88
333.88
299.88
388.88

α.17.2.25
BET1 1859.88
535.88
MOPC 104E 346.88
282.88
299.88
BET1 1359.88
B1-8 382.88
355.88
355.88
271.88
AF6 78 219.88
381.88
B1-8 287.88
356.88
292.88

α.17.2.25 BET
1 18963.69
2 11894.11
S2.15.6 3 12681.81
4 12417.28
5 11364.77
L3.10.6 (8) 3551.88
S2.14.4 6 1861.88
S2.15.6 12 942.88
L3.5.2 778.88 (2)
L3.4.4 576.88 (2)
745.88 (5)
L3.14.5 465.88 (5)
S2.8.2 291.88
L4.13.2 521.88 (1)
L3.4.4 375.88 (1)
L3.9.4 114.88 (4)
L4.5.2 147.88 (6)
S2.14.4 257.88 (6)
L3.10.6 185.88 (8)
BSA 385.88
131.88
117.88

AF6 78
354.88
396.88
386.88
516.88
383.88
378.88
347.88
353.88
2153.88
1724.88
1508.88
1383.88
1268.88
568.88
614.88
561.88
584.88
827.88
163.88
144.88
BSA 266.88
214.88

BET.1 used: 30μl/ml of stock
AF6. used 20μl/ml of stock

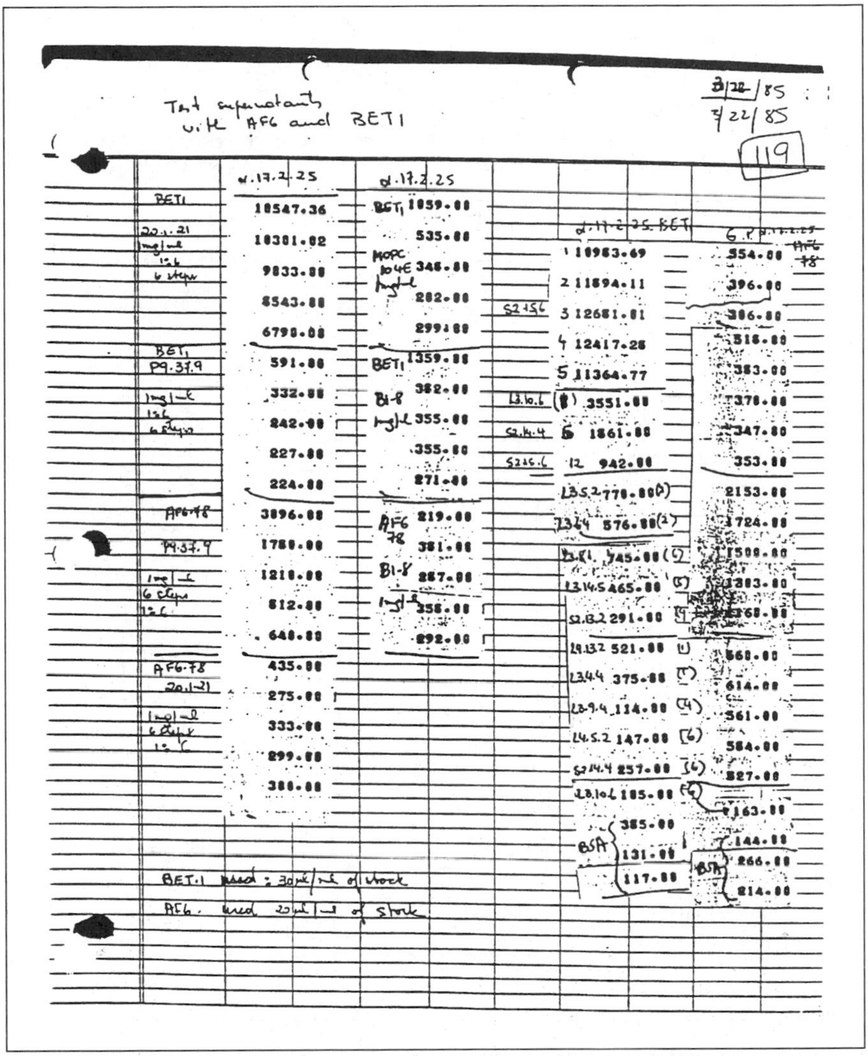

FIGURE 5-B

B, C. Each of the two batches of lymph cells and two batches of spleen cells that Weaver and Imanishi-Kari managed to grow were originally deposited on a plate with 24 wells. Each well contained a number of hybridoma lines deriving from different individual cells. The two lymph plates were designated *L3* and *L4*; the two spleen plates, *S1* and *S2*. A hybridoma used in the experiment was identified, first, by the letter and number indicating its cell type and the plate from which it came, then by a number indicating the well from which it was taken, and then by the clone—that is, the particular hybridoma line—that had been isolated from the several growing in the well. For example, the hybridoma L3.10.6 in the third column from the right in *B* is the sixth clone from the well number 10 on plate L3.

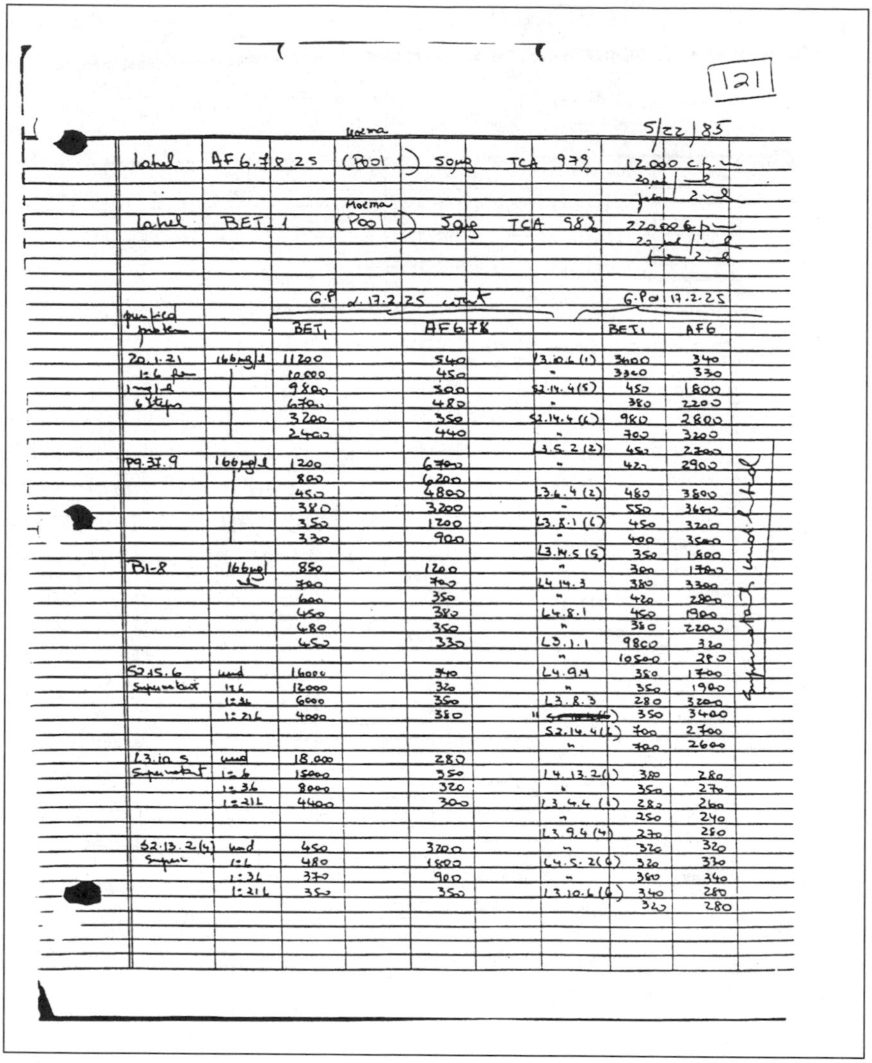

121

5/22/85

label AF6.78.25 (Pool 1) 50µg TCA 97% 12000 c.p.m

label BET-1 (Pool 1) 50µg TCA 98% 22000 cpm

purified protein		G.P. α.17.2.25 wt BET1	AF6 78		G.P.α 17.2.25 BET1	AF6
20.1.21	166µg/l	11200	540	13.10.6 (1)	3400	340
1:6		10000	450	"	3300	330
		9800	500	52.14.4(5)	450	1800
		6700	480	"	380	2200
		3200	350	52.14.4(6)	980	2800
		2400	440	"	700	3200
				13.5.2(2)	450	2300
P9.37.9	166µg/l	1200	6700	"	420	2900
		800	6200			
		450	4800	13.6.4(2)	480	3800
		380	3200	"	550	3600
		350	1200	13.8.1(6)	450	3200
		330	900	"	400	3500
				13.14.5(5)	350	1800
B1-8	166µg/l	850	1200	"	380	1700
		700	700	L4.14.3	380	3300
		600	350	"	420	2800
		450	380	L4.8.1	450	1900
		480	350	"	380	2200
		450	330	L3.1.1	9800	320
				"	10500	280
52.15.6 Supernatant	und	16000	340	L4.9.4	350	1700
	1:6	12000	320	"	350	1900
	1:36	6000	350	L3.8.3	280	3200
	1:216	4000	350	"	350	3400
				52.14.4(6)	700	2700
				"	400	2600
13.10.5 Supernatant	und	18000	280			
	1:6	15000	350	L4.13.2(1)	380	280
	1:36	8000	320	"	350	270
	1:216	4400	300	L3.4.4(1)	280	260
				"	250	240
				L3.9.4(4)	270	250
52.13.2(4) Supern	und	450	3200	"	320	320
	1:6	480	1500	L4.5.2(6)	320	330
	1:36	370	900	"	360	340
	1:216	350	350	L3.10.6(6)	340	280
					320	280

FIGURE 5-C

(ribonucleic acid) that complements the coding regions in the gene's DNA. Weaver thus determined the characteristics of the RNA that the hybridomas were producing. He used a standard technique of joining the RNA in question with a radioactive molecule likely to resemble it, stimulating the ensemble to migrate through a gel several inches long under the force of an electric field, then taking a picture—a *radiograph*—of the gel (*Figure* 6). The radiograph revealed the distance the radioactive ensemble had moved through the gel. He could then compare that distance with the distance traveled by RNA whose originating DNA was known—in this case the transgene DNA. The comparison would indicate whether the transgene or some other gene had expressed itself.

Weaver's early radiographic probes indicated—here was another of the surprises—that while the transgene was expressed in some of the hybridomas, it was not expressed at all in many others. Its DNA was definitely present, a point that Weaver troubled to make sure of, but it was quiescent. Weaver's analysis also revealed the presence of RNA that traveled the distance characteristic of the DNA for a *gamma* rather than a *mu* isotype—a result consistent with Imanishi-Kari's serological isotyping.[47]

Weaver then sought to identify by molecular means just which genes were generating antibodies in a selection of the transgenic hybridomas he had grown with Imanishi-Kari. He analyzed thirty-one such hybridomas and Albanese examined another three, making a total of thirty-four. Weaver determined that in many of these hybridomas, the gene that was being expressed belonged to the native repertoire of the Black/6 mice, which also complemented Imanishi-Kari's serological findings. Albanese's molecular data reinforced Weaver's. "No data was thrown out because it didn't fit a story," Albanese later told federal investigators, adding that "in many cases, what you find that's surprising is very interesting."[48]

Weaver did his work mainly in the new Whitehead Institute, of which Baltimore was the director and to which he had moved his laboratory and the two dozen or so people then working in it in the summer of 1984. Affiliated with M.I.T., the Whitehead Institute was only a couple of blocks from the cancer center, and Weaver visited Imanishi-Kari regularly while Baltimore discussed the developing data with her in telephone calls and meetings in his laboratory and hers.[49] Baltimore later told several N.I.H. staff investigators that he took special care to figure out what Imanishi-Kari was telling him, explaining, English is "about her third or fourth [language], and by the time things got translated from Portuguese through Japanese into English, with an occasional foray into Finnish and German . . . some things are not perfectly clear. So I had, all through the time I have dealt with her, times when I was a little uncertain about exactly what was being said, and then we would just sit down and I would go over it until I felt comfortable."[50]

The Collaborators' Conclusions

The collaborators wanted to be sure that normal Black/6 mice did not generate the kind of NP-sensitive antibodies with the idiotypic birthmark

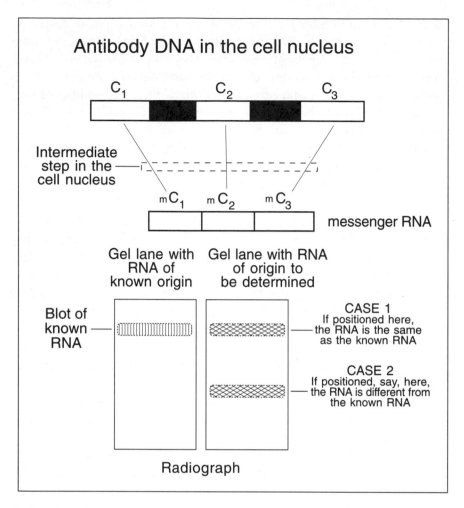

FIGURE 6

Molecular Characterization of Antibodies

The coding regions—C_1, C_2, C_3—in the DNA of the B cell are transcribed through an intermediate step in the cell nucleus and then into a molecule called *messenger RNA* that carries the genetic information out to the cell, where it is used to generate an antibody. Molecular characterization of the antibody involves comparing the messenger RNA to a known sample of antibody RNA. The characterization is accomplished by joining the known RNA and the unknown RNA with radioactive molecules like themselves, then allowing the hybrids to migrate under the influence of an electric field in parallel lanes on a gel. In this figure, the known RNA is in the left lane and the unknown RNA is in the right one. If the unknown RNA comes to rest at the same distance along the gel as the known sample—case 1—it is likely the same as the known RNA. If it comes to rest at a different distance along the gel—case 2—it differs from the known RNA. A photograph of the gel reveals where each molecule of RNA has come to rest, since each emits radioactivity that creates an image resembling a blot on the film.

that showed up in their transgenic siblings. Imanishi-Kari's characterization of antibodies in the blood of the normal mice had already indicated that they did not. Now, during the course of the experiment, she and Moema Reis each separately constructed and tested hybridomas with lymph cells and spleen cells taken from normal mice. Between them, they found that only one among 143 of these hybridomas produced antibodies with the distinctive idiotype.[51]

Late in the summer of 1985, Weaver drafted the paper about their results. Taken together, its findings were remarkable. Many of the hybridomas produced antibodies that had the idiotypic birthmark resembling the transgene's but that derived from genes native to the transgenic mice. It seemed that the introduction of the transgene did not inhibit rearrangement in the Black/6 mice's B cells. On the contrary, it seemed that somehow the transgene stimulated the abundant production of antibodies that their B cells would have otherwise produced infrequently or not at all. O'Toole, who by then had been in the laboratory for several months, was asked to read the draft critically. She gave it "a very careful review," she says. She supplied a number of editorial suggestions, including a rewording of the title into the one that was actually used. "It was a beautiful paper, beautiful data, dramatic findings," she remembered her reaction.[52]

During the fall of 1985, it was passed back and forth between the two laboratories, undergoing several revisions and rewriting by Baltimore. The collaborators, thinking that their experimental results raised important questions about how immune genes were rearranged to produce antibodies, submitted their paper to *Cell* on December 13, 1985. They subsequently dealt with the comments of referees—the reports were enthusiastic, praising the data as "very convincing and properly detailed" and calling the research "an important study"—and submitted their revised article on February 10, 1986. It appeared in the journal's issue for April 25.[53]

While the collaborators agreed that the foreign gene stimulated the abundant production of antibodies in the Black/6 mice that they did not ordinarily produce, they disagreed about why the phenomenon occurred. Imanishi-Kari was inclined to think that the responsible mechanism was a process that Niels Jerne had proposed in 1974. Called *idiotypic mimicry*, it figured in a kind of network of antibody responses that mobilized the immune system. She says that when she left Germany she doubted that any idiotype network was significant in the actual functioning of the immune system but that the data from the transgenic mice compelled her to reconsider.[54]

Baltimore, skeptical of idiotypic mimicry, thought that some kind of molecular mechanism within the transgenic mouse cells accounted for the unexpected antibody production. But whatever his interpretive disagreement with Imanishi-Kari, he was convinced that the high incidence of native antibodies to NP reported in the *Cell* paper was genuine. Two independent lines of analysis led to the same conclusion. Baltimore later told a congressional subcommittee, "For [Imanishi-Kari] to elaborate fraudulent data would have been most unlikely because the redundancy in the study would so likely have shown it up."[55]

T w o

■

Tough Customers

ON JUNE 1, 1985, Imanishi-Kari welcomed Margot O'Toole to her laboratory, a brightly lit, well-equipped room on the ground floor of the M.I.T. cancer center, spacious enough to accommodate the five to seven people who worked in it. Imanishi-Kari intended to have O'Toole start as soon as possible on work deriving from the study discussed in the *Cell* paper. Contrary to later erroneous press reports, her task was not to confirm the observations reported in the paper but to extend them.[1] Imanishi-Kari wanted to see whether the odd antibody production in the Black/6 mice could be provoked by interactions between cells, which would supply evidence that some kind of network actually figures in the process of immune response. To that end, O'Toole's assignment was to isolate two types of white blood cells—normal B cells and T cells containing the transgene— and then to transfer a mixture of both into the bone marrow of a Black/ 6 mouse whose own immune system had been destroyed by radiation.[2] O'Toole, who was experienced in this procedure, was to test whether the transgenic T cells would stimulate the normal B cells to produce antibodies against NP with an idiotypic birthmark resembling that of the transgene. She worked with Moema Reis. Reis did the radioiodinations of the reagent for O'Toole, because O'Toole was pregnant again and wanted to avoid risking exposure of her fetus to radioactive iodine. Reis, an expert in mouse genetics, was also in charge of breeding the lab's mice, including those for the venture in cell transfer.[3]

During her early months in the laboratory, O'Toole was somewhat side-tracked, later telling an N.I.H. scientist named Walter W. Stewart that she was busy with her pregnancy and with battling the Boston police. On May 1, 1985, while hurrying back to Tufts through Boston's Chinatown, on Kneeland Street, she had seen a police detective punch a man in the course of attempting to arrest him. Outraged, she collected the names of several people in the crowd that gathered, declaring, according to a written statement that she provided at her home that night, that she would bear witness to "police brutality." She also notified the *Boston Globe*, which published a story on the incident the next day. The incident became a cause célèbre in Boston's Asian community. O'Toole, testifying at a police inquiry in July, faced aggressive cross-examination that challenged her credibility and brought her to tears, forcing a recess in the proceedings.[4] (The inquiry found the detective guilty and he was suspended from the force, but a judge later concluded that the finding was without merit and reinstated him with back pay.)[5]

Imanishi-Kari tolerated O'Toole's distraction with the police, but she could be harsh and mercurial—"a tough customer," Henry Wortis noted. A friend and immunologist at Brandeis University named Joan Press reflects that "Thereza is very intense . . . [with] high expectations. . . . She's a very sweet person, but she calls them as she sees them. She calls me a twit all the time." Mary White-Scharf, a postdoctoral fellow, had met Imanishi-Kari in Rajewsky's laboratory and accompanied her to M.I.T. She later told federal investigators that she found the work environment "miserable," Imanishi-Kari manipulative, and her research practices hap-hazard. She confided her unhappiness to Herman Eisen, a senior professor in the cancer center; he helped her obtain another position at M.I.T., in Malcolm Gefter's lab. Eisen himself says that he was aware of "unhap-piness" and "personality conflict" in Imanishi-Kari's laboratory.[6]

One of O'Toole's mentors, knowing Imanishi-Kari's impatient frank-ness, later reflected that Margot's relationship with Thereza was doomed from the start.[7] When O'Toole arrived in the laboratory, she told Ima-nishi-Kari that she thought the mice were being bred to contain comple-ments of immune genes inappropriate for the cell-transfer experiment. She urged using mice with a different complement. Imanishi-Kari declares that she rejected the idea because it would becloud the issue of whether the antibodies in the mice came from the transgene or a native one. O'Toole says, "That was probably the first scientific argument I had with Thereza. I had it the very first day I was there."[8]

O'Toole later confided to Walter Stewart that during the summer she told Imanishi-Kari that certain hybridomas she was working with had been contaminated by a parasite, that Imanishi-Kari was furious at the news, and that she attacked her for getting too little work done. According to Stewart's notes of the conversation, an undergraduate in the group named Philip Cohen disparaged Imanishi-Kari to O'Toole and advised her to leave. She announced to her husband that she was quitting, but Imanishi-Kari apologized the next day and so she stayed on. Unlike Moema Reis, O'Toole found Imanishi-Kari's laboratory, as she told Stewart and other N.I.H. staff, "weird," marked by a "complete lack of camaraderie or friendship," with "a lot of unhappiness" and even "open hostilities." Hopeful at the time, O'Toole says that she resolved that she would be friendly with everyone.[9]

Not enough mice were available from the breeding program during the summer to start on the cell-transfer experiment, so Imanishi-Kari put O'Toole onto an interim piece of research with the transgenic hybridomas. She was to investigate whether the transgene mutated when it was introduced into the normal mouse, a change that would alter the immunological characteristics of the resulting transgenic mouse. O'Toole's preliminary results suggested that it did. She says that she wanted to probe the apparent change more deeply to check that it was genuine before going into print but that Imanishi-Kari insistently urged her to publish the preliminary results and that she refused. "We fought about it literally from September to May," she recalls.[10]

O'Toole felt that she needed to know more about the reagents she was using with the hybridomas. She says that Imanishi-Kari provided them from her freezer in vials labeled only with a number, neglecting to explain their origins or much about what they were, saying she could look in a laboratory notebook for more information.[11] In August, when Imanishi-Kari was away on vacation, Charles Maplethorpe, a graduate student in the laboratory who was putting the final touches on his doctoral thesis, told her that Philip Cohen had devised a reagent that would help her figure out whether she was detecting transgenic antibodies or native ones in her hybridomas. She says that she was "astonished" that she had not been told that such a reagent was available. She got some of it from Cohen, used it on the hybridomas, and concluded that her caution about the results was well founded because they came from the transgene.[12]

O'Toole says that when Imanishi-Kari returned from vacation, she told her what she had done and that Imanishi-Kari "became immediately hostile to me." O'Toole added that Imanishi-Kari insisted that Cohen's reagent was useless and that "Philip doesn't know what he's doing," that she decided who used which reagent, and that O'Toole should henceforth restrict discussions of her work to Moema Reis and herself. According to

Imanishi-Kari, the trouble was that Cohen's reagent recognized antibodies with the idiotype but did not determine whether or not they were transgenic in origin. She says, "I told [O'Toole] that these experiments don't tell us very much unless you find out whether this is actually transgene or not transgene by other means. And other means would have involved using Bet-1 or a molecular approach." Imanish-Kari adds, "So, yes, it is probably true that I was very mad, but not because she did the experiment but because of the uselessness of it. And also there were no controls. . . . It was badly planned." O'Toole says that she reported the imbroglio to Henry Wortis and that he advised her just to be a team player and see to her own experiments.[13]

In September, when enough mice were available from the breeding cages under Reis's supervision, O'Toole started on the cell-transfer experiment, destroying the bone marrow in the target mice and injecting them with transgenic T cells and normal B cells. Her initial results were arresting. Using the same reagents that Imanishi-Kari had employed, she detected antibodies to NP circulating in the mice—which suggested, remarkably, that the transgenic T cells had actually stimulated the normal B cells to produce them.[14] "I was *sure* that I was onto something hot. I was on cloud nine. I thought this was the most important scientific thing that I had ever done," O'Toole says, pointing out that in October, "when I got that first result, I was called over for a conference with Dr. Baltimore." Baltimore scrutinized her data, questioned her closely about the experiment, and was impressed enough to permit mention of her cell-transfer results in a note in the *Cell* paper.[15]

In the summer, O'Toole gave an informal talk at Tufts on her cell-transfer findings and, in October, Imanishi-Kari presented the results at a New England meeting on immunology, fully crediting O'Toole, who accompanied her. Some time in the fall of 1985, O'Toole drafted a revised bibliography of her research, writing in three new articles that she expected to publish from her work in Imanishi-Kari's laboratory, including one on the investigation into transgene mutation and another on the cell-transfer experiment. "It was a very big deal in my life," she said later. "I was going to be able to be a scientist, which was a very iffy proposition at that point. . . . I was really excited and delighted about the whole thing."[16]

The delight was short-lived. O'Toole had to obtain confirmation of the cell-transfer experiment before it could be published, and, despite repeated attempts, she was unable to reproduce the original results. She did detect the presence of antibodies to NP in the blood of the normal mice, but she could not satisfy herself that these antibodies were being produced by normal B cells. In principle, the only B cells transferred into the mice should have been normal ones, but she worried that the preparation of transgenic T cells might in practice be accompanied by a contamination of transgenic B cells. O'Toole says that it was "imperative" to show that none of the NP-sensitive antibodies came from transgenic B cells contaminating the T-cell preparation. She thought that otherwise, the experiment would not necessarily mean anything.[17] If the transfer mouse's blood was not contaminated with transgenic B cells, then the result of a Bet-1 test of the serum would be negative. To O'Toole's consternation, the test result was positive.[18]

O'Toole believed that her problem centered on Bet-1. The positive test result could have meant that the mouse blood contained transgenic B cells—or that the particular batch of Bet-1 employed was no good. Reagents like Bet-1 can degrade when they are radioiodinated and Bet-1 itself was tricky to use. Imanishi-Kari relied on it because it was the only reagent available at the time that would detect *mu-a*, the signature of the transgenic antibody. A batch prepared for use in an experiment had to be assayed against known standard proteins to make sure that it was working properly—that is, detecting what it was supposed to detect and ignoring what it was supposed to ignore. O'Toole checked the reliability of her Bet-1 against such standards. Imanishi-Kari had provided a graph in the *Cell* paper, identified as "Figure 1," that summarized the sensory characteristics of both Bet-1 and AF6-78.25 and that was glossed in the text as showing that Bet-1 reacted "only" with antibodies of transgenic origin (*Figure 7*).[19] But in O'Toole's hands, Bet-1 did not behave in conformity with the figure; it reacted positively to both native and transgenic antibodies.

Imanishi-Kari and Moema Reis responded that the proteins used as standards in the tests, which Reis had prepared, had been inadvertently contaminated. As soon as Reis purified them, O'Toole would be able to determine the reliability of her Bet-1. Even with the purified standards, O'Toole could not get Bet-1 to work the way Imanishi-Kari said it did. Without a successful assay of the reagent, she says, "I had no way of knowing whether the cells I transferred were contaminated [with transgenic ones] or whether Bet-1 was not working."[20]

O'Toole felt that she "*had* to get the assay to work." Otherwise, the cell transfer experiment could go nowhere.[21] She asked Moema Reis about Bet-1 but says that she had difficulty understanding her responses because Reis did not speak English very well. She does recall Reis's telling her at one point that Bet-1 did work. Reis later testified to federal investigators, "This reagent was no problem for us. It was very effective. . . . It did not

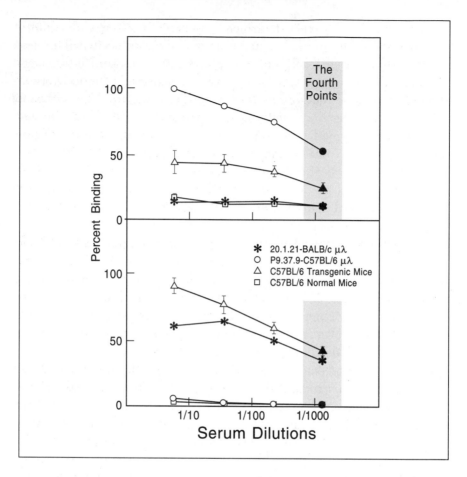

FIGURE 7
Figure 1 of the Cell *Paper*

The lower section of Figure 1 shows tests of Bet-1 on different types of supernatants; the upper section shows similar tests with AF6-78.25. The antibodies tracked in each of the curves have the idiotypic birthmark of the transgene, except that very few of the antibodies tested in the curve joined by squares carry that idiotype. The curves indicate the reactivity of each reagent running from high concentrations of supernatant on the left to low concentrations on the right. In the bottom half of the figure, the two upper curves show that Bet-1 reacts strongly with antibodies carrying the *mu-a* allotype and the two lower curves show that it reacts very little with antibodies carrying the *mu-b* allotype. In the upper half of the figure, the two upper curves indicate that AF6-78.25 reacts strongly with antibodies carrying the *mu-b* allotype; the lower curve joined by asterisks indicates that it reacts very weakly with antibodies carrying the *mu-a* allotype. The fourth point on each of the six curves later came to be disputed, even though it is evident that each of the curves is well established without its fourth point.

function well in only one experiment."[22] When O'Toole read the draft of the *Cell* paper, she raised no objection about the graph or related text concerning Bet-1. When later asked why in a congressional hearing, she testified, "I naturally assumed it was I who was making the error, because I wasn't getting the results that other people were getting."[23] Nevertheless, she spent months continuing to struggle with the reagent. "Bet-1 became absolutely an obsession with me," she says.[24] Frustrated, she began to grow suspicious of Imanishi-Kari's data, although she later told federal investigators that she thought the problem was just "miscommunication, eccentric personalities, lack of understanding of what my data actually said, extraordinary pressure on Dr. Imanishi-Kari at that point in her career."[25]

Imanishi-Kari says that, despite her early disagreements with O'Toole over matters like the mouse breeding, she "never thought there was anything serious involved," adding, "Normally, when there is something wrong and I perceive it, I go to talk to people. I may mumbo and jumbo a little for a while, but then I do it. I didn't perceive anything wrong in Margot." Moema Reis left the laboratory in mid-January 1986 to return to Brazil. Imanishi-Kari assigned O'Toole the important job of sharing with herself the responsibility for the care and breeding of the laboratory's mice, although by then she was becoming increasingly irritated and dissatisfied with the younger scientist's work and attitude. She complained to several colleagues that she wondered how O'Toole expected to make a career in science, that she simply didn't work hard enough.[26] Imanishi-Kari recalls suggesting that O'Toole go out to Stanford for a short time to collaborate with two scientists who were engaged in research similar to some of hers and that O'Toole adamantly refused, declaring, "No. This is my experiment." She remarks that O'Toole also complained that a postdoctoral fellow named Evelyn Rabin was working in Wortis's laboratory on a project closely related to hers.[27]

Imanishi-Kari felt that O'Toole was wasting the laboratory's valuable and limited resources. She disagreed with O'Toole's insistent claim that it was important to be sure that the blood results did not arise from the presence of transgenic B cells. What O'Toole had to pin down, she says, was that antibodies against NP derived from native genes—that the allotypes in their constant regions were *mu-b* rather than *mu-a*. The crucial reagent for that identification test was AF6-78.25, not Bet-1. Imanishi-Kari notes that, in any case, O'Toole used preparations of Bet-1 that were drawn from the same batch that Reis used and that, at one point, she actually did get the result with the reagent that other people got. Imanishi-Kari reflects that O'Toole seemed plagued by "some kind of insecurity" about experiments, continuing, "She feared that some little thing, some detail was wrong, or not quite right. They were little details. They were not on the big picture."[28]

Imanishi-Kari holds that O'Toole "was supposed to know more than I did in adoptive cell transfer" but that many of the mice that she used in her cell-transfer experiment died. Consulting an expert in bone-marrow transfers about the problem, Imanishi-Kari learned that O'Toole was handling the mice poorly, irradiating them too soon after they arrived in the laboratory. Imanishi-Kari complained that O'Toole was not subjecting her reagents to suitable tests on known standards to make sure that they were working properly. Imanishi-Kari says, "I kept telling her, 'You have to put on all these controls,'" declaring that it was O'Toole's lack of adequate controls that led to "her big frustration." She adds that it was not easy to tell O'Toole what to do. Despite Imanishi-Kari's seniority in age, she was not much senior in science, and O'Toole, she says, was not the kind of person whom it was "very easy to tell what to do."[29]

Imanishi-Kari had advised O'Toole at the outset that she had enough grant money to employ her for only one year. Once O'Toole's grant with Imanishi-Kari ended, she would be ineligible for any further N.I.H. postdoctoral training support under the rules of the agency. Imanishi-Kari says that she was "very worried about her" and that she repeatedly urged her to apply for grants but that O'Toole did not seek any.[30] O'Toole remembers Imanishi-Kari's telling her in the summer of 1985 that O'Toole might move with her to Tufts, if Imanishi-Kari herself was hired there, so long as O'Toole obtained her own research support. (Imanishi-Kari doubts that she made such a firm tender to O'Toole, pointing out that she did not feel "comfortable with her" and that, in any case, O'Toole planned on an independent research career.[31]) O'Toole recalls Imanishi-Kari's urging her to apply for grants to pursue the cell-transfer research but also screaming at her to publish the preliminary results of another experiment whose results O'Toole doubted. O'Toole says that Imanishi-Kari berated her constantly, saying, "At least I publish something. You're so smart, but all you can do is not publish. You publish nothing." And that she would go on, "You'll never amount to anything. You'll never get a job. You'll be just one of those women the husband has to support."[32]

Imanishi-Kari, separated from her husband for four years, was in the process of getting a divorce, and was meanwhile rearing her daughter, then ten. She knew all about the conflicts women scientists face. She says, "You have to work very late, you have to work Saturday and Sunday without any distractions, let alone when you have distractions. It's very hard."[33]

Yet she was a survivor in science. Vivien Igras, a technician in a neighboring laboratory in the M.I.T. cancer center, noted that Imanishi-Kari had been forced "to give a lot up to get to where she was" and continued, "I think she expected the same kind of commitment from other women. Ironically, instead of sympathizing with you as a woman because of what she's given up, she expects you to sacrifice for your career as well."[34]

Imanishi-Kari wondered whether O'Toole had the necessary tenacity and devotion. She says that during the year O'Toole often had to be away from the laboratory to take care of her child, and that she "was trying very hard to get another child and was having a lot of difficulties." Igras recalls that, at a party in late 1985, O'Toole mentioned that her pregnancy had miscarried and that she was having time-consuming day-care troubles with her son. She also said then that she wanted another child and for that reason was thinking about giving up science. Imanishi-Kari says, "She kept telling me, Don't worry about me. I don't know what to do with my life."[35]

O'Toole seems in truth to have wanted a life in science, but not the seven-day-a-week life exemplified by Imanishi-Kari. Imanishi-Kari was probably right about her job prospects, O'Toole feels, but she would not do science the way she perceived Imanishi-Kari to be practicing it.[36] Nothing alienated her more from Imanishi-Kari than her frustrations with Bet-1. She says that at one point Imanishi-Kari told her just to use the data that Moema Reis had gotten in her control assay of the reagent, sometimes yelling at her across the room to take the shortcut. O'Toole adamantly refused. By early spring of 1986, the two scientists were barely speaking to each other.[37]

In March, when Imanishi-Kari returned from a trip of several weeks, O'Toole greeted her with the news that she was still unable to repeat the cell-transfer experiment and that her initial apparent success had arisen, she believed, from a mistake in the genetic typing of the mice she had used. O'Toole says that Imanishi-Kari was uninterested in the explanation. She told O'Toole to do no more experiments and to confine herself to the care and breeding of the mice.[38]

Imanishi-Kari notes that Moema Reis was gone and that she had originally asked O'Toole to look after the breeding and testing of the mice in Reis's place on the understanding that she would help her. She adds, however, that she realized that O'Toole "had done lots of experiments that were completely unacceptable," elaborating, "It wasn't that the answer was bad and therefore the experiment was bad. That was not the case." In her judgment, the way O'Toole did experiments it was impossible to determine whether the experiment was no good or the assay was just not working. She was reluctant to see O'Toole go on wasting valuable laboratory resources to no productive purpose. Imanishi-Kari points out that her mouse bill alone was high, running $40 to $50 a day for the care of the two hundred or so animals bred and kept for experiments. She thought it better for O'Toole to concentrate on caring for the mice for the time being, thinking that she might resume research later.[39]

Imanishi-Kari had declared that she would pursue the cell-transfer experiments herself, telling O'Toole that she did not want her to do them anymore because, O'Toole put it to herself, "I had lost faith."[40] In 1992, O'Toole told me that shortly thereafter Imanishi-Kari called her in and said, " 'Margot, look. There's really something here,' " showing her data that suggested the results O'Toole had been unable to repeat. Then Imanishi-Kari, sitting at her desk, went over the data with a pen. To O'Toole, who was looking over her shoulder, she appeared to be crossing out high measurements in mouse groups that she wanted to be low, and low measurements in those that she wanted to be high. " 'See, there's a real trend here,' " O'Toole recalls her saying. "She's really happy and perky. I'm just astonished."[41]

"Until that moment," O'Toole continued, "I was in complete turmoil. I was frantic trying to make myself get the data. I was frantic trying to understand why I am not able to be a scientist. Watching her, I just had this utter feeling of tranquility that I was not partaking because I would not partake. She said, 'Bring me your data.' And then she went through my data and made them conform. She turned around to me and she said, 'So, what do you think?' Then she turned back. And this word came out of my mouth, spontaneous and genuine; it just escaped in a whisper out of my lips: 'Fascinating.' And she turned around to me and she looked in my eyes, and her eyes were smiling at me. And she liked me. And she pitied me. And she welcomed me back into the fold."[42]

O'Toole interpreted Imanishi-Kari's handling data differently on other occasions. For example, in an interview with N.I.H. investigators in 1988, a time much closer to the incident, she said nothing about manipulation or crossing out of data. She said only that Imanishi-Kari "called me in and she said, this experiment worked." O'Toole continued:

It only worked if you threw out two out of three of each of the groups of five, and you selected the high groups here and the low groups here. . . . I was through fighting. And I looked at her and I said, "Thereza, I feel you are being overly optimistic. . . ." And she said, now bring me your own data, I bet all your experiments worked and you have been sitting down whining, saying they didn't work. So I brought her all my own data, which was just as shitty as hers—excuse my language—and she did the same thing. She called me in and she said, look the pattern holds, you have seven experiments here that worked, and you are telling me they didn't work. And I looked at her and I said, "That's fine Thereza."[43]

Imanishi-Kari, who went over data with her several graduate students and postdoctoral fellows all the time, says that she has no specific recollection of the encounter that O'Toole found so significant. "What I recall is that I went through her data myself, and then I made an ordered sheet of her own data, keeping order my way. I did not change her data." Nor, she says, did she ask O'Toole to manipulate her data. "I never told her to write any paper, because I didn't think she had enough data."[44]

O'Toole later declared that she resolved simply to "stop arguing and just not do what [Imanishi-Kari] told me," adding, "I decided to leave Dr. Imanishi-Kari's laboratory and seek employment elsewhere. I gave my notice but agreed to continue taking care of the mice through May." Since O'Toole's appointment at M.I.T. expired at the end of May anyhow, it is difficult to know what she meant by "notice." She may have been alluding to the suggestion that she says Imanishi-Kari had made the previous summer of accompanying her to Tufts if she raised grant money of her own, but she had not obtained any such funding nor had she applied for any. O'Toole nevertheless says, "We were at each other's throats, so she was very relieved that I was the one that said, 'Forget it. I'm not going with you to Tufts.' "[45]

O'Toole did expect to go to Tufts, but to work with her husband. She says that she had been promised a renewal of her appointment there as an assistant research professor if she raised the money to support it. (Martin Flax, the chairman of the department at the time, says that he had not made an explicit commitment to give her such an appointment but that she would have been eligible for it if she had obtained the necessary grant money.) In any event, O'Toole hoped that the collaboration with her husband would lead to the kind of results that would attract funding. She recalls that Imanishi-Kari, who knew her plans, turned very friendly, coming to chat while she worked on the mice. "There was this peace between us." O'Toole continues, "I no longer railed against having to go to work. I thought I'll just do what she wants me to do and I'll leave."[46]

But O'Toole continued to brood over the failed cell-transfer experiments. "I just couldn't let go of trying to figure out what it was that I couldn't figure out. So one day, despite my absolute determination to keep the peace and not rock the boat and get out with my tail in one piece, without even thinking about it I said to her, 'Well, Thereza, why do you think that Bet-1 worked for you and it didn't work for me?' And she laughed and she said, 'It works the same for us as it does for you.' " Imanishi-Kari says that, the remark, if she made it, was an offhand declaration that Bet-1 was working for her and that it was working for O'Toole, too—if she would only handle it properly. O'Toole, however, heard it as an admission that the reagent did not work at all.[47]

For a number of weeks O'Toole had been seeking out Charles Maplethorpe, the graduate student who had steered her to Philip Cohen and his reagent the previous summer. She had seen little of him since he had finished his doctorate, in August, but she remembered him as one of the unhappy people in the lab. She wanted to meet with him because she knew that he didn't get along with Imanishi-Kari and she wanted to know why, so that perhaps she could figure out the source of her own troubles in the lab.[48]

I talked with Charles Maplethorpe one evening in March 1993 at a cafe near DuPont Circle, in Washington, D.C., where he works as an examiner

CHARLES MAPLETHORPE
*Photo Credit: Joe Runci/*Boston Globe

for the Food and Drug Administration. He is a lean, handsome man of high intelligence and guarded reserve. A country doctor's son, he had done his undergraduate work at M.I.T. and had gotten excited about immunology while he was in medical school in his native Iowa. In 1977, while completing his M.D., he returned to M.I.T. to earn a doctorate. The next year, he enlisted for research under Malcolm Gefter, who was then interested in getting into immunology. Gefter assigned him to work in the laboratory of Michael Bevan, a junior faculty member and immunologist, where he was to learn how to grow T-cell hybridomas with the ultimate aim of developing an experimental system that he and Gefter could exploit. Gefter says that Maplethorpe got mired in the details of the work and that his relationship with Bevan steadily slid into "an irretrievable breakdown."[49]

Bevan notes that Maplethorpe struck him as a smart guy but that he "didn't take advice very well and was much too cocksure of himself." Maplethorpe says that Bevan "was having an affair" with one of his graduate students. The couple eventually married, but Maplethorpe saw their relationship as "prostitution" and reportedly complained about it to women's consciousness-raising groups on the campus. He contends that the young woman got "all the attention, all of the experiments, all of the favorite treatment." (Bevan declines to comment on how his relationship with her affected his lab except to say brusquely that "Charlie is screwed up.") Maplethorpe later testified that Bevan "called me a queer." He told me, with evident emotion, that his time with Bevan amounted to "the worst years of my life," that the affair made the spaceship-like circumstances of the lab into a "nightmare."[50]

Maplethorpe left Bevan's lab and Gefter's supervision in December 1980, and in January 1981, at Herman Eisen's suggestion, he joined up with Imanishi-Kari, who was just coming to M.I.T. and who, Eisen thought, would be eager for a graduate student already familiar with immunology.[51] Maplethorpe says that he helped set up Imanishi-Kari's laboratory and establish its capacity to use techniques in molecular biology. Some of his work involved isolating the antibody genes activated by NP, determining their size and ascertaining the sequence in which the core constituents of their DNA—chemical arrangements called *base pairs*—occur. Size and sequence identify a gene so that it can be compared with other genes; the sequence also encodes the DNA's genetic information, indicating the kind of protein—in this case, antibody—it produces. Maplethorpe says that he wasted a year extracting a gene from a hybridoma and then identifying it, only to discover that it was a gene from a viral contaminant. He holds that Imanishi-Kari refused to acknowledge the merit of his work and he decided that she was "ridiculous." He considered Imanishi-Kari "domineering" and unsupportive, averring, "No one succeeds when the bastards don't want you to succeed."[52]

Imanishi-Kari recalls that she had been told that Maplethorpe was very smart and deserved another chance, but she says that "in spite of his brightness, he would not do experiments." After a couple of years, he seemed far from accomplishing enough research for the doctorate, and at Imanishi-Kari's initiative, the biology department, which does not like to see its graduate students fail, formed an ad hoc committee to help Maplethorpe qualify for his degree.[53] Imanishi-Kari says that, in her view, "Charlie has an essential problem, which must have come from long ago," explaining, "He's an unhappy person. . . . He thinks everybody's after him, that everybody is doing things to damage him . . . I really thought, for many years, that poor guy, I should help him. Even if he is not getting his data . . . I should be helpful. Because some people need some opportunity. But I don't think he ever saw that."[54]

Eventually, Imanishi-Kari wanted nothing more than to see Maplethorpe leave. She sensed his contempt and gathered that he did his best to turn new members of the lab against her. Nancy Hopkins, a member of the ad hoc committee, recalls that during visits to her office, Maplethorpe at times seemed despairing but boasted that he maligned Imanishi-Kari to first-year students. She says that he seemed to hate Thereza, an impression that he gave others. (Maplethorpe says that he didn't hate Imanishi-Kari, declaring, "You have to respect someone to hate them.")[55] Christopher Albanese told federal investigators that the relationship between Imanishi-Kari and Maplethorpe was marked by off-scale animosity, noting, "It was as bad as it got between Thereza and Charlie."[56]

In the fall of 1984 Imanishi-Kari completed a research paper with a doctoral student in her lab named Martina Boersch-Supan that was based mainly on Boersch-Supan's thesis work but that used some of Maplethorpe's sequence data and that credited his contribution in a footnote. On November 15, he complained to Imanishi-Kari that she had not made him a coauthor of the paper. According to Maplethorpe's record of the conversation, he said that she had not consulted him at all in drafting the paper and, indeed, had made it available to him only in the last day or two. Although the footnote acknowledged him for having given permission to publish his data, he insisted he that had not given any such permission and had not been asked for it. Imanishi-Kari responded that she intended to give him authorial credit on another paper that used his sequence data. Maplethorpe argued that he was entitled to coauthorship by academic tradition and for a "moral" reason. His sequence data comprised "the very first bit of molecular biology to come out of our lab. . . . Every bit of that data was produced through an arduous struggle and should now be given due recognition." According to Maplethorpe's notes, Imanishi-Kari insisted that she alone was responsible for all decisions regarding papers from her lab and that "she will not follow the academic tradition." The notes add that Imanishi-Kari was "directed to accommodate me" by Gene Brown, the chairman of the M.I.T. biology department, and by Salvador Luria, the Nobel laureate who then headed the cancer center, but that "she refused."[57]

Imanishi-Kari says she does not remember a fight with Maplethorpe over the matter. She recalls telling him at some point that she saw no reason to add him to the paper as a coauthor because his contribution to it was "pretty small, just one little piece of what Martina had found." Maplethorpe's sequence data in fact concerned only one type of DNA fragment out of sixteen that were analyzed, many of the rest also by sequence; it thus formed only a minor fraction of the material presented in the publication. Boersch-Supan was reluctant to have him included as a coauthor, partly, she says, because "it always looks better if you have not too many authors on the paper" and partly because his data was very incomplete. Imanishi-Kari appears to have accommodated Maplethorpe in an appropriate way, crediting him in the text of the paper as well as in the footnote.[58]

Maplethorpe felt cut out of the scientific life of the lab, denied access to materials, and shunted away from important developments. He says that he found Imanishi-Kari generally "very secretive" with data, reporting, by way of illustration, an encounter he had with Albanese in the spring of 1985. Hearing about Imanishi-Kari's work with David Baltimore, Maplethorpe was curious to know more about the extraordinary immune response of the transgenic mice. He asked Albanese for a look at the radiographs he had gotten—they were the same kind that Weaver had obtained—bearing on which gene produced the antibodies against NP. Maplethorpe later testified in Congress that Albanese said "that he wasn't permitted to show me the data, that she [Imanishi-Kari] specifically told him not to show me the data."[59] Albanese, however, later said to federal investigators, "I can't remember ever being told, 'Don't show anybody your research, your results,' but on the other hand, it wasn't encouraged that you go around and show everybody your data." Albanese said that he was friendly with both Imanishi-Kari and Maplethorpe, but he "didn't want to be caught in the middle of two clashing personalities," one of whom was his boss. He continued, "Now, I'm not saying that Charlie might not have had, in his own mind, a reasonable reason to see my data, and I'm not saying that I didn't refuse to show him my data, but I'm trying to stay outside this controversy as much as possible."[60]

Maplethorpe recalls that in late May 1985, at the regular Thursday-afternoon immunology seminar held on the sixth floor of the M.I.T. cancer center, Imanishi-Kari spoke about the remarkable results of her collaborative research with David Baltimore and David Weaver. He says that, in response to a question following her talk, Imanishi-Kari mentioned that Moema Reis had recently found evidence that the transgene might be expressing itself at a low level in the hybridomas. Such evidence would have implied that the transgene rather than native genes might have been responsible for the production of antibodies displaying idiotypic birthmarks related to the transgene. It would thus have undercut the remarkable results of the experiment. Maplethorpe says that the news "shocked the people in the room because it seemed to go against everything that they had just heard in the hour presentation." Maplethorpe thought that Imanishi-Kari's alleged remark "in particular . . . must have shocked David Weaver, who was present."[61]

One evening about a week later, Maplethorpe continues, Weaver, accompanied by an Israeli postdoc who had been at the seminar, visited Imanishi-Kari in her lab. Maplethorpe, who happened to be there working on his thesis, eavesdropped on the conversation. He says that Weaver and the Israeli questioned Imanishi-Kari, particularly about the specificity of Bet-1, the reagent that detected the presence of transgenic antibodies, trying, it seemed, "to milk information out of her but at the same time, not threaten her and not to seem suspicious." Maplethorpe later testified in Congress that what he heard revealed that Imanishi-Kari "was obtaining the same results that Dr. O'Toole subsequently obtained"—that is, Bet-1 reacted regularly with *mu-b* as well as with *mu-a*.[62]

However, Imanishi-Kari says that she was the organizer of the weekly immunology seminar that year and thought it inappropriate to designate herself a speaker. She does not recall giving a talk, and neither Weaver nor Nancy Hopkins remembers her giving one, let alone in some way shocking the audience. Weaver has no memory of the subsequent meeting in Imanishi-Kari's lab and points out that it was his custom to come to chat with her at the end of the workday to catch up on the progress of their experiment. And what Maplethorpe heard Imanishi-Kari tell Weaver that evening was the opposite of incriminating. She said that the cell line from which Reis had obtained the Bet-1 used in the recent experiments appeared to be contaminated, making some of the reagent improperly cross-react with antibodies produced by genes native to the transgenic mice. She also said that the Bet-1 generated by the uncontaminated part of the cell line reacted only with the antibodies from the transgene, as it was supposed to.[63]

Maplethorpe says that the air of secrecy in the lab, the incident with Albanese, and the overheard conversation all made him suspicious of the *Cell* paper.[64] Since data was not being shared with him, he says, he took it upon himself to search it out on occasion in Albanese's notebooks and Imanishi-Kari's.[65] He secretly tape-recorded Imanishi-Kari's conversation with Weaver and the Israeli postdoc—perhaps without being aware that to do so was illegal—even though in June 1985 he could not have been sensitive to O'Toole's problems with Bet-1 because they had not yet occurred and no one else had complained about the reagent the way she would. Vivien Igras says that Maplethorpe was intent on a "vendetta," against Imanishi-Kari, that he declared at a party, " 'that bitch; I'm going to get her somehow.' " Maplethorpe later categorically denied that he had made any such statement to Igras.[66]

Late during the winter of 1985, Maplethorpe had gone to the M.I.T. ombudsperson, a woman named Mary Rowe, to complain that Imanishi-Kari was harassing him. Rowe was supportive, he says, counseling and encouraging him through a number of visits and talking with various faculty to facilitate the completion of his doctorate. At his final meeting with Rowe, after finishing his thesis in the summer, he mentioned that he suspected Imanishi-Kari was committing scientific fraud. She promptly handed him a copy of M.I.T.'s guidelines on what to do in such cases, but Maplethorpe told me that he raised the issue only in passing, because he thought Rowe should know about it. According to Nancy Hopkins, Maplethorpe successfully convinced another professor at M.I.T. that Imanishi-Kari had committed fraud, but Maplethorpe declined to press the matter when he saw Rowe, believing that universities lacked the moral capacity to deal with it and fearing that he would be worse off for doing so.[67]

Maplethorpe was worried about getting a job. Hopkins says that his doctoral thesis was beautifully written but of marginal scientific importance. He told me that one day after finishing his degree, he was with Imanishi-Kari in her secretary's office and that she said to him, "What are you going to do now, Charlie?" He thought that, since she knew full well he had no job, she merely intended to humiliate him in front of the secretary. He said that he didn't know, but asked whether he could count on her for a recommendation. She called him into her office and, with the door closed, told him that she had succeeded against the odds, implying, he thought, that if she could make it on her own, so could he. He says that she warned him: I have to tell you one thing, Charlie. If you put me down for a recommendation, you can be sure that you won't get the job.[68] Imanishi-Kari remembers Maplethorpe's asking her for a letter recommending him for a postdoctoral fellowship and telling him that she could not write a letter on his behalf. She explains, "Our relationship had been very bad, he never had any respect for me particularly, and I didn't think I would have been telling the truth if I wrote a positive recommendation letter. So I said that the best thing would be for him to ask somebody who . . . had a much better appreciation of his work than I had."[69]

Maplethorpe sought a job for more than a year, finally obtaining one in a new biotechnology firm in the Boston area in December 1986. (He remained there until December 1989, which was when he joined the Food and Drug Administration.) During the months he was looking—a period that included Margot O'Toole's time with Imanishi-Kari—he frequently returned to the M.I.T. cancer center to visit friends. Imanishi-Kari asked the administrator of the center to request that he leave, which she did, but Maplethorpe kept showing up anyway.[70] Maplethorpe says that he "really didn't have much to do with O'Toole." He found her "hopeful" and "kind of innocent and naive" about the situation in the laboratory. He didn't want "to harm her chances," but he adds that he "couldn't really respect anyone that much who had become a member of Thereza's lab."[71]

In the early spring of 1986, several members of the lab told him that O'Toole was having trouble with Imanishi-Kari and that he should speak with her. He says that at first he brushed off the idea, not least because he had more important things to do, like getting a job, but he got in touch after a friend told him twice that O'Toole wanted him to call her and gave him a small yellow Post-it with her phone number. At lunch in late April, she appeared distraught, telling him how Thereza would question her competence and scream at her when she repeatedly failed to obtain the experimental results she expected. He says that he responded in a "jaded" but reassuring way, explaining that what happened to her would happen to anyone working with Imanishi-Kari, that she shouldn't let it bother her, that she should just try to leave. He also remembers informing O'Toole that he found Imanishi-Kari secretive and unwilling to share the results of experiments and that he smelled something fishy about her work. He later said of the meeting with O'Toole, "I told her all of my suspicions."[72]

About the same time, O'Toole talked with David Weaver about what light his molecular results might cast on her observations in the cell-transfer experiment. He had heard that she was having difficulties with

Imanishi-Kari, but she said nothing about them to him.[73] She did remonstrate with Wortis. He was well aware of Imanishi-Kari's dissatisfactions with O'Toole and he says that O'Toole had previously grumbled to him not only about Imanishi-Kari's alleged tolerance of parasitic contamination but also about the way that Donald Mosier at Fox Chase, and even Sidney Leskowitz, at Tufts, did science. He declined to get involved, thinking that Margot and Thereza's clash boiled down to personality differences and that O'Toole's complaints were nothing different from what he had heard from her before.[74]

O'Toole also spoke with Herman Eisen, who was in charge of the grant to the cancer center that was funding her. Later, she declared that she told Eisen that Imanishi-Kari was pressuring her to misrepresent her experimental results but that Eisen appeared "aloof" and took no action. Eisen says, "I don't easily get angry, but that really angered me." He calls it "absolutely inconceivable" that he would do nothing if a postdoc complained, "My mentor wants me to lie about my results." Eisen recalls that when O'Toole came to him she was "very unhappy and distraught" and seemed to want advice. He got the impression that Imanishi-Kari was "very hard on her . . . the way a mentor can be hard on a postdoc that is not functioning well in the mentor's view." Eisen knew that Imanishi-Kari would shortly be moving to Tufts and that O'Toole's postdoctoral position would end in a few months. He advised O'Toole that she had little choice but to "hang in there and just look for another job."[75]

O'Toole told Mary Rowe, the M.I.T. ombudswoman, about her problems with Imanishi-Kari, including Imanishi-Kari's having confined her to the management and breeding of the mice.[76] She says that Rowe urged her to file a charge of misconduct against Imanishi-Kari on grounds that she was being mistreated. "I refused," O'Toole says, emphasizing that her quarrel with Imanishi-Kari was not over mistreatment but over the science and that Mary Rowe only wanted her to complain "about how I was not being mentored, a whiny thing that had nothing to do with my professional standard, had nothing to do with what I was trying to accomplish as a scientist." Besides, she says, "I had the utmost sympathy and empathy for Thereza. She was going through a divorce. She was clearly extremely physically ill. . . . She was having to move." O'Toole says that she did not see how she could file charges of mistreatment against a woman who was "barely getting by." She rather wanted "somebody to step in and say, 'There's a scientific issue here that needs to be resolved.' "[77]

Around the beginning of the last week in April, such a somebody appeared by chance in the M.I.T. laboratory. She was Brigitte Huber, an immunologist on the Tufts medical faculty and one of O'Toole's mentors and friends, who had come to pick up something on a day when Imanishi-Kari happened to be away. O'Toole recalls that Huber asked how things were going and that she replied, flatly, "Not so good," continuing, "She sort of waved her hand and said. 'Yes. Yes. I've heard all about it. I don't want to know about it. You and Thereza have a personality conflict.' I said, 'No. That's not what's going on here.' She said, 'You mean there's something wrong with the science.' I said, 'That's exactly what I mean.'" Huber urged O'Toole to talk with someone about the matter, and O'Toole said, "What about you?" Huber reluctantly agreed, but said that she was busy then. O'Toole should come see her in two weeks.[78]

T H R E E

■

Assertions of Error

ON WEDNESDAY, May 7, 1986, shortly before she was to see Huber,
O'Toole opened a blue notebook of Moema Reis's to look up some mouse-
breeding data and stumbled on a cluster of pages—seventeen of them—
that appeared to be a record of key experiments published in the *Cell*
paper. By then she knew the paper in detail. "I had studied this paper
better than I had studied anything before in my life," she says. "I would
weep over this, saying why can't I get the same result." But when she
came across columns of numbers spread across Reis's pages, she suddenly
believed she understood why and exclaimed, "It's not me!" The data she
saw suggested to her that much about the paper was unfounded, including
its representation of the behavior of Bet-1 in Figure 1 and its central claim
that the introduction of the transgene somehow induced the Black/6 mice
to produce from their own genes abundant antibodies to NP with an
idiotypic birthmark resembling that of the transgene.[1]

O'Toole xeroxed the pages and studied the copy at lunch. One page
recorded tests of Bet-1 indicating that the reagent did not distinguish
between native and transgenic antibodies; it reacted equally with the *mu-
a* and *mu-b* allotypes. Others called into question parts of the experimental results reported in two tables—Table 2 and Table 3—in the *Cell* paper
(*Figures* 8 and 9). Table 2 summarized many of the serological tests that
Imanishi-Kari and Reis had carried out using the 340 hybridomas from

Table 2. Frequency of 17.2.25 Idiotype-Producing Hybridomas in Normal and Transgenic Mice

Organ	17.2.25 Idiotype-Positive	17.2.25 Idiotype-Positive Plus:			
		Anti-NP (κ)	Anti-NP (λ)	μ^a	μ^b
Normal Spleen	1*/144 (<1%)	1†/144	2/144	0/144	0/144
Normal Lymph Nodes	0/100	0/100	0/100	0/100	0/100
Transgenic Spleen	43/150 (28%)	0/43	7‡/43	9§/43	1/43
Transgenic Lymph Nodes	129/190 (68%)	6/129	12/129	33/129	10/129

B cell hybridomas were isolated from spleen and lymph nodes of transgenic and normal C57BL/6 mice. Secreted Ig from hybridomas was assayed for binding to the anti-17.2.25 idiotypic antibody or for NP-binding with either λ or κ light chains. The portion of the hybridomas that contained the μ isotype was assayed by the anti-allotype antibodies (see Figure 1) to associate the anti-17.2.25 idiotype binding with either the μ^a or μ^b allotype.
* Weak reactivity of hybridoma to anti-17.2.25.
† Not 17.2.25 idiotype.
‡ 2/7 were observed to have μ^a allotype.
§ 2/9 were found to be λ-bearing Ig, and 7/9 were found to be κ-bea. Ig.

FIGURE 8
Table 2 of the Cell *Paper*

The data in the box on the upper left show that idiotypically birthmarked antibodies were produced by only 1 out of 144 hybridomas from normal spleen cells and by none of 100 hybridomas from normal lymph cells. The data in the box on the bottom left show that 172 (the sum of 43 and 129) of 340 (the sum of 150 and 190) transgenic hybridomas generated idiotypically birthmarked antibodies. The data in the box on the lower right indicate that only 53 (the sum of 9, 1, 33, and 10) of these transgenic hybridomas produced antibodies with a *mu* isotype.

transgenic mice and the 144 from normal ones. Table 3 delineated the serological and molecular data that Imanishi-Kari, David Weaver, and Chris Albanese had jointly obtained from the thirty-four specific hybridomas. Some of the data in the seventeen pages appeared to indicate that antibodies to NP with the transgene-like idiotypic birthmark had been produced in abundance by a Black/6 mouse identified as No. 56 and "NORMAL" (*Figure 10*).Since the animal appeared to be capable of producing such antibodies even though it was said to lack the transgene, the data implied that the introduction of the transgene made no difference to the immunological characteristics of the mice used in the experiment.

Table 3. RNA and DNA Analysis of Transgenic Hybridomas

Hybridoma	Ig	mRNA 17.2.25 V$_H$ Probe	S1 Nuclease Probe	Transgene DNA	V$_H$ Sequence
L3.10.5	μ, κ, NP	+	+	+	Transgene
L4.4.1	μ, κ, NP	+	+	+	Transgene
S2.15.6	μ, λ, NP	+	+	+	Transgene
L4.14.3	μ, κ, NP	+	−	−	
L3.14.5	μ, κ, NP	+	+/−	+	
L4.13.2	γ₁, κ, NP	+	+	+	
L4.7.2	γ₁, κ	+	−	−	
S1.2.6	γ₂ₐ, κ	+	−	−	
L3.1.4	γ₂ᵦ, κ	+	−	+	
L3.3.2	γ₂ᵦ, κ	+	−	−	
L3.4.4	γ₂ᵦ, κ	−	−	+	81X
L3.5.2	μ, κ	−	−	+	
L4.5.2	γ₂ₐ, κ	−	−	+	81X
L3.10.6	γ₂ₐ, κ	+	−	+	
S2.13.3	μ, κ	−	−	+	
L4.8.1	μ, κ	−	−	+	
L4.3.4	γ₂ᵦ, κ	−	−	+	81X
L3.9.4	γ₂ₐ, κ	−	−	+	
L3.1.1	μ	+	+	+	Transgene
L4.9.4	μ, κ	−	−	+	
L3.7.3	γ	−	−	−	81X
L4.2.6	γ₂ᵦ, κ	−	−	−	
L3.6.3	n.d.	−	−	+	
L3.8.3	μ	−	−	+	
L4.4.3	γ₂ᵦ	−	−	+	
L4.8.2	γ₂ᵦ	−	−	+	
L4.10.1	α	−	−	+	
L3.13.6	n.d.	−	−	+	
L3.16.3	α	−	−	+	
L3.18.5	n.d.	−	−	+	
L3.17.3	α	−	−	+	
S1.3.1	γ₂ₐ	−	−	+	
S2.14.4	μ	−	−	+	81X
S1.3.2	n.d.	−	−	−	

The transgenic hybridomas were examined by S1 nuclease and Northern blot analysis of expressed Ig heavy chain RNA as depicted in Figure 3 and Figure 4. In addition, inheritance of the microinjected DNA was determined by Southern blot analysis. As indicated, only a fraction of these hybridomas were also positive for μ transgene transcription by the S1 nuclease assay. Similarly, not all of these clones retained the DNA insertion of the microinjected DNA, as assessed by Southern blots.

C   B   A

FIGURE 9
Table 3 of the Cell *Paper*

In column A, the plus signs (+) indicate that the hybridoma carries the DNA of the transgene and the minus signs (−) indicate that it does not. The plus signs in column B indicate that the transgenic DNA was expressed in at most six of the hybridomas. The Greek letters on the left in column C—α (*alpha*), γ (*gamma*), or μ (*mu*)—identify the isotype of the antibody produced by the hybridoma. (*Alpha* is the isotype associated with the immunoglobulin IgA. The notation "n.d." means "not determined.") Fifteen of the hybridomas produced antibodies with variants of a *gamma* isotype and three produced antibodies with *alpha* isotypes. Seven of the hybridomas produced antibodies with a *mu* isotype that did not express the transgene DNA, which indicated that their antibodies derived from native genes, too. Although the hybridomas were not selected for generating antibodies with the idiotypic birthmark of the transgene, Imanishi-Kari had unpublished data showing that all but the four designated "n.d." in column C did so. Although hybridoma L4.13.2 expressed transgenic DNA, molecular analysis revealed that the variable region of the antibodies it produced came from native DNA. Thus, a total of twenty-five (15 plus 3 plus 7) of the hybridomas generated idiotypically birthmarked antibodies that were native in origin.

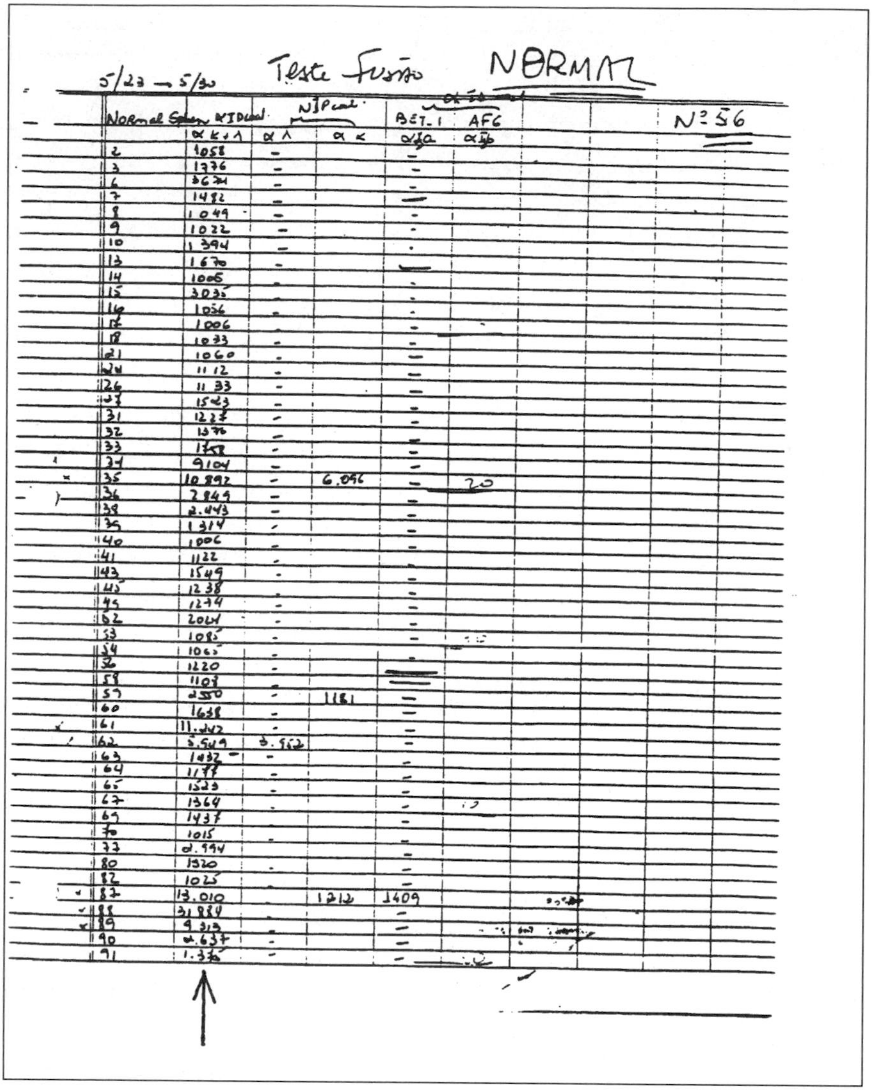

5/23 → 5/30 Teste Fusão NORMAL

Nº 56

Normal Spleen	αIDcal αK+λ	NIPcal αλ	αK	BET-1 α3a	AFG α3b
2	1058	-		-	
3	1776	-		-	
6	3624	-		-	
7	1482	-		-	
8	1049	-		-	
9	1022	-		-	
10	1394	-		-	
13	1670	-		-	
14	1005	-		-	
15	3035	-		-	
16	1056	-		-	
17	1006	-		-	
18	1033	-		-	
21	1060	-		-	
24	1112	-		-	
26	1133	-		-	
27	1523	-		-	
31	1227	-		-	
32	1376	-		-	
33	1758	-		-	
34	9104	-		-	
35	10 892	-	6.056	-	20
36	2849	-		-	
38	2.443	-		-	
39	1314	-		-	
40	1006	-		-	
41	1122	-		-	
43	1549	-		-	
44	1238	-		-	
45	1274	-		-	
52	2024	-		-	
53	1085	-		-	
54	1065	-		-	
56	1220	-		-	
58	1108	-		-	
59	2550	-	1181	-	
60	1638	-		-	
61	11.242	-		-	
62	3.549	3.562		-	
63	1432	-		-	
64	1177	-		-	
65	1523	-		-	
67	1364	-		-	
69	1437	-		-	
70	1015	-		-	
77	2.994	-		-	
80	1320	-		-	
82	1025	-		-	
87	13.010	-	1212	1409	
88	31 834	-		-	
89	9 313	-		-	
90	2.637	-		-	
91	1.375	-		-	

FIGURE 10

Among the Seventeen Pages: the Mistyped Mouse

This page from among the seventeen discovered by O'Toole in Moema Reis's notebook shows data obtained from a test of the spleen cells of the putatively normal mouse No. 56. The antibodies produced by these cells were captured by a reagent that selected for the idiotypic birthmark. The presence of such antibodies is indicated by the radiation counts above 1,000 in the column to which the arrow is pointing. According to the *Cell* paper, a normal mouse would not produce so many idiotypically birthmarked antibodies. Thus, O'Toole's discovery of the data on this page prompted her to think that the paper might be wrong. Another possibility, however, was that the mouse had been mistyped, that it was really transgenic.

It struck O'Toole as a major discrepancy from the published results and encouraged her to think that the *Cell* paper's remarkable central claim was entirely without merit.[2]

Molecular tests of antibodies produced by the hybridomas listed in Table 3, showed that the transgene was unexpressed in twenty-six or twenty-seven of them, but the notebook records of Imanishi-Kari's serological tests showed high radiation counts from Bet-1 assays done on thirteen of those same hybridomas. The high counts implied that at some point in their lives, the thirteen hybridomas had produced transgenic antibodies. O'Toole suspected that perhaps Weaver's tests were not as sensitive as the Bet-1 assays. Still, the published data did show that fifteen of the thirty-four hybridomas generated antibodies with *gamma* isotypes, which could not have come from the transgene. O'Toole says that her doubts about the paper would have been satisfied by seeing data showing that the *gamma* antibodies carried the distinctive idiotypic birthmark.[3]

The seventeen pages did not contain such data, so after lunch that Wednesday, O'Toole told Imanishi-Kari that she would like to see the record of idiotype tests done on the Table 3 hybridomas. When asked why she wanted them, she concealed her real reason and said nothing about the pages that she had discovered. Imanishi-Kari said that the data were in "that brown spiral notebook" that was "around here somewhere," O'Toole recalls. She looked around the laboratory and in her office but was unable to find it either then or the next day, when O'Toole asked again. Imanishi-Kari casually mentioned that it might have been stolen, according to O'Toole, who continues, " 'Well,' she said, 'It's disappeared. Other things have been disappearing. I have been wondering about this.' I said, 'What else has disappeared?' She said, 'Well, Charlie Maplethorpe's thesis has disappeared. He didn't want me to have it, and now it's disappeared along with my notes.' "[4]

O'Toole says that she was sure Imanishi-Kari was going to accuse Maplethorpe of theft, so she telephoned on Thursday to warn him. She also told him about the seventeen pages, though not to ask for advice, she notes, because she knew he hated Imanishi-Kari.[5] Maplethorpe recalls that she sounded angry and said that she was going to tell Huber about her discovery of the deviant data. He urged her not to, indicating that he did not think much of Huber as a scientist, that she would not get a fair hearing at Tufts, and that she would only get herself in trouble. O'Toole says, "From day one, Charlie was beside himself with worry about what these people were going to do to me. He knew they were going to trample me." Maplethorpe was worried about getting trampled himself. He says that she asked him if "I would come forward and tell the truth" about what he knew, and he remembers replying, "Yes, I will," but only "at the appropriate time" and in "an appropriate forum"—by which he meant that, since he was looking for a job, not then and certainly not in Boston.[6]

On Friday, May 9, two days after finding the seventeen pages in Reis's notebook, O'Toole brought her photocopies of them to her previously scheduled meeting with Huber. She pointed to the discrepancies between the columns of numbers and the data reported in the *Cell* paper. Huber, a Swiss native with a Ph.D. from the University of London, was in her late thirties. She had joined the Tufts medical faculty in 1977 and had been appointed an associate professor on the Tufts medical faculty only three years earlier. She was reluctant to involve herself in the dispute. She telephoned Robert Woodland, a friend since they had both been post-doctoral fellows at Harvard and now an immunologist at the University of Massachusetts Medical School in Worcester. He knew Imanishi-Kari and Henry Wortis from their joint participation in a monthly discussion group comprising mostly junior immunologists from the area. He suggested that O'Toole discuss her concerns with Wortis. Huber remembers that O'Toole was "very happy" with the idea, although O'Toole says she objected to Wortis because he had appeared unsympathetic when she complained previously to him about Imanishi-Kari and because she knew he was collaborating with her on a grant proposal.[7]

O'Toole nevertheless did go over the seventeen pages with Wortis that same day. As she recalls the conversation, he seemed "crestfallen" but unwilling to act on the matter beyond averring that he would confront Imanishi-Kari. If her data were problematic, he would quietly "rehabilitate her." Wortis insists that he never said any such thing. By O'Toole's own testimony, he did not seem determined to keep the matter confined to Tufts: She says that when she objected to what he proposed, he urged her to take up the issue with David Baltimore. She declined to do so, testifying that going to Baltimore "did not make any sense to me," since he was a coauthor of the paper and was neither her supervisor nor Imanishi-Kari's.[8]

O'Toole later said she interpreted Wortis to be intent on a cover-up and, at the urging of a colleague, she promptly paid a visit to Martin Flax, the chairman of the Tufts pathology department. (Wortis says he had nothing to gain by covering up bad science or by collaborating with someone who did bad science.)[9] O'Toole apprised Flax about the seventeen pages of data and their implications and about her meeting with Wortis. She also told him that Imanishi-Kari dealt poorly with students, recounting some of her experience with Imanishi-Kari and declaring that her own difficulties were not unique. She invoked Maplethorpe as a witness, calling him at home and putting him on the line with Flax. Maplethorpe, embarrassed, agreed to speak with Flax in the future about the details of the issues raised by O'Toole (Flax never got in touch with him). Flax is white-haired, solid-looking, and unflappably circumspect. He soon asked Sidney Leskowitz to prepare a report on Imanishi-Kari's relationships with students and colleagues, but he told O'Toole at their meeting that he declined to intervene in what Wortis was doing and advised that her proper course was to bring the matter to the attention of M.I.T., which was ethically and legally responsible for the work. She says that his advice was "just devastating," explaining that Wortis, Huber, and Imanishi-Kari "represented my entire support in the scientific community" and "I knew that I was saying goodbye to it if I [went to M.I.T.]."[10]

That evening, O'Toole reported the events of the day to her husband, Peter Brodeur, as they were driving home. Two days earlier, he had told her that he never wanted to see the seventeen pages of data she had just found, fearing that he would be placed in a hazardous position at Tufts if her assessment was right. Now, hearing that she had gone to Flax, he was upset. Brodeur worried that her action would be interpreted "as a frontal assault on Thereza," O'Toole says. He "begged" her just to "let Henry and Brigitte deal with it." During the weekend, O'Toole telephoned Flax at home, declaring that bringing the matter to M.I.T. "would cause immense tension personally for me" and that she was unlikely to do it.[11]

Martin Flax judged that O'Toole's "initial motive" in coming to him was to "right a wrong." Wortis says she told him she had a "moral obligation" to protect her friends at Tufts "from someone who might not be a competent scientist," but he was angry at O'Toole for bringing the matter to Flax's attention. Wortis, experienced in radical politics, is sensitive to the nuances of politics in the lab. It struck him that, in turning to Flax, O'Toole had completely "muddied the waters" about just what issue she wanted resolved. If it was the scientific dispute, he explains, then "we were dealing with it. If it was a formal complaint about fraud or misconduct, then go to M.I.T." Tufts was then formally considering Imanishi-Kari's appointment to its faculty. For Wortis, O'Toole's approach to Flax "raised the specter" that she aimed "precisely to threaten Thereza's being appointed here."[12]

On Monday, May 12, Wortis and Huber reviewed with O'Toole the problems that she discerned in the data, and Wortis further discussed the matter with her in telephone calls over the next couple of days. O'Toole insisted that she was not charging fraud—only that the paper seemed marked by serious errors. Wortis and Huber resolved to look into the matter informally on behalf of O'Toole as her friends and colleagues, mindful, all the same, that even though Tufts had no official standing in the matter her claims bore on the merits of Imanishi-Kari's pending appointment. They invited O'Toole to attend a meeting that they would arrange with Imanishi-Kari, but she declined.[13] They formed an ad hoc committee of three comprising themselves and Robert Woodland, from the medical school in Worcester, who was an expert in idiotypes.

O'Toole claims that Wortis and Huber agreed that if they learned that her assertions about the paper were well founded, they would urge that Imanishi-Kari submit corrections to *Cell* or withdraw the publication. Neither Wortis nor Huber recalls having made such a commitment, but Wortis acknowledges that he might have done so before he talked with Imanishi-Kari, when the key issues were the characteristics of the normal Black/6 antibodies and whether Bet-1 worked at all. Huber notes that the data on the No. 56 mouse made it look as though what was reported in the paper was "falsified." She says that she was "tremendously" concerned by that implication and was willing to take on the investigation for that reason.[14] A retraction would have been warranted by discrepancies in the data that were so fundamentally destructive of the paper's central claim.

On Tuesday or Wednesday, Wortis telephoned Imanishi-Kari to tell her that he wanted to talk with her about some of her work. On Thursday morning, May 15, he visited her at M.I.T., told her that questions had been raised about her *Cell* paper data, and set up the informal inquiry for the next day. Imanishi-Kari was hurt and angry. Wortis had promised O'Toole that he would not reveal the source of the questions and he declares that he did not tell Imanishi-Kari about the seventeen pages or mention O'Toole. However, in a congressional hearing in 1988, O'Toole stated that within a half hour of the scheduled beginning of that meeting, Imanishi-Kari telephoned, accused her of "vindictive motive and threatened to sue me." (At a later hearing, O'Toole testified that Imanishi-Kari had said that "there were laws to protect people like her from people like me" and that she had interpreted the remark as "a threat to sue me.") In 1988, O'Toole declared that Imanishi-Kari "ordered" her back to the laboratory, that she refused to return without a mediator present to discuss her responsibilities, and that Imanishi-Kari then told her not to return. Imanishi-Kari says that she got the impression from Wortis that it was O'Toole who had raised the questions, but she says, "I never called her and I never would threaten to sue anybody."[15]

On Friday, May 16, Robert Woodland, unable to drive because his leg was in a cast, had a graduate student chauffer him the forty miles from Worcester to Cambridge and joined Wortis and Huber some time in the late afternoon in Imanishi-Kari's office at M.I.T., which was just across the hall from her laboratory. He says that Huber had informed him about the gist of O'Toole's quarrels with Imanishi-Kari's reported results but that she had not mentioned the seventeen pages as such and that no one brought a copy of them to the meeting. He knew O'Toole casually, from various immunology occasions, and says that he considered her "very conscientious, very, very intelligent." The overarching issue that afternoon was the merit of the research in the *Cell* paper. In pursuit of the matter, Woodland says the trio of inquirers wanted to see original data.[16]

They asked for and saw a great deal of data. Woodland later wrote that they "examined the original notebooks and verified all conclusions from the primary data" and that Imanishi-Kari cooperated during the data analysis in a "full and open manner."[17] They quickly learned that the No. 56 mouse had been mistakenly mistyped. It was not normal but transgenic, which is why the antibodies from its hybridomas reacted strongly with Bet-1. Once the mistake was discovered, Imanishi-Kari had obtained hybridomas from a genuinely normal mouse and it was the report on them, pooled with those from other normal mice, that appeared in the *Cell* paper. Huber says, "The mistyping was very straightforward. She showed us why. She showed us the evidence."[18] Reviewing data on the behavior of Bet-1, the Wortis committee discovered that the *Cell* paper's sentence elaborating on Figure 1 was erroneous. The reagent did not react "only" with transgenic antibodies; it also reacted with native antibodies if they were present in sufficiently high concentration. However, the Bet-1 data showed that under the right conditions, it did distinguish between native and transgenic antibodies.[19]

In the course of looking at the data on the idiotypically distinctive hybridomas reported in Table 2, Woodland noticed an odd fact: When Moema Reis had checked some of the transgenic hybridoma wells during the last week of May 1985, they did not show evidence of producing transgenic antibodies; but when they were checked some days later, on June 6, 1985, such antibodies appeared to be present in some of them. The anomaly occurred in the seventeen pages, and O'Toole had called Huber's and Wortis's attention to it, arguing plausibly that the idiotypic birthmark that had been detected and reported in the table came from the transgenic antibodies rather than from the native ones (*Figure 11*). Woodland, unaware of O'Toole's views, was prompted by the data he saw to ask Imanishi-Kari for an explanation of the delayed expression of the transgene. She replied that the wells might contain two different lines of B-cell hybridomas, each deriving from a different parent cell and each

BRIGITTE HUBER
Tufts University Medical School
Photo Credit: Courtesy Brigitte Huber

ROBERT WOODLAND
University of Massachusetts Medical
School at Worcester
Photo Credit: Courtesy Robert Woodland

HENRY WORTIS
Tufts University Medical School
Photo Credit: Janet Stavnezer

HERMAN EISEN
M.I.T.
Photo Credit: Courtesy Herman Eisen

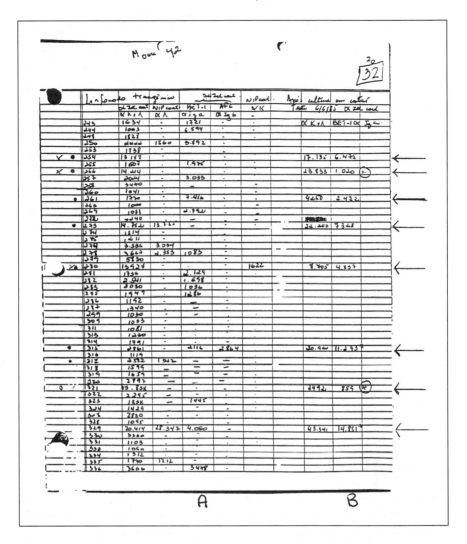

FIGURE 11
Among the Seventeen Pages: Latent Transgene Activity

This page and another raised questions for O'Toole and for Robert Woodland about whether the antibodies with the idiotypic birthmark truly derived from native genes. The hybridomas designated by the arrows produced antibodies that carried the idiotypic birthmark; as column A shows, some initially displayed no reactivity with Bet-1, but some did exhibit such reactivity, which meant they expressed the transgene at a low level. Column B shows that, when tested several days later, on June 6, 1985, all these hybridomas, including those that at first had no reactivity with Bet-1, displayed very high levels of transgene activity. The data raised the possibility in O'Toole's mind that the idiotypically birthmarked antibodies really derived from the transgene but that the analytical methods used in the experiment were not sensitive enough to reveal the fact.

producing different antibodies. Scientists call such lines *clones*. The antibodies from one clone could derive from native genes while those from the other could come from the transgene.[20]

Woodland knew that the presence of antibodies from two clones would make the serological test results ambiguous. For example, the positive response to the idiotype test might come from the antibodies produced by the transgenic clones while the negative response to the Bet-1 test might be a feature of the antibodies produced by the native clones. The tests would thus falsely indicate that the antibodies carried the idiotypic birthmark yet were produced by native genes. For that reason, Woodland asked Imanishi-Kari if she had subcloned the wells—that is, if she had isolated the cell lines by growing each in a separate well and then serologically tested the antibodies each produced. She had, in fact, carried out this procedure beginning in June 1985 with Moema Reis, partly to obtain clones generating transgenic antibodies, partly to check out the wells that showed delayed expression of them. (Questions about the veracity of these data, later dubbed the "June subcloning data," would play a key role in the fraud investigation.) Imanishi-Kari told Huber, Woodland, and Wortis that she had done the subcloning. They later wrote, "We asked if we could see the data. She asked if we didn't trust her. We were silent, and she began to cry as she brought out the data for us to examine."[21]

In the view of Wortis, Huber, and Woodland, Imanishi-Kari responded to their questions satisfactorily. The matter of the No. 56 mouse was settled decisively, and the data on Bet-1 indicated that the reagent did its job well enough to serve the purposes of the experiment. In fact, when Bet-1 erred, it did so by mistakenly identifying antibodies produced by the native genes as having come from a transgene, thus overestimating the production of idiotypically distinctive antibodies by the transgene, which was the opposite of the result Imanishi-Kari was looking for. The Wortis committee recognized that Imanishi-Kari thus had no reason to overstate the discriminatory power of Bet-1 and that the overstatement in the text was inconsequential to the point of the paper.[22]

The Wortis committee asked to see the radiographs that Chris Albanese had obtained in making his contribution to the table; they bore on whether the transgene had been expressed in the hybridomas he had analyzed. Imanishi-Kari looked all over the lab for the pictures but was unable to locate them. She said that she thought Albanese had probably taken them from the laboratory. He was looking for a new job and wanted to use the radiographs to show his work. She said that she would telephone Albanese to ask for them. She would show them to Wortis and Huber—

Woodland saw no point in getting someone to drive him eighty miles round trip just to see a few pictures—when she got them back.[23]

The next morning, in a telephone conversation, Huber told O'Toole the results of the ad hoc committee's inquiry, reporting that they had seen all they asked for except the radiographs, which Imanishi-Kari was retrieving. O'Toole says that she accepted "completely" the explanation that the No. 56 mouse had been mistyped, but she resisted the conclusion that nothing was seriously wrong with the *Cell* paper. Huber recalls that she couldn't understand why O'Toole wasn't "overjoyed" that the crucial issue of the mouse had been resolved. She says that O'Toole complained "on and on, But you haven't looked at this and you haven't looked at that." O'Toole holds that she was particularly distressed that the Wortis committee had not asked Imanishi-Kari for the idiotype data from the Table 3 hybridomas—that is, the data that she had requested of her after lunch on the day she discovered the seventeen pages. O'Toole says that some time during the next several days she telephoned Huber to protest that the investigation "was being mishandled." Huber puts it that O'Toole "harrassed me with phone calls" and that she finally told O'Toole that if she considered the investigation inadequate, then a second meeting might be held, but only if she agreed to appear. Otherwise, Huber explains, O'Toole would "come again afterwards and say, What about this, and this, and this."[24]

On Thursday, May 22, O'Toole telephoned Wortis, declaring that she now thought the central claim of the paper completely wrong. Wortis jotted down a brief note of the conversation in red marker because, he says, "all of a sudden I had a feeling that we're in a totally new ballpark." According to the note, O'Toole said that elevated production of idiotypically birthmarked antibodies by genes native to the transgenic mice was supported neither by Imanishi-Kari's data nor by the data that she herself had obtained (presumably in the research on cell transfer). O'Toole felt herself facing a "moral dilemma." She wanted to explain her position to "Brigitte, Bob, and Henry" or to those three joined by Imanishi-Kari and an "outside[r]." Part of her position was that she had arrived at an explanation different from that in the *Cell* paper of the apparent presence of idiotypically telltale antibodies in the transgenic mice. Wortis thought her explanation was dubious but believed it should have a hearing.[25]

Wortis called Imanishi-Kari, who by now had gotten back the radiographic records missing the week before, to arrange a meeting at her laboratory for the following evening. The next afternoon he telephoned

O'Toole, inviting her to the colloquy and informing her that Woodland, who had no one to drive him, was unable to come in from Worcester. She thought that someone else ought to replace him, but Wortis rejected the request, not wanting to put off the meeting by taking the time to find somebody. O'Toole had been calling Martin Flax repeatedly—the more that, in her view, Huber and Wortis "messed things up." Now she telephoned him to ask whether he thought she should proceed with the meeting, and he advised her to attend.[26]

On Friday, May 23, Wortis, Huber, and O'Toole met with Imanishi-Kari in her office at M.I.T. She produced the radiographs, which seemed in order to Wortis and Huber. Imanishi-Kari says that she had not known O'Toole was coming and was surprised to see her because she had understood that the radiographs were the sole item on the agenda. Imanishi-Kari remembers O'Toole's turning to a point made in the *Cell* paper about the idiotypically birthmarked antibodies from the 119 hybridomas reported in Table 2 whose isotypes were neither *mu-a* nor *mu-b*. The text said that antibodies from these hybridomas had other isotypes and mentioned parenthetically that a majority of them belonged to a subclass of *gamma* called *gamma 2b*. O'Toole asked whether Imanishi-Kari had done this additional isotyping. Imanishi-Kari recalls suddenly realizing that the parenthetical statement was incorrect and saying, "Oh, Lord, this is a mistake." She said she had not done any further isotyping tests on the Table 2 hybridomas. Neither had she caught the misstatement in reviewing the successive drafts by Weaver and Baltimore.[27]

However, the fact that the antibodies from the 119 hybridomas had isotypes other than *mu-a*—the isotype of the transgene's antibodies—was strong, if indirect, evidence that their idiotypically birthmarked product derived from genes native to the Black/6 mouse. Moreover, Imanishi-Kari had a good deal of evidence to show about the isotypes of the idiotypically birthmarked antibodies that were reported in Table 3 of the *Cell* paper. Indeed, the parenthetical statement had been intended to apply partly to them rather than to those in Table 2. She had isotyped antibodies produced by the Table 3 hybridomas, fifteen of which were *gamma*. She had also isotyped some hybridomas used in a parallel experiment, with comparable results.[28]

O'Toole asked for some of the data, as she had done on May 7, when Imanishi-Kari had been unable to locate it. She was particularly eager to see data from tests carried out on purified versions of the Table 3 antibodies; such data might show conclusively that antibodies displaying the idiotypic birthmark were nontransgenic in origin. Imanishi-Kari offered the data, in the form of two loose-leaf sheets with radiation-counter printouts taped to them. According to O'Toole, she expressed irritation that she had been compelled to spend time in the last week gathering material to satisfy her when she had so many other things to do, including preparing for her move to Tufts.[29]

"The only data I saw were these two sheets," O'Toole told federal investigators in 1988, adding that she thought the sheets looked "fresh." She took Imanishi-Kari's outburst of irritation to mean that she had been compelled "to generate" those two sheets in particular during the last week. O'Toole said she believed then that the data were "real" and not "made up," but she held that they "weren't original data." She recalled that she asked Imanishi-Kari for the original data that she assumed the two sheets summarized, continuing, "I can't tell you the courage it took for me to say that in that environment. . . . I cry when I think of how hostile it was. . . . There was dead silence in the room. Nobody spoke for several seconds, and then Henry Wortis said to me, 'Margot, you will deal with the data you are shown,' and so I dealt with the data I was shown." Imanishi-Kari declares that she does not recall saying it took her a week to generate the sheets or being asked for original data and that, in any case, the data on the sheets were original and not freshly generated. Wortis says, "I did not feel that she was saying that she was producing new experiments or producing [data in a] . . . newly written form. . . . The data . . . looked like they were from . . . a typical laboratory notebook."[30]

The sheet that most interested O'Toole dealt with a selection of hybridomas in Table 3 that produced antibodies with a *gamma* isotype. The tests recorded on this sheet showed that these *gamma* antibodies also carried idiotypic birthmarks related to the transgene's. O'Toole looked over the sheet and jotted notes on it. She was not satisfied. She says she told Imanishi-Kari that the degree of reaction with the reagent used to test for idiotype was too weak to mean anything, that it was not much different from what was observed in hybridomas from normal mice. Imanishi-Kari says that she responded, "You must be crazy." The reactions were somewhat weak because the antibodies had idiotypes that were related to the transgene but not identical to it.[31]

O'Toole pressed her alternative interpretation of the data. As Wortis and Huber remember it, she proposed that many of the antibodies that Imanishi-Kari had observed were "heterodimers," a term for a particular kind of hybrid molecule. She argued that the hybrid combined, on the one hand, a *mu* constant region and an idiotypically birthmarked variable one, both generated by the transgene, and a *gamma* constant region, which was produced by a native gene. Such a molecule might account for what she considered to be the valid experimental facts: The transgene half of the molecule would support the detection of the idiotypic birth-mark. The native half of it would explain why the birthmarked antibodies carried the *gamma* isotype. O'Toole's hypothesized hybrid allowed for Imanishi-Kari's data without accepting the production of idiotypically dis-tinctive antibodies by native genes. However, it ran contrary to what physical chemistry and laboratory experience indicated—that no such hybrid was likely to form. O'Toole later insisted that she never proposed such a hybrid molecule and that to attribute such a scientifically implau-sible suggestion to her was "professionally damaging." She said that what she had in mind was a hybrid of *mu-a* and *mu-b*. But the counter-recollections of both Wortis and Huber aside, the data she saw on May 23 meant that she had to account for the production of antibodies that carried the idiotypic birthmark yet showed the *gamma* isotypes. Hybrids of *mu-a* and *mu-b* would not accomplish that trick.[32]

Wortis later testified that overall "the meeting was intense and at times heated," though "not unlike other scientific meetings and discussions I have witnessed." At the end of the meeting, according to Wortis, Ima-nishi-Kari told O'Toole that if she wanted "to believe that the data can be explained on the basis of hybrid formation between *gamma* and *mu* chains . . . you are free to believe it. Whereupon Dr. O'Toole stood up and said, I'm satisfied and offered to shake hands with Dr. Imanishi-Kari." But Imanishi-Kari refused.[33]

O'Toole says she left the meeting "under the impression that the paper would be retracted" and that she telephoned Huber the next morning, May 24, to say that she wanted to see the letter that would be sent to *Cell*. She says that Huber told her that she was "right scientifically" but went on to declare that no letter would be sent, a position that Wortis affirmed to her shortly thereafter. According to O'Toole, Huber explained that a retraction would be "devastating" to Imanishi-Kari and that the errors did not have to be corrected because they had not been made with "fraudulent intent." Huber added, O'Toole says, that if Imanishi-Kari "didn't get grants, it would be bad for everybody in the [Tufts] depart-ment, because . . . there would be less money for everybody, including junior people."[34]

Wortis and Huber deny that their inquiry had concluded with an agree-ment to retract or to ask for corrections. "Zero. Zero," Wortis exclaims, alluding to the probability that either conclusion had been reached. Their inquiry had uncovered only two errors—the overstatement concerning the discrimination of Bet-1 and the misstatement concerning which hybrid-omas had been isotyped. Neither matched the magnitude of the issues that O'Toole had initially raised, particularly that of the No. 56 mouse.

Both errors were minor. In the judgment of Huber and Wortis, neither warranted a correction, let alone a retraction, because neither affected the central claim of the paper. They also thought that, whatever the merits or lack of merits of O'Toole's alternative interpretation of the experiments, her differences with the authors of the *Cell* paper on that score amounted to a substantive scientific dispute, not an issue of misrepresentation. Such disputes were properly settled by further research, not by letters of correction.[35]

On May 30, the Tufts pathology department met to review Imanishi-Kari's appointment. Flax says that since "a cloud had been hanging over" the prospect, "I wanted to be sure that everyone had a chance to speak out." Wortis reported orally on the outcome of his committee's inquiry, and Sidney Leskowitz discussed the results of his investigation of Imanishi-Kari's merits as a colleague and teacher, pronouncing them sufficient to bring her in as an assistant professor. The department voted to approve her appointment. Wortis remembers that the endorsement was unanimous except for one abstention and that the "ayes" included O'Toole's husband, Peter Brodeur. He says that Brodeur had made clear, "I've got to have a professional life and a private life. The only way I can survive is, I try to deal with all of this within the department on a professional basis."[36]

On the afternoon of May 29, the day before the vote at Tufts, O'Toole paid a visit to Mary Rowe, the ombudswoman at M.I.T. She says that earlier in the day she had talked with Wortis and that he had "totally closed the door to any kind of correction." She went to M.I.T. without telling her husband and despite his objections, she later told a reporter, because she interpreted Huber's warning about the risk of reductions in money for junior people to be a veiled threat against Brodeur. The Tufts people were holding him "hostage" and trying "to turn my love and marriage against me," she claimed. (Huber denies saying anything to O'Toole about jeopardizing grant money by blowing the whistle on Imanishi-Kari.) Nicholas Yannoutsos, then a graduate student of Imanishi-Kari's, reports that some time during the succeeding months he bumped into O'Toole and "asked her how she could be doing all this," continuing, "To my astonishment she answered that she had to do it to protect her husband whom she was sure Dr. Imanishi-Kari was going to destroy out of hatred for her."[37]

O'Toole provided Mary Rowe a substantial account of her quarrels with Imanishi-Kari's science, the conditions in her laboratory, and the outcome of the Tufts inquiry, stressing her irritation that no correction would be published. Rowe urged her to bring formal charges of fraud against Ima-nishi-Kari. O'Toole declined, partly, she later testified, because of "my strong belief that a formal charge of fraud was not warranted by the infor-mation available to me at that time."[38] O'Toole explained to me that she shied from a charge of fraud because she was bending "over backwards to be accommodating, to be responsible, to be non-confrontational"; also, she could "see nothing deliberate about [Imanishi-Kari's] behavior—just sloppy, desperate, panic." She continued, "You must remember: The things that are in that paper, she stood and screamed at me to do in my own paper. This is not the behavior of fraud. This is the behavior of somebody that doesn't understand," adding, "I was definitely sure she was guilty of self-deceit. But is that fraud? I certainly didn't think that I was supposed to be the one to make that call."[39]

O'Toole told Rowe that she felt she had a professional responsibility to see to it that the incorrect statements were corrected but that she was worried about antagonizing Henry Wortis, no doubt having in mind her husband's prospects at Tufts and, perhaps, her own. According to O'Toole, Rowe assured her that M.I.T. could handle the matter ethically without jeopardizing her Tufts connections. She said that "coming for-ward was the right thing to do" and that she would enlist the chairman of the biology department and a professor of biochemistry named Gene Brown, who was the dean of science at M.I.T., "in making sure that a position would be found" for O'Toole in an M.I.T. laboratory if her oppor-tunities at Tufts disappeared. Then and there, Rowe called Brown and arranged for O'Toole to see him at once about her challenge to the *Cell* paper. Rowe, who had made notes of O'Toole's account of the matter, urged her to follow them in talking to Brown and sent her off to see him.[40]

That afternoon, O'Toole says, she told Brown that some of the exper-iments that Imanishi-Kari reported in the *Cell* paper had not been done and that others were misrepresented. He seemed impatient to her, press-ing whether she wanted to charge fraud and, when she said she didn't know what to call it, insisting, "either charge fraud or forget about it entirely." O'Toole concludes, "I said, well, then, I'd forget about it entirely, and he showed me the door, and I went home."[41] Brown protests that she said nothing about experiments not being done and that he did not tell her to charge fraud or drop the matter. He recalls that he asked her if she thought that fraud had been committed and that she said, adamantly, no; she just "felt a mistake had been made in the interpre-tation of the data." In Brown's recollection, O'Toole declared that "the source of the real dispute" between herself and Imanishi-Kari was that Imanishi-Kari "had said to her that she thought she had done the assays wrong and that if she would just do them right she would probably get the same results. . . . That was the thing that really bothered her more than anything else."[42]

Gene Brown had never heard of the imbroglio before that afternoon, and O'Toole told him nothing about the inquiry at Tufts. O'Toole's quarrel with Imanishi-Kari struck him, he says, as possibly "a simple dispute between a supervisor and a student or a postdoc"—the kind that he had seen and successfully mediated before. He told O'Toole that, since she was not charging fraud, he was unable to appoint a committee to examine the matter. However, since she had raised a scientific issue, he asked whether she would mind if he requested Herman Eisen to look into it. She seemed happy to have him do so, he says (which adds weight to Eisen's version of his earlier discussion with O'Toole: She would likely have objected to Eisen's involvement now if he had been unresponsive before to a claim that Imanishi-Kari was pressuring her to misrepresent her data). Shortly after O'Toole left, Gene Brown called Eisen.[43]

Eisen was then in his mid-sixties, a prominent immunologist who had joined the M.I.T. faculty in 1973 after almost two distinguished decades at Washington University Medical School, in St. Louis. He is a tall, white-haired man with a grandfatherly manner and a reputation for meticulous research, generosity to colleagues, and even-handed judgment. He is the kind of professor to whom administrators turn for resolving disputes, especially those marked by high feelings. He had refereed a number of such disagreements over the years, and now he agreed to look into this one. He thought well enough of Imanishi-Kari's work to write a letter on her behalf to Tufts, but, since they worked in different branches of immunology, his scientific relationship with her was not close. His scientific interests overlapped those of David Baltimore, whom he greatly admired, but he was not personally or socially close to Baltimore either.[44]

The morning of Friday, May 30, the day after he heard from Gene Brown, Eisen telephoned O'Toole from his vacation home in Woods Hole, Massachusetts, on Cape Cod, where he was spending the Memorial Day Weekend. He asked if she could see him at M.I.T. on the following Monday. However, O'Toole had cancelled her day care as of the coming Sunday, so that Friday, at Eisen's invitation, she drove the seventy miles to talk with him on the Cape. They spoke for a couple of hours while sitting on the outside porch of Eisen's gray-shingled house, the calm of Oyster Pond visible through the trees and the rumbling of the Atlantic Ocean audible from beyond it. O'Toole showed Eisen the seventeen pages and tried to explain her quarrels with the *Cell* paper. Eisen, who had not yet read the paper, looked through the pages, unable at this first sight to appreciate what the numbers or the notes, which Moema Reis had written in Portuguese, meant. He wanted to know if O'Toole was charging fraud. She said that she was not. He found it difficult to figure out what O'Toole was getting at scientifically; he later said that she was "incoherent." In 1988, O'Toole herself told federal investigators that she didn't remember entirely what she said. "I was like blubbering, trying to defend myself." Eisen asked her to set down in writing the issues that were bothering her. O'Toole says that she was unsure during the weekend whether she would write anything for Eisen because her husband was upset but that by Sunday evening, partly as a result of telephone calls from Mary Rowe, she had decided to proceed.[45]

On June 6, in a memorandum for Eisen that exceeded four single-spaced pages, O'Toole wrote that the *Cell* paper data were marked by "serious weaknesses" that made the paper's central claim unsupportable. In elaboration of her challenge, she pressed the issues she had raised with the Wortis committee, bolstering them with a close analysis of what she had found in the seventeen pages and somewhat enlarging on them with what she had learned in the Tufts inquiry. She identified four sources of error in the paper. The first involved Bet-1, which, according to data "from all my experiments . . . and from others in our laboratory," failed to discriminate between native and transgenic antibodies. In the second, O'Toole referred to hybridomas from normal mice "other" than those reported in Table 2 and contended that a "significant" number of them produced antibodies with an idiotypic birthmark resembling that of the transgene. She also quarreled with the data from the normal mouse that *was* reported in the table—that is, the mouse that had been substituted for the one that had been mistyped. Most of the ground covered in the remaining two sources of alleged error concerned the use of analytical methods that, O'Toole said, were inadequate to recognize that many idiotypically distinctive antibodies derived in whole or part from the transgene.

O'Toole summarized the paper's weaknesses as she saw them: The data did not show that idiotypically birthmarked native antibodies were produced at significantly higher levels by transgenic mice than by normal ones. They lent themselves far "more likely" to an alternative explanation—that in many of the hybridomas the apparently higher levels of such antibodies resulted "from *low-level* expression of the transgene" and by the formation of "heterodimers"—that is, the hybrid molecules she had in mind. What O'Toole had written could be interpreted to mean that each of the *gamma* hybridomas in Table 3 was a double producer—that is, each produced an antibody that derived from the transgene, which would account for the detection of the idiotypic birthmark, and another that came from a native gene, which would account for the evidence of a *gamma* isotype. Later, O'Toole claimed she had proposed that such double producers were at work in the *Cell* paper experiments. At the time, however, she evidently had in mind heterodimer formation alone, judging by the summary of the alternative interpretation with which she concluded her memo. In any case, she argued that her interpretation better explained the data obtained from at least thirteen, and possibly six more, of the thirty-four hybridomas in Table 3. All thirteen produced antibodies that were idiotypically birthmarked but displayed the *gamma* isotype of a native gene.[46]

O'Toole's memo did not specify the composition of the heterodimers. Perhaps because of the Wortis committee's rejection of her argument that they could be *mu-gamma* hybrids, she asserted specifically of the hybrid molecules only that they might account for the data represented in the *Cell* paper's Figure 1 and Figure 2, neither of which had anything to do directly with *gamma* isotypes. Her later claim that what she had in mind was the formation of *mu-mu* heterodimers was to an extent consistent with the data in the two figures. It was, however, not uniquely determinative of the data shown in Figure 2, which concerned idiotypically birthmarked antibodies circulating in the blood of the transgenic mice. The isotypes of those antibodies were unknown and were thus not necessarily limited to *mu*.[47]

Soon after completing the memorandum, O'Toole discussed it with Eisen in Cambridge. She asserts that he heatedly asked how she could make such claims about the data without charging fraud. However, O'Toole says that Mary Rowe had vetted the memorandum for her, sanitizing it of any suggestions that data were misrepresented in the *Cell* paper because "she was concerned that I would be sued." If Rowe did perform such a service, the result was effective. Eisen says that the nature of the memorandum dispelled any lingering uncertainty he may have that it was fraud that troubled her. The document spoke only of "error" in connection with Bet-1 and of an invalid assay of the Table 2 mouse cells as a "mistake." It mentioned the *Cell* paper's misstatement that the antibodies for Table 2 had been isotyped only matter of factly, pointing out that they had been isotyped for Table 3 and that she would analyze those data. Most important, O'Toole's memorandum parsed the data in a way that accumulated to a case for her alternative explanation of them. Eisen says that if she had believed the data were "fake," she wouldn't have tried "to devise a new alternative explanation." One makes the kind of case she did only if one thinks "the data is really there but that people are just not looking at it the right way."[48]

Although in the memorandum O'Toole said that she had discussed the issues with Imanishi-Kari, she said nothing in the document about the Tufts inquiry. Eisen recalls that by now he had heard rumors of the Wortis investigation, but he scrupulously avoided getting in touch with Wortis because he wanted to come to his own, independent judgment. Eisen did not ask to see Imanishi-Kari's notebooks. He explains that O'Toole was not charging fraud, and that such an examination would have comprised a huge undertaking, one that he could not possibly shoulder himself. Besides, O'Toole's memorandum did not refer explicitly to the seventeen pages. O'Toole says that she did not attach them on the advice of Mary Rowe. The memorandum itself actually discussed the data in sufficient detail to make most of its case. Eisen was impressed by the overall cogency of the document. He sent it to Baltimore, Imanishi-Kari, and Weaver and arranged for O'Toole to meet with them and himself to discuss what he says he cared about and believed O'Toole cared about—the merits of the *Cell* paper's science.[49]

On June 16, around 4 P.M., the five scientists sat down around a coffee table in a second-floor conference room at the Whitehead Institute to hash out O'Toole's dissents. The seventeen pages figured little in the colloquy, since they were not attached to the memorandum and no one other than O'Toole had a copy of them. Weaver was characteristically quiet; the others did most of the talking. Weaver notes that the air was tense but that Baltimore and Eisen were solicitous of O'Toole, recognizing that it was no doubt difficult for her to press her quarrels with the paper on its senior authors, one of them a Nobel laureate. Eisen let the debate range free-for-all over the scientific issues. He recalls that a lot of time was spent on Bet-1, but by 6 P.M. or so, when the meeting ended, the interchange had pretty much covered all the points in O'Toole's memorandum.[50]

Baltimore says that he came to the meeting unfamiliar with the experimental intricacies of Imanishi-Kari's serological analyses, certainly not at the level of detail of O'Toole's memorandum. "It was the kind of work I didn't know how to do, had never done, and I had collaborated with Imanishi-Kari for that reason." He was nevertheless upset to learn that Figure 1 and the text of the *Cell* paper appeared to be misleading about the ability of Bet-1 to discriminate between *mu-a* and *mu-b*; according to several accounts, he spoke sharply to Imanishi-Kari about the textual error. She responded that Figure 1 rested on data obtained by Moema Reis and that the reagent worked well enough for the experiment. Imanishi-Kari's response did not entirely satisfy Baltimore, according to O'Toole, but Baltimore and everyone else—except O'Toole—were satisfied that the difficulties with the reagent were not scientifically consequential.[51]

O'Toole pressed her assertion that normal Black/6 mice produce idiotypically distinctive antibodies at a higher frequency than what the *Cell* paper reported. Eisen acknowledged that she might have a point, but he was unconvinced. The claim contradicted everything known about the immune response of normal Black/6 mice, particularly the fact, Eisen says, that antibodies in blood taken "from these mice don't show any idiotype activity." To him, O'Toole's view just did "not make any sense."[52] O'Toole's objection to the hybridomas from the normal mouse that had been reported in Table 2 did not make much sense to Eisen, or to Baltimore either. In her memorandum, she pointed out that those hybridomas had been "generated at a different time from those of the transgenic" mice, and at the meeting she apparently contested the data from the mouse because it was not a sibling of the transgenic animals. But so far as anyone else knew, the difference in times made no difference to the outcome of the experiment, and neither did the fact that the transgenic and the normal mice were not members of the same immediate family. Both belonged to the highly inbred Black/6 strain. One inbred mouse varies very little from another, which is why scientists use them. "You can take a mouse today, or here, or in Hong Kong next year," Eisen says, "if it is a C57 Black/6 mouse you have got it. That is the power [of inbred mice]."[53]

O'Toole argued strongly and at length for the double-producing explanation of the data she had advanced in her memo to Eisen, giving particular attention to the Table 3 hybridomas. Here, in the meeting, she seems also to have left vague the kind of heterodimers she thought formed in the *gamma* hybridomas of Table 3. Rather, she pointed to the apparent discrepancy in a number of these hybridomas between the results of Weaver's molecular tests and—according to the seventeen pages—those of Imanishi-Kari's serological tests with Bet-1: Weaver's assay did not find expression of the transgene and Imanishi-Kari's did. O'Toole explained the discrepancy by claiming that Weaver's tests were insufficiently sensitive to detect low-level expression of the transgene. However, Weaver sharply disagreed with her assertion about the relative sensitivity of his molecular tests. Imanishi-Kari contended that the Bet-1 used to test the Table 3 hybridomas in the experiment recorded in the seventeen pages was no good, that the seventeen pages themselves showed it behaved inconsistently. Imanishi-Kari declared that a retest of the same hybridomas with good Bet-1 had showed that they did not express the transgene. At one point, according to O'Toole's recollection, Imanishi-Kari brought out some charts of data that she had obtained in Germany and discussed them with Baltimore, arguing from them that hybridomas in her NP system did not generate hybrid antibodies.[54]

O'Toole says she was skeptical that the retest had been done but that she chose not to press the matter. She was sitting next to Baltimore and tried to impress on him the reasons for her alternative explanation of the results by showing him data in the seventeen pages. She focused his attention on the data for Table 2, particularly, it seems, the column that showed evidence of transgene expression in the assay done on June 6, 1985, a week or so after the hybridomas had originally tested negative for the transgene. At one point later, she testified that Baltimore brushed her evidence aside, and at another she declared that he took the seventeen pages "and looked at them carefully." Whatever the case, she recalled consistently his saying, " 'You can't tell anything at all, one way or the other, from this.' " She also remembered that Baltimore "said that the published claims could not be based on" the seventeen pages. "This was precisely my point," she stated.[55]

But it was not precisely Baltimore's point. He meant that the data on the seventeen pages, like those that O'Toole analyzed elsewhere in her memorandum, were inconclusive with regard to her view of the Cell paper. They represented only part of Imanishi-Kari's data, and, in any case, as with much experimental data, the use and interpretation of the overall body of results demanded the exercise of judgment and imagination. For example, Imanishi-Kari held that the late expression of the transgene in some of the Table 2 hybridomas was irrelevant to the detection in them of idiotypically birthmarked antibodies a week earlier. To be sure, she had subcloned selected hybridomas to check what was happening in them; but independently of the subcloning, she reasoned that since the telltale antibodies appeared when the transgene was not expressing itself, the transgene had not been responsible for them. They must have arisen from native genes.[56]

Baltimore recognized that Imanishi-Kari's judgment calls might be questioned, but he also held that the Cell paper was not a claim of absolute truth. Like any other scientific paper, it was a statement in a process of evolving understanding of how the matter looked to the coauthors at the time of its completion. He remembers that on reading O'Toole's analysis and listening to the conversations, it seemed to him that at issue were "questions of detailed interpretation and alternative interpretation," adding, "I was satisfied that what we knew up to that point was appropriately represented in the paper. Maybe some day it was going to turn out that there was a wrong interpretation on the basis that Margot was suggesting, and maybe not. But you weren't going to answer the questions by arguing about them." Like the Wortis committee, Eisen, Baltimore, and Weaver all thought that O'Toole's challenge could only be answered by further research, and Baltimore suggested experiments that might be done to test her ideas.[57]

O'Toole nevertheless hung on. At the meeting, Baltimore declared that he was responsible for the misstatement about the isotyping of the Table 2 hybridomas, explaining that he had gotten some general information about transgenic mouse cells from Imanishi-Kari over the telephone, thought it referred to specific cells in Table 2, and inserted the point into the text. O'Toole found Baltimore's account credible but says that she still thought the misstatement should be corrected. Baltimore disagreed, seeing no more reason than had Wortis's ad hoc committee to correct it. He later observed, "I think large numbers of things are known by people to be incorrect later, and not publicly corrected because they're minor enough that they would not have a significant impact on understanding the paper. Most corrections are either when a central fact has turned out to be incorrect or when the paper was somehow misprinted." O'Toole says that Baltimore's resistance to publishing a correction was not mean-spirited. She likens his attempt to explain why such a correction was unwarranted to the head cop's instructing the rookie cop that a certain level of corruption has to be accepted and not to bother about it.[58]

Baltimore suggested that O'Toole write a letter to *Cell* that outlined her alternative interpretation; he expected that the editors would contact him and that he would respond. He was proposing a friendly, civil exchange, the kind in which scientific disputes are often aired, but O'Toole later said that she thought he would oppose her views, that the editors would believe him, and that his opposition would in effect thwart the publication of her critique. She declined to write a letter. Baltimore said, "Well, then there's no problem" and asked Eisen to write a report on the meeting. Imanishi-Kari recalls that O'Toole went on to say that she had brought the issue to Eisen only "because there were rumors going around that she, Dr. O'Toole, was harassing me because I didn't want to take her to Tufts" and that she only wanted to stop the rumors and clear the air. Imanishi-Kari asked whether O'Toole had raised all the issues on her mind, saying she didn't want to have to respond to questions "over and over again." O'Toole "said that, no, that was it. And I thought that was it."[59]

On the way down the stairs afterwards, Weaver and O'Toole stopped to talk. According to O'Toole's later congressional testimony, "Weaver was very upset" and concerned about her, seeing that she was shaken. Apologizing to her, he said that he "greatly admired" her courage and that he thought Imanishi-Kari's data were faulty. She said that she asked him how David Baltimore could "think this is all right" and that Weaver responded "he doesn't think it's all right." O'Toole also told federal investigators that Weaver said of Baltimore, "What he says and what he thinks are completely different." Weaver disputes O'Toole's account, declaring that he never said he suspected the data or offered an apology; he did not feel he had "anything to apologize about, especially to her." He notes that he could not have known what Baltimore thought about her challenge because he had not talked with Baltimore about it before the meeting and that, in any case, in his experience Baltimore rarely said one thing when he thought another. Weaver recalls that he expressed "some feeling of identification with her as being courageous," being "willing to sit in a room with three senior scientists and defend a position." He "thought she did a good job." "But ultimately," he adds, "the other position could be defended as well. So they weren't necessarily required to back off the position of the paper just because there was an alternative explanation."[60]

Eisen praised O'Toole in a memorandum that he composed about the discussion the next day, writing, "I do not think that I or anyone else present at the meeting felt that Margot O'Toole's disagreements were frivolous. They are indeed based on pretty carefully thought out ideas of the limitations of the analytical methods." Eisen doubted that the experimental methods used had been too insensitive to detect low-level expression of the transgene, but he did not address O'Toole's argument about heterodimer formation. It had not been explicitly discussed at the meeting; Eisen nevertheless understood her to have meant *mu-gamma* hybrids. Although he believed that such a hybrid was chemically improbable, her use of the term *heterodimer* suggested to him that she was talking about *mu-gamma*; a *mu-mu* hybrid warranted a different technical term. Besides, the *mu-mu* combination would not explain the detection of idiotypically birthmarked *gamma* isotypes. Between the improbability of *mu-gamma* and the inadequacy of *mu-mu*, Eisen doubted that hybrid molecules better explained the data in the *Cell* paper. However, he was certain that O'Toole's challenge to the paper was unresolvable by argument; it could only be decided by experiments. He expected that further research would soon reveal whether the coauthors were correct or O'Toole was correct about antibody production in the transgenic mice. If the interpretations of the coauthors "are incorrect and require revision," Eisen wrote in his day-after memo, "then so be it."[61]

O'Toole says that the day after the meeting at the Whitehead Institute, she talked by telephone with Mary Rowe, telling Rowe that she wanted a copy of whatever report Eisen produced. According to O'Toole, Rowe replied that she had urged Eisen not to write one and explained that he had "told her unflattering things about me and that my charges were unsubstantiated." Rowe advised, "Believe me, it is not in your interest to have the report filed." O'Toole says that she then reminded Rowe of her earlier offer to help her get a job at M.I.T. and that Rowe told her that the time when she might have helped her was past, declaring that the entire matter was now "in the hands of God." Rowe says that, as an ombudsperson, she is duty bound not to reveal what transpired between herself and O'Toole. She adds, though, that "an ombudsperson can't defend herself," perhaps suggesting that her version of the matter differs from O'Toole's.[62]

In fact, Eisen's memorandum actually raised only one point against O'Toole—that the coauthors other than Imanishi-Kari had first learned about her quarrel with the paper via painful rumors because she had brought the matter to the attention of Tufts faculty rather than to them. Otherwise, Eisen commended O'Toole, declaring that "it took a rather considerable amount of courage to face a senior scientist whose scientific judgment she was questioning in a serious way."[63] He later told federal investigators that the coauthors agreed that a letter should be sent to *Cell* correcting their paper's exaggerated statement about the specificity of Bet-1, but Baltimore thought Imanishi-Kari should sign it whereas she thought it should be sent with all their signatures. The dispute "just dribbled away," Eisen recalled. Imanishi-Kari was busy moving her lab to Tufts, her collaboration with Baltimore was over, and Eisen doubted anyway that the editor of *Cell* would publish a correction on such a minor matter.[64]

Thus O'Toole did not get the correction she demanded nor was she made aware of the respectful attention that her interpretive critique of the experiments had earned. Having left M.I.T., she was no longer around the cancer center. Eisen does not recall seeing or speaking with her after the meeting on June 16, and he was neither required, instructed, nor asked to provide her or anyone else a copy of his report on the dispute. Although a kind and well-meaning man, he confined his praise of her to his memorandum and confined the memorandum to his files. He was occupied with what further research might reveal about immune responses in the transgenic mice and thought that O'Toole ought to be, too. Some years later he found a remark of Mark Twain's that he tacked onto his bulletin board. "Supposing is good, but finding out is better," it reads. Eisen says that Margot O'Toole thinks that because "she supposes there might be another explanation, that's sufficient," continuing, "She doesn't realize supposing is the beginning. It's where you start, not where you end. Finding out the flaws, or what you think are flaws, in a paper's data—how it's collected, how it's analyzed, how it's interpreted—those are just a springboard to go on and try to find out more about the system."[65]

O'Toole does appear to have expressed little interest in further investigation of the immune response in the transgenic mice. If she had a lab to work in, she might have displayed more. But what she seemed at heart to crave was recognition as an insightful scientific critic and, more important, legitimation as a practicing scientist who was not incompetent because she could not get Bet-1 to work. She wanted a public correction of the *Cell* paper's overstatement that Bet-1 discriminated absolutely between the native and transgenic antibodies. She believed that it did not discriminate between the two at all, and that its inadequacy had cost her months of fruitless work and dashed her professional hopes. To everyone else, the issue of a correction concerning Bet-1 was weighted with no ethical obligation, not least because its technical importance seemed minimal, inadequate even to gain space in a scientific journal. To O'Toole, publishing the correction was virtually a moral imperative. Although Bet-1 did work well enough for Imanishi-Kari, it did not, in O'Toole's hands, suffice for what she thought she needed to do the cell-transfer experiment. From her perspective, the paper demanded correction because, as she later put it, other scientists would "waste effort and time attempting, as I had for almost a year, to extend it."[66]

O'Toole's one-year appointment with Imanishi-Kari had ended on May 31. She was ineligible for another N.I.H. traineeship. (When Eisen's secretary tried to get O'Toole to sign the standard form that she had completed her trainee year, she ignored the request, explaining later that she had not received any real training in Imanishi-Kari's lab.)[67] She still had in mind joining her husband's lab at Tufts, perhaps reviving her position as an assistant research professor, but it was rumored that Imanishi-Kari had been guaranteed that O'Toole would not be permitted to work there. O'Toole telephoned Martin Flax, the department chairman, to check out the rumor. She later testified that "he confirmed for me what I had already heard . . . ," elaborating that Imanishi-Kari "had asked for his assurances that I would not be allowed to return to my position" and that "he had assured [Imanishi-Kari] that he understood how she felt."[68]

Flax recalls that Imanishi-Kari had told him that "she would not enjoy having Margot as a colleague." He says he did tell her he understood how she felt, but that he "did not in any way suggest this was going to be determinant." Imanishi-Kari insists that she had no such conversation with Flax and wouldn't have thought about having one, noting that as a beginning assistant professor whose research had just been investigated, "I had very little leverage at that point with anybody." Henry Wortis told Brodeur, "Anyone who thinks that Thereza and Margot could coexist in the same department without fireworks would have to be crazy." Wortis intended to discourage Brodeur from bringing his wife into his laboratory. O'Toole understood the message.[69]

O'Toole was now pregnant again, and she went to work for her brother at his Gentle Giant Moving Company, setting up a computerized dispatching system for him. She later told a congressional subcommittee that after her year with Imanishi-Kari and the inquiries that ended it, she left science "saddened and disillusioned."[70] Her mother provided solace, giving her a copy of William Butler Yeats's poem "To a Friend Whose Work Has Come to Nothing," which opens: "Now all truth is out / Be secret and take defeat." But O'Toole's defeat rankled. She remained in touch with Brigitte Huber, who had conveyed the news that no correction would be published. Huber's two children were close in age to O'Toole's son, and she remembers that on a visit to the Science Museum in Cambridge that fall, with their children in tow, O'Toole "blasted me for what I had done to her career. I felt attacked by her continuously."[71]

F O U R

■

Misconduct in America

AROUND THE time that O'Toole told Charles Maplethorpe about the seventeen pages, some friends gave him an article that had appeared in the *New York Times* in April about two scientists, Ned Feder and Walter W. Stewart, who had written a controversial paper about a case of scientific misbehavior. Both were staff members of the N.I.H. in Bethesda, Maryland. Their paper was as yet unpublished, but they were saying that further studies of scientific fraud and misconduct were warranted.[1] On May 16, the day of the first Tufts inquiry, Maplethorpe found out how to reach Stewart and Feder by telephone in their laboratory. He talked with them several times during the next few days, without informing O'Toole, who had asked him to keep her doubts about the *Cell* paper in confidence. "I had nothing to lose," he says. According to Stewart's notes of the conversations, Maplethorpe identified O'Toole only as "Margot," outlined her quarrels with the paper and vouched for her "honesty," describing her as "an activist in police brutality." He declared that Imanishi-Kari was untrustworthy and "manipulative" and said, "I have xeroxes proving fraud." He also told Stewart and Feder that one of the paper's coauthors was David Baltimore.[2]

Stewart later claimed that he and Feder paid little attention to Maplethorpe's report, explaining, "We had plenty else to do, we simply assumed that the matter would run its course and become public in due time." However, on May 20, they told Maplethorpe that they had alerted a journalist about the dispute and urged that the reporter call Henry Wortis to ask what was going on. Maplethorpe, furious, remonstrated that they had "just ruined any chance that anything will come of this," that "a journalist can only fuck things up." He worried that "outside interest may harm Margot," to which Stewart replied, "No matter what happens Margot will be hurt." Maplethorpe managed to dissuade them, though with difficulty, indicating, "Given the choice reporter or not—I'd say turn it off." Maplethorpe says of this, his first encounter with Stewart and Feder, that "it was like grabbing a shark."[3]

Feder and Stewart comprised the sole staff of the section on biophysical histology at the National Institute of Diabetes and Digestive and Kidney Diseases. Feder, then in his early sixties and seventeen years Stewart's senior, was tall, white-haired, and reserved, measured in most of his opinions. He was formally boss of their section, but Stewart, I discovered on a visit to them one late-summer day in 1992, was the spark plug of the pair. Feder drove me out to Stewart's home, a sprawling house at the end of a wooded lane in Potomac, Maryland—a neighborhood for rich people, Feder remarked matter of factly. Stewart, of middling height and clad in shorts, his eyes bright behind black-rimmed glasses, popped around his kitchen and talked torrentially for several hours. Feder nodded with a kind of paternal pride, saying little, volunteering to do small chores like letting out the cat. He urged me to see the technology that Stewart had devised in an adjacent den. The room was empty except for an array of five or six Macintosh computers on the rug that Stewart had wired together and programmed to compare texts for plagiarism.[4]

Feder grew up in New York City, where his father had started as a janitor in a synagogue and eventually became its executive secretary and ultimately, head of a national association of synagogue secretaries. Feder entered Harvard University at the end of World War II, when the university was opening its doors far more widely to Jews, eventually graduating from Harvard Medical School and then doing postdoctoral research in biology. In 1955, he joined the staff of the N.I.H., but in 1961 he returned to Harvard under N.I.H. sponsorship for an extended joint stay in the medical school and the biology department, where he was a junior member of the teaching faculty.[5]

Stewart went to Harvard, too, arriving in 1963 from Phillips Academy Andover. He says that in a sense he grew up Jewish, even though he is not, explaining that he also came from New York City and that his father was a psychoanalyst. Little of what the elder Stewart knew about people apparently rubbed off on the son. "I was a very unpeople person . . . ," Stewart once told a reporter. "People are quite a mystery to me. I don't understand them and I never have. Certainly one of the reasons I went into science . . . is that it's about things you *can* understand—like why the sky is blue." At Harvard, he tended to a boyish dishevelment, wearing torn clothes and eyeglasses kept together with adhesive tape. His professors were impressed not only by his irrepressible talkativeness but also by the insatiable curiosity and academic brilliance that the talk revealed. He majored in physics and chemistry, but he sidestepped formal course work as much as possible in favor of independent study. In his junior year, wanting to dissect a cat and learn some basic biology, he hooked up with Feder, who gave him a pickled cat and a table to dissect it on. In his senior year, he completed a paper with two Harvard scientists, one of whom says that Stewart did most of the work, that was published in the *Proceedings of the National Academy of Sciences*.[6]

In 1967, awarded his Harvard degree *summa cum laude* and elected to Phi Beta Kappa, Stewart went to Rockefeller University for graduate work. The unstructured nature of the curricular program there suited him. He tidied up his dress, occasionally donning a velvet jacket and a decent shirt to attend a concert. Nevertheless, he thought that the Rockefeller professors looked like businessmen and he disliked the button-down ambience of the place. He was also disturbed by the war in Vietnam and was in jeopardy of being drafted. Rollin Hotchkiss, who was then a professor at Rockefeller and an admirer of Stewart's, recalls that Stewart's friends were distressed at the prospect. "They doubted that he could be taught to hold a gun and expected that, if he could, he would march forward with it, smiling at the enemy," Hotchkiss says. Wanting him saved for science, the friends urged him to seek a commissioned officership at the N.I.H., which after two years would fulfill his obligation to selective service. The officerships were highly coveted at the time and relatively few were awarded in science, as compared with medicine or even sanitary engineering. Nevertheless, Stewart managed to obtain one. In 1968, giving up his quest for a doctorate—permanently, as it turned out—he left Rockefeller for Bethesda and the laboratory of Ned Feder.[7]

Stewart's many interests included how scientists arrived at their understanding of nature, particularly the relationship of the data they obtained to the conclusions they reached. While an undergraduate, he had attempted to re-derive Kepler's laws of planetary motion to see how Kepler had accomplished his triumph. At Rockefeller, he had absorbed himself in the work of a deceased member of the faculty named D. W. Wooley on a toxin called wildfire that attacked tobacco plants. Wooley had published the chemical structure of the toxin, but his conclusions proved to be wrong. Stewart examined Wooley's notebooks with the aim of figuring out where he had gone off the track; he continued to struggle with the problem after he went to Feder's lab and, in 1971, published the likely correct structure, showing in addition how and why his results differed from Wooley's.[8] The work was remote from Feder's, but Feder and Stewart had become pals. Feder indulged him the holiday from the wildfire work that Stewart began at the end of 1970 to act as a consulting editor for the scientific journal *Nature*. For Stewart, who had volunteered for the task, it was another way to explore how scientists reached conclusions. He was soon refereeing twenty papers a week.[9]

In 1971, a remarkable report of research on brain chemistry crossed Stewart's desk. It had been submitted by three scientists in Houston, Texas, who claimed that learned behavior could be transferred from the brain of one animal to another by a chemical called *scotophobin*. The three researchers had been discussing their findings at scientific meetings for several years and their work had been heralded in the national press. Stew-

WALTER STEWART and NED FEDER (standing)
Staff Scientists, N.I.H.
Photo Credit: Marty Katz/NYT Permissions

art, suspicious of the claim, demonstrated that it had not been proved by the experiment its authors had described. He scored something of a coup when *Nature*, in mid-1972, published his rebuttal alongside the article. In a reply, the hapless authors wondered why Stewart had gone beyond the normal duties of a referee. He later explained that "the model of science is supposed to be free and open debate," adding that there was "much too little debate"—"too little of anybody criticizing anything"—and that he had resolved to supply it.[10]

For several years, Stewart's enthusiasts at Harvard had been nominating him for election as a junior member of the university's select Society of Fellows. The junior fellowship was made to order for Stewart, since it put a premium on independent work. "I would say give him rope, he's an excellent investment," one of his nominators urged. "He will enliven dinners enough to make him worth it on that count alone." Stewart was flattered by the prospect and was awarded the fellowship in 1971. But once back at Harvard, he wanted his own laboratory and complained that no one would give him the necessary space. After a short time, much to the irritation and disappointment of his sponsors, he returned to the N.I.H. and Feder, pursuing his fellowship in absentia as a guest scientist and, when the fellowship ended in 1974, receiving a regular appointment in Feder's section.[11]

In the late 1970s, Stewart devised a dye called *Lucifer Yellow* for biological research that was one hundred times more sensitive than those previously available and that came to be widely used. It could be introduced into living cells without damaging them, would fluoresce with an intense yellow light, and would also pass from one cell to another. He patented the dye but gave it away to other researchers without condition, including the courtesy, typical in biomedical research, that those who used it grant Stewart collaborative credit in their publications.[12] For several years, Stewart and Feder tried together to develop a red-fluorescing dye, which could be used in conjunction with a dye like Lucifer Yellow to study pairs of neighboring neurons, but they evidently did not succeed. In the meantime, they began collaborative research using the yellow dye to study nerve cells in inbred snails. They accumulated data on the genetics and nervous systems of thousands of snails, filling numerous notebooks with observations, but they kept what they learned pretty much to themselves. By the early 1980s, Stewart had only nine research papers to his name and Feder, twenty-two to his, the last original one in 1976, outputs that were low by N.I.H. standards of productivity. At this time, however, Stewart was drawn to revive his interest in scientific practices—in a sense, a return to his scotophobin triumph—by the emergence of scientific fraud and misconduct as a public issue in the United States.[13].

The issue was a peculiarly American phenomenon, compelling attention at the time in no other scientifically vital nation, a product in part of the political culture of the day. The war in Vietnam and Watergate had generated a deep distrust of public authority and institutions in the United States that was sustained and perhaps deepened by the exposure of the sale of political favors in scandals such as Abscam in the late 1970s and the corrupt administration of environmental policy in the early 1980s. (By 1984, some two dozen officials at the Environmental Protection Agency had been removed from office or forced to resign, and one of them eventually went to jail for lying to Congress.[14]) As an ally of business and a ward of government, science, too, was vulnerable to suspicion, and the kind of investigative reporting that had blossomed in politics had spread to the coverage of science in both the technical and the lay press. By the early 1980s, a spate of press reports had appeared on cases of scientific fraud, enough to provoke talk of a "crime wave" in science. Virtually all the cases were in biomedical research and at leading institutions of research, including Stanford University, Yale University, Boston University, and Massachusetts General Hospital. The public exposure of the cases led Congressman Albert Gore, Jr., of Tennessee, to hold hearings on fraud in biomedical research in April 1981—the first inquiry into the subject on Capitol Hill.[15]

Whether fraud, or merely the reporting of it, was rising was unclear, but when it came to science, especially biomedical science, just a few cases of demonstrated misconduct were enough to elevate distrust. In its early days, many practitioners of natural philosophy were gentlemen and expected their word to be taken as such.[16] Many were also clerics, faithful intermediaries between the works of God and general human comprehension of them. In the modern world, scientists acquired the image and standing of a kind of secular priesthood—virtuous in its eagerness to discover the truths of nature and formidable in its ability to exploit them.[17] In the United States after World War II, physical scientists were recognized as indispensable to the nation's military security and economic growth. The Cold War advanced their fortunes in American society, but also rendered the public ambivalent towards them as progenitors of the nuclear arms race, and in the wake of the Vietnam War the ambivalence sharpened considerably. Dissidents indicted physicists and engineers for complicity in the military-industrial complex, attacked scientific technology for despoiling the environment, and called its very reliability into doubt, taking the nuclear-plant disaster at Three Mile Island, in 1979, as exemplary of its fallibility.

People associated biomedical scientists, in contrast, with healing, with unalloyed benefits such as the discovery of antibiotics and the conquest of diseases like polio. Life scientists had not been prominently involved in the arms race and had taken the lead in the movement for environmental protection. During the era of Vietnam and Watergate, they prospered steadily, more than did physical scientists, obtaining enormous support from public and private sources, including, by 1976, more than a doubling of federal obligations for basic biomedical research, and by 1980, more than a tripling of them, to roughly $2 billion. From 1971 to 1976, life scientists trained some 27,000 new Ph.D.s, roughly a third more than from 1965 to 1970, and some built academic research empires.[18] Biomedical scientists marched from one technical triumph to another, using molecular biology to gain major purchases on the genetics of the immune system, the nature of viral infection, and the mechanisms of cancer. They expected still greater progress to come from the techniques of recombinant DNA—the ability to isolate and manipulate genes that permitted the kind of experiments in basic research that David Baltimore would perform in establishing the transgenic mice for his experiment with Imanishi-Kari.

However, recombinant DNA technology promised not only to extend the range of experimental biology but also to make a reality of genetic engineering by enabling the alteration of genes or the combining of the genes of one species with those of another. It created the power "to join duck DNA with orange DNA," in the remark of a British wag.[19] It opened the door to the transformation of organisms—plants, animals, and possibly even human beings—at the core of their hereditary essence. Many people, including many scientists, found these prospects unsettling. A number of biologists worried that recombinant microorganisms might threaten life or health or whole ecosystems. Some questioned the reconfiguring of life itself as an act of hubris that would lead to unpredictable and dangerous consequences.

In 1974, eleven biologists, led by Stanford University Professor Paul Berg and including David Baltimore, called for a moratorium on most research in recombinant DNA pending a review of its potential hazards at a meeting at Asilomar, a conference center near Pacific Grove, California. The moratorium achieved virtually unanimous voluntary compliance, and some 140 biologists showed up at the conference, in 1975. A number said they wanted to avoid the censure increasingly visited upon physicists for their responsibility at Hiroshima. In a sense, the meeting at

Asilomar was about the assertion of morality in biological science. The conference produced a set of recommendations for the conduct of recombinant research that formed the basis of the guidelines that the N.I.H. issued in mid-1976 to govern federally sponsored work using the new techniques.[20]

In the later 1970s, however, the dangers explored at Asilomar provoked widespread lay apprehension, especially in areas housing universities engaged in recombinant research. In 1976, the city council of Cambridge, Massachusetts, voted to impose a moratorium on recombinant DNA research until its hazards could be reviewed. In May 1977, Alfred E. Vellucci, the Cambridge mayor, formally asked the head of the National Academy of Sciences whether the recent sightings of a "strange, orange-eyed creature" in Dover, Massachusetts, and "a hairy, nine-foot creature" in Hollis, New Hampshire, might have something to do with recombinant DNA experiments taking place in the New England area. Critics inside and outside the scientific community attacked the way the biomedical community had chosen to deal with the dangers, pointing to the N.I.H. guidelines as tantamount, in the charge of a professor of biology at M.I.T., to "having the chairman of General Motors write the specifications for safety belts."[21] Local and state governments and the U.S. Congress geared up to legislate tough, mandatory restrictions on such research. A bill in the New York State legislature proposed to empower local health commissioners to levy a fine of $5,000 a day against any scientist who did experiments with recombinant DNA that the commissioners found hazardous.[22]

Robert Pollack, a microbiologist on the faculty of the State University of New York at Stony Brook, had endorsed the civic responsibility of science expressed at Asilomar. Now he declared himself "very discouraged and disturbed" that it had unleashed a "medieval prescription for punishing witches and sorcerers." Norton Zinder, a biomedical scientist at Rockfeller University, denounced some of the recombinant regulatory bills in Congress for setting up "vast bureaucracies, cumbersome licensing, harsh penalties and tedious reporting procedures." They included "search and seizure provisions," he noted. "They read like a narcotics bill." James D. Watson, a signer of the call for a moratorium and famed for his co-discovery of the structure of DNA, thought that the biologists should "head down to Washington and tell Congress that these scaremongers are an odd coalition of spaced-out environmental kooks and leftists who see genetics as a tool for enslaving the masses."[23]

Watson's colleagues ultimately convinced Congress that the hazards of recombinant DNA were no greater than those characteristic of traditional work with virulent organisms. They assured the public that the production of monsters was neither their aim nor within their powers. And they promised that recombinant techniques would lead to revolutionary practical benefits in medicine and agriculture. By the 1980s, they had beaten back the threat of intrusive bureaucratic controls and obtained even a major relaxation of the N.I.H. guidelines. Nevertheless, the recombinant DNA wars left many biomedical scientists ruing Asilomar. "The political dynamite of this issue escaped us!!" Zinder said. Now that they fully recognized its explosiveness, they were, to say the least, skittish about anything that smacked of government intrusion in the practice of research. Paul Berg himself put it for many of his colleagues: "This exercise in science and public policy has become nightmarish and disastrous. . . . I now believe that Society has more to fear from the intrusions of government in the conduct of scientific research than from recombinant DNA research itself."[24]

The molecular biologists prevailed partly because recombinant research had been pursued for several years without apparent mishap and partly because the rapidly increasing industrial interest in biotechnology added weight to their predictions of practical utility. But they won also because Congress and much of the public took them at their word. They secured the right to develop their powers on the implicit understanding, as the social critic Edwin Yoder later put it, that they would "take vows of uncompromising openness, integrity and self-scrutiny."[25]

Congressman Gore, a sturdily moral liberal who had gotten involved with policy making for recombinant DNA when he entered the House in 1977, opened his inquiry into fraud in 1981 with the pointed reminder that the American people's huge investment in science rested on their "trust" in it and on "the integrity of the scientific enterprise" itself. Gore held that the corruption of fraud raised serious questions about the ability of scientists to deal reliably with the "ethical judgments" that now confronted biological science "in great magnitude."[26] For Gore and for many others, fraud in the biomedical sciences was akin to pederasty among priests. In a country that more than most others often expects its public— and, especially, publicly funded—shamans to conform to high standards of trustworthiness, scientific fraud invited the kind of censure associated with a morals charge.

The Gore hearings exposed several disturbing tales of sin. There was John Long, of the Massachusetts General Hospital, in Boston, who had faked biochemical studies of cells taken from patients with Hodgkins' disease, and Vijay Soman, who, while an assistant professor at the Yale Medical School, had misrepresented data in at least one article on anorexia nervosa and possibly committed plagiarism. Both had been discovered and forced to resign, but authorities at both institutions had not detected any evidence of their respective frauds for some time and had been slow to act when it did appear. Although Long testified that the responsibility for his fraud was entirely his, he conceded that his commission of it was "made under great pressure . . . for a grant application," meaning that the failure of his experiments could have jeopardized his funding. Soman's senior collaborator at Yale, a distinguished professor of medicine named Philip Felig, observed, "The desire for success may in some individuals override principles of professional ethics." Felig had attempted to place announcements that Soman's work was in doubt in two scientific periodicals where his work had been published. Neither would publish a letter—one, the *Journal of Clinical Investigation*, because it had no letters section; the other, *Nature*, because its editor, John Maddox, was concerned for Soman's well-being. Maddox wrote that he admired the steps Yale was taking to audit Soman's data but that he was "a little uneasy about the effect of too much publicity on Dr. Soman himself, who in spite of his obvious reprehensible behavior, must presumably be in a state of some distress."[27]

Felig concluded from the recent revelations of research fraud that science, too, included people "prone to unethical behavior" and that research institutions needed "mechanisms to minimize" its occurrence. Yale's had been inadequate to deal with Soman's transgressions; the university had tacitly relied on Felig's willingness to check his own collaborator's veracity. In collaborative research, trust in the honesty of coworkers "may be misplaced," Felig noted. "Consequently, when a challenge has been raised concerning the authenticity of collaborative or single-authored research, a mechanism should exist within the institution for a review process which takes the matter out of the hands of the involved investigators." Yale had now established fraud procedures conforming to that principle, he testified, including the requirement that the outcome be reported to the challenger. Whether the new procedures would suffice, only time could tell.[28]

Apart from Yale, few if any institutions had formal, well-articulated procedures for handling allegations of scientific fraud or misconduct. According to news accounts of the cases that had been cropping up, scientific whistle-blowers tended to suffer retaliation and most had a hard time getting institutional attention. University investigations were often shallow and desultory, and they might well result in quiet, slap-on-the-wrist punishments that permitted the malefactor to continue in science. The N.I.H. had no formal fraud procedures either and exercised little if any oversight of academic inquiries. The agency assumed that the incidence of fraud was "vanishingly small," as an official later put it, and that the matter could be handled on an ad hoc, case-by-case basis. Gore suspected that one reason for the persistence of scientific misconduct was "the apparent reluctance of people high in the science field to take these matters very seriously."[29]

While Felig, burned by the Soman case, obviously took it seriously, the hearings were not otherwise reassuring on the point. The principal scientific witnesses were Philip Handler, the head of the National Academy of Sciences, and Donald Fredrickson, the head of the N.I.H. Handler acknowledged that the scientific community, both at large and in its local institutions, had "never adopted standardized procedures of any kind to deal with these isolated events [of misconduct]. We have no courts, no sets of courts, no understandings among ourselves as to how any one such incident shall be treated." Both Handler and Fredrickson implied that such courts or understandings were unnecessary and might even be dangerous. They insisted that fraud in science was rare and that false claims were exposed by the scrutiny that scientists gave each other's work. Fredrickson contended that the perpetration of fraud shook scientists "to our very core": When fraud did occur, the penalties were harsh, the equivalent of "excommunication" for the sinner.[30]

The biomedical community's post-1970s supersensitivity to government controls was evident in the testimony. Fredrickson called it "frightening" that the larger, lay culture might intrude "roughly" upon scientific affairs while "failing to understand the requirements of the scientific method or the fact that its own correctives are in place." He defended caution in responding to charges of misconduct, declaring, "We want to be sure . . . because human beings are involved; ambitious human beings, seeking honor and prestige and we can easily injure them; we can destroy them for a whole career. We can cast them out of science. . . ."[31] Although Fredrickson acknowledged that the N.I.H. had to be sensitive and responsive to changing public perceptions of science, he held that the agency "certainly cannot guarantee the behavior of scientists, or certify the quality of their work through a whole system of independent analyses or fraud squads, or even special statutes." Ronald Lamont-Havers, the director of research at Massachusetts General Hospital, insisted that American science faced several dangers, including cutbacks in research support and "the imposition of official dogma in place of free scientific inquiry," but "the problem of falsification of research data is not one of them."[32]

Congressman Robert Walker, a Republican member of the Gore subcommittee from Pennsylvania, detected "a certain amount of arrogance" in "a lot of the testimony." He doubted that a policy of "self-policing" was adequate for oversight of the rapidly burgeoning biomedical community. Someone might gussy up data in support of a scientifically sound conclusion, and yet nobody would necessarily be the wiser, he suggested. To Walker, fraud, even if aberrant, was a "major issue" if phony data were "used to justify the expenditure of [public] funds."[33] He warned, alluding to Abscam, "We in politics would like to think that the people who stuck $50,000 into their pockets at some townhouse here in Washington are an aberration in our profession, too. It doesn't mean . . . that we should not be conscious of the need to clean up problems of that kind. . . ." Walker predicted that if the press continued to report these aberrations and nothing is done, the "credibility [of science] will decline pretty quickly" and the public will think of everybody in science as they think of everybody in politics—that is, as somehow a little crooked."[34]

Gore had wondered whether the several fraud cases that had come to public attention were not " 'the tip of the iceberg.' " The testimony suggested to him that fraud might occur at a higher frequency than scientists were willing to concede. Indeed, during the months following the hearing, more cases of scientific fraud surfaced in the press, and in a book titled *Betrayers of the Truth* that was published early in 1983, William Broad and Nicholas Wade contended that crookedness was virtually endemic in science.[35]

Broad and Wade had worked in the news department at *Science* magazine, had recently moved to the *New York Times*, and were respected and influential science journalists. Between them, they had written several stories on recent cases of scientific fraud, some of which had apparently helped prompt Gore to hold his hearings. In their book, which melded narratives of the cases with analysis of why they occurred, they said that writing the stories had made them doubt the "conventional wisdom" of science, especially that the scientific system inevitably roots out fraud because scientists check each other's results. In fact, the evidence is to the contrary: In the cases they examined, the fabrication of data had been exposed by the forger's arrogance or carelessness, or by whistle-blowing in the forger's own laboratory. They concluded, "The chances of getting caught in committing a scientific fraud are probably quite small. Replication in science is a philosophical construct, not an everyday reality. It . . . is almost never the prime cause of suspicion."[36]

Broad and Wade had also discovered that, in acquiring new knowledge, scientists were not wholly governed by logic and objectivity but were guided "by such nonrational factors as rhetoric, propaganda, and personal prejudice." They insisted that fraud was encouraged by the demanding incentives of the research enterprise, an issue that Gore had raised at his hearings. Noting that modern science was highly competitive, fostering ambitious careerism, Broad and Wade contended that to publish papers, get grants, obtain positions, or win prizes, "some researchers yield to the temptation of cutting corners, of improving on their data, of finagling their results, and even of outright fraud."[37]

Broad and Wade argued that "an ambiguous attitude toward data was present from the very beginning of Western experimental science," holding, "On the one hand, experimental data was upheld as the ultimate arbiter of truth; on the other hand, fact was subordinated to theory when necessary and even, if it didn't fit, distorted." Citing historical studies, they said that Galileo may not have done some of his experiments exactly as he reported them; that Isaac Newton had trimmed some of his data; that Gregor Mendel's pea-plant numbers were too perfect; and that the American physicist Robert Millikan, who won the Nobel Prize in 1923 for measuring the charge on the electron, had selected only the experimental readings that fitted his convictions.[38]

Broad and Wade recognized that "self-deception" might shape the handling of data, that a scientist might simply "see what he wants to see." But they argued against the existence of a "clear distinction between conscious and unconscious manipulation of data," suggesting that "the two phenomena probably lie at opposite ends of a spectrum." To their minds, scientists who fabricated data were guilty of "major fraud." Scientists committed "minor fraud" if they selected or distorted "data from real experiments so as to make them appear smoother or more convincing." Broad and Wade expected "that for every case of major fraud that comes to light, a hundred or so go undetected," extrapolating, "For each major fraud, perhaps a thousand minor fakeries are perpetrated." They estimated that "every major case of fraud that becomes public is the representative of some 100,000 others, major and minor combined, that lie concealed in the marshy wastes of the scientific literature."[39]

Knowledgeable observers agreed with Broad and Wade that scientists were overly confident of their self-policing and that the scientific system responded to fraud weakly at best, but otherwise opinion on the issues they raised was mixed. Donald Fredrickson had told Congressman Gore that, while competition might be generating adverse effects in biomedical science, "an increase in deceit was not one of them"; David Baltimore conceded that unquestionably "the pressure on research workers grows because of the limitation of funds and the increasing formalism of the academic world, with its demands to produce and appear successful." Baltimore said he was "sure that everyone has a cracking point."[40] Some critics accepted Broad and Wade's characterization of minor fraud, but at a special session titled "Fraud and Dishonesty in Science" held at the 1983 annual meeting of the American Association for the Advancement of Science, Norton Zinder argued that "science would not have survived" if cheating were as prevalent as they claimed and, citing sensational medical headlines, he went on to indict journalism as "the only profession in which fraud is prevalent." Several other critics lambasted Broad and Wade for attacking common scientific practices with what amounted to moralizing naivete.[41]

Broad and Wade in fact overlooked many of the realities of the laboratory. Data does not always speak for itself, lucidly declaring its meaning. No honest scientist fabricates or falsifies data, but neither do most— especially the innovative ones at work on the cutting edge, where experimental results are likely to be contradictory or ambiguous—respond to data mechanically. In an exquisite autobiography titled *In Praise of Imperfection*, the neurophysiologist Rita Levi-Montalcini recounts the work that led to her Nobel Prize and quotes approvingly "the law of disregard of negative information" that a Russian psychologist had once advanced: "Facts that fit into a preconceived hypothesis attract attention, are singled out, and remembered. Facts that are contrary to it are disregarded, treated as exceptions, and forgotten."[42]

Einstein defied another physicist whose experimental findings were incompatible with his newly minted special theory of relativity, telling him that his laboratory results simply counted for little. Linus Pauling published his groundbreaking structure of the alpha helix protein, brilliantly and rightly casting aside certain data that were inconsistent with it as likely inconsequential. If Einstein or Pauling had been wrong, either might have paid a price in his professional reputation or scientific pride (Pauling's *amour propre* took a hit when he published an incorrect model for the structure of DNA that relied more on intuition and guess than on extensive data).[43] Galileo, Newton, and Mendel operated within their day's standards of faithfullness to data. Millikan may have once stretched the boundaries of full reportage, but the reading of data by all of them figured in the creative judgment and force of conviction that were integral to their achievements.[44]

The critics pounced on Broad and Wade's estimate of the incidence of fraud in science, finding it extravagant and, in the judgment of one physicist, "wildly inconsistent with the personal experience of most active scientists." Broad and Wade supplied no data for their estimate. Reviewers were surprised that, given the subject of their book, they should advance a claim that was important yet that appeared to have been fabricated out of whole cloth.[45] Nevertheless, the defenders of science had no hard evidence either. They deployed arguments of plausibility—for example, a scientist would have to be mad to falsify data because he or she would ultimately get caught and suffer professional ruination. The fact of the matter was that reliable data on the incidence of scientific fraud did not exist.[46]

It was in the interest of helping to fill that gaping lacuna that, in the spring of 1983, not long after the publication of Broad and Wade's book, Walter Stewart and Ned Feder turned to the issue of professional misconduct in science. Their attention to the subject was prompted specifically by the case of John R. Darsee, a young researcher in cardiology who had done a good deal of his work at their alma mater, Harvard, as a postdoctoral fellow sponsored by the N.I.H. Darsee was an exceptionally prolific scientist, so much so that several junior scientists were motivated one evening in May 1981 to observe him secretly in his Harvard laboratory. They saw him forge raw data. Confronted with their testimony, Darsee confessed to his mentor, Eugene Braunwald, a prominent cardiologist and professor at Harvard. Darsee said that he had never faked data before. Braunwald believed him. He had Darsee stripped of his postdoctoral fellowship, but he permitted him to continue working in the lab on a major collaborative project. Although the project was sponsored by the N.I.H., Braunwald declined to inform the agency of Darsee's misconduct. "Public disclosure would have ruined him for life," Braunwald later explained. However, an official at the N.I.H. responsible for overseeing the project grew suspicious of other data attributable to Darsee. An investigative panel appointed by the N.I.H. ultimately found him guilty of multiple fakeries in his research, and in February 1983, the agency announced that Darsee, then 34, had been barred from eligibility for grants for ten years.[47]

Stewart and Feder had the idea that they might be able to measure how commonly scientific misconduct occurred by studying Darsee's coauthors. Darsee had published 109 papers with a total of forty-seven coauthors, including his Harvard supervisors, Braunwald among them, a group that Stewart and Feder considered a sample of biomedical scientists large enough to reward scrutiny. John Maddox, the editor of *Nature*, expressed interest in publishing the study. Stewart and Feder gathered and studied the coauthored publications, which comprised eighteen original research papers, eighty-eight abstracts, and three book chapters, intending to compare them against common standards of scientific conduct. They found that no generally accepted code of standards existed, so they devised one, basing it on what they held to be "an unwritten code of which scientists are aware and to which they profess adherence." They defined the code in terms of "lapses" from it, which they divided into two categories, Type A and Type B. Type A lapses, four in all, included allowing factual errors to remain in papers or failing to check a research group's recent data against those in its earlier publications; Type B, six in all, included misleading statements, failure to credit others for their research data, and publishing "very similar abstracts under different titles." They did not discuss what they said might be referred to as Type C practices—"wholesale forgery of data." Type C crimes were almost universally condemned, appeared to be rare, and seemed "in their aggregate effect" to "do little harm to science."[48]

Stewart and Feder concluded that more than half of Darsee's coauthors had lapsed from the Type A standards and that a quarter of them were guilty of the Type B variety. They said that five papers from Harvard "contained statements which we believe the coauthors . . . knew or should have known were inaccurate." They spotlighted a family pedigree in a publication from Emory—their "most striking example" of error—that included four children, one of them eight years old, whose father, only seventeen, would have conceived her when he was nine. Such lapses tended to debase the scientific literature, Stewart and Feder claimed, suggesting that journals and their referees were at fault for not catching such errors before publishing them. They held that further systematic studies of scientific publications were in order, and they suggested that the task might be accomplished by subjecting scientific papers to an "external audit" like the one they had just conducted or, better yet, to an "internal audit," close examination of the original data on which a paper rested.[49]

Stewart and Feder speculated on the reasons for scientific misconduct, advancing an explanation seemingly colored by their own institutional position and mode of work. The N.I.H. is divided into "intramural" and "extramural" research programs. The intramural enterprise is composed of the numerous individual institutes that are devoted to particular areas of biomedical research—for example, cancer—and are housed mainly on the N.I.H. campus in Bethesda. The campus is huge, and several thousand scientists work on it, but the individual research groups in the different institutes tend to operate on a small scale, much as Stewart and Feder had done for twenty years. Intramural research at the N.I.H. is on the whole "Little Science." The extramural part of the agency primarily awards research grants and training fellowships to scientists in colleges, universities, and other research institutions. While much of the work it supports comprises cottage-industry efforts like those on its own campus, some of its largesse goes to biomedical research groups that are large-scale, hierarchical, and expensive—a version of "Big Science."

From their Little-Science perspective, Stewart and Feder discerned several objectionable features in Big-Science biomedicine. Their critique was partly social: Such large-scale research "created a new class of scientific bureaucrat—the research czar—who is a prominent and influential (although often feared) member of the scientific community." Although his contribution to the work of his group might consist only of raising money for it, he took credit in the form of "honorary authorship" for its research publications—another of their Type A misdemeanors. Yet the principal burden of their critique was that Big-Science biology not only allowed Type A and Type B malpractices; it also encouraged them. The powerful lab chief might choose poor research projects but insist that subordinates carry them to successful completion despite inadequate support from the data, forcing a "hard choice" on the team's junior scientists. The chief's prestige "may lull or intimidate other coauthors, the reviewers, or the readers into uncritical and inappropriate acceptance of the work as valid." To Stewart and Feder, it was "probably no accident that many of the discovered instances of fraud have occurred in large research groups," the "perfect setting for a forger" as well as exemplars of emphasis on high publication rates and intense competition for grants.[50]

Stewart and Feder completed their paper in the fall of 1983 and sent it to Maddox, at *Nature*. Maddox realized that he had been handed a legal Molotov cocktail. Even though Stewart and Feder had not named the coauthors—Stewart later testified that their intention was not to expose individual scientists—they were not difficult to identify from the paper's references. Maddox was apparently worried by Stewart and Feder's implication that a number of the coauthors were complicit in Darsee's fraud by reason of inadequate diligence. (In 1987, they went beyond implication, writing that "about 30 percent of the 47 scientists in the group we studied had been personally involved in misconduct in their research.") In March 1984, Maddox circulated the paper to eighteen people, including several of Darsee's coauthors. Two of them, including Braunwald, responded through their lawyers to *Nature*, Stewart and Feder, and the N.I.H. that they would regard publication of the paper as willfully and maliciously defamatory on the part of both the authors and Maddox's magazine. Maddox asked Stewart and Feder to revise the paper, which they did, but he kept postponing publication. In February 1985, they withdrew the study hoping to find an alternative outlet.[51]

During the next twelve months, Stewart and Feder sent their paper to some 100 fellow scientists for comment and submitted it to more than a dozen other journals. A number of the respondents and referees found it serious, responsible, sobering—an article in the public interest that ought to be published. However, even some of the enthusiasts had reservations about the study, particularly its suggestion that the misconduct that Stewart and Feder had discovered among the coauthors might be representative of misconduct in the biomedical community at large. A number of critics derided it—one, for example, declaring its code of scientific conduct arbitrary, and still another slamming the paper as "petty, vindictive, gratuitous, self-serving, and of no useful purpose." Elizabeth Neufeld, a biochemist at the University of California at Los Angeles (UCLA) and one of the people to whom Stewart and Feder sent their paper, later commented: "There are instances of sloppiness that we would not like to excuse professionally, but if you put it in terms of crimes, you don't equate murder with running a traffic light."[52]

Stewart and Feder had specified few of the errors they detected, nor had they bothered to ask for explanations from Darsee's coauthors or, in the case of the teenager reported to have an eight-year-old child, from the *New England Journal of Medicine*, which had published Darsee's coauthored article with the seemingly ludicrous pedigree. (Several years later, the editor of the journal wrote to Stewart that, if he had troubled to check, "we would have told you that the ages given in the figure were intended to be the ages *at diagnosis*," adding, "We had simply neglected to include that point in the legend and had overlooked our mistake. . . . No rational charlatan could possibly have expected readers to believe that a 17 year old father had an 8 year old daughter."[53]) But whatever the mixed reviews of the paper on its merits, none of the journals was willing to publish the article because they feared libel suits from the coauthors.[54]

Stewart and Feder, neglecting their snails for their Darsee paper, increasingly raised eyebrows at the N.I.H. They had already earned a reputation as irritants on the Bethesda campus, and embarrassingly unproductive ones at that. To their many critics, they now seemed to be misusing their time and lab space, not to mention N.I.H. stationery, on which they wrote numerous letters in pursuit of their fraud studies. Their letters included the postscript that they were "not acting as spokesman for N.I.H." They were pursuing the matter voluntarily "in our official capacity as scientists." Magda Gabor, Rollin Hotchkiss's wife and a scientist herself, says that she asked Stewart how he could reconcile pursuing scientific misconduct with being paid by the N.I.H. to do experimental research, and that he would tell her, This is a free country. I do what I have to do.[55]

Some high-ranking scientists at the N.I.H. thought Stewart and Feder should conduct their inquiries into fraud on their own time, but high agency officials held that their academic freedom included the right to pursue their studies of scientific misconduct. N.I.H. officials did insist that freedom had its obligations. They allowed Stewart and Feder time away from snails to work on their Darsee paper on the understanding that once it was completed they would return to regular research duties, holding their studies of scientific wrongdoing down to a fraction of their time. Around the beginning of 1984, the director of their institute, Jesse Roth, told Stewart and Feder in a memorandum that the continuation of a technician who worked for them was contingent on their producing articles drawn from their research for submission to scientific journals. "There is no demand that these be literary masterpieces in first line journals," the director advised, "Journeyman works for publication in second, third or fourth line archival publications will be quite satisfactory."[56]

Stewart and Feder considered Roth's quid-pro-quo condition for keeping their technician outrageous. (They also considered it hilarious, a caricature of the pressure to publish no matter what the significance of the work. They retaliated, in their Darsee paper and later congressional testimony, by coyly quoting this "official memorandum" to scientists, whom they did not name, from the director, whom they also left unnamed, "of one of the world's leading research institutions.") They neglected to meet Roth's criteria of productivity, were given poor performance ratings, and lost their technician. When everyone in their building was moved out to allow for renovation, they were assigned to a small basement laboratory in another facility. Stewart says that neither of them had a desk and that they did much of their writing "either standing up or sitting on the floor." Stewart and Feder held that N.I.H. officials were retaliating against them for their misconduct research, but sources at the agency later pointed out that space was always tight and less of it was given to unproductive scientists.[57]

The new space was inadequate for all their equipment, some of which was put into a storage room. Stewart and Feder were told several times that the room was to be cleared, with the equipment in it to be removed to surplus. They were asked to tag whatever equipment they wanted to keep, but they did nothing. All of it was eventually moved to surplus, which determines what equipment is to be discarded. Stewart and Feder later informed a congressional committee that one morning in October 1985 they learned that much of their electronic and dissecting apparatus was being hauled out of surplus to a dumpster. It was "heartbreaking,"

Stewart said. He did not inform the committee how the equipment had gotten to surplus in the first place.[58]

In mid-1985, Stewart and Feder wrote to Floyd Abrams, a lawyer in New York, about their problems getting the Darsee paper published, asking whether he could assure the editor of a small scientific journal "that he will be protected against the possibly ruinous costs of a legal defense if he should be sued." Abrams specialized in constitutional issues and frequently represented major publications and broadcasters. He could give no such assurance, and reported that he could not give it in an article on the need for changes in libel law that discussed Stewart and Feder's publishing difficulties, among others cases, in the *New York Times Magazine* for September 29, 1985. The article evidently caught the attention of the House Judiciary Subcommittee on Civil and Constitutional Rights. In February 1986, Stewart and Feder testified at the subcommittee's hearing on the conflicts between a free press and the laws of libel, reporting on the substance of their misconduct study, detailing the threat of libel suits that had greeted it, and providing for the hearing record copies of the correspondence, by now substantial, from the coauthors' various lawyers.[59]

Not long after the hearings, a journalist named Daniel S. Greenberg walked into Stewart and Feder's laboratory. During the 1960s, Greenberg, then in his thirties, had been news editor of *Science* and developed a stable of young reporters who turned a cocked, iconoclastic eye on the affairs of research, irrepressibly exposing the politics that shaped science, pure and otherwise. Greenberg made many scientists uncomfortable and earned the enmity of some. In 1971, he left *Science* magazine for full-time work at *Science and Government Report*, a biweekly newsletter that he had founded two years earlier. His newsletter, by turns sardonically muckraking and thoughtfully informative, won an influential following. Now here he was, in Stewart's recollection, looking big and hulking, announcing—Greenberg disputes this—that he told the *New York Times* what was important in science, and, in all, making Stewart anxious.[60]

Greenberg wanted to hear about the problems with the Darsee paper. Stewart says that he admired the *New York Times* for having published the Pentagon Papers in 1971 but suggests that he generally distrusted the press. He and Feder were committed to changing the ethics of science in a strictly scientific fashion, confining the expression of their views to professional journals and never talking to the media. Nevertheless, having already testified publicly to Congress, Stewart spoke with Greenberg. In his newsletters for March 15 and April 1, Greenberg reviewed what he called "The Tale of the Fraud Study That's Too Hot To Publish," describing Stewart and Feder as "undramatic, cautious, and credible witnesses."[61]

Two weeks later a reporter named Philip Boffey, a veteran of the *Science* news department under Greenberg and now with the *New York Times*, arrived at the lab. According to Stewart, he said that he had read about Stewart and Feder in *Science and Government Report*. Stewart says they told him some of their story and, having generally been rebuffed for several years in trying to peddle their claims, were pleasantly shocked when Boffey asked to see their evidence. He not only went through all of it; using the Freedom of Information Act, he soon got the N.I.H. to release to him Stewart and Feder's paper on Darsee and hundreds of pages of the accumulated legal and scientific correspondence pertaining to it. Boffey published an account of the entire mess, accompanied by a picture of Stewart and Feder, in the *New York Times* of April 22, 1986—the story that drew Charles Maplethorpe's attention to the pair.[62]

John Maddox told Boffey that Darsee's coauthors had been "careless" rather than "dishonest in any active sense," but that even so, Stewart and Feder's analysis contained "enough horrors to make people sit up on the edge of their chairs." Maddox was in fact reconsidering the publication of the paper, and a number of scientists—David Baltimore among them— welcomed the spotlight they threw on corrupting practices such as honorary authorship. A growing number of people outside of science were also sitting up and taking notice of the issues it raised. Even if fraud was rare, the publication of false results in biomedical science might jeopardize consumers and be costly in time and money to other scientists who tried to follow them up.[63]

The ethics and responsibilities of authorship had become a subject of debate in the scientific press, of formal discussion at scientific meetings, and of policy pronouncements by technical journals and societies, including the *New England Journal of Medicine*, the *Journal of the American Medical Association, Annals of Internal Medicine*, the Council of Biology Editors, the Association of American Medical Colleges, the American Psychological Association, and the American Chemical Society. The N.I.H. now received fifteen to twenty allegations and reports of misconduct every year, almost half of them from individuals closely associated with the suspected work. Of those handled by the agency itself, more than half were substantiated. Still, according to a survey conducted at a major research university, while most scientists thought that fraud was rare, many also believed it was unlikely to be detected.[64]

In 1982, following the Gore hearings, the N.I.H. had begun developing policies and procedures governing misconduct for the Public Health Service and its grantee institutions. The agency established an ALERT system to provide orderly, confidential sharing of information about investigations under way and sanctions imposed. Colleges and universities devised their own mechanisms for dealing with allegations of fraud, and in a revision of the health services act in 1985, Congress required that any organizational applicant for federal funds in aid of biomedical or behavioral research have a process in place for responding to charges of scientific misconduct. In June 1986, in a follow-up to the new law, the Public Health Service codified the rules it had devised on the basis of its experience since 1981, promulgating a policy that defined misconduct in science as "serious deviation from practices which are reasonable and commonly accepted within the scientific community for proposing, conducting, or reporting of research." The deviations included but were not limited to "fabrication, falsification, plagiarism, and deception." The chief misconduct official at the N.I.H. at the time says that the policy was "the Bible" for all grantee institutions.[65]

A month earlier, on May 14, Stewart and Feder were back on Capitol Hill, testifying about their Darsee study and its reception before a House Task Force on Science Policy. They entered their paper into the record of the hearing, which made it available to the entire world, and they declared that it "does raise the possibility that misconduct may be far more common among biomedical scientists than is realized."[66] Stewart noted that they had initially intended "just to do this study and get right back to our work" and that further studies of fraud were "the last thing in the world we would like to do personally." "Once was more than enough," he said. But two days later Maplethorpe initiated his first round of telephone calls to Stewart and Feder. And on July 10, he called them again.[67]

F I V E

■

A Demand for Audit

MAPLETHORPE SAYS that he had "held out some hope that at the M.I.T. meetings, David Baltimore, for whom I had a lot of respect, would be furious and renounce Thereza." Now that Baltimore had not, Maplethorpe revived contact with Stewart and Feder to recount the outcome of the Tufts and M.I.T. investigations. In successive conversations, he reported that a scientist named "Margot" had written a critique of the *Cell* paper based on the laboratory notebook pages she had copied. While he said that its message was "defects, not fraud," he reiterated his own suspicion that fraud might be involved. Stewart recalls, "I asked if the woman bringing the charges had been satisfied and he said no, she had not, but she had decided not to pursue the matter further." Stewart adds that, having become "extremely curious about the circumstances behind these very interesting events," he asked Maplethorpe for the name of the woman. Maplethorpe at first declined, saying, "Any contact with her— she'd know it came from me." However, probably around the end of July, Maplethorpe told O'Toole about his talks with Stewart and Feder. Although angry at the violation of her confidence, she permitted him to give them her full name and her mother's telephone number, and Stewart got in touch with O'Toole early in August.[1]

Stewart asked if she was satisfied with the results of the investigations. She said no. Stewart remembers telling her that if she knew "about defects in a published paper, it was our belief . . . that she had a professional obligation to bring that knowledge forward." She said that she was reluctant to pursue the matter beyond the actions she had already taken, but, pressed by Stewart, she soon provided an extensive account of her dispute with the *Cell* paper and its coauthors. Stewart asked her to send a copy of the seventeen notebook pages so that he and Feder could analyze the data themselves. O'Toole worried that reopening the matter would jeopardize her husband's career at Tufts. Stewart pressed her for weeks, striking up a telephone acquaintance with her mother in the course of his entreaties. O'Toole had entrusted the seventeen pages of data to her, and Stewart says that she would have sent them if Margot had not. Several years later, O'Toole recalled her mother's saying one day, in allusion to O'Toole's concern for her husband, "You know, no man can be blamed for not controlling his mother-in-law." O'Toole, taking her to mean that she was going to send the pages, said she decided to do so herself, thinking, "My God, there's nothing more embarrassing than to have my mother do it."[2]

Stewart's contemporary notes of several telephone conversations with O'Toole late in the summer of 1986 reveal, however, that she was forced to take other considerations into account. Stewart and Feder told her that the data, having been "paid for by [the] public," was in the "publ[ic] domain" and "sh[ou]ld be accessible." They added that if the data were not turned over, they "must go to N.I.H.." (Apparently they had prepared a memo about the dispute.) They had the agency's legal adviser, Robert Lanman, call to urge that she "come forward." O'Toole said that she was "worried about getting sued" and that she was "frightened." She insisted on "maximum protection for Peter and me before we send [the data]." On September 10, she asked Stewart whether he would "promise" to say that "you called me and said if I didn't give you the data, you'd give the memo to N.I.H." Stewart replied, "Of course," whereupon O'Toole agreed to send a copy of the data, adding, "Right now I want my role to be passive—leave me out of conf[erence]s with Balti[more] and Theresa [sic]. They'll lie to you."[3]

On September 16, 1986, Stewart dashed off a note of evident satisfaction: "We received a copy of d[a]ta from Margot O'Toole. Postmark Sept. 11. Postage due: 2¢."[4]

O'Toole says that Stewart went at mastering the arcana of the *Cell* paper experiments "like a tiger," calling her at home and at the Gentle Giant Moving Company for advice and guidance. Stewart and Feder soon convinced themselves that the data in the seventeen pages undercut the *Cell* paper's claims. They proposed that O'Toole write a paper using their analysis. She remembers telling Stewart, "I can't. I have to paint my mother's house," and his replying, "Well, I'll come up and paint and you can write the manuscript."[5] In the end, Stewart and Feder analyzed the data in a paper of their own.

Their study reinforced the issues O'Toole had raised and added a new one by calling into question the way Imanishi-Kari had decided what constituted an indication that, say, Bet-1 had bound to an antibody. The indicator of binding was the level of radiation, measured in counts per minute, generated by the sample in a well. Imanishi-Kari had set the level of significance at 1,000 counts per minute, which she said was two and a half times the background level. (Background radiation came from the tubes that carried the sample through the radiation counter. Although the tubes were washed out after each run, they tended to retain small residues of radioactive material.) Stewart and Feder argued that the cutoff of 1,000 counts per minute was misleading because, in the data they saw, the wells assessed did not all produce sharply different counts per minute compared with background; many counts fell along a continuous distribution between background and 1,000, making the determination that a count of a 1,000 meant something seem arbitrary.[6] Stewart and Feder recognized that the exaggeration of the discriminatory powers of Bet-1 was inconsequential to the central claim of the *Cell* paper, but Stewart says that he believes that no paper should contain untrue statements of any inherent consequence.[7] In preparing their analysis, he and Feder also rigidly excluded any information that was not in the seventeen pages, including what O'Toole herself had learned from the Tufts inquiry about the misstatement concerning the isotyping of the Table 2 mouse cells and even the mistyped mouse. Having adopted such criteria, they concluded that the coauthors' "experimental results not only failed to support their main conclusions, but in many cases actually contradicted these conclusions."[8]

Stewart and Feder hoped to publish their paper in *Cell*. This time, departing from their procedure in the Darsee case, they intended to ask the coauthors of Weaver *et al.* for comment on their analysis before submitting it to the journal.[9] In mid-October 1986, however, they were told that the N.I.H. Office of Extramural Research did not want them to contact the coauthors or in any way to publicize their critique of the *Cell* paper. Since the extramural office ran the agency's programs of research and training grants to scientists in colleges and universities, it had jurisdiction over misconduct investigations in academia. Stewart and Feder had alerted the extramural staff member in charge of misconduct inquiries, a woman named Mary Miers, that fraud might have been committed. Miers, claiming control over any inquiry into the *Cell* paper that might be mounted, held that if the authors and their institutions were to be contacted, it was extramural officials who rightly should do it.[10]

At the end of October 1986, Stewart and Feder, following standard procedure at the N.I.H., requested permission from their immediate supervisors to submit their article to *Cell*. Such requests were routinely granted, but this one was bumped by their immediate bosses up to "Building 1"—N.I.H. shorthand for the building that houses the agency's central administration—where it landed on the desk of the deputy director for intramural research and training. He was Joseph E. Rall, a physician and former scientific director of the institute where Stewart and Feder were employed. Rall, citing academic freedom, had supported Stewart and Feder's request to submit their Darsee paper for publication. Now retired from administration and unabashedly frank about its ways, he explains that Stewart and Feder's superiors were "a little nervous about making any decision," continuing, "They regarded Stewart and Feder as a big pain in the ass, but I think they didn't want to censor things." Since he had hired the pair, they shuffled the matter to him, telling him to handle it.[11]

Rall considered Stewart "a brilliant scientist" and thought it "dumb" for him "to be grubbing around in the sewers of scientific stupidity, sloth and fraud when he could be doing something positive." He says he told Stewart and Feder as much any number of times, but he always defended their right to investigate whatever they wanted. He was devoted to the principle of academic freedom and thought that the N.I.H. should operate in the mode of a university rather than of a government laboratory. Besides, part of Rall sympathized with Stewart and Feder's purposes. An enthusiast of the predominantly Little-Science culture of intramural N.I.H., he felt that they had done a good service by exposing practices such as honorary authorship in their Darsee study. He says that "too much money goes to too few people having too large an organization" and that the scale of such activities, more than any other factor, encourages the rare event of "falsification, fabrication." For that reason, he was "interested" in the issues that Stewart and Feder were raising while "the extramural people were uninterested." He notes that "Baltimore had always run a huge operation," adding, "Now Baltimore's very smart and he does it full time and he's very good. [But] how could he be totally on top of everything Imanishi-Kari did? Obviously he couldn't."[12]

Stewart and Feder had argued for Rall's prompt assent, pointing out that other scientists were building on the paper's results and that, if they "are wrong, perhaps hundreds of thousands of dollars are being wasted every month."[13] Rall, recognizing that he had a hot paper on his hands, postponed a decision until he could obtain an evaluation of what Stewart and Feder had written from three internal referees. The referees each independently advised that the paper was based on information—the seventeen pages—that resembled "an anonymous phone call," as one of them put it. They said that the paper should not go out until Stewart and Feder had checked with the coauthors whether the seventeen pages were genuine, and whether the coauthors might possess other unpublished data that substantiated what was reported in the paper. Rall decided to withhold approval of the paper until Stewart and Feder resolved the matter with the coauthors. He suggested that they send them a copy of the seventeen pages, so that they will know "exactly what you are referring to," and ask for an explanation of the discrepancies with the published data. He added that it might be unwise to ask for "access to all the authors' data so that you can perform an 'internal audit.' "[14]

On December 18, 1986, Stewart and Feder wrote to all the coauthors, saying that they had conducted "a partial internal audit" of the *Cell* paper using copies of seemingly relevant laboratory records that they had obtained from "a scientist," that the paper seemed problematic by this reckoning, and that they wished access to most of the records so as to "perform a more thorough analysis." They insisted that they were "not alleging or implying wrongdoing" but were only concerned with the *Cell* paper's "verifiability" in relationship to its data."[15] (Stewart says that submitting to internal audits is a scientific obligation, a duty imbedded in "the very meaning of science," explaining that if someone, "for whatever reason," wants to see data that you've published, "you are obligated to supply the data.")[16] They told the coauthors that they would take their comments into account before "deciding whether to attempt to publish" their analysis. They assured them that they were not agents of the N.I.H., that they were mounting their inquiry "as scientists" eager to get the facts solely in pursuit of their "scientific interests."[17]

Stewart and Feder had not approached the coauthors in the way Rall had suggested. They had not specified the problems they had found with the paper, had not sent the seventeen pages, and had not listed the data they wished to examine beyond referring to several of the paper's tables and figures. Rall remarks that in cases such as Stewart and Feder's inquiry into the *Cell* paper, he has always been "torn between freedom of expression . . . freedom to say what you think and the boundaries of decency," adding, that in many cases, but "particularly with Walter and Ned," he found it difficult to decide between the two.[18] *Nature*, in its issue for January 15, 1987, published Stewart and Feder's article on Darsee, in a form sanitized to avoid libel suits. The article was accompanied by an angry attack from Darsee's coauthor and Harvard mentor, Eugene Braunwald, on their guilt-by-association reasoning and by an editorial observing that Stewart and Feder "have not understood that the unfettered right to publish scientific data does not equate with a right to denigrate others' character."[19]

In the months preceding the arrival of Stewart and Feder's letter at M.I.T., Herman Eisen had been intermittently troubled by several of the issues that had been probed at the meeting with O'Toole and the coauthors in mid-June. He reread the *Cell* paper and O'Toole's memorandum a number of times, trying better to understand the science. "I was disturbed by O'Toole's memorandum, which really made a lot of sense," he recalls. He telephoned William Paul at the N.I.H. in whose laboratory Bet-1 had been developed; Paul agreed that the reagent could perform as Imanishi-Kari said it did. On August 4, he spoke with Weaver to check out a technical objection raised by O'Toole: It turned on her suspicion that the hybridomas might have been sampled for DNA sequencing by Weaver and for antibody characterization by Imanishi-Kari at different times. Weaver assured him that the objection had no merit: O'Toole was "wrong" in claiming that the samplings had not been done simultaneously, Eisen jotted in a note of the conversation. According to Eisen's note, he and Weaver agreed that a correction of the overstatement about the discriminatory powers of Bet-1 ought to be submitted, but Eisen later recalled they also agreed that *Cell* would not publish a correction of such a small error.[20]

However, Eisen was concerned not only about the specificity of Bet-1 but also about whether it was reliable—that is, whether it worked as it was supposed to most of the time. The way it was represented in the *Cell* paper suggested that it did, but O'Toole's objections and Eisen's own inquiries suggested that it might not. Eisen quizzed Imanishi-Kari on the point in a hurried hall conversation one day in early September, when she happened to be at M.I.T. picking up some of her frozen hybridomas for transportation to Tufts. He thought he heard her say that Bet-1 was highly problematic, that she had "known it all along," and that she told anyone who asked her about it. Eisen, startled and upset, mentioned to Baltimore that Imanishi-Kari had admitted that Bet-1 rarely worked properly, conveying to him that more was wrong with the reagent than he had thought.[21]

In a letter to Eisen on September 9, Baltimore, furious, called Imanishi-Kari's statement that she knew about the unreliability of Bet-1 "all the time . . . a remarkable admission of guilt." It was beyond him why she had chosen to use the data generated with the reagent and to "mislead" Weaver, himself, and anyone who might read the paper. He declined to withdraw the paper, however, holding that Weaver's sequence data sufficed to support its central claim and that he would "hate" to see Weaver's "integrity questioned," as senior author, "for something he accepted in good faith and where his contribution is what makes the paper strong." In his summary judgment, "a retraction would harm the innocent and raise doubts about quite solid work." He advised simply acknowledging "to colleagues that the Bet-1 results are not reliable," adding that he, "for one, will be skeptical of Thereza's work in the future."[22]

Baltimore's letter, smacking of cover-up, eventually became public, and in a congressional hearing he would express his "profound regret" for having written it, declaring, "I fully understand that when a serious error has been made, it must be fully acknowledged." At the time, he quickly forgot about the letter because a day or two later, in a discussion with Imanishi-Kari, Eisen learned that he had misunderstood her: Although various batches of Bet-1 failed to discriminate between *mu-a* and *mu-b*, enough of them worked to make the reagent reliable if it was used with care.[23] Eisen was relieved, and so was Baltimore. It was plausible to Eisen that some batches of Bet-1 would fail as a result of having been treated with radioiodine; he calls such disruptions "an automatic issue" for people in the field. Nevertheless, in a later, rueful note to himself, he wrote of the difficulties with the reagent: "Why oh why was this not indicated everywhere, and especially in the paper? It would have saved so much grief and cost to so many people and institutions." All the same, in the fall of 1986 Eisen did not urge that an erratum be submitted to *Cell* about either of the problems with Bet-1—that is, with its specificity or reliability—judging that the journal "would not deign . . . to publish . . . a trivial point."[24]

Stewart and Feder's approach to the coauthors, which Eisen knew about, revived his misgivings about the way the *Cell* paper had characterized Bet-1. It also prompted him to think differently about the dispute, shifting somewhat defensively to frame it in terms of fraud and misconduct rather than scientific disagreement. In late December, the chairman of the M.I.T. biology department asked Eisen for a written report of the inquiry he had conducted in June. Instead of retrieving the contemporary summary from his files, Eisen responded on December 30 with a new memorandum titled "Allegations of Misconduct" that was devoted almost entirely to Bet-1. He had not mentioned the subject at all in the report he had written the day after the June meeting; now he wrote that O'Toole's assertion about Bet-1 was "disturbing because it raised serious questions about deliberate misrepresentation of data." Countering that possibility with what he had learned since about the reagent from both Imanishi-Kari and William Paul, he concluded that O'Toole was right to claim that the paper contained an error, but he added that the error was "not . . . flagrant." It was "too minor to rate a letter to the journal," and it "certainly does not warrant a retraction, especially because the paper contains a substantial body of other data that is clear and impressive."[25]

Baltimore, on his part, had thought that the matter of the *Cell* paper was closed. After Stewart and Feder's letter arrived, he assumed that O'Toole had insisted on reopening it. As the coauthors' spokesman, he replied that he saw no point in cooperating with Stewart and Feder, declaring that he was satisfied that the data in the *Cell* paper fell "within the norms of scientific evidence" despite the questions raised about them by "a discontented postdoctoral fellow."[26] Baltimore rejected Stewart and Feder's "notion of doing an internal audit," contending that it was unwarranted except in cases of suspected fraud, which the *Cell* paper was not; that otherwise it "would tie up the scientific community in continuous wrangles"; and that he did not recognize their "right to set yourselves up as guardians of scientific purity."[27]

Baltimore says that Stewart and Feder appeared "adversarial," far more interested in exposing scientists than in genuinely assessing the accuracy of the scientific literature. They seemed to expect to protect scientific purity by deciding which scientists were lying about data and which were not. (Stewart once told a reporter that what they aimed to do was to "take liars one at a time, and expose them, and have public opinion disapprove of them.")[28] In Baltimore's view, what they called their "research" really amounted to indictments of scientists; a critique from them might put a scientist's reputation at risk without providing the person accused with the protections of fair procedures.[29] To Baltimore, Stewart and Feder resembled vigilantes riding under the cloak of science. While they invoked the cry of free and open debate, their practices could amount to free and open slander. Rumor had already prompted telephone calls to Baltimore from reporters and colleagues wondering about the merits of Stewart and Feder's analysis.[30]

Stewart and Feder would not be denied. Imanishi-Kari and Eisen had told them in telephone calls that the powers of Bet-1 had been overstated in the paper. For weeks they bombarded the coauthors, especially Baltimore, as well as Eisen and Henry Wortis with letters and calls, asking, unsuccessfully, for data and documents, then for responses to a long list of technical questions, and ultimately, in March, to their analysis of the *Cell* paper. Wortis, in a letter he hoped would end their correspondence, told Stewart and Feder that he would not cooperate because they were "not conducting a scientific study to answer a specific question, such as 'What is the frequency of mistaken conclusions in scientific papers?'" Eisen held their idea of auditing science in contempt, believing it implied "that you can do science like you run a bank." He wrote Stewart and Feder that he wanted nothing more to do with them, that O'Toole's belief in an alternative explanation of the data reported in the *Cell* paper amounted "to an obsession," and that the question of who was right would be decided "through additional laboratory work, and not through endless conversation."[31]

Baltimore discussed Stewart and Feder's analysis of the *Cell* paper with Imanishi-Kari, scrutinizing the seventeen pages, and probing the issue of the cutoff at 1,000 counts per minute. She assured him there was no substance to their principal points, particularly the matter of the control mouse. To Baltimore, the mistyped mouse seemed "the most dramatic" argument Stewart and Feder had raised. The rest of their points—for example, how cutoffs are determined—struck him as "clearly a matter of interpreting scientific data"; "they had their own way of doing it," one that he did not necessarily agree with. Baltimore says that he was highly irritated by Stewart and Feder's seeming "arrogance" that they could analyze a complicated scientific study on the basis of merely seventeen pages of data, a small fraction of the hundreds generated in the course of the experiment.[32]

In a telephone conversation with Stewart, Baltimore lost his temper. According to Stewart, he characterized O'Toole as pursuing a "personal vendetta," accused Stewart and Feder of having written a "vicious manuscript," and threatened that he might sue both of them for "harassment and libel." He contended that O'Toole had "stolen" the seventeen pages and that most scientists, if given such unauthorized materials, would have returned them to their owners with an apology. Baltimore wrote ruefully to Wortis and Eisen that he had let himself "get sucked into a telephone conversation with Stewart," conceding, "Some of the quotes are accurate." Stewart wrote to Baltimore that his outburst was "not helpful," that threat of suit was "not in the scientific tradition."[33]

Rall remembers Baltimore's being "furious" at him for letting Stewart and Feder go so far. In mid-March, in a letter to Rall, Baltimore suggested a course of action that he had urged on Stewart during their tempestuous telephone conversation—that Rall appoint "a couple of immunologists" to examine Stewart and Feder's assessments, on the understanding that if the *Cell* paper passed muster again, Stewart and Feder would apologize and henceforth shut up about the subject. "The apology is absolutely necessary to counter the publicity that the issue has already received,"

Baltimore explained. "The reputations of young scientists (never mind an older one) have been impugned by Stewart and Feder's activities and this wrong must be righted by them."[34] On April 1, senior officials at the N.I.H. asked the agency's extramural misconduct office to look into the case. While they declined to muzzle Stewart and Feder forever, they forbade them to submit their manuscript to any scientific journal until the matter was resolved.[35]

To Stewart and Feder, that restriction sounded intolerably long. In a memorandum to Rall they protested at length, insisting that it violated the N.I.H.'s own precedents and policies, not to mention the law governing the agency. They also argued, in evident allusion to O'Toole's experimental difficulties with Bet-1, that "a person unable to replicate the published results will be uncertain whether the problem is with the original report or with his own techniques." "Scientific careers may be harmed by the wasted time and effort" and the "harm being done by the *Cell* paper greatly outweighs any potential embarrassment to the authors."[36]

Rall's balance of principles had shifted against Stewart and Feder. He was concerned that they were forcing Baltimore to waste his time. He also worried that Baltimore and Imanishi-Kari might, with justification, sue them for stealing the seventeen pages. Rall refused them permission to publish their manuscript, holding that "a resolution among scientists would be preferable to airing what may not be a correct analysis of the data in the public press." Nevertheless, since Stewart and Feder had revised their study of the *Cell* paper, Rall gestured to academic freedom by sending it for evaluation to a new set of referees.[37]

By July 1, the new round of refereeing had only reiterated the problematic nature of Stewart and Feder's analysis. In the meantime, however, they had gone public, broadcasting the suppression of their paper to scores of scientists through a chronological summary of events at the N.I.H. and of their telephone conversations and correspondence with the coauthors. In an extensive covering letter, they asked for "advice" about how to resolve the "real professional dilemma" they faced—between, on the one hand, jeopardizing their jobs by defying the suppression and, on the other, abiding by it and thus violating the "generally accepted standards of research" that scientists were obligated to report errors about which they had unique knowledge.[38]

Toward the end of June, the American Civil Liberties Union (ACLU) learned of Stewart and Feder's difficulties, including that, in mid-June, Feder's supervisor had given him an unfavorable performance rating. (Rall says that Stewart should have had an unsatisfactory performance rating, too. Stewart, however, was always rated by Feder.)[39] In mid-July, the Washington law firm of Morrison and Foerster, acting at the behest of the ACLU, intervened on their behalf, protesting the denial of permission to publish their paper and suggesting that Feder's unsatisfactory review might have been connected to his efforts with Stewart to study and publish on scientific misconduct. Almost immediately, in a move that effectively denied the extramural misconduct office exclusive rights to the case, Rall gave Stewart and Feder the green light to publish, pointing by way of explanation to "the overriding importance of permitting free and untrammeled investigation and reporting thereof." On September 9, Feder was told that his job performance was satisfactory.[40]

Early in September, Stewart and Feder's attention had been caught by a multiply authored letter in *Nature* that further explored the peculiar immune response of the transgenic Black/6 mice.[41] One of the coauthors was David Baltimore, who had helped analyze the experimental data, but the lead coauthor was Leonore Herzenberg, who had carried out the principal experimental work reported in the letter with her husband, Leonard Herzenberg, and a postdoc named Alan M. Stall in their laboratory at the Stanford University Medical School. The Herzenbergs had met at Brooklyn College in the early 1950s, when he was a senior and she a freshman, and they married soon after he moved to California to study at the California Institute of Technology (Caltech) for a Ph.D. She left school before completing her degree to join him in California, got permission to sit in on courses—women were not then allowed into the graduate program at Caltech—and, with encouragement from several professors, fashioned herself into a research biologist. Like many such scientific couples, the Herzenbergs had collaborated for many years. Now he held a professorship at Stanford and she, a senior research associate, would soon hold one, too. The team of Lee and Len Herzenberg, as everyone called them, was highly regarded among immunologists on both sides of the Atlantic.[42]

On a visit to the Whitehead Institute one day a couple of years or so before the publication of the *Cell* paper, Lee Herzenberg stopped by the laboratory of a young pathologist named Jonathan Braun, who was then doing postdoctoral work with Baltimore. He told her, "Lee, I took these mice apart and I sectioned them. They're really screwed up." They had "funny pathology," she recalls, funny spleens and funny lymph nodes. From then on, she wanted to analyze the B cells from the mice using special equipment available in the Stanford laboratory that would permit identification of the type of B cells they were from markers on their surface. Later, after reading the article by Weaver *et al.*, she decided that she just had to get the transgenic mice. Baltimore, who judged the project interesting, provided some through Rudy Grosschedl, his former postdoc who was now at the medical school of the University of California at San Francisco and who, with Braun, by then at UCLA, would appear among the coauthors of the letter in *Nature*.[43]

Normal mice produce mostly conventional B cells but also a small number that are anomalous and termed *Ly-1*. The Herzenbergs discovered that the transgenic mice generated virtually no conventional B cells but did generate the normal complement of the Ly-1 variety. The news was exciting back in Cambridge because Ly-1 B cells produce abundant antibodies with the distinctive idiotypes related to those characteristic of antibodies from the transgene. Part of what Baltimore contributed to the letter was the suggestion that the disproportionately high expression of the Ly-1 cells might account for Imanishi-Kari's detection of idiotypically birthmarked antibodies that were native to the Black/6 mice. These antibodies might well have arisen just from the Ly-1 B cells, not from some form of idiotypic mimicry. David Weaver, also a coauthor of the letter in *Nature*, later declared to federal investigators, "It was in complete agreement with the *Cell* paper."[44]

In fact, the agreement was tentatively incomplete on one important point. A large fraction of the cells the Herzenbergs analyzed produced IgM antibodies—that is, antibodies with the isotype *mu*—that derived from both the transgene and native genes. Thus, contrary to what Imanishi-Kari and Weaver had found, many of the Herzenbergs' transgenic mouse cells appeared to be double producers. The Herzenbergs suggested that the discrepancy between the results of the two experiments might be rooted in the fact that they had scrutinized cells taken directly from the mouse while the coauthors of the *Cell* paper had used hybridomas. Whatever the reason, they intended to pursue the subject further in collaboration with their postdoc, Alan Stall.[45]

The Herzenbergs knew O'Toole but at the time were completely unaware of her dispute with Imanishi-Kari, let alone her claim that double producers could better explain the *Cell* paper data. In the fall of 1986, Lee Herzenberg innocently called Henry Wortis, told him about their preliminary double-producer results, and asked him to pass on the news to Imanishi-Kari. She apparently also sent him material prepared for the letter that would be submitted to *Nature*. On November 11, Stewart jotted notes of a telephone conversation with O'Toole about the draft: "Highly conf[i]d[ential]. Smuggled to Peter by grad student in Henry

Wortis' lab. Afraid to make copy. Margot not supposed to know. Henry just rec[eive]d from Herzenbergs at Stanford. They have Thereza's mice. . . . Thereza all wrong."[46]

Lee Herzenberg remembers that the following September she received a telephone call from Walter Stewart. He asked a lot of questions about the kind of laboratory techniques used in the *Cell* paper, especially those involving Bet-1. He gave no indication that he was investigating the publication. Herzenberg says she initially got the impression that he was a postdoc somewhere who was about to start work at the bench on the analysis of antibodies.[47] Following up the conversation, Stewart and Feder sent the Herzenbergs a copy of their critique of the *Cell* paper, noting that the recent letter in *Nature* referred to the paper and claiming that some of its results "are in fact contradicted by the original data." They included a copy of the memorandum that O'Toole had prepared for Herman Eisen in June 1986 arguing for double producers as a preferable explanation of the experimental results. Stewart had mentioned the document in his telephone conversation and expected that the Herzenbergs might want to cite it in articles to which it might be relevant. Stewart and Feder hoped that once the Herzenbergs had studied their critique, they would "not cite the [*Cell*] paper for any propositions that are not supported by the original data in any of your future publications in this area."[48]

Now well aware of the dispute, the Herzenbergs aired their views on the matter in a letter to Baltimore. "Margot O'Toole was probably quite correct in raising questions about the serological data from the transgenic hybridomas." They had experienced "nothing but trouble" with Bet-1 and had ruled it out for use in the kind of experiments they were doing. They nevertheless had achieved some "partial successes" with the reagent and, given them, could "see where strong arguments in favor of the serological data as presented could have been made." They wondered what Baltimore thought about why Imanishi-Kari had not detected the kind of double producers they were finding. Although they had not been in touch with O'Toole, they considered her memorandum to Eisen "well presented and quite reasonable." They pointed out that what they were doing—producing a new set of hybridomas from the transgenic mice—would "clearly help to resolve this issue," adding, "If Margot really has a legitimate claim to some credit for her insights and for some relevant data that she produced herself, we could consider inviting her out here to participate in the production of the new hybridomas and thus in the publication of the new results."[49]

The Herzenbergs added summarily:

If we had to make a guess right now, with hindsight we would say Margot is probably correct in suggesting that double-producing hybridomas can be obtained from the transgenics. On the other hand, we think that Margot and/or Feder and Stewart are extraordinarily incorrect in attempting to turn this into an issue of scientific dishonesty or improper data reporting. The arguments in this case rest on subtle data interpretation and represent the typical kinds of questioning that lead eventually to scientific progress. We can perhaps fault the original investigators for being wrong or even being pig-headed, but we cannot fault them for being dishonest.[50]

On November 6, Baltimore replied, declaring that the data in the seventeen pages "is not all Margot's" and adding parenthetically, "I do not know if any is hers." He acknowledged that Bet-1 was "not as great as it seems in the paper," that it was *"relatively* specific, not absolutely." He did not doubt that direct examination of the transgenic mouse cells showed double producers but did not know whether a repeat of the hybridoma tests would show the same and would be glad to have the Herzenbergs look at Imanishi-Kari's originals. He noted in closing, "I've spent more time on this than I would like, but it won't go away. I'm glad you see it as the usual scientific uncertainties."[51]

The Herzenbergs learned from Henry Wortis, too, that O'Toole had not worked on the project and had no personal claim to the data. They concluded, as they wrote to Baltimore two days before Thanksgiving, that they did not see "any necessity for her name to appear on a publication concerning this data" nor "any particular reason for inviting her here to work on the problem."[52] In the succeeding weeks, they continued, with Alan Stall, to develop and analyze their own hybridomas from the transgenic mice. They soon completed the work and reported the results in a paper that they submitted at the end of December for publication in the *Proceedings of the National Academy of Sciences*: The transgenic hybridomas were double producers.[53]

In mid-January 1988, the Herzenbergs sent Stewart and Feder a copy of their new paper on the transgenic hybridomas. In an accompanying letter, they said that O'Toole's criticisms of the *Cell* paper were "novel and creative" and suggested "an error of significant proportions was made." They noted that nevertheless the further lab work they had pursued had been started independently of O'Toole's critique of Weaver *et al.* and without knowledge of it. They saw no legitimate way to credit O'Toole. In any case, they doubted that the errors in the *Cell* paper merited the attention that Stewart and Feder were giving them. They stressed "that science is better served by the findings we have recently presented than by the publication of a critical manuscript such as the one you have sent to us."[54]

By now, Stewart and Feder were deeply absorbed in fraud busting. After the publication of their Darsee paper early in 1987, mail had begun flooding into their laboratory telling them about incidents of suspected fraud and asking for their help. They had gone, for the better part of several months, to work in Madison, Wisconsin, as private, unpaid experts on behalf of a party in Cambridge, Massachusetts, that was attacking a patent obtained by a scientist at the University of Wisconsin. On their return, they found that six boxes of letters had accumulated—more pleas for help than they had time even to read. With regret, they discarded them unopened. Their phone was ringing frequently. They estimated that allegations of scientific fraud were coming to them at the rate of about one hundred per year, five times the rate they were arriving at the N.I.H. People interested in scientific fraud were interested in them. Eventually, they threw out their snails.[55]

Through the fall of 1987, Stewart and Feder lectured on college campuses about fraud in science, suggesting that it was rampant, and with the N.I.H. lid off them, aired their version of the *Cell* paper's claims and Baltimore's response to their inquiries, including what they took to be his threats. ("It wasn't slander because it was absolutely accurate," Stewart says.) Patricia Woolf, a sociologist of science who heard them speak at Princeton, remembers that they seemed to misrepresent matters and suggested fraud. After their Princeton lectures, they decided to donate the one-thousand-dollar honorarium they received to a scientist who had served the public interest by disclosing scientific misconduct.[56]

Meanwhile, they kept trying to publish a revised version of their analysis of Weaver *et al.*, submitting it to *Cell*, which turned it down, and then to *Science*, which also rejected it. The reasons at both journals were the same, and the same that Rall's referees had advanced—that "the manuscript is not a scientific paper" suitable for peer review, as the managing editor of *Science* put it in a letter to Stewart and Feder in mid-December, adding, "The type of allegation you make is best handled by an investigating committee."[57]

Their allegations, in fact, figured in the follow-up by Mary Miers's office to the request that had been put to it in April 1987 to determine whether an investigation of the *Cell* paper was warranted. In May, Miers's office asked the coauthors how the objections raised by Stewart and Feder might be resolved, and it requested Tufts and M.I.T. to report on the inquiries they had conducted in 1986. Baltimore, in a response that Imanishi-Kari appeared to think would suffice for her, too, conveyed his opinion of Stewart and Feder's "accusations" by sending Miers's office copies of his earlier letters to the pair and to Rall.[58] M.I.T. promptly provided an account of Eisen's review. The Tufts administration had first to obtain a written report of the Wortis committee's deliberations, because in 1986 nothing had been set down on paper. In the course of preparing a document, Wortis performed an assay on Bet-1 himself, with results, he told Eisen on the telephone, that "completely confirm" what Imanishi-Kari had always claimed. Wortis's report covered the principal technical issues reviewed the year before, was endorsed by his co-inquirers, Brigitte Huber and Robert Woodland, and concluded:

NO EVIDENCE OF DELIBERATE FALSIFICATION
NO EVIDENCE OF DELIBERATE MISREPRESENTATION.
ALTERNATIVE INTERPRETATIONS OF THE EXISTING DATA CAN BE MADE, BUT THAT IS THE STUFF OF SCIENCE.[59]

In late June 1987, Tufts sent the N.I.H. a summary of the Wortis inquiry's outcome rather than the report. In July, Miers's office asked for the report itself, and at the beginning of October Tufts finally mailed it to Washington.[60]

With the help of N.I.H. scientists, Miers and her small staff reviewed the Tufts and M.I.T. assessments against Stewart and Feder's critique of the paper, approaching the matter as an issue of scientific error rather than as one of misconduct. Neither assessment responded to the critique because, of course, neither had been written to respond to it. And neither could be used to address the critique: M.I.T.'s assessment did not deal at all with the original data; Tufts's invoked it at several points but not sufficiently to resolve Stewart and Feder's challenges. Miers, realizing that a determination of who had said what to whom would not resolve the scientific questions, concluded that the agency should mount an independent inquiry into the experimental data. Since the substance of the dispute turned on intricacies of molecular and cellular immunology, the N.I.H. decided to appoint an independent panel of experts competent in those fields. By early 1988, Miers's office had obtained the services of three respected biologists—Ursula Storb, of the University of Chicago; Frederick W. Alt, of Columbia University; and James Darnell, of Rockefeller University.[61]

Baltimore was surprised at the choice of Alt and Darnell. He had done postdoctoral work in Darnell's lab at M.I.T., and Darnell had later joined him in writing a major cell biology textbook. Alt had worked with him as a postdoctoral fellow and was a coauthor on more than a dozen of his papers. Baltimore says that when Alt and Darnell called to tell him that they had been asked to serve on the N.I.H. panel, he remarked that they had "a conflict of interest." He adds that he nevertheless refrained from telephoning Mary Miers to suggest that they did not belong on the panel, thinking that for him to judge "either who should or who shouldn't be on the committee . . . immediately taints the whole situation."[62]

Walter Stewart found out about the panel's membership and in a letter to Mary Miers in mid-February 1988, he protested the inclusion of Alt and Darnell. He also volunteered that he and Feder should be given an opportunity to see the data that the N.I.H. panel would obtain, noting that the information on which they had based their critique was "incomplete" and that having access to all of it would permit them to improve their critique before giving it to the investigators. He emphasized that he and Feder were competent to do the job, calling attention to their "unusual and, I would suggest, unmatched record in detecting and analyzing scientific error and misconduct."[63]

In mid-March, Miers responded that her office would have no objection to the panel's seeking the assistance of Stewart and Feder and that she saw no conflict of interest in Alt and Darnell's participating in its deliberations, explaining, "It is well known in peer review that one's peers are often colleagues as well." Miers says she considered the conflict of interest only "a perception," not a reality. Besides, she later declared, "By that time I was sick of Stewart and Feder. I wouldn't have done anything they asked me to do. If they had said, 'Black,' I would've said, 'White.' "[64] But no matter what Miers, or for that matter, the Herzenbergs, the editors of *Science* and *Cell*, or anyone else said, Stewart and Feder intended to gain satisfaction, one way or the other.

S I X

■

"A Perfect Object Lesson"

LATE IN the winter of 1988, Stewart and Feder went down to Capitol Hill to discuss scientific fraud with Bruce Chafin and Peter Stockton, who worked for Representative John Dingell. A moderate Democrat from the Detroit area, in Michigan, Dingell chaired the House Energy and Commerce Committee, which had jurisdiction over the N.I.H., and also its subcommittee on oversight and investigation. He was normally a good friend of the N.I.H. Indeed, his father, a congressman before him and a New Deal stalwart, had helped foster its development. His younger brother was a scientist at one of its institutes, his wife, a wealthy member of the Fisher Body family, actively supported the Children's Inn, a special facility on the N.I.H. campus for sick children and their families, and Dingell himself had majored in chemistry at Georgetown University before earning a law degree there. He was interested in scientific fraud because, he said, of his concern for the N.I.H. and the magnitude of its budget, $6 billion in 1988 and rising. He held that his committee could not "afford to divert precious dollars into areas of meaningless or fraudulent work."[1]

Dingell aggressively pursued people and institutions he thought were violating the public trust that came with access to taxpayers's money. After taking over the Energy and Commerce Committee in 1984, he enlarged its scope and authority with hard work, high intelligence, and exceptional mastery of House rules and procedures. The committee now had some 140 staff members and handled a significant fraction of the

domestic legislation that went through the House. A specialty legal practice known as the "Dingell bar" had sprung up in the Capitol, a corps of
lawyers who represented clients subjected to probes by his oversight subcommittee. Dingell was not known primarily as a legislative initiator, and
he did not hold hearings mainly to gather legislative information. He professed a special affection for whistle-blowers, and he used the hearing
room to spotlight an issue by probing some one, or some case, or some
practice exemplifying it. "You have to use dramatic examples that mom
and pop will understand," Stockton said. Dingell had gone after extravagant defense contractors, corrupt bureaucrats, and illegal influence ped-
dlers with ferocious and, usually, successful tenacity. Many people during
the Reagan years saw him as a hero for the targets he hit.[2]

To others, not all of them conservatives, Dingell was a bully. He was
six feet, three inches tall, broad-shouldered, and imposing, and he pursued
witnesses relentlessly, interrogating them from the high dais of the hearing
room as they sat at a table below. Congressional insiders liked to say that
he inspected the seats after a hearing to see how much sweat his witnesses
had left. His staff were likened to "junkyard dogs"—scrappy, intimidating,
at times vicious. They courted favored reporters, exceeding the normal
standards of the sievelike Capitol in leaking confidential documents and
information. Stockton, who held a masters in economics, and Chafin, one
in business, were hard-nosed, irreverent, and smart. Both had extensive
experience in governmental oversight and sharp eyes for phony balance
sheets, financial or otherwise. Chafin dressed in business suits and was
expansive enough to belong in a faculty club. Stockton was casually up-
scale, inclining to blue jeans. He laced his conversation with obscenities
and, according to some people who encountered him, behaved like a
"thug."[3]

Now Dingell intended to turn his investigative resources on the subject
of fraud in federally sponsored biomedical research. Whatever the N.I.H.
and its grantee universities had done to establish reasonable procedures
for dealing with fraud and misconduct, Dingell and his staff considered
the efforts inadequate. More cases of misconduct had surfaced in the
press, along with accompanying evidence of dilatory institutional
responses. The most notorious of them, widely reported in 1987, was Stephen Breuning's fraud. Breuning, a member of the Western Psychiatric
Institute at the University of Pittsburgh, was an influential researcher in
psychopharmacology for children institutionalized by reason of severe
mental retardation. Most scientists held that the appropriate drug therapy
for such children was tranquilizers. Breuning, drawing on his own research,

argued that they should be given stimulants such as Ritalin, not least because it would elevate the IQ of a mentally retarded child by as much as a factor of two. At the end of 1983, Robert Sprague, his mentor, senior collaborator, and a member of the University of Illinois faculty, reported to the National Institute of Mental Health, which was funding their research, that he deeply doubted the legitimacy of Breuning's data. It took more than three years and the persistence of Sprague for N.I.H. and the University of Pittsburgh to determine that Breuning had faked his results. During that time, Sprague found himself investigated by the mental health institute for supervising Breuning ineffectively. His application for renewal of his long-standing research grant was deferred, then funded at a reduced level.[4]

The Breuning case widened interest in scientific fraud on Capitol Hill. The issue captured the attention of some moderate Republicans but largely of middle-of-the-road to liberal Democrats like Dingell himself. In the view of James Wyngaarden, the director of the N.I.H., some members of Congress were concluding "that scientists are no more honest than businessmen, and that universities are no more honorable than corporations." Scientific fraud fell in the category of corruptions—notably, the savings and loan scandals, defense profiteering, and influence peddling—that plagued the mid-1980s. Congressman Ron Wyden, an Oregon Democrat and a member of Dingell's subcommittee, called it "more serious than most ripoffs in Government," explaining, "Fraud in scientific research is like using substandard materials in the foundation of a skyscraper. The entire structure is compromised by the unsound foundation."[5]

It did not escape the notice of Dingell or of Representative Theodore Weiss, a highly respected liberal Democrat from the west side of Manhattan, that while Breuning's false claims remained unrepudiated, Ritalin, a potentially dangerous drug for mentally retarded children, had likely been administered to untold numbers of them. On April 11, 1988, at hearings before the subcommittee on government operations that Weiss headed, Sprague declared that culpability rested with the University of Pittsburgh and the National Institute of Mental Health, testifying that universities and agencies of government "act first in their own interest and slow down, or, sometimes, stall" investigations of scientific fraud. Congressman John Conyers from Detroit, a key figure in the Congressional Black Caucus, learned enough to say "there may be more fraud going on in scientific activity than we—certainly myself—thought," adding, "It is shocking and disturbing to hear that we don't really have a good system in place for dealing with it." Conyers suggested that scientific fraud might be deterred by making it a criminal offense, part of a white-collar crime bill the House Judiciary Committee was then considering.[6]

A salient theme in the congressional shock was the degree to which science relied on self-policing. The N.I.H. was supposed to oversee fraud investigations conducted by its grantee institutions, but its misconduct office comprised only three people—Mary Miers and two part-time staff—a small complement given the volume of misconduct that Stewart and Feder believed to be occurring.[7] As a result, universities where allegations of fraud arose were essentially left to investigate themselves. Congressman Dennis E. Eckart, of Ohio, another member of Dingell's subcommitee, noted that defense contractors tell us they have systems to catch fraud, but the rest of us know those systems don't work. "What is at issue here, much in the same way [as] within the defense industry, over at NASA, over at the accounting profession, over at savings and loans, are the adequacy of safeguards which will give you and me and the public confidence that waste, fraud, abuse or misconduct do not occur with taxpayers' dollars."[8]

To Stockton and Chafin, scientific corruption appeared to be going unpunished and research institutions were covering up to protect themselves. The system was failing to deter wrongdoers or to take whistle-blowers seriously. They explain that Congressman Dingell wanted to expose the flaws in the system and get the academic community to heal itself. In keeping with the subcommittee's *modus operandi*, they sought an ongoing high-profile case that would show the system still in need of repair. Stockton says, "We were going to whop the thing; kick 'em around a little and go back to Defense. One goddamn hearing and they'd do the right thing."[9]

Then, they heard from Stewart and Feder about the challenge that Margot O'Toole had raised against M.I.T., Tufts, Imanishi-Kari, and David Baltimore. Chafin says that back then he "didn't know David Baltimore from Adam," noting, "I had never heard of him. We don't travel in those circles." On a trip to Boston, Stockton and Chafin did not acquaint themselves with either Baltimore or Imanishi-Kari, but they did look up Margot O'Toole.[10]

O'Toole and her family were living with her mother on Clark Road in Brookline while, for a mixture of personal and financial reasons, they rented out a house they owned elsewhere in Boston. O'Toole was still working at her brother's Gentle Giant Moving Company. Early in 1987 she had given birth to a second child and around May she had put her new baby in day care and begun looking for a job in science.[11] The only offer she received came from a physician who proposed to pay her out of his own pocket to do research in his office. She says she kept looking for a job, answering advertisements from biotechnology companies while she continued to live with her mother and put in time at her brother's moving company.[12]

O'Toole attributed her difficulty in finding a job to her whistle-blowing. She says that Imanishi-Kari angrily disparaged her to colleagues in the greater Boston biomedical community. She had it from Stewart and Feder that Baltimore, during a telephone conversation in January 1987, had termed her a "disgruntled postdoc" who was engaged in a "personal vendetta" and that, in a telephone conversation with Eisen's secretary in early March, they learned that Eisen thought O'Toole's critique of the *Cell* paper "pretty incoherent." O'Toole also knew from Stewart that Eisen had written a response to her critique following the meeting at M.I.T. in June 1986. She had not yet been sent a copy. On March 17, she wrote to Eisen requesting one, expressing surprise at his characterization of her analysis. Five weeks later, having not heard from him, she wrote again, noting that it was "most important to me that this matter be resolved, particularly in view of the characterizations of me and my conduct of which I have heard."[13]

Eisen replied towards mid-May, declaring that he had mentioned "incoherence" only in reference to her oral account of the issue at Wood's Hole and that he found her written critique "entirely coherent and took it seriously." He added that he would send her his report so long as she agreed to keep it out of the hands of Stewart and Feder.[14] (Several days later, Eisen telephoned Imanishi-Kari for further clarification of the powers of Bet-1 against Stewart and Feder's claim that the reagent did not adequately discriminate between *mu-a* and *mu-b*. She told him that she had data showing it was in fact between ten and twenty-five times more responsive to *mu-a* than to *mu-b*, and emphasized that Stewart and Feder were basing their analysis on a bad assay that appeared in the seventeen pages that O'Toole had xeroxed, an incomplete record of the experiment.)[15] In early July, O'Toole assured Eisen that she would not share his report with Stewart and Feder. Eisen promptly sent her—without thinking about it, he recalls—the rhetorically defensive report that he had written to the biology chairman at M.I.T. in December 1986, which dwelled on Bet-1, rather than the generous assessment of O'Toole's overall critique that he had composed the previous June.[16]

O'Toole took Eisen's report to be a professional smear. She read its "Allegations of Misconduct" heading to mean that she had leveled an unfounded charge of fraud against Imanishi-Kari even though she had scrupulously confined her challenge to error. She found unwarranted its characterization of the issues she had raised as turning mainly on "matters of judgement." She held that to dismiss them as such implied that they were somehow in error, yet no one had brought any error to her attention. In October, in an angry reply to Eisen, she detailed her objections to his report and asked that he send a memorandum of correction to anyone who had received it. Eisen, in notes jotted on her letter's margins, declared some of her account of events in 1986 "a total fabrication!" and her requests for corrections "imperious." He sent out no memos and filed her letter without answering it. Although Eisen had not sent his "Allegations of Misconduct" report to anyone except the biology chair at M.I.T. and, now, O'Toole, she was sure that his failure to do as she had asked further disadvantaged her in seeking employment. She later told federal investigators, "You're not going to get a job when people are accusing you of having made a baseless accusation against somebody for no good reason."[17]

O'Toole was still looking for a job in science when Stockton and Chafin visited her in Boston in the early part of 1988. They interpreted her circumstances as sufficient evidence that she had been victimized for her whistle-blowing. "We physically saw her living in her mother's house," Chafin recalls, offering support for their view. Stockton and Chafin thought they had an exemplary case of a cover-up of scientific misdeeds. "Here was someone chewed up by the system," Chafin says. "Here was a perfect object lesson."[18]

Stockton and Chafin arranged for Stewart, Feder, O'Toole, and Maplethorpe to appear at a hearing that they scheduled for Tuesday, April 12, 1988—the day after the inquiry by Weiss's subcommittee—on fraud in N.I.H. grant programs. Congressman Weiss's inquiry had not even touched on the Baltimore case, perhaps out of regard for the fact that it was in process. Stewart says that the Weiss subcommittee had actually wanted O'Toole to testify, but behind a sheet to protect her anonymity. The proposal seemed silly to Stewart, Feder, and O'Toole, since she was reluctant to hide, her Irish accent was unmistakable, and she would testify openly the next day at the Dingell subcommittee hearings.[19] Dingell's witnesses also included several officials from N.I.H., among them Mary Miers. Dingell and his staff knew, probably from Stewart, that Miers had appointed two of Baltimore's coauthors, Alt and Darnell, to the panel that was to investigate the *Cell* paper, and on March 22, they spoke about the matter with Stewart and Miers. The next day, Dingell asked Stewart and Feder to review Wortis's account of the Tufts inquiry "to provide an interpretation for the Subcommittee staff." In the two weeks before the hearing, the N.I.H. removed Alt and Darnell from the panel.[20]

David Baltimore first heard of the imminent hearing from reporters who called him for comment in the several days before it took place. Stewart and Feder had briefed a number of journalists on the dispute, and stories on the pending inquiry appeared in newspapers across the country. Over the weekend, Baltimore opened his *Boston Sunday Globe* and was mortified by an article about the coming proceedings. The piece prominently displayed both his picture and opinions from Peter Stockton about the Tufts and M.I.T. investigations—"piss-poor"—and about the *Cell* paper: "It's hard to tell if it's error or fraud. At certain times, it appears to be fraud and other times, misrepresentation."[21]

At the hearing on April 12, O'Toole said that she was reluctant to testify and was doing so at the subcommittee's "direction." (Dingell's staff reportedly had told her that she would be subpoenaed if she declined to appear.) She said later that she was scared, and she doubtless remained worried about endangering her husband's career at Tufts. Her voice quivering at times, she provided the subcommittee a detailed narrative of the development of her quarrels with the *Cell* paper, declaring that at the inquiries in 1986 her "assertions that the data did not support the published claims were completely confirmed"—although both investigative bodies had decided that the data she disputed was more than adequate. Despite the merits of her challenge, she said, the Tufts committee had not wanted to publish corrections of the errors in the *Cell* paper because the admission would injure Imanishi-Kari. Baltimore, for his part, had given as his reason that "part of the study was valid."[22]

O'Toole testified that "not one" of the senior faculty and officials at either Tufts or M.I.T. had offered her help or support. Alluding to Eisen's report of December 1986, she said that for her troubles she had been portrayed as "a person who had raised ridiculous and trivial and unsubstantiated allegations against my supervisor for vindictive reasons." She noted that she had written to Eisen asking him to "correct" his report so that she could "clear my name" but that he had not replied.[23] She declared, to Dingell's expression of agreement, that "lab notebooks that are paid for by the public are the property of the public and . . . belong to everybody that's interested in the subject." She complained, likely with Baltimore's remark in mind that she had stolen the seventeen pages, that "implications were made that I had done something obscene to look at somebody's data." Asked whether N.I.H. had offered her any protection, she answered, "The N.I.H. has never contacted me," forgetting the telephone call that she had received from the agency's legal adviser, Robert Lanman, in the summer of 1986, urging her to come forward. She added that she was fortunate to have a husband and family willing to support and comfort her until "I got on my feet."[24]

O'Toole only alluded to her personal frustrations with Bet-1, noting that "the time and efforts of scientists [are] too valuable to waste our precious resources discovering errors that were known to others from the start." Dingell's staff believed they knew what she meant. If Baltimore and company had shown that they took her seriously by publishing a correction for Bet-1, Stockton told me, with strong support from Chafin, "they could have been out of the whole fucking thing. Absolutely nobody would have ever heard of this dispute. They could have made O'Toole probably happier than shit."[25]

Stewart and Feder entered their analysis of the *Cell* paper into the hearing record, thus achieving a kind of publication of it.[26] In testimony, they selectively summarized their findings: "In a number of crucial cases" the paper gave "an inaccurate and, in fact, misleading picture of the underlying data." The reports of the investigations at Tufts and M.I.T. were "seriously defective," with certain issues "ignored or treated evasively." And when they tried to discuss O'Toole's allegations with Baltimore, he "attacked her character and motives" and went on to assail them "personally" and threaten them with legal action. Charles Maplethorpe testified that at any university a whistle-blower would "pay the price." Congressman Ron Wyden asked Maplethorpe whether he had seen data from the research that Imanishi-Kari had done for the *Cell* paper that was "falsified and misrepresented." Maplethorpe said, "Yes."[27]

After Stewart, Feder, O'Toole, and Maplethorpe had finished, Dingell angrily turned to the several witnesses from the N.I.H., lambasting them for the way they were handling the investigation of the *Cell* paper. He demanded an end to "cooked figures and . . . cooked panels judging the people who cook the figures." He joined Representative Ron Wyden in castigating the agency for taking so long to act in the matter and for not getting in touch with O'Toole. Her life had been "hanging in the balance," yet the N.I.H. had not been "going to bat" for her, Wyden said. "She's gotten the back of the hand from all the people who should be congratulating her for wanting to come forward in the name of scientific truth."[28]

Mary Miers had a masters degree in political science, had been with the N.I.H. for some eighteen years, about half of them in the agency's legislative office, and had spent many hours in the front row at congressional hearings. She says that she had expected to be browbeaten at the hearing, that such performances are staged on Capitol Hill to achieve a purpose. She tried to explain that her office was following an orderly procedure and that the expert panel, once it was reconstituted, would be talking with O'Toole. Dingell hammered at her anyhow:

MR. DINGELL: When do you propose to talk to Dr. O'Toole?

MS. MIERS: Sir, I'd be happy to talk to her tomorrow.

MR. DINGELL: Is talking to Dr. O'Toole a part of a complete and thorough investigation?

MS. MIERS: Yes, sir.

MR. DINGELL: It is? Are you sure?

MS. MIERS: Yes, sir.[29]

A moment later Wyden interrupted to ask what she had "learned new on the matter of Dr. O'Toole this morning?" Miers responded that she had "been pointed in the direction of much more specific issues that [O'Toole] raised, that were not as apparent from the materials that I had in hand. As well as some questions about the response of the institutions and other individuals involved."[30]

Congressman Dingell left no uncertainty about his purpose. Toward the end of the session, which ran nonstop for six hours, he admonished Miers and her colleagues from N.I.H.: "Now I don't want you to get the impression that this Committee is going to stage a one-day hearing, bring you up, make you miserable, and then let you go on about your business. When we do these things, we try to see to it that the pain and suffering go on for a greater period of time, until the abuse that is obvious is taken care of."[31] For the subcommittee, the abuse comprised not only inadequate scrutiny of the *Cell* paper but also cruel mistreatment of the whistle-blower. Dingell's message was unambiguous: The N.I.H. had better speed up the investigation—and pay solicitous attention to Margot O'Toole.

Soon after the hearing, the Dingell subcommittee, declining to wait for the N.I.H., initiated its own in-depth investigation of the *Cell* paper and the responses of Tufts and M.I.T. to O'Toole's complaints. It arranged to borrow Stewart and Feder from the N.I.H. on an as-needed basis to conduct what Stewart would have called a complete internal audit of the *Cell* paper. It asked Tufts and M.I.T. for copies of all documents relevant to the case and sent investigators, including Stewart, to Boston to interview several of the principals. Chafin was sure that the subcommittee was onto something with its investigation of how institutions responded to allega-

CONGRESSMAN JOHN D. DINGELL
Chairman of the House Committee on Energy and Commerce
and of Its Subcommittee on Oversight and Investigations

Photo Credit: © Ken Heinen

PETER STOCKTON
Investigative Staff, Dingell Subcommittee

Photo Credit: © Anne Adjchavanich

BRUCE CHAFIN
Investigative Staff, Dingell Subcommittee
Photo Credit: Herb Parsons, Cold Spring Harbor Laboratory Archives

MARGOT O'TOOLE, WALTER STEWART (middle), and NED FEDER
Testifying at the Dingell Subcommittee Hearing on Fraud in
N.I.H. Grant Programs, April 12, 1988
Photo Credit: AP/Wide World Photos

tions of scientific misconduct. He says that within twenty-four hours of the hearing the subcommittee received word of thirteen cases from people saying, " 'Thank God. I finally have somebody I can call and trust.' " Stewart and Feder told a *New York Times* reporter just after the hearing that their work on fraud and misconduct was "absolutely exhilarating," "really a lot of fun." Dingell declared that they had "really done a great public service by bringing to the fore what the National Institutes of Health haven't been willing to face."[32]

Across the country, press accounts of the hearing tended to take the testimony of Stewart, Feder, Maplethorpe, and O'Toole at face value. A point made by the *Washington Post*, in an editorial titled, "When Scientists Fudge . . . ," was commonplace: "Like other such whistleblowers, Dr. O'Toole suffered ostracism and has now left the field."[33] However, the hearings infuriated a number of scientists. Bruce Chafin recalls that Dingell and his staff expected scientists to acknowledge they had a problem and pledge to repair it. Instead, "they acted like the Rockettes," he says. "They all turned and kicked right in formation," putting the subcommittee on the defensive, complaining that "people outside of science were trying to judge what's going on inside of science."[34]

Many scientists questioned the subcommittee's reliance on Stewart and Feder. Even those who thought that they had performed a useful service in the Darsee case now held that the pair had become messianic. Arnold Relman, the respected editor of the *New England Journal of Medicine*, wrote, "They've taken a good principle too far. They have arrogated to themselves a mission that nobody has given them. They have set themselves up as more or less grand inquisitors." Lee Herzenberg told a reporter what she had earlier advised Stewart and Feder: The findings of the Herzenbergs' lab at Stanford lent weight to the approaches suggested by O'Toole but were part of the normal "self-correcting" process of science. The dispute between O'Toole and Imanishi-Kari had been blown out of all proportion to its scientific significance. It had "no business being aired in public."[35]

Several scientists faulted the subcommittee for politicizing a scientific dispute and for putting on such a one-sided hearing—giving a forum to the accusers but not to the accused, who had not even been invited to testify. David Baltimore's personal lawyer, Normand Smith, called the hearing "rather a surprise. . . . We kind of heard on the grapevine that we were being tried in absentia." In a well-attended public meeting held in the Sackler Center at Tufts University Medical School on April 20, 1988, Huber, Wortis, Woodland, and Flax each outlined the events and issues of May 1986 and then took questions from the floor. According to a summary of the meeting, Imanishi-Kari's former and present graduate students and colleagues "all stressed the openness of the atmosphere in her laboratory, and her rigorous requirements with regard to reproducibility of data." In a letter to *Science*, Wortis, Huber, and Woodland protested that, contrary to O'Toole's testimony, in 1986 they "did not concede that her criticism was sound." They insisted that they had done nothing to impede her career, and David Baltimore said publicly that he had done nothing to impede it either.[36]

The Dingell subcommittee's behavior touched political nerves in Baltimore that grew out of his personal history. His parents, the children of poor Jewish immigrants to New York, had prospered enough from his father's work in the garment district to move to Great Neck, Long Island, to take advantage of the excellent public schools. Politics and public affairs were a staple of family conversation. Both his parents had strong left-wing sympathies, Baltimore says, although "they were both proud of the fact that they had not involved themselves in the Communist Party in the thirties." David Baltimore spent part of his sophomore year in high school watching the congressional hearings in which Senator Joseph McCarthy questioned whether the army was adequately vigilant against communists within its ranks. While neither his parents nor their friends thought of themselves as likely targets of McCarthyism, they saw the witch hunts as an attack on left-leaning people like themselves.[37]

Baltimore's mother, an experimental psychologist, prompted him to learn about physiology and, while in high school, he became passionately interested in biology. When he was a student at Swarthmore, he taught himself molecular biology, since it was a new field and the college offered no courses in it. He impressed visiting scientists with his masterful knowledge of cutting-edge work in the subject. He whizzed to a doctorate at Rockefeller University in 1964, at age twenty-six, and went to work at the Salk Institute in La Jolla, California, drawn there by its research emphasis on the molecular analysis of viruses, particularly those that cause polio. He says that he liked the outdoor life in the sun and deserts, but that he found the political conservatism increasingly irritating. With the escalation of the war in Vietnam, he became politically active, helping to organize protests, taking to the streets, and speaking out. In 1967, when several pieces of art were removed from an exhibit at the Salk Institute because they were said to use the American flag improperly, he quit, protesting censorship. He says that his ambitions were outrunning his laboratory space anyway, he had an offer in his drawer to join the M.I.T. faculty, and so he took it.[38]

Back in Cambridge, Baltimore grew increasingly absorbed in the problem of how viruses like those that cause polio interact with the cells they infect—the line of research that led to his Nobel Prize. The genetic material of such viruses is not DNA but RNA. Through the 1960s, a biologist at the University of Wisconsin named Howard Temin had been arguing from his studies of RNA tumor viruses that the viral RNA generated DNA complementary to itself and that this DNA integrated itself into the DNA of the host cell. The integrated viral DNA then used the machinery of the cell to produce new copies of the original RNA virus. Many scientists scoffed at Temin's idea because it was widely believed that, while DNA could yield RNA, the reverse did not—indeed, could not—occur. In 1970, however, Temin demonstrated that RNA viruses contain a special type of enzyme that catalyzes the synthesis of DNA from RNA.[39]

Baltimore made the same discovery simultaneously and independently. He had met Temin, who was four years his senior, when he was in high school, at a summer program at the Jackson Laboratory in Bar Harbor, Maine. He was well aware of Temin's theory of RNA tumor viruses. He later recalled that although Temin's logic "was persuasive, and seems in retrospect to have been flawless, in 1970 there were few advocates and many suspects," adding, "Luckily, I had no experience in the field and so no axe to grind—I also had enormous respect for Howard dating back to when he had been the guru of the summer school." Baltimore obtained his first evidence of the special enzyme during the opening days of May 1970. He recognized its enormous significance and set to work to reproduce the results, but about two days later, the news flashed that President Richard Nixon had invaded Cambodia. Baltimore recalls, "I stopped everything that I was doing and joined everybody else out on the streets and didn't go back to the laboratory for a week."[40]

At the end of June 1970, in the same issue of Nature, Temin and Baltimore reported in independent articles the discovery of the enzyme, which was promptly dubbed reverse transcriptase, in recognition of its ability to transcribe RNA back into DNA. RNA viruses came to be termed retroviruses because they are equipped with the enzyme to accomplish the transcription. The achievement cast a brilliant light not only on how retroviruses replicate but also on how they interact with cells.[41] It had important implications for cancer research and, within a decade, for research on the acquired immunodeficiency syndrome (AIDS), too, since the human immunodeficiency virus (HIV) is a retrovirus.

In the years after his Nobel Prize—the award, which came in 1975, was shared with Temin and another scientist, who had worked in a related area—Baltimore was drawn increasingly into the world beyond the lab, contributing to the ongoing debates over genetic engineering, agitating against biological warfare research, and in 1981, joining a delegation of scientists designated by Pope John Paul II to urge a slowdown in the nuclear arms race on President Ronald Reagan. He nevertheless continued to produce brilliant research and to train numerous graduate students and postdoctoral fellows. He was encouraging to young women scientists, and a number of those he trained now hold academic positions in universities across the country. One of them is his wife, Alice Huang. She was his postdoctoral student at Salk, collaborated in some of his retroviral research, and became a full professor at Harvard Medical School. Baltimore says that he likes "to catalyze the developing careers of younger people, often by helping them find a fruitful direction in their research," adding, "I was kind of doing that with Thereza. I thought she had more potential than she'd shown."[42]

In 1980 Edwin C. Whitehead, who had made a fortune in a family medical instrument business, proposed to establish the Whitehead Institute for Biomedical Research with a multimillion-dollar gift, and Baltimore was his first choice for director. The Whitehead Institute was to be a partner of M.I.T. and many of its members would hold teaching appointments there, but its research program and appointments would not come under M.I.T. control. Many M.I.T. faculty openly resisted the move, calling it a sellout of the university and expressing suspicions that it would be unduly influenced by Edwin Whitehead's commercial interests. Through a year of discussions, Baltimore was influential in persuading the M.I.T. faculty to accept the venture, emphasizing how biomedical research at the university would benefit from the munificent gift and stressing that the fears were baseless. After the Whitehead Institute was established—a faculty motion to oppose it was defeated by a margin of three to one—Edwin Whitehead endowed the new institute with $135 million. Baltimore promptly brought in stellar young faculty, helped design a building that would foster collegial interactions, and made the Whitehead an overnight success.[43]

The Whitehead directorship added to Baltimore's prominence in the rapidly expanding arena of the life sciences. He provided a key link between basic molecular biology and the burgeoning biotechnology industry, advising several successful start-up companies and becoming wealthy in the process. He continued to defend the importance of protecting autonomy in basic biological research against political interference, a position that he had begun advancing during the battles over recombinant DNA in the 1970s, but he increasingly struck broader themes as well. In February 1988, in a major address at the annual meeting of the American Association for the Advancement of Science, he castigated activists who put "animal rights ahead of the human right to optimum health"; denounced the Strategic Defense Initiative as a "cruel hoax"; and attacked federal science policy for putting too much money into the Human Genome Project—a biological "moon shot," he said—and too little into AIDS research.[44]

It was just about two months later that John Dingell gaveled to order his hearing on fraud at the N.I.H. The Dingell subcommittee's accusation by staff leak and the one-sidedness of its hearing struck Baltimore as a kind of latter-day McCarthyism. Its attack on Imanishi-Kari, a powerless immigrant who lacked the fluency in English to defend herself effectively, offended him more deeply. He was also mindful of the impact of the widespread press coverage. "Much of it [was] inadequate and some of it downright wrong," he judged, with the result "that my own reputation and those of my coauthors were being sullied."[45]

To Baltimore, both the hearing and the coverage expressed a distorted idea about how science really works. He says that "all of our notebooks, and the notebooks of all of [our] students and postdocs, [contain] individual experiments that show quite the opposite from the conclusions that are published. . . . And there is some explanation of it because . . . there are too many parameters in biologic[al] experiments to have them all under control every time perfectly." Margot O'Toole felt that Imanishi-Kari's belief in idiotypic mimicry had led her to overlook data that raised difficulties. But in fact O'Toole's charge that the *Cell* paper misrepresented its experimental undergirdings in reality called Imanishi-Kari to account partly because she simply decided that those ambiguities were not of consequence. Imanishi-Kari had solid analytical reasons for doing so, but O'Toole had not conveyed that possibility to Congressman Dingell.[46]

To Baltimore, the Dingell subcommittee appeared to be discouraging the play of judgment and creativity in science. Dingell himself appeared to be interested only in the fact that the paper contained errors. It was his opinion, as he later said in a lecture at a meeting of the Massachusetts Medical Society in Boston, "that when errors are recognized scientists have a duty to make them known." He seemed not to understand that errors vary in their scientific significance, that evaluating their meaning involves critical judgment, and that discrimination in the use of data is a feature of scientific inquiry.[47] If some analysts had found scientists such as Gregor Mendel and Robert Millikan candidates for fraud charges, perhaps given the chance Congressman Dingell would have gone further.

All things considered, including Dingell's enlistment of Stewart and

Feder for his subcommittee's investigation of the *Cell* paper, Baltimore felt that the coauthors needed to defend themselves. In mid-May 1988, he sent a lengthy, open letter to some four hundred colleagues around the country, explaining that he wanted to clear the names of himself and his coauthors but, more important, to warn the scientific community that "a small group of outsiders" threatened to use "this once-small, normal scientific dispute" to "cripple American science."[48]

Baltimore acknowledged the overstatement concerning the discriminatory powers of Bet-1. Otherwise, he defended the *Cell* paper, reviewing its science and castigating Stewart and Feder, not least for drawing sweeping conclusions about it on the basis of seventeen pages out of the thousand or so on which it was based. He also denounced the tactics of the Dingell subcommittee, including its apparent attempt to make truth or falsehood in science a matter for determination in congressional committees. Baltimore stressed that O'Toole had not brought any kind of formal accusation against the coauthors, that her critique of the *Cell* paper was carefully developed, but that the coauthors had not found good scientific reasons to credit her alternative interpretation of the data. Reiterating his conviction that the resolution of the dispute demanded further experimentation, he called attention to how his recent work with the Herzenbergs had further illuminated the immunological response of the transgenic mice.[49]

Two weeks before the Dingell subcommittee hearing, Imanishi-Kari had provided the N.I.H. a long, detailed account of the experiments reported in the *Cell* paper, noting her eagerness to help "solve this unbearable situation." Two weeks after it, Baltimore had complained to Wyngaarden, at the N.I.H., that the lack of an official inquiry by the agency had "greatly prolonged an atmosphere of doubt and suspicion" harmful to the principals in the dispute and had created a "vacuum" into which professionally unqualified people—an allusion to Stewart and Feder—have moved, "freely making judgments which do not reflect the factual circumstances surrounding our work." Now, in his public letter, he told the many friends and colleagues who had volunteered to help that they write to Dingell and Wyngaarden encouraging the N.I.H. to proceed to a "rapid and complete review" of the dispute.[50]

Baltimore warned that new laws and regulations could have a chilling effect on intellectual life. He saw "this affair as symptomatic, warning us to be vigilant to such threats, because our research community is fragile, easily attacked, difficult to defend, easily undermined." He added in closing, "What is now my problem could become anyone else's."[51]

■

A Moment's Vindication

SHORTLY AFTER the Dingell subcommittee hearings in April 1988, the N.I.H. enlarged its misconduct office by two professional staff members. It also began searching for two scientists to join Ursula Storb on the investigative panel in replacement of Alt and Darnell. In mid-May, with the aim of laying the groundwork for the panel, the agency sent Mary Miers, its legal adviser, Robert Lanman, and the two new misconduct officers to Boston to interview all the principals in the dispute, including Charles Maplethorpe and, first, Margot O'Toole. Thus began a full investigation by the N.I.H. to determine whether the *Cell* paper scientifically accorded with the original data and, if it did not, whether the discords revealed misrepresentation or misconduct.[1]

O'Toole reviewed her research and progressive falling out with Imanishi-Kari, the inquiries at Tufts and M.I.T., and the consequences of the imbroglio for her scientific career. She emphasized that "at no point was I suspecting any fraud or misconduct." She indicated some of her resentments, pointing out that when at the meeting of the Wortis committee she asked Imanishi-Kari for certain data, Imanishi-Kari had declined, saying "you're so picky, it's not good enough for you, tough luck, at least I publish." She warned the N.I.H. staff, "I know that everything I have said will be called a lie. . . . I still feel my best protection is to tell the truth because only the truth will fit the facts of a proper investigation."[2]

O'Toole said that she accepted "completely" Imanishi-Kari's explanation of the mistyped mouse. She conceded that if the overstatement concerning Bet-1 "was all that was wrong with the paper, it wouldn't be much," noting, "it wouldn't bring down the theory of the paper." She allowed it was "probably true" that the misstatement about the isotyping of the Table 2 hybridomas had resulted from a telephone miscommunication between Imanishi-Kari and Baltimore, but she held that "it should still be corrected." More important, she argued that, since the isotyping had not been done, Table 2 was "completely invalid, absolutely 100% invalid," claiming that Baltimore and Imanishi-Kari both "agreed to it." Table 2 was shown to be invalid by the seventeen pages, O'Toole insisted, adding that even though Imanishi-Kari had obtained considerable data beyond what those pages contained, they were "like the Rosetta stone" for the experiment. They "revealed what was wrong with the thinking."[3]

O'Toole recounted her version of events in extensive detail, interweaving anecdotes, conversations, and science. Bruce Maurer, who was on loan to Miers's office from another part of the N.I.H., had come to Boston. He was a level-headed microbiologist and had been appointed executive secretary for the investigation of the *Cell* paper. He recalls that O'Toole was "intense" and "convincing," adding, "When she left, we looked at each other and said, If only half of what she says is true, it's horrible."[4]

Maplethorpe, his manner furtive, asked for some form of witness privilege, by which he meant either immunity from suit or a promise that the N.I.H. would provide for his legal defense if he were subjected to one. "He acted like he was a major player in a spy novel," Maurer says. Maplethorpe later claimed that he needed the privilege because of "the hostile environment for witnesses to scientific misconduct." Lanman told him that the N.I.H. could not offer such protection because of budgetary restrictions. Questioned whether he had told Stewart and Feder about any other problems in Imanishi-Kari's lab, Maplethorpe refused to answer, and he similarly refused when asked whether he had provided documents to the subcommittee in support of his claim that she had committed fraud. In the main, he limited his response to what he had said publicly in his testimony before the Dingell subcommittee. The N.I.H. staff found him so unforthcoming as to think it "unlikely," as the agency put it in a later report, "that the scientific panel would learn anything new from Dr. Maplethorpe."[5]

The N.I.H. staff talked with all the *dramatis personae* in the dispute, now numerous in accumulation—the principal coauthors: Baltimore, David Weaver, and Imanishi-Kari; the scientific and institutional reviewers at M.I.T.: Gene Brown and Herman Eisen; and their counterparts at Tufts: Henry Wortis, Brigitte Huber, and Robert Woodland as well as

Martin Flax, Sidney Leskowitz, and Joseph J. Byrne, who was Tufts's associate provost for research. Before the hearing of the Dingell subcommittee, Mary Miers's office may have seen no point in finding out who said
what to whom and when; now, after the hearing, the N.I.H. interlocutors
devoted much of their probing to that subject. They obtained what struck
them as candid accounts of the Tufts and M.I.T. inquiries, including a
number of rebuttals to O'Toole's version of events. David Weaver's
responses summarized how those close to the science viewed O'Toole's
challenge: Some of the issues she raised had "no validity" while some
"weren't relevant" to her point. "A fraction . . . had some validity," were
extensively discussed, and were either of minor significance or resolvable
only by further experiment.[6]

By the end of May, the N.I.H. had reconstituted the scientific panel
that would investigate the dispute, keeping Ursula Storb and adding two
new members. One was Hugh McDevitt, a Harvard-trained member of
the Stanford University Medical School faculty, a ruddy man of quick,
voluble intelligence who says he agreed to serve out of a sense of citizenly
duty. (O'Toole, asked by the N.I.H. staff whom she would recommend
for the panel, had said that McDevitt might be all right, noting, "I know
that the Stanford people are 100% against me, but even so.") The other
was Joseph Davie, thoughtful, precise, reticent, a likable, upright scientist.
He held a Ph.D. from Indiana University and an M.D. from Washington
University, St. Louis, where he had succeeded Herman Eisen and spent
twelve years as chair of his department. The year before, he had left the
university to become president for research and development of the large
Illinois pharmaceutical corporation, G. D. Searle and Company. Davie was
designated chairman of the panel.[7]

Storb, a native of Germany and a medical graduate of the University
of Freiburg, is tall, intellectually forceful yet reserved in manner, with
deep-set eyes that McDevitt says some times looked "haunted." She had
risen through the ranks of the University of Washington and become head
of its division of immunology before moving to the University of Chicago,
in 1986. Maurer says, "She's a gem . . . extremely bright . . . probably the
best panel member," explaining that her "specific knowledge" of what
was at issue in the dispute was deeper than McDevitt's and Davie's. All
the panel members were distinguished scientists of independent mind. As
a group, they commanded the molecular and cellular subjects, including
idiotypes, that the Cell paper covered. Each knew the other well. They
were acquainted with the coauthors through professional encounters, but
none was close to any of them, and Storb says that in one of her areas of
research she and Baltimore were competitors.[8]

The panel was to review all the relevant data from the laboratories of Weaver and Imanishi-Kari, the seventeen pages, and the memorandum critical of the paper that O'Toole had drawn up for Eisen in June 1986. It was also to be given the reports of the inquiries at Tufts and M.I.T. and the critical analysis of the paper by Stewart and Feder. The N.I.H. arranged for the panel to examine the data and talk directly with the coauthors in Boston in the latter part of June. In preparation, Davie, McDevitt, and Storb were provided with the promised documents, the records of the mid-May interviews with the coauthors, and several other items, including the Herzenbergs' papers and a lengthy, detailed account of the experiment that Imanishi-Kari had submitted on March 28. McDevitt recalls that he spent a long time poring through Stewart and Feder's critique of the *Cell* paper and that it impressed him.[9]

Baltimore told the N.I.H. staff in mid-May that the investigating panel would be "inundated with data" from Imanishi-Kari. Her data included numerous notes on her experiments, mouse-breeding records, tapes and transcriptions of readouts from the radiation counters, and graphs or compilations of results. Many of the records had been generated on the run, with, for example, radiation counts being recorded in one experiment while hybridomas were being tested in another. Imanishi-Kari recorded some of the data in a spiral notebook or placed it in manila folders, in anticipation of collating and analyzing the results later, when she had time. Her notebooks and folders as well as the notebooks of assistants such as Moema Reis moved from one part of the lab to another as they might be needed for reference.[10] Imanishi-Kari had plenty of data to show the N.I.H., but it was disparate, unorganized, and scattered around her laboratory.

Several days before the N.I.H. panel's visit, Imanishi-Kari collected her records into boxes and discussed what to do about presenting them all with Baltimore and his personal lawyer, Normand Smith. The Dingell subcommittee also wanted Imanishi-Kari's records relating to the experiment and it would later express acute interest in the arrangement of her data. Baltimore joked, as Smith recalls, that Imanishi-Kari "should just dump it on the doorstep of the committee and let them try and sort their way through what . . . was incomprehensible." Smith advised that better she should put the data in order, catalog it, and make it comprehensible. Heeding Smith, Imanishi-Kari organized her records of the experiment and, with the help of several secretaries, xeroxed multiple copies of them for the use of the scientific panel. She says that the secretaries complained that it was too laborious to xerox the pages in the spiral notebook one by one. To facilitate the process, she ripped them out so they could be machine fed, then collected those pages and others into a loose-leaf note-book. She says that in composing that notebook, she did not add dates or any other information to the individual sheets. "I had very strict orders from Normand [Smith] not to touch anything, not to write anything. . . . So, that is exactly what I did."[11]

The panel, accompanied by Robert Lanman and Bruce Maurer, spent Thursday through Saturday, June 23 to 25, 1988, interviewing the principals and examining data in a conference room in the Tufts Medical Center, in Boston. Although the sessions were not taped by the N.I.H. staff and notes of them are not available, some of what transpired can be reconstructed from recollections of the participants. Storb recalls that she came to the panel meetings with the impression that the *Cell* paper was not "such a milestone in the first place" or that "the accusations were terribly hair raising." Davie says that they approached their task "as most scientists would," disinclined to think "that anybody really would falsify information" but with "a very open mind." He adds, "Nobody to my knowledge really came in with any preconceived idea."[12]

The panel talked first with O'Toole, going over her quarrels with the paper in detail and at length.[13] Davie recalls that they spent a long time convincing themselves "that this was not just a personal dispute between a disgruntled postdoc and a faculty supervisor." O'Toole apparently argued for her alternative interpretation of the data, insisting, McDevitt remembered a year later, that the cells produced "lots" of double anti-bodies and that the doubles were hybrids, "mu gamma molecules."[14] She struck Davie as "very smart . . . very aggressive . . . absolutely certain about herself," an impression that she also gave McDevitt, who says that she made "inflammatory statements" and that he got into several arguments with her. O'Toole, on her part, says she felt that McDevitt demeaned and dismissed her. She says that the panel asked her a great deal about Stewart and Feder's manuscript and that McDevitt "repeatedly and insistently" called on her to derive its numerical results on the blackboard. She holds that she told the panel that Imanishi-Kari had pressured her to misrepresent her data and that Storb "dismissed my assertions, saying that I probably had not understood Dr. Imanishi-Kari's English." She adds that "Davie was unable to be present for the second half of my interview" and that "he informed me of the major conclusion of the panel during a break." (Davie says he does not recall having told O'Toole about any conclusion and finds it improbable that he did.)[15]

At the end of the session, Maurer remembers, "Margot was at the elevator, waiting to go down by herself. And she had this kind of down look. I guess I was concerned. And I asked her, 'What's the matter?' And she

said something to the effect, 'They didn't hear me. They didn't listen.' "
Maurer was sure that they had listened, but he adds, "I think she was
deflated by getting the sense that she didn't win the arguments. I said,
'You wait.' And I went back in and I said, 'I want you to really talk with
her again.' And they did."[16]

The panel spoke next with David Weaver, Baltimore, and then, beginning apparently some time on the second day, extensively with Imanishi-
Kari. Weaver had already sent Mary Miers his data. Imanishi-Kari brought
hers to the Tufts conference room, arriving with a postdoc who wheeled
in a cart with several boxes on it, including the newly constructed loose-
leaf notebook. She provided the originals and enough photocopies so that
each member of the panel could have a set to examine as the group went
over the contested issues. She remembers Joe Davie keeping the originals
in front of him. Bruce Maurer says, "She was to the point, she never
ducked a question. She would take them through the data and show them
what was what and why. And when she didn't know the answer to something, she didn't try . . . to make up something. She said, 'I just don't
remember why I did that.' I thought that she was honest, detailed, forthright, and I think someone who was telling the truth." Maurer notes that
at times, the panel members shook their heads in trying to understand
her English and "they'd come back at it in another direction. But if you
kept at it, you finally understood."[17]

Imanishi-Kari showed how she had checked out the putatively normal
but anomalous control mouse, reviewing the extensive tests she had done
to determine that it had been mistyped. She demonstrated that Bet-1 had
worked many times with sufficient specificity, offering in support not only
her own data but also a record of the reagent's behavior from O'Toole's
own notebooks. She rebutted Stewart and Feder's contention that she
had been arbitrary in taking a result as positive if the radiation exceeded
a thousand counts per minute and negative if it did not: She explained
that the thousand-count cutoff represented a count rate at least two and
a half times that of background. She reviewed the data and reasoning that
led her and Weaver to conclude that their hybridomas were not double
producers. She acknowledged, as she had done before, that the hybridomas reported in Table 2 had not been isotyped, but she pointed in her
records to isotype data for other idiotypically birthmarked hybridomas,
including those in Table 3.[18]

However, as the end of the day approached, the panel members grew
increasingly puzzled by certain features of the data for Table 2. McDevitt
says that their attention had been drawn to the table because Stewart and
Feder so severely questioned its legitimacy in relationship to what was

JOSEPH DAVIE, G.D. SEARLE AND
COMPANY
Member, Scientific Panel of N.I.H.
Investigations of Imanishi-Kari
Photo Credit: © Ken Heinen

HUGH MCDEVITT, STANFORD
UNIVERSITY
Member, Scientific Panel of N.I.H.
Investigations of Imanishi-Kari
Photo Credit: Elson Alexander

URSULA STORB, UNIVERSITY OF
CHICAGO
Member, Scientific Panel of N.I.H.
Investigations of Imanishi-Kari
Photo Credit: Courtesy Ursula Storb

BRUCE MAURER, N.I.H.
Executive Secretary, N.I.H. Investiga-
tive Panel
Photo Credit: Courtesy Bruce Maurer

recorded on the seventeen pages. The panel was particularly curious about the data undergirding the contribution of Table 2 to the central claim of the paper. According to the numbers, a high proportion—roughly 50 percent—of what was labeled in the table and referred to in the text as transgenic "hybridomas" or "clones" produced antibodies with the distinctive idiotype (see *Figure 8*). The panel members pressed Imanishi-Kari for details. She told them that, no matter the terms in the table or text, the figure actually referred to the percentage of wells that contained those cell lines, not to the percentage of hybridomas or clones themselves.[19]

To the panel, that news seemed to invalidate the results reported in Table 2. Wells were not clones or interchangeable with them. Wells were physical bowls in the plastic plate that contained the hybridomas. Clones were the hybrid cell lines that grew in the wells. Any well might contain two or three or more hybridoma lines, each of which had originated in a different cell from the mouse. Made aware of that fact, Storb seems to have been concerned by what had bothered Robert Woodland during the first session of the Tufts inquiry: Tests that were positive for the idiotype and negative for the transgene would be ambiguous with regard to whether the two results revealed features of the same antibody. Storb in particular was concerned that one hybridoma might have produced the antibody that displayed the idiotypic birthmark while another in the same well might have generated one that arose from a native gene.[20]

All three panel members, but McDevitt and Davie especially, felt that the possibility of multiple hybridomas per well raised an even more disturbing point. The *Cell* paper's central claim, after all, was that the introduction of the transgene into the otherwise normal mouse prompted the animal to produce idiotypically birthmarked antibodies from its initial stock of genes at a far higher frequency than it would if it had been left alone. Imanishi-Kari's remark that she had used the word "hybridomas" when she meant "wells" indicated that the frequency of such production had been calculated by dividing the number of hybridomas that generated such antibodies by the number of wells. The panel reasoned, however, that if the wells contained multiple hybridoma lines, the denominator for calculating the frequency should properly have been the total number of such lines. To the panel, it was as though Imanishi-Kari had picked some green apples from several barrels full of unknown numbers of reds and greens, then determined the percentage of greens in the mixture with a calculation that divided the number of greens picked by the number of barrels instead of by the total number of apples. Since the number of hybridoma lines was likely larger than the total number of wells—just as the total number of apples would be larger than the number of barrels—

the actual frequency of hybridomas producing the antibodies with the distinctive idiotype would be smaller than what she had reported; and the actual frequency of those producing such antibodies deriving from native genes would be smaller still—in the panel's view, possibly so small as to be scientifically inconsequential.[21]

To Imanishi-Kari, the panel's objection was without merit. The hybridomas reported in Table 2 had been generated precisely to obtain the frequency with which transgenic mouse cells occurred that produced native antibodies with the distinctive idiotype. It aimed, in short, to estimate the strength of the transgene's impact on the immune system of the mouse into which it was introduced. Baltimore and Imanishi-Kari knew that what they were using in the denominator of the frequency calculation was the number of wells rather than the number of hybridomas, but they also had convinced themselves that for their purpose counting wells instead of hybridomas didn't matter. They had been able to reason to that conclusion from knowing the fraction of the wells in which hybridomas had failed to grow, which was about 37 percent of the wells containing mouse spleen cells. They factored that number into a standard statistical formula—it is called the Poisson distribution, after the eighteenth-century French mathematician Simeon Poisson, who devised it— which immunologists commonly employ to calculate the distribution of hybridomas on the wells of a plate. Their Poisson calculation revealed that on such a plate many of the wells in which hybridomas did grow likely contained only a single hybridoma line, fewer likely contained two lines, and fewer still three. Thus, although the denominator in the frequency calculation was in fact larger than the number of wells, it was not larger by enough to have affected the point of the arithmetic—which was that the transgenic hybridomas generated idiotypically birthmarked antibodies from native genes at a rate at least three times higher than did hybridomas from normal mice.[22] The frequency of 50 percent was not exact, but Baltimore and Imanishi-Kari felt confident in claiming that the data strongly supported their conclusion of a remarkable qualitative difference in the kind of antibodies produced by the two types of hybridomas.[23]

Imanishi-Kari says that she told the panel about the reasoning behind the qualitative claim. Storb remembers that she "had a very good statistical argument when we went over this in Boston," but McDevitt did not buy it and Davie says he was bothered that the frequencies of the different types of antibody production reported in the *Cell* paper were imprecise and that the actual frequencies could have been much different. Discussion of the issue was cut off shortly before 6 P.M. because the building was about to be locked up and the panel had to adjourn. Its three members went to eat at Jacob Wirth, a German restaurant that McDevitt knew from his medical school days and that was around the corner from their hotel in downtown Boston. McDevitt recalls that the food was "lousy" and they were glum. They feared "not so much that there's fraud but that Table Two's wrong, that Stewart and Feder are right . . . that we're really going to have to say this is a royal fuckup."[24]

Imanishi-Kari recognized from the tenor of the questions at the close of the session that the panel thought the claims based on Table 2 were unsupportable. She felt that McDevitt had not understood the statistical argument. She thought about what evidence might convince the panel and found a strong candidate in the subcloning data that she showed to the Wortis committee at its first meeting in May 1986. She had done the subcloning for several reasons and had discussed it in the detailed account of the experiment that she had sent to the N.I.H. in late March and that had been distributed to the panel. None of the reasons bore directly on the issue of frequency, but the process of subcloning yielded only one hybridoma line per well rather than several, and the characteristics of the antibodies they generated were consistent with the elevated frequency with which, according to Table 2, the transgenic mice produced idiotypically birthmarked antibodies of native origin. Between 20 and 40 percent of the subclones—the fraction depended on which plate of subclones was used—produced such antibodies, and between 80 and 90 percent of those odd antibodies derived from native genes.[25]

Imanishi-Kari, wanting to tell McDevitt about the subcloning data, telephoned him at his hotel that evening and left a message. When McDevitt returned from dinner, he declined to return the call, preferring to hear what she had to say when the inquiry resumed in the morning. Imanishi-Kari arrived at Tufts first thing the next day, found the panel already assembled, and called its attention to the subcloning data. (Contrary to some later reports, she did not bring the subcloning data with her that morning; it was already there, among the cartload of records she had brought to the session the day before.) The panel members were surprised to learn about the data, Davie says, declaring that before that morning none of the them knew Imanishi-Kari had subcloned any of the Table 2 hybridomas. (Neither Davie, McDevitt, nor Storb recall having read the letter of March 28 in which Imanishi-Kari discussed it.) However, once they saw the subcloning data, they changed their minds about the merits of the *Cell* paper's central thesis. Storb, like Woodland before her, was convinced that the tests positive for idiotype and negative for the transgene had detected characteristics of the same antibodies. The panel as a whole took the high fraction of subclones producing such antibodies as strong evidence in support of the *Cell* paper's frequency claim. The subcloning data "really demonstrated to us," Davie recalls, that the paper's conclusions rested on "a big foundation of data," not just its reported results.[26]

Davie, McDevitt, and Storb completed their work in Boston by mapping out a draft report and agreeing who would compose its different parts. During the next several weeks they developed the document, meeting by conference calls. In late July, the panel asked for and obtained from Imanishi-Kari further illumination of various technical points. Charles Maplethorpe read in *Science* for July 15 that the panel "is said to have found no evidence of fraudulent research." At the beginning of August, he volunteered new information to Mary Miers that he had not supplied during his interview with the N.I.H. staff in May, suggesting that Miers might want to give his new material to the panel "so that they might better perform their task." Maplethorpe's addenda included statements that he had been "a constant and vigilant observer of the experiments" in Imanishi-Kari's laboratory, that the data in Albanese's notebook were inconsistent with the paper's central claim, and that he had heard Imanishi-Kari tell Weaver that Bet-1 responded to *mu-b* as well as to *mu-a*. Maplethorpe's new reports were passed along to the panel with the understanding that its members could pursue the matter with him if they deemed it necessary. Apparently, they did not.[27]

However, in a subsequent telephone conference call the panel evidently raised questions about how it came to be that the laboratories of Baltimore and Imanishi-Kari had both permitted misstatements in the *Cell* paper. They seem to have been particularly concerned about the paper's erroneous indication that the Table 2 hybridomas had been isotyped. While they could comprehend the misunderstanding that led Baltimore initially to insert it, they were somewhat puzzled that it had remained in the paper through successive drafts. Davie worried a lot whether the persistence of the misstatement revealed an intent to deceive. The panel members were concerned still more that the frequencies of antibody type had not been calculated on the basis of individual hybridomas rather than of wells. They were sufficiently troubled to make another trip to Boston, where on September 28, 1988, accompanied by Robert Lanman and Bruce Maurer, they again interviewed, each separately, Baltimore, Imanishi-Kari, and Weaver.[28]

Baltimore no doubt told them what he had mentioned in the inquiry at M.I.T. in 1986—that the misstatement about the isotyping arose from a misunderstanding in a telephone conversation with Imanishi-Kari. The isotyping misstatement was, in fact, elaborative rather than direct, an added remark in a discussion of the Table 2 hybridomas: "The remaining 119 clones produced other Ig . . . isotypes, the majority being γ[gamma]$_{2b}$ (data not shown)." Imanishi-Kari explained that she had not recognized its import on reading successive drafts of the paper. She held that before the panel brought up the issue of terminology, it had never occurred to her or, so far as she knew, to anyone else that her use of "hybridomas" in place of "wells" could be misleading. Neither the Wortis committee, the Eisen inquiry, nor O'Toole herself had raised the point. In her lab when hybridomas were first cloned, it was the practice to refer to them as "clones" no matter whether they were individual clones or wells. Besides, to her mind the numbers in Table 2 were so large that anyone who did this kind of research would know that they represented wells.[29]

The panel found Imanishi-Kari's explanation of the isotyping misstatement plausible because of her poor English; and, mindful that the Table 3 antibodies had been isotyped, they concluded that Imanishi-Kari had not intended to deceive. They nevertheless remained unhappy that the paper misstated the isotyping facts, and they held that Imanishi-Kari's terminological interchange of "clones" with "wells" was misleading enough to constitute a serious problem with the paper.[30]

Baltimore asked what the coauthors could do, and, according to Lanman, the N.I.H. officials at the meeting mentioned that one option available to the coauthors was to write a letter to *Cell* detailing the misstatements in their paper. They promptly composed one, providing N.I.H. a copy of it in mid-October and publishing it in the issue of the journal for November 18, 1988. Taking the opportunity to correct all the errors in the paper that they knew about, they acknowledged the overstatement concerning the discriminatory capabilities of Bet-1. They declared that the Table 2 data represented "primary hybridoma wells, not isolated clones" but added, pointing to their statistical considerations, that "the number of clones per well . . . was likely to be only one or a few." They noted that the isotyping reported to have been performed on the Table 2 hybridomas had actually been done in "separate experiments" with similar results. They closed with the assurance that "these are not material alterations and do not affect the conclusions of the paper."[31]

The publication of the letter apparently angered Congressman Dingell and his staff. In late August, Dingell's staffer Bruce Chafin had discussed the N.I.H. investigation with a member of the N.I.H. misconduct office named John Butler, an accountant who had previously investigated the mismanagement of grants and who now wrote a memorandum to files about the conversation. Chafin said that if the N.I.H. would inform the subcommittee of the panel's tentative findings, the staff "could let us know if there was additional information that the panel might have missed, and we could avoid the embarrassment of publishing a report that was incomplete and possibly incorrect." Butler declined to reveal the panel's findings, emphasizing that it was N.I.H. policy not to disclose the results of an ongoing investigation and that if the subcommittee wished otherwise, it should talk to Lanman.[32] Now, with the correction appearing in *Cell*, it seemed to Dingell that the N.I.H. had leaked investigative information to the coauthors so that, as he complained to Secretary of Health and Human Services Otis R. Bowen, they could "preempt the findings" of the N.I.H. investigation and "defuse the N.I.H. report." Dingell remonstrated that the coauthors' correction gave the impression that the errors in their paper had come to their attention only recently, continuing, "It is quite clear that this is not what actually happened: the authors were informed of these same misrepresentations two and a half years ago, but were only willing to admit them publicly when they realized they could no longer be kept secret."[33]

Baltimore had, of course, publicly acknowledged the overstatement concerning Bet-1 in his "Dear Colleague" letter six months earlier; the coauthors had not regarded the misstatement of the isotyping as sufficiently significant to warrant a correction; and they had only just learned of the problem with the terminology of wells and clones. Lanman reportedly advised Dingell's staff in a telephone call that the N.I.H. had not told the coauthors about the panel's findings; the coauthors had inferred them from the questions put to them at the meeting on September 28. In a follow-up letter, Wyngaarden assured Dingell that the correction sent to *Cell* "was neither directed nor encouraged by the panel." Whatever the N.I.H. had or had not revealed to the coauthors, it was evident that Dingell's subcommittee remained strongly interested in the investigation. By mid-November, the panel's draft report on the *Cell* paper was ready. On November 18, the N.I.H. misconduct office sent by Express Mail copies to the coauthors and O'Toole for their information and comment. All were asked to maintain the report in strict confidence—and all were told that the report would also be made available to Congressman Dingell "based on his request," which it was, the same day.[34]

The report announced unambiguously that "no evidence of fraud, conscious misrepresentation, or manipulation of data was found." It concluded that the normal mouse had indeed been mistyped; that Bet-1, as even some of O'Toole's data showed, was specific enough for the job; that the 1,000-count-per-minute cutoff, though not fully backed, was defensible. It judged that the subcloning studies "support the trend seen in

Table 2"—which was to say that the transgenic hybridomas produced an elevated frequency of native, yet idiotypically birthmarked antibodies. It concluded that Table 3 added weight to that central claim because roughly half the hybridomas reported in it generated antibodies that displayed both the distinctive idiotype and isotypes other than *mu*. In a passage that adds weight to the view that O'Toole had once again proposed the formation of *mu-gamma* hybrids to account for some of the data, the report explicitly considered the idea and rejected it, observing that the occurrence of such hybrids was "unlikely" on grounds of physical chemistry. The panel declared itself "impressed" by the amount of work the coauthors had done for the study, "by the completeness of the records, and by the abilities of both Drs. Imanishi-Kari and Weaver to find, accurately interpret, and present data on experiments that were performed as much as three years earlier."[35]

The draft report nevertheless held that the panel had discovered "significant errors of misstatement and omission, as well as lapses in scientific judgment and interlaboratory communication," by which was meant the overstatement concerning the specificity of Bet-1, the misstatement concerning the isotyping, and the ambiguous references to "hybridomas" and "clones" in Table 2. Despite what the panel said about the support that the subcloning provided for the trend in Table 2, its members seemed somewhat uncertain as a group over whether the central claim of the paper was, scientifically, altogether correct. Maurer recalls that the panel was "somewhat divided," with Joe Davie thinking something about the *Cell* paper was "fishy." Davie says that he just saw a lot of "sloppy science" in the paper and attributes any inconsistencies in the panel's report to the fact that it was written by a committee. Whatever the case, the panel declared itself persuaded that the transgenic mice did display "unusual characteristics," the "most salient" of which was that their B cells produced a high level of antibodies derived from their native genes, not from the transgene. But for several reasons, among them problems it discerned with Bet-1, the panel found it "difficult to be certain" that the idiotypically birthmarked antibodies were not transgene products. It ventured the thought that some of the transgenic hybridomas might be, if not producers of *mu-gamma* hybrids, double producers of some other kind, pointing as warrant for the idea to the paper that the Herzenbergs and their student Alan Stall had by now published in the *Proceedings of the National Academy of Sciences*.[36]

The panel concluded that the paper's inaccuracies were sufficiently serious to merit public correction and that the coauthors' letter to *Cell* did not go far enough. It recommended that the coauthors submit another letter to the journal that would correct various clerical errors in the paper, clarify that Table 2 represented numbers of wells rather than hybridomas or clones, and assess how the problems with the reagents they had used, including Bet-1, affected the interpretation of their data. The panel also held that the problems with Table 2 were so serious that the authors should offer "in replacement, the data obtained from the subclones of the wells represented in this table."[37]

The coauthors, although happy with the overall thrust of the panel's report, were displeased with a number of its technical conclusions and the tentativeness of its support for their paper. In a response on November 29, they contested both, pointing to inconsistencies in the report and bolstering their rebuttal with a seventeen-page memorandum of data and argument. They dissented from the panel's deployment of the Herzenbergs-Stall paper. They called it "appropriate for new approaches to uncover new data and to require the rethinking of previous conclusions—this is the stuff of science!"—but they suggested it was inappropriate to question the scientific legitimacy of the claims made in 1986, given what they then knew, on the basis of results that were published long afterwards and were based on a study of different materials.[38]

The coauthors objected strongly to writing an additional letter of correction that included everything the panel asked for. They had seen no reason when they had published the paper for an elaborate discussion of the reagents used—meaning Bet-1—noting journal space was limited and that they had followed the common practice of relying "upon the general understanding of the scientific community." They said that they saw no reason for it now but were nevertheless "prepared to describe them in whatever detail is recommended."[39]

They vigorously protested substituting the subcloning data for that in Table 2. Their objection rested in part on the statistical argument for the table's utility which they had already advanced and on which they elaborated now, but only in part. They also believed that invoking the subcloning data in support of the paper's central claim would be scientifically inappropriate. The claim was about the kind of antibodies produced by the overall population of B-cell hybridomas from the transgenic mice. The subclones represented only a subset of the population; and in the process of subcloning, the characteristics of the B cells tended to diverge from those of the population (because, for example, they lost chromosomes as the cells multiplied). The original population was no more represented by the subclones than was, say an original population of immigrants represented by the descendants of several families, some of whom had intermarried. Thus, one could conclude little directly from the subclones about the frequency with which the original hybridomas produced one kind of antibody or another. For that reason, the coauthors said that they would be willing to have the subclone data used as a "supplement to Table 2, not as a replacement."[40]

Margot O'Toole, on her part, angrily protested the panel's draft report, charging in an eleven-page letter that it was "wholly inadequate." She faulted the panel on its procedures, pointing out that it had declined to interview Maplethorpe or Stewart and Feder, and on a number of technical points, especially for reaching "important conclusions" on the basis of the subcloning data. She was suspicious that the panel had found nothing significantly wrong with the claims made for Bet-1 and asked to see the original data for Figure 1, saying that she had a right to it. She charged that the subcloning had not been performed, declaring that "no such data were ever produced during the previous investigations." She asserted on page 2 of her letter that at her meeting with the Wortis committee in May 1986 "Dr. Imanishi-Kari stated emphatically that, besides the tests recorded in the 17 pages already in my possession, no other test had been done on the Table 2 cells"; on page 3 she went further: "I state categorically that Dr. Imanishi-Kari told me in May 1986 that no subcloning analysis of Table 2 hybridomas had been done."[41]

Despite the injunction of confidentiality, the draft report was leaked, probably by someone at the Dingell subcommittee, to Daniel Greenberg's *Science and Government Report*. Greenberg promptly published the report's conclusions in an article in his newsletter's issue for December 1 that seemed colored by Stewart's views. Stewart had told a reporter that he was dissatisfied by the letter of correction to *Cell*, declaring, "It doesn't appear to address all of the scientific issues we raised." Although Greenberg's story noted that no evidence of fraud had been discovered, it emphasized the errors detected by the panel, holding that "the findings on scientific points in large part vindicate the assertions of . . . Margot O'Toole." It added that the panel had not touched on "the squalid treatment O'Toole received from her mentors when she questioned the work of several superstars of contemporary science."[42]

The *New York Times* picked up the article, and *Nature*, which had also obtained a copy of the draft report and whose editor, John Maddox, was a fan of Stewart's, weighed in with a story on the matter. In an accompanying editorial, Maddox faulted Baltimore's laboratory for its "error-prone" work, declared that O'Toole's objections had been "at least partly validated," and reported that Stewart and Feder's varied criticisms had been found "to have a force not previously acknowledged." What counted now, Maddox argued, was not who was right or wrong but how whistle-blowers were handled. He lectured Baltimore and his coauthors that they might make a useful start on that score "by responding frankly, and more generously" than they had in their letter to *Cell* the previous month.[43]

Baltimore was infuriated by Maddox's swipe at the quality of work in his laboratory, Maddox's seeming alliance with Stewart and Feder, and his stated disregard for exactly who was right and who was wrong in the dispute. An analysis done in Baltimore's office concluded that the panel had actually dismissed most of the specific challenges that O'Toole and Stewart and Feder together had raised against the *Cell* paper. Baltimore drafted a heated rebuttal to what he took to be the editorial's distortions of fact, expressing astonishment that Maddox had written it without even talking with the coauthors beforehand and saying that he would have been pleased to discuss the coauthors' extensive response to the draft report with the editor "had he inquired." Baltimore apparently thought better of sending the letter, but in the wake of the press's seemingly one-sided treatment of the leaked draft, he telephoned the head of the N.I.H., James Wyngaarden, urging him to have the scientific panel consider the coauthors' response in their final report.[44]

The panel, however, had no obligation to modify its report or its recommended letter of correction in light of the responses from either the coauthors or O'Toole, and it did not. Davie and McDevitt at least found unpersuasive the coauthors' statistical defense of the legitimacy of talking about hybridomas when they meant wells. Davie says, "You had very competent scientists . . . arguing that it was okay to talk about percentages when, in fact, they had no reason. Frankly, we could have been much tougher. . . . They had defined a percentage of B lymphocytes that were expressing this abnormal idiotype . . . We're saying, 'The data that you presented in that paper do not allow you to extrapolate to that point.' The extrapolation is big-time extrapolation. . . . This was really a landmark paper that had great implications." Davie adds that the panel continued to insist on the coauthors' publishing the subcloning data as a substitute for Table 2 because they did not want to "leave the impression that . . . the precision of what they had done in Table 2 was sufficient." McDevitt was particularly adamant on the issue. He did not recognize the coauthors' argument that the subclone data was an inappropriate basis on which to stake a claim about the population of hybridomas. To McDevitt, certain features of the subclone data indicated that each well had originally contained many more hybridomas than the statistical analysis concluded. He says he simply thought the statistical defense was "bullshit."[45]

The panel felt that they had already carefully examined most of the issues O'Toole had raised, and they apparently did not credit her subcloning charge because it was new. O'Toole had actually not heard of the subcloning data until she saw the draft report, but to the panel she seemed to be changing her story. In an earlier version of the draft, the N.I.H. staff had included a balancing statement praising O'Toole: "The panel was impressed by Dr. O'Toole's response to questions and her intellectual grasp of a very complex system." Maurer says that McDevitt had insisted on taking it out. Although McDevitt says that he doesn't recall the matter specifically, he declares that "for whatever reason, she had an axe to grind, and I was not going to go out of my way to say she was great."[46]

In early December, the panel's draft report cleared the standard review given such evaluations by a group of senior N.I.H. staff and, later that month, it was subjected to scrutiny at a high-level meeting in the office of the agency's director that was headed by Katherine Bick, the deputy director of extramural research. O'Toole's claim that Imanishi-Kari told her she had not done the subcloning gave Bick's group pause. Bick's group might have wondered why, if Imanishi-Kari had in fact admitted such an experimental omission, O'Toole had not mentioned it either in her 1988 congressional testimony or in the memorandum that she prepared for Herman Eisen in June 1986, only about two weeks after her encounter with Imanishi-Kari in the Tufts inquiry.[47]

Bick, instead, went directly to Imanishi-Kari, putting several questions to her in a letter that she answered on December 30: Had she shown the Wortis committee the subcloning data in May 1986? "YES." Did she tell O'Toole in May 1986 that she had not subcloned any of the Table 2 hybridomas? "NO. At no time did I say to Dr. O'Toole that the subcloning analysis of wells in Table 2 was not performed."[48]

About the same time, the N.I.H. sent query letters to members of the Wortis committee: Had they seen the subcloning data in May 1986? If so, did it support Table 2? Had they discussed it with Imanishi-Kari? In independent responses during the second week in January, Wortis, Huber, and Woodland responded yes to all three questions.[49]

In mid-January, Bick endorsed the recommendations of the scientific panel in a lengthy memorandum for the head of the N.I.H., James Wyngaarden. She also reviewed the history of the controversy, attempting to account for why it had dragged on so long and caused so much trouble for the agency. She declined to censure M.I.T. or Tufts for failing to bring about a speedier resolution of the dispute, noting that Tufts had no authority over it, since the research had not been performed there, and that M.I.T., faced with an allegation of error rather than fraud, had gone "beyond commonly understood requirements of academic integrity and stewardship of public funds." She declared that it had been "extraordinarily difficult" to identify the issues clearly and to focus on the agency's "legitimate . . . interests." Bick blamed the difficulty on the intense publicity given the case by Stewart, Feder, and Dingell and on the coauthors' "initial failure" to deal with the matter "in a collegial manner." She called it "significant" that the coauthors had "only recently . . . acknowledged that some correction is warranted" and held it "unfortunate that despite the growing challenge to the validity of their research, the coauthors apparently did not undertake a comprehensive review of their data until they met with the N.I.H. scientific panel."[50]

Bick included a letter for Wyngaarden's signature that on January 31, 1989, he signed and sent to the coauthors and O'Toole with the panel's final report, unmodified from the draft version save for minor clarifications and changes. Wyngaarden noted the panel's exoneration of Imanishi-Kari from any finding of serious wrongdoing but insisted that the coauthors submit to *Cell* the correction letter it had recommended. Following Bick's analysis, Wyngaarden chastised them for neglecting to deal fully with the allegations of O'Toole, Stewart, and Feder, despite having known about them since 1986. If they had done so, he added, they might well have corrected the "serious" inaccuracies in the paper and might "well have made a full investigation unnecessary."[51]

The final report, accompanied by the responses of the coauthors and O'Toole as well as by Bick's long account of it for Wyngaarden's decision, was promptly released to the public. *Science and Government Report* pointed to Bick's review of the case, "including the stonewalling that Feder and Stewart encountered from Baltimore and company when they sought to come to O'Toole's assistance." The newsletter called Baltimore's early proposal that Stewart and Feder agree to cease discussing the subject after the investigation "outrageous" "a new high in arrogance." Shortly before the release of the report, in an article in the *Chronicle of Higher Education*, O'Toole had castigated the biological community for self-interested timidity in dealing with allegations of misconduct, declaring, "Anyone who points to the problems of misrepresentation is automatically perceived as a troublemaker." Now she publicly asserted her dissatisfaction with the report, reiterating her claim that the panel had drawn its conclusions from experiments—the subcloning—that had not been performed. Baltimore told a reporter, "I feel vindicated." He declined to say, however, whether the coauthors would comply with the directive for an additional letter of correction. "This is a matter of scientific judgment," he said.[52]

In a letter in mid-February, Baltimore, Imanishi-Kari, and Weaver told Wyngaarden that they had, in fact, conferred about the allegations when they were made and had found "no substance" to those of Stewart and Feder. They added that they had deemed the errors in the paper insufficiently serious to warrant a formal correction, a judgment that was in consonance with the conclusions of the Tufts and M.I.T. inquiries, and they noted that when the confusion in terminology about the wells was called to their attention, they had promptly submitted a clarification to *Cell*. They emphasized that they had good reason to think that Bet-1 was "more than up to the job." They also pointed out that "no scientific argument" had been offered against their contention that it would be inappropriate to substitute the subcloning data for the hybridoma data in Table 2. They resolutely concluded:

> We believe that it is unfair—and quite damaging to all of us—to suggest that we have been remiss in responding to the allegations. You conclude that had we been more aggressive in reviewing the allegations a full investigation might not have been necessary. It is equally possible to conclude that since the original allegations have not found scientific support, it was solely the zeal of our accusers that forced the convening of all of these investigations.[53]

Baltimore nevertheless was willing to accommodate Wyngaarden. He says that he figured Wyngaarden's admonitions were "political," intended to show Congressman Dingell that the N.I.H. was calling the coauthors to account. Baltimore continues, "Since I believe in his agency, and I believe it's important to protect it, I didn't go shouting about it. I responded to him quietly and precisely. As long as he was willing to accept what I said, I was willing to accept what he said." He also felt that unless the coauthors abided by Wyngaarden's directive "our heads would go."[54]

Baltimore, Imanishi-Kari, and Weaver thus enclosed in their reply to Wyngaarden a letter that they proposed to send to *Cell*. It supplied data on the specificity of Bet-1 and recorded the subcloning data in the way they had offered at the end of November—as a supplement to Table 2 rather than as a replacement for it. The coauthors wanted to know if Wyngaarden considered it sufficient and, "if not, how and for what scientific reasons it should be modified to satisfy you." Wyngaarden, retreating somewhat from his chastisement of the coauthors, conceded in a reply to them on March 10 "that we have the advantage of hindsight, and a certain distance from the scene." He accepted their letter, having obtained approval of it from Davie, McDevitt, and Storb. "Maybe it's not 100 percent responsive, but it's okay," he told *Science and Government Report*.[55]

The letter, which appeared in *Cell* in its issue for May 19, opened with the notice that it had been sent "at the explicit request of the Director of the National Institutes of Health" and concluded: "Nothing in this clarification affects the conclusions or interpretations in the original paper." In an accompanying editorial, Benjamin Lewin, the editor of the journal, observed that the necessity of publishing the statement provided "a measure of the lack of a sense of proportion" that had come to govern the dispute over the paper. He pointed out that the statement covered detailed technical matters that would normally concern people close to such work and be most appropriately resolved "in individual discussions with the authors." *Cell* would not usually publish material of such recondite nature but had decided to print it "in view of the intense political concern about this paper."[56]

Imanishi-Kari says that when Maplethorpe had publicly accused her of fraud in the 1988 hearing, she had felt personally violated, smeared as a scientist and, because of the stories in the *Boston Globe*, distressed for her thirteen-year-old daughter. Throughout the months of the dispute she had difficulty taking pleasure in the little things of life. "You always have this feeling in the back of your head," she recalls. When the N.I.H. issued its final report, she went with her daughter and niece to the Tufts Athletic Center where they had a swim and then sat together in the sauna, chatting about nothing special. "I felt so good. I said, 'My God. It's such a difference. It's been three years.' "[57]

E I G H T

■

Baltimore v. Dingell

TO CONGRESSMAN Dingell and his staff, the N.I.H. report was far from the final word. Since the spring of 1988, Walter Stewart and Ned Feder had been pressing ahead with the subcommittee's own inquiry into the case. Stewart was one of the investigators—Peter Stockton was another—whom the subcommittee sent to Boston in May 1988 to interview the principals. Baltimore, Weaver, and Imanishi-Kari declined to submit to interviews with the subcommittee's investigators, not least because they thought Stewart did not measure up as a scientifically qualified or unbiased investigator. Henry Wortis, Brigitte Huber, and Robert Woodland spoke with them on May 9 and 10.[1] Herman Eisen and Gene Brown met with them on May 11, accompanied by Robert E. Sullivan, a lawyer at Palmer & Dodge, a Boston law firm that was outside counsel for M.I.T. At one point Sullivan interrupted Stewart to tell him, "I know the difference between questions that are trying to find out as much as possible and those that are just trying to confirm pre-held assumptions. You are asking the latter type and I am very annoyed." Eisen said that he was irritated, too, and Stewart apologized.

A few minutes later Stewart queried Eisen, "Anything else I should ask about the science [of the *Cell* paper]?"

Eisen replied, "Yes. You should ask what's good about it."

Stewart responded, "I don't think so. I would have published a correction based on the 17 pages."[2]

In mid-May, M.I.T. and Tufts sent Peter Stockton documents that had figured in their respective inquiries into O'Toole's allegations, and Weaver sent his notebooks on June 21; but as of late June Imanishi-Kari had not yet sent the subcommittee the laboratory records it wanted from her. By then, she had obtained a lawyer through a roundabout process that began with the help of local friends, led to an attorney in Washington who held John Dingell in opprobrium but who could not take her case because of a conflict of interest, and came full circle to Bruce Singal in Boston. In his early forties, Singal was a former assistant attorney general of Massachusetts, director of the state's consumer advocacy agency, and member of the U.S. attorney's office in the city, where he had fought cases of political corruption. He was more than familiar with the intricacies of government power and its abuses. Now in private practice, he agreed to take Imanishi-Kari's case pro bono.[3]

Singal told the subcommittee that it could have copies of everything that she had provided to the N.I.H. panel and that it was welcome to review the originals in Boston. The subcommittee deemed such an arrangement unacceptable. On June 30, it subpoenaed everything in her possession that was even remotely related to the *Cell* paper, including all original data, calculations and notes, all the laboratory notebooks and data produced by Moema Reis and Chris Albanese, all correspondence and memoranda, and copies of all grant applications. Imanishi-Kari, in a small act of defiance, shipped the materials to the N.I.H., which promptly turned them over to Dingell.[4]

In the meantime, Baltimore had enlisted lawyers in the Washington offices of Akin, Gump, Strauss, Hauer & Feld, a firm experienced at the Dingell bar, to get the coauthors' point of view before the subcommittee and its investigators. "Up in Boston, we're very provincial," Baltimore's lawyer, Normand Smith, told a reporter. "When you want to know what's going on in Washington, you hire a lawyer in Washington." Baltimore later elaborated, "I didn't know anything about John Dingell then. I was looking for someone to give me advice." On June 28, in a meeting arranged by Akin, Gump lawyers, Baltimore detailed his concerns to Representative Thomas Bliley of Virginia, the subcommittee's ranking Republican, who subsequently conveyed them to Dingell. Baltimore seems to have stressed in particular that it was inappropriate for the subcommittee to rely on Stewart and Feder, since they had a stake in the outcome of the investigation. A lawyer at Akin, Gump named Joel Jankowsky, who was a former assistant to Speaker of the House Carl Albert, put it to a reporter that Stewart and Feder were acting as "judge as well as witness." Baltimore's concerns were said to have been duly noted by Dingell.[5]

In early August, Dingell asked Baltimore and Weaver for all of their documents relevant to the *Cell* paper, including all original data and all correspondence and memoranda since 1985 that even mentioned any of the scientists in the case or connected to it—for example, the Herzenbergs, Stewart, or Feder. Baltimore supplied all the materials pertaining to the substance of the *Cell* paper, but in a letter to Dingell objected to the blanket request for correspondence. Honoring it would "violate my privacy and the privacy of those who have sent personal letters to me." He added, "I have nothing to hide, but I cannot determine what it is that the Subcommittee is seeking. Couldn't we sit down together and discuss your concerns?" Baltimore received no reply. He was eventually forced to turn over the letters to the subcommittee; warning his private correspondents, he expressed the hope that their acts of friendship in writing to him would not cause them embarrassment or hardship.[6]

Stewart started devoting full time to the subcommittee at the beginning of August, roughly two weeks after Imanishi-Kari had sent her original notebooks to the N.I.H. He pored through the material at a small desk in Dingell's staff office while Feder, with whom he kept in touch by fax and telephone, provided backup at the N.I.H. He labeled the notebooks, calling the one that Imanishi-Kari had constructed "I-1." Stewart says that when he finally got to examine all the documents, he knew "instantly" that something was funny about them because dates had been rewritten and some of the radiation-counter tapes mounted on the same page differed from each other in shade and print intensity. The June subcloning data struck him as an especially sloppy counterfeit. He says that Imanishi-Kari must have prepared the data in haste the night before she called it to the attention of the N.I.H. panel.[7] (He was mistaken, of course. She had already given the panel the subcloning data along with all the rest of her notebooks that morning. Imanishi-Kari only asked them to turn to it the next morning.)

Stewart says that he told the staff he thought much of her data "was forged" and that they told him, " 'Figure out a way of proving it.' " He pursued the task assiduously, obtaining considerable help from O'Toole, to whom he described some of the materials and spoke with for hours on the telephone. Stewart recalls that he managed "to construct really tight arguments" to prove his case. Dingell's staff found Stewart's arguments too scientific and arcane to sell to the public, so at the end of August they called on the Questioned Document Branch of the Secret Service for help—to see, Bruce Chafin says, if the Secret Service could confirm the fraud "in a way that you could do in a hearing."[8]

John Hargett, the chief of the branch and the senior document exam-

BRUCE SINGAL
Thereza Imanishi-Kari's Lawyer
Photo Credit: Courtesy Bruce Singal

NORMAND F. SMITH, III
David Baltimore's Lawyer
Photo Credit: Courtesy Normand F. Smith

iner for the Secret Service, took charge of the work in collaboration with Larry F. Stewart, who was the lead document examiner for the agency's Instrumental Analysis Section. Both had extensive experience in the forensic analysis of papers and inks. The Secret Service has some seven thousand samples of inks in its International Ink Library, the world's largest collection of its kind. The agency has considerable experience in the detection of falsified documents, since its duties include the enforcement of laws against counterfeiting cash, identification documents, and food stamps. The Secret Service rarely did this sort of work for Congress (legal experts say it raises a question about the separation of powers if an agency of the executive branch takes a committee of the legislative branch as a client). The Service also had little experience in dealing with the type and volume of material that was in the notebooks, but Dingell's staff said that the issue was potentially a criminal matter.[9]

The Secret Service laboratory staff first worked with two of Imanishi-Kari's notebooks, including the one she had constructed in the spring and that Walter Stewart had designated I-1; they were soon given Moema Reis's notebooks, too. Hargett was concerned to make the Secret Service's work as objective as possible. To that end, he preferred to know nothing about the significance of any of the data for the experiments. He wanted to deal only with the ink and paper of the laboratory records. "We had no idea about the research," he later said.[10]

However, a good deal of the agency's analysis worked differently in practice. The volume of materials was so large that the Secret Service staff did not have the time to give equal attention to all parts of the laboratory records. After an initial examination of various pages in one of the notebooks, Hargett or Stewart would confer with Walter Stewart at the subcommittee about items that appeared questionable. Walter Stewart would then tell them to pursue the analysis of those items further or drop it. One of the Secret Service agents explained in a meeting with N.I.H. officials, "We used their direction to lead us to what was important. . . . We reported the forensics to them and let them determine if it was something they wanted to bring up to the subcommittee."[11] Thus, the Secret Service analysis did not comprise an entirely independent effort of forensics. At the beginning it was guided by Walter Stewart's suspicions of the *Cell* paper's science.

Stewart, who continued with his own analysis of the notebooks, says that all the while it began to dawn on him that the three principal coauthors, the three scientists at Tufts, and Herman Eisen must have decided to "lie about the facts," "that seven people had conspired to conceal a really important thing to the enormous detriment of O'Toole." He holds that Henry Wortis probably helped Imanishi-Kari with some of her fabrications.[12] That January, the Cold Spring Harbor Laboratory, on Long Island, hosted a meeting on scientific fraud at the nearby Banbury Conference Center that was attended by a number of scientists and several congressional staffers, including Stewart. At one point in the discussion, Stewart wrote "Holocaust" on the blackboard. Outraging many in the audience, he said, in so many words, that people who stood by while misconduct occurred and whistle-blowers suffered were tantamount to good Germans.[13]

O'Toole, meanwhile, had been waiting for the N.I.H. to send her the data for Figure 1 of the *Cell* paper, which she had requested when she first saw the panel's draft report in November. In early February 1989, after two dissatisfying telephone conversations with an N.I.H. official, she wrote a letter asking for it again; but now that the final report from the N.I.H. had been released, she also questioned the scientific panel's reliance on the Wortis committee's assertion that it had seen the subcloning data in May 1986. Several weeks later, in another letter to the N.I.H., she complained that the agency's decision on the panel's recommendations had praised Dean Gene Brown at M.I.T. for "going beyond commonly understood requirements of academic integrity and stewardship of public funds." She declared that, so far as she knew, Brown "did nothing more than wash his hands of the matter" and that the N.I.H.'s praise of his actions raised questions about how junior scientists might expect to be treated if they brought allegations of misconduct. She added that at the meeting with Eisen in June 1986 Baltimore had "decided . . . [on] a cover-up." O'Toole sent copies of her letters to Congressman Dingell. Early in March, the N.I.H. provided her the data for Figure 1. At the end of the month, O'Toole replied that the data for the fourth point in each of a series of curves on the figure were missing and that she wanted to see those results[14] (see *Figure 7*).

The data had been obtained by Moema Reis. Imanishi-Kari pored over Reis's notebooks looking for it and, unable to find it, telephoned her in São Paulo, Brazil. In April 1989, she sent the N.I.H. a letter that she had just received from Reis: Reis thought she had gotten the data for the fourth point the day after she had taken it for the first three and had entered it directly onto the original figure in her notebook. Imanishi-Kari told an N.I.H. official what she had learned, and also mentioned how she had reorganized her data in preparation for the inquiries by the panel and the subcommittee.[15]

The news of these "unorthodox data handling practices," as Wyngaarden termed them, was unsettling to the N.I.H. In a letter to Wyngaarden at the end of January, Dingell had taken note of O'Toole's dissatisfactions with the draft report, including her claim that certain experiments reported to have been done were not, and he had concluded, "I assure you that the final report will receive a careful and searching review for accuracy and completeness by the Subcommittee." At the beginning of April, Dingell had asked the N.I.H. for the complete records of its investigation into the *Cell* paper. Now, at the end of April, apparently with a worried eye to Capitol Hill, the N.I.H. told Dingell's staff and the coauthors that the agency was reopening its investigation of Imanishi-Kari with a detailed audit of all her data.[16]

By then, the Secret Service had told the subcommittee that it had come up with what seemed to be the kind of evidence Dingell's people were looking for. Its document analysts had found a number of date alterations in Imanishi-Kari's I-1 notebook and determined that some radiation-counter printouts in it had been moved from one place to another. They went beyond what the naked eye could discern by using a technology that detected the indentations made on one page when someone wrote on another page directly on top of it. Their analysis indicated that, although a number of pages in the I-1 notebook came from the same lined tablet, the pages ordered by the dates on them did not match the order in which they had occurred on the original tablet. In short, pages with later dates appeared to have been written on top of—that is, before—pages with earlier dates. In all, twenty-five pages in the I-1 notebook were suspect.[17]

So was a composite photo illustration of fourteen radiographs that had been prepared for Figure 4 of the *Cell* paper by David Weaver (*Figure 12*). Weaver had prepared these radiographs to test whether transgenic DNA, which produces antibodies with a *mu* isotype, was active in fourteen of the hybridomas summarized in Table 3. In these radiographs, small horizontal bands represented the type of DNA—transgenic *mu* or native *mu*—that had been expressed in each of the fourteen. Recall that the type was revealed by the distance the band traveled. If no band appeared on the radiograph, that hybridoma had not expressed transgenic DNA; its antibodies had to derive from a native gene. If one did appear, its location on the radiograph revealed whether the DNA generating the antibodies in that hybridoma came from native DNA or transgenic DNA. Ten of the radiographs had no band. Three of them had one that derived from a native gene with *gamma* DNA. Only one had a band that came from the transgene. Figure 4 thus showed that the transgene was expressed in only

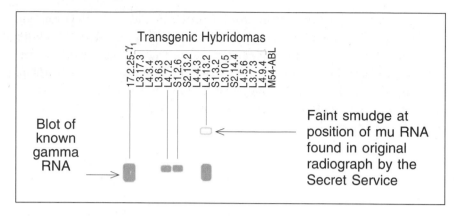

FIGURE 12

The Smudge in Figure 4 of the Cell *Paper*

The suspect smudge occurred in the radiograph lane for hybridoma L4.13.2. The dark blot, which appeared in Figure 4, was at a location in the lane characteristic of the RNA for an antibody with a *gamma* isotype. The faint smudge, which did not appear in Figure 4, was at a location characteristic of the RNA for an antibody with a *mu* isotype, which gave rise to the possibility that the hybridoma was a double producer. Note: Figure 4 is drawn rather than reproduced here so as to make the point at issue clear.

one of the fourteen hybridoma lines, which comprised important under-girding for the paper's conclusion that an abnormally high fraction of the idiotypically birthmarked antibodies came from native genes.[18]

However, multiple exposures had been made of the radiographs for each of the fourteen hybridomas, and some of the original exposures for one of them raised questions. That hybridoma—numbered L4.13.2—was one of the three that appeared to express a native gene. Although the exposures showed an intense band characteristic of such a gene, they also exhibited a separate faint smudge or blur at a point characteristic of the transgene. A Secret Service agent named Steven Herzog had discovered that the photo illustration was a composite and that, in preparing it, Weaver had chosen an exposure for the hybridoma in question on which the band for the native gene showed clearly but on which the blur was so faint that it became invisible in the publication process. To the sub-committee staff, both discoveries were grounds for suspicion. Walter Stewart took Weaver's choice of radiograph to be suppression of evidence that the hybridoma, in expressing the transgene, was a double producer.[19]

Late in April, the subcommittee called down several N.I.H. officials, including Bruce Maurer, for a private briefing on the Secret Service results. Maurer remembers that they were seated at a long table, with Stewart about eight feet away from him on the other side. The conversation got to Baltimore, and Maurer said something in support of him. "Stewart came at me, climbing right across the table, like a cat," Maurer says. "He went on and on about how wrong I was and what a terrible scientist Baltimore was." The N.I.H. soon decided that its renewed investigation of Imanishi-Kari would place special emphasis on forensic analysis of her notebooks.[20]

Congressman Dingell proposed to showcase the Secret Service results in hearings on scientific fraud that he scheduled for May 4 and May 9, 1989. His witnesses would include the coauthors, Wyngaarden, the Davie panel, and all but one of the scientists and administrators involved in the investigations at Tufts and M.I.T. The missing scientist was Robert Woodland, from the Tufts inquiry. He told me matter-of-factly over the telephone why he thought he had not been included: "I'm black, and one of the subcommittee's lawyers said, 'Dingell's from a district where it wouldn't look too good for him to be beating up on black people.' And I said, 'No. It's that bad, is it?' And she said, 'You'd be surprised. This is a political procedure and that's a political fact.' "[21]

Baltimore says that the coauthors first learned about the Secret Service's involvement in the affair via John Crewdson, a reporter for the *Chicago Tribune*, who was warning scientists against defending the *Cell* paper because the Secret Service had discovered data tampering in Imanishi-Kari's notebooks. Shortly thereafter, Dingell, who later said he wanted "to be fair," invited the coauthors to come to Washington a week ahead of the hearing so they could be briefed on the Secret Service findings. In separate meetings on Tuesday, April 25, Baltimore and Weaver heard about the results from Stockton, Chafin, Walter Stewart, and several Secret Service agents, and on Thursday Imanishi-Kari got her briefing from the group.[22]

The coauthors considered the Secret Service findings pointless. The construction of composite photo illustrations was standard practice in the presentation of radiographs. And it made no difference that the faint smudge had been omitted because the aim of the radiograph was to show that the native gene was expressed in the hybridoma, not that the transgene was not expressed, and that it was the native gene that was responsible for the idiotype of the antibodies produced by the hybridoma. David Weaver says that what Walter Stewart and the rest of the subcommittee staff tried to do with the photo illustration was "really ridiculous," adding, "They had no objectivity at all. They weren't interested in 'the truth.' "[23]

Imanishi-Kari was accompanied at her session by her lawyer, Bruce Singal. He asked the Secret Service agents for a written lab analysis and report of their findings, which, in Singal's prosecutorial experience with the U.S. attorney in Boston, it was the agency's common practice to provide. He says he was "stunned" to discover that they had no such report (they later said the subcommittee had told them not to bother preparing one). Singal continues:

> I asked for further data at that time relating to the particular pages and the particular notation and particular handwriting which they were questioning, and other than one or two isolated examples, they were not willing to share that with us. What they did do was go through and generally talk about the laboratory processes . . . from which they derived certain conclusions, but for the most part without showing us the particular evidence.[24]

Singal adds that when the Secret Service finished their "skeletal recitation," they attempted to ask Imanishi-Kari questions but that as her lawyer he advised her not to answer because they had only just been given the information and had not been provided with the data necessary to respond properly.[25]

Baltimore recalls that the evidence he was shown of the changed dates in Imanishi-Kari's notebooks was "laughable," explaining that "for anybody to create a fraud by obviously changing a number, so you can see the old number and the new number, is not a fraud." He says that he did not know what to make of the indentation evidence that some of Imanishi-Kari's notebook pages had been written out of chronological order. "Knowing how chaotically she maintained her records, I was of the belief that it was very difficult to distinguish between messiness and conscious deception." He found the Secret Service's conclusions on the matter difficult to credit. Without the kind of written report that Singal had asked for, he had to rely on what he was told orally about a complicated forensic analysis. During his briefing, Baltimore declined to answer questions about the Secret Service presentation because under the circumstances it was difficult to follow. Shortly after the briefings, Bruce Singal told a reporter, "This has all the classic trappings of a witch hunt."[26]

The same day that Imanishi-Kari was briefed, the N.I.H. convened Joseph Davie, Hugh McDevitt, and Ursula Storb in Washington, probably in anticipation of the upcoming hearing, to go over the data reviewed during its investigation, particularly that for Figure 1. Several days earlier, the subcommittee staff had asked Davie to come to Washington, saying they had new data he ought to hear about. He arrived in the evening and spent until midnight with Chafin, among others, learning what the Secret Service had discovered. He said that the revelations were important enough that McDevitt and Storb should be told about them, too. McDevitt remembers that now, during the meeting at the N.I.H., the subcommittee staff invited the panel down to Capitol Hill to hear about the Secret Service results. The panel members changed their plane reservations, and at a briefing that began the next morning, Friday, McDevitt and Storb heard, while Davie heard again, about the findings, including the impression analysis, the repasted tapes, and the altered dates. McDevitt says, "Clearly, Peter Stockton and Stewart and Feder expected that we would walk in, see this evidence, and lie down and say, 'They're guilty.'" The panel stayed all day and into the early evening trying to comprehend the forensic analysis, questioning its details, and debating its import with the subcommittee staff. Storb says that the staff seemed genuinely eager to have the panel's opinion about the forensic data and to ensure that all their questions were answered. Stewart was notably outspoken in his conviction that the evidence showed fraud. McDevitt recalls, "I said, 'I don't see anything that convinces me that there's fraud here.' And they said, 'Well, but she wrote this over this.' And I said, 'So what?'"[27]

Davie, McDevitt, and Storb were informed that the coauthors had been given a look at the findings of the Secret Service. McDevitt says that they asked, "What kind of look?" and were told a half hour or so each. "We said, 'That's ridiculous. It's [taking] us a whole day to begin to make sense of it, with access to the Secret Service data.'" (Dingell later claimed that the subcommittee staff had spent ten hours with the coauthors; even so, since each was briefed separately, each got only a portion of that time.) Davie remembers the circumstances somewhat differently. He says he got the impression that the subcommittee was going to surprise the coauthors by springing the full Secret Service data on them at the hearing, in front of reporters. Whatever the variance in recollections, Davie and McDevitt agree that the panel was "outraged," to use Davie's word, by their perception of the subcommittee's tactics.[28]

The coauthors asked to meet with the panel the day before the hearing, and the panel was more than agreeable. "N.I.H. was leary of the idea," McDevitt says, "but we were adamant." The panel wanted to make sure that the coauthors would not be blindsided the next day by the Secret Service testimony. Davie, for one, also was eager to hear just how they would answer the implications of the agency's discoveries. Whatever McDevitt thought, Davie and Storb found them unsettling and wanted to see how Imanishi-Kari responded to them.[29]

At the meeting, Singal jumped in immediately to refute what he had heard the N.I.H. had been told—that Imanishi-Kari had no explanation for the Secret Service and was uncooperative. "She was dying to answer questions because she felt that what they were saying was of no merit whatsoever," he said. She had remained silent on his advice but that she would not remain silent now. Imanishi-Kari told the N.I.H. panel members that she often left the radiation-counter tapes in different parts of the laboratory—some times on table tops or window sills, sometimes in drawers or file folders—until she could collate, record, and analyze them. By the time she entered them in her notebooks, they had often faded to different shades. She said that she was not at all surprised that the Secret Service had discovered that her notebook pages had been prepared out of chronological order. "I did record old data after I had recorded much newer data. It all depended on how busy I was." She said that she worried about the substance of her experiments, not the dates on which they were performed, insisting, "I am not an accountant."[30]

McDevitt and Storb seemed to find Imanishi-Kari's explanations plausible. Davie allowed that he understood how she organized her data, but that the subcommittee preferred an alternative account of her notebooks and that "it's very difficult for you until you prove otherwise."

You mean the "accused person has to prove [herself] innocent?" she asked.

"That's right," Davie said.[31]

Baltimore explained to Imanishi-Kari that the Secret Service was saying that she had "made up something that never existed" and then changed the date to allow time for the experiment actually to have been done. Indeed, the word "fakery" had been used during Imanishi-Kari's briefing. Imanishi-Kari said angrily, "I understand all the inference. No, that's the inference that they want to make, and the only thing that I can say is that I know which experiments I did, I know exactly how these experiments were done, and I know how much fucking shit work I did for this. Right? And if I like to fake, I would do a much better job than this. There are easier ways to do this."[32]

The coauthors and the panel members exchanged notes about their respective briefings and concluded that they had not all been given the same information. The panel, in fact, appeared more knowledgeable than the coauthors about the Secret Service results. Normand Smith, Baltimore's lawyer, remarked that the discrepancy "gives us great pause for what we're going to hear tomorrow." Baltimore commented that, according to Stockton, the subcommittee had been under no obligation to tell them anything before the hearing, that they could have just "blindsided" the coauthors. Bruce Singal interjected that, so far as he knew, Stockton was right. "But what's happened here," he continued, "is that they're trying to convey the pretense of fairness by making it appear as though they've shown the information and given us a chance to respond, when in fact that is not really the case."[33]

Phillip Sharp, a friend of Baltimore's and the director of the M.I.T. cancer center, had suspected that Dingell intended the hearing to be a show trial. Sharp concluded that biomedical scientists had to play political hardball. "You can't fight City Hall in the ordinary way if City Hall is Congressman Dingell," he later said. In mid-April, he sent a "Dear Colleague" letter and a fact sheet to hundreds of scientists, warning that Dingell and his staff were bent on rejecting "the judgment of qualified scientists" and at the hearings would likely try again "to prove that misconduct occurred." He urged them to use the fact sheet as the basis for letters of protest to their own congressional representatives or op-ed pieces in their local newspapers. Paul Berg of Stanford, since 1980 a Nobel laureate in biology, too, wrote to a member of the Dingell subcommittee that its "staff has used innuendo and rumor to harass, embarrass and vilify Dr. David Baltimore, one of this nation's most accomplished and respected scientists," adding, "Anyone who values fairness and due process should be outraged by the direction and tone this investigation has taken."[34]

Like Bruce Singal, many scientists considered Dingell to be on a witch hunt that threatened not only Baltimore but also science itself. In an op-ed piece published in the *New York Times* two days before the hearing, Robert E. Pollack, now a professor of biology and the dean of the college at Columbia University, warned that Dingell was jeopardizing the very "process of scientific investigation itself." As Pollack understood Dingell's case, it was "that published science must be free of error, and that error itself indicates bad faith and fraudulent intent." But the case was "wrong," Pollack continued. "Published error is at the heart of any real science. We scientists love to do experiments that show our colleagues to be wrong and, if they are any good, they love to show us to be wrong in turn. . . . If we are asked to foreswear error, or worse, to say that error means fraud, then we cannot function as scientists." Lee Herzenberg wrote Baltimore a private letter praising the *Cell* paper for pioneering the use of transgenic mice in the area of immunology it addressed and attesting that the modifications in its conclusions produced by their own recent work represented nothing more than the normal scientific process at work.[35]

The day before the hearing, a seasoned Washington political consultant named Michael Berman provided Herman Eisen and several other faculty at M.I.T. guidelines for testifying before the Dingell subcommittee:

> In all of your exchanges with Members, it is important to treat them with the respect to which they are entitled, *even* if they don't seem to be treating you with the respect to which you are entitled. . . . Do not attempt to one-up a Member. It may make you feel good for a moment, but it works against you in the long run.[36]

Baltimore may or may not have seen the guidelines, but, if he did, he apparently paid little attention to them. He had once again turned to the law firm of Akin, Gump. He says, "Had I really asked them for advice, they would have said, 'Run and hide.' But I said, 'I'm not going to run and hide. We've got to stand up and face this thing.' They were happy, because they'd virtually never had a client who wanted to confront the situation." On Tuesday, May 2, two days before the hearing, Baltimore led a public relations blitz in the capital, addressing a meeting of more than one hundred science writers and talking separately with members of the *New York Times*, the *Wall Street Journal*, and the *Washington Post*. A *Post* writer observed, "By the time the hearing opened on Thursday morning, every reporter who cared knew the general outline of the case, the allegations and Baltimore's defense."[37]

Dingell opened the proceedings at 10 A.M. on May 4, more than a year after his first public inquiry. The hearing room was packed with journalists and scientists, including two who had won the Nobel Prize and two who would soon win it. Dingell set the tone for the occasion, announcing that he was "well aware of the attacks on the subcommittee's activities" and the suggestions "that we do not know the difference between fraud and error." He wished to "assure all that the jail houses have had a number of people who can testify that this subcommittee very well understands the difference between fraud and error." He declared that the proceedings would focus on "the ability and the will of major research institutions and the N.I.H. to police themselves when concerns are raised about potential misconduct," but he quickly turned to the discrepancies in Imanishi-Kari's notebooks and David Weaver's radiograph, insisting, "We must inquire whether they were altered with an effort to conceal or to confuse or to deceive."[38]

Agents Hargett, Herzog, and Larry Stewart testified to what their analyses had revealed about the notebooks and the radiograph. They explained their findings using huge displays of the seemingly cropped illustration and out-of-chronological-order notebook pages. They laid out their results in extensive detail, far more extensive than what they had provided the coauthors in their briefings. They had been apprised of Imanishi-Kari's explanations of the dating discrepancies from the statement she had submitted to the subcommittee in anticipation of her testimony. Congressman Ron Wyden asked Hargett whether, in his professional opinion, she could have written the pages in the chronological order she described. Hargett replied that "it would not be plausible." Wyden inquired whether in Hargett's professional opinion various dates had been written over "with an attempt to conceal." Hargett responded that "it was done with the attempt to conceal the initial date, yes."[39]

Hugh McDevitt, who was seated next to James Wyngaarden, recalls that Wyngaarden was sweating under the pressure. Wyngaarden assured Dingell that future N.I.H. panels, when faced with questionable data, would be provided with the assistance of someone "with forensic experience." He testified that the N.I.H. was concerned with the fate of whistleblowers, particularly with O'Toole, whose "scientific career has been damaged simply because she has pursued her convictions." Dingell observed that M.I.T., acting in accord with its policies, had not examined O'Toole's challenge more extensively because she had not alleged fraud. Then he asked, "Well, is that a good result—that to get a serious review you have to charge fraud, and anything else gets a lick and a promise?" Wyngaarden allowed that was not a good result, whereupon Dingell asked whether he agreed with the policy that led to it. Wyngaarden, for the moment, appeared to dig in his heels. "Scientific disputes—differences of interpretation are common. They're everyday occurrences in the laboratory. The institution cannot convene a formal inquiry for every disagreement between two scientists."[40]

The subcommittee's interest in the possibility of deception prompted questions about the *Cell* paper's presentation of its science. Bruce Chafin pressed Davie on the salient points—the use of "clones" to mean "wells," the reliability of Bet-1, the misstatement about the isotyping. Davie responded unflinchingly on most points but wavered on some. The frequency calculations in the paper constituted a "serious error," but the coauthors disagreed with the panel on the point and Imanishi-Kari herself had made "no attempt to hide" what her terminology meant. Bet-1 was a problem reagent, but "frankly, lots of reagents are problems in science, and we do not insist on descriptions of the idiosyncracies of all reagents." Besides, Bet-1 worked "many times," even though Margot O'Toole said that "it never worked"—or at least that it never worked in her experience, Davie corrected himself under prodding from Chafin.[41]

Chafin, referring to the isotyping, spoke of "experiments that were reported on but not done." Davie said such reporting could be called "misconduct," but pointed out that the panel had not so termed it in their report because they had not found evidence of "intent to deceive." He added a few minutes later, "We are really talking not about something that was made out of whole cloth. We're talking about experiments that were done, but not in precisely the same way that they're described in the text." Ursula Storb intervened to say that the isotyping had been done on other hybridomas with the results reported in the paper. "Therefore, the statement in the text is correct as far as the conclusions of the paper are concerned, and is supported by the evidence that we have inspected." Dingell engaged Davie on the faint band missing from Weaver's radiograph. Davie said that the coauthors might have acknowledged in the paper that the smudge could represent the transgene and he conceded that their presentation of the radiograph without the smudge was in some sense "misleading." But when Dingell pressed whether the radiograph was "a matter of some concern," Davie responded, "It would not be a surprise to anybody on the *Cell* paper or any of the panel, if transgene were expressed; the claim of the paper is not that transgene is never expressed" simultaneously with native genes.[42]

Davie admitted that Imanishi-Kari's way of keeping data was unusual, but Imanishi-Kari testified that although her notes might seem "messy," she knew "where they are and how to read them; and that's what's important." She accounted for the out-of-sequence ordering of her notebook pages, telling the subcommittee what she had explained to the N.I.H. panel the night before. She said that the Secret Service analyses made "no sense." They seemed to be charging "that I am not a neat person," which, she added, was true. She also pointed out that the pages whose dating the Secret Service called into question appeared to contain no data that had been published in the *Cell* paper—a surmise that later proved largely correct. She noted that changes in dates in her notebook were "clearly visible to the naked eye," adding, "I do not understand why you needed the Secret Service to tell you something that even a child could see was written over. . . . I am not . . . a sophisticated person. But if I wanted to trick somebody, I would have known better than to just change the date by writing over it in such an obvious way."[43]

Imanishi-Kari concluded by noting that her research might contribute to finding cures for autoimmune diseases such as lupus, continuing, "Mr. Chairman, I have lupus. My sister died from lupus. That was in my mind all the time I was doing my research. . . . If I had fabricated data, it would have misled scientists, wasted their precious resources and retarded their efforts to cure the disease that killed my sister and threatens me."[44]

In a statement submitted to the subcommittee, David Baltimore acknowledged the errors in the *Cell* paper and declared that O'Toole's raising questions about it in 1986 was "healthy and proper" but that "what was needed to decide the merits of her arguments were new experiments." Additional experiments had been done by scientists other than O'Toole on immunologically transgenic mice, Baltimore averred. Just before the hearing, he had received letters not only from Lee Herzenberg but also from a biologist named Erik Selsing, at Brandeis University, who had recently published the results of an experiment with transgenic mice similar to that reported in the *Cell* paper. Selsing said that his mice, which were somewhat different from Baltimore's, produced antibodies that derived from native genes but that displayed a variable region characteristic of the transgene. Although he doubted that his laboratory's work supported an interpretation of idiotypic mimicry, the peculiar features of the antibodies he detected were "clearly consistent" with those observed in the experiment by Baltimore, Imanishi-Kari, and Weaver. Baltimore entered Selsing's letter as well as Lee Herzenberg's in the record of the hearing. He also testified that key features of the *Cell* paper had been confirmed in other laboratories and that its claims had not been contradicted anywhere.[45]

Baltimore made no secret of his dissatisfaction with the subcommittee. He said that he understood its right to "demand accountability for government funding of scientific research," and he acknowledged that research institutions must "respond quickly to any allegations of irregularities in the scientific process." But he attacked the subcommittee's press leaks and its "prosecutorial style." He disparaged its recent apparent message for the treatment of scientific notebooks: "Never overwrite a date or add a page or the Secret Service will catch you." He conceded that with the heightened sensitivity to scientific fraud, it was "more crucial than ever" that scientists maintain records whose substance could be easily assessed; yet he opted for finding ways "to encourage individuality and not to straitjacket science with a preconceived notion of how notebooks should be kept."[46]

Baltimore insisted that the Secret Service's briefings a week before amounted to "a charade of helpfulness," since its agents had presented a good deal more information at the day's hearing than it had earlier provided the coauthors. The briefings were designed more "to terrify" than to inform. He contended that a congressional hearing was an inappropriate forum for assessing the merits of a scientific paper. (Congressman Dingell unwittingly corroborated the point in a question indicating that he thought the specificity of Bet-1 was "a major thrust of the article in *Cell*," to which Baltimore responded, "No, not at all. . . . The paper . . . was in no sense a discussion of the Bet-1 reagent; we merely used it.") Baltimore denounced the subcommittee's substitution of Walter Stewart's "pernicious" data audits for the assessments of qualified immunologists. He called Stewart, who was seated with the subcommittee staff and looking busy with its affairs, a man of "significant analytic skills but poor judgment" and censured him for "the loathsome comparison of scientific fraud, of which he accuses me, to the Nazi Holocaust."[47]

Around 7 P.M., after nine hours of testimony, Dingell, summing up, chided the *Cell* paper's principal coauthors for not cooperating and, in particular, for refusing to respond to the findings of the Secret Service prior to the hearings. He chastised Baltimore for his "rather ringing attack upon this committee," pointing to his claim that he had been charged with fraud—a claim that Dingell said was "untrue"—and to his "allegation" that "some of the persons involved in these matters were behaving in a fashion worthy of Hitler."[48]

Subcommittee watchers say that Dingell arranges to have the last word with closing statements that are intended to be unanswerable and unanswered. However, as Dingell raised his gavel to end the proceedings, Baltimore asked for a chance to respond. "I *was* charged with fraud," he said, passing up to Dingell a copy of the *Boston Globe* story with its headline on the 1988 hearing and its quotation from Stockton. He picked up an article from *Science* that described Stewart's behavior and read it aloud, his voice low and shaking with fury. Stewart's comparison of scientific fraud to the "Holocaust" and his likening of scientists who looked the other way to good Germans were, the journal noted, "not the best analogy to use before an audience of scientists, where more than a few are Jewish." As for the Secret Service briefing, Baltimore said that he was given only two pieces of paper about the agency's analysis, and not at the briefing itself but three days later. He would "not respond in a situation in which I've been presented with material I have never seen before and asked to analyze it on the spot." He gave Dingell a copy of the letter he had sent the previous August, in which he had offered to come to Washington to discuss the congressman's concern, and pointedly noted that he had received no response. Dingell soon closed the hearing and hurriedly left, brushing aside questions from reporters.[49]

Daniel Joseph, one of Baltimore's Akin, Gump lawyers and long practiced in the Dingell bar, told him that he'd never seen such a performance and that he was sure that people all over the hearing room had been cheering to themselves. As the hearing broke up, many of them hugged Baltimore and Imanishi-Kari. Baltimore went off through the corridors to a press conference, trailed by a crowd of reporters. On the way, the entourage bumped into Dingell. One of Baltimore's associates remembers that the congressman was "very red in the face . . . clearly angry."[50]

After Baltimore's blast on Thursday, the Dingell subcommittee asked Margot O'Toole to testify again at the hearing they had scheduled for several days later, on May 9. O'Toole says that she had no babysitter over the weekend but managed to draft a statement for the subcommittee while holding a child in her lap. She appeared the following Tuesday as the lead-off witness, opening her testimony by praising the subcommittee for its intervention in the dispute.[51]

O'Toole reviewed her dissent from the *Cell* paper, covering much of the ground that she had in the 1988 hearings but departing from her initial version in key respects. In 1988, she had stressed that her quarrel turned on her belief that the paper contained errors that warranted correction and that its conclusions were not supported by the data. Now she insisted that the errors in the paper were material and "made it wrong," that the paper made "false statements" and "false claims." Now she flatly declared, "I challenged the paper because it presented evidence that simply did not exist, period. . . . This is the crux of my dispute with the authors." She contested David Baltimore's testimony, holding that "the part [of the paper] I challenged has not been confirmed." More important, improving on the charge she had leveled against the draft report of the N.I.H. panel, she testified that she had "asked" Imanishi-Kari explicitly "if these subcloning experiments had been done" and that Imanishi-Kari had said they had not been done.[52]

Dingell questioned O'Toole about how she had fared as a result of her challenge. "I don't think I can advise any young scientist to come forward," she said. She had received "absolutely no support" from the either M.I.T. or Tufts. Because the matter involved her mentors, "I was left without a job." Dingell pointedly observed that Dr. O'Toole's career is "in shambles and a number of other people are prospering mightily."[53]

Congressman Dingell soon turned to the M.I.T. and Tufts faculty and administrators who had come to testify, hammering at the two institutions for their handling of O'Toole's complaint and questioning the merits of the conclusions reached in the inquiries conducted by Herman Eisen and Henry Wortis. John Deutch, the provost at M.I.T., insisted that his institution did not require a formal allegation of misconduct to look into a scientific dispute—its procedure did not "make the hurdle to raise a question as high as the Wall of China"—and that O'Toole's criticisms had, in fact, been given a hearing. Dingell asked Louis Lasagna, the dean of the Graduate School of Biomedical Sciences at Tufts, why it had been necessary to call in the N.I.H. panel and the Secret Service to get to the bottom of some of the issues O'Toole had raised with the Wortis committee. Lasagna answered that the necessity had arisen "in response to the pressures from a number of quarters, including Congress," adding that he was unaware that any of the additional reviews had modified the original judgment about the thrust of the Cell paper.[54]

Brigitte Huber rejected suggestions that the Wortis committee had been unable to evaluate O'Toole's criticisms objectively because Imanishi-Kari had been about to join the Tufts faculty: "Why should we have wanted to recruit as a colleague someone who commits fraud and whose science is not trustworthy? It was in our interest as much if not more than anyone's to investigate the matter thoroughly and assure ourselves of the integrity of Dr. Imanishi-Kari's work." Henry Wortis said that he knew of no data "that conflicts with" the central claim of the paper. Although he said at one point that the claim itself had not yet been independently tested by "any other lab" that he knew about, he indicated a few moments later that it had been confirmed by two unpublished experiments at Tufts. He also suggested that quite possibly most scientists had stayed away from the research topic "for fear that they might wind up spending 3 years enmeshed in trying to defend themselves."[55]

Dingell pressed Huber and Wortis on O'Toole's charge that crucial experiments had not been done at the time the Cell paper was published. Huber insisted that she and Wortis and Woodland had seen the subcloning data for Table 2 in May 1986. Dingell probed further, asking whether she knew then that the subcloning had actually been performed. She replied that she did not know when it had been done but, taking him to mean that perhaps Imanishi-Kari had done the subcloning in the week between the first and second meetings at Tufts, she added, "I can say, sir, that these data could not have been produced in a few days. . . . It's technically not possible." Congressman Doug Walgren, a Pennsylvania Democrat, wanted to know about data reported in the paper that had been alleged not to exist. Wortis explained that it was wrong to "infer that there is no data at all" because the isotyping had not been done for the Table 2 hybridomas. Such data did exist for Table 3 and in relation to Table 2 scientifically, and it was "sufficient," he testified. "We don't have to go around in circles saying yes, but where is the data for Table 2?"[56]

A few minutes later Dingell nevertheless took Wortis around the circle once again. Wortis had already told Dingell that their differences over whether the published data was supported by the laboratory data seemed to express "the difference between a lawyer and a scientist."[57] Dingell, as though eager to sharpen the impression of the alleged difference, asked why his Tufts committee had not challenged the coauthors' "conclusions" when the N.I.H. panel had successfully demanded that they publish "corrections." "That's not the same as challenging conclusions," Wortis responded.

Dingell asked, "What is the point of having a correction or suggesting a correction, which in fact is made, if there's no error?"

Wortis replied, "If there is a belief here that science, in order to be correct, has to be absolutely free of error, then we're stuck."

Dingell interrupted to declare that he was "not so foolish as to say that it has to be absolutely correct." He thought it had to be "procedurally correct," and "factual." Was not science "essentially a search for truth?"

Yes, it was, Wortis replied, noting that "the conclusions of this paper have been so far found to be true" and that what the N.I.H. panel had requested were "clarifications."

"We get clarifications out of the White House all the time," Dingell remarked. "We got them during Watergate."[58]

Dingell's fellow subcommittee Democrats, notably Wyden and Walgren, manifestly shared his suspicions, but Wortis, Huber, Eisen, Baltimore, and the N.I.H. panel were treated gently by the subcommittee's Republicans. Representative Thomas Bliley declared that the evidence showed "only the complexity and uncertainty of pure science and the difficulty of imposing external judgments on scientific work." Congressman Norman Lent, the ranking Republican on Energy and Commerce, judged that Margot O'Toole had gotten "the attention and consideration of almost a score of eminent scientists." They "have reviewed her complaints and have found them wanting." Lent rebuked O'Toole for copying Imanishi-Kari's laboratory notes and turning them over "to a man who was not qualified, who was not an immunologist, in the hopes that he would somehow bring this whole process into review." He said that if someone in his office did something similar, the person would "be out of there in a flash" and "wouldn't be rehired by anybody that I could call up."[59]

Many observers called the hearing, especially its dramatically confrontational last half-hour on May 4, a triumph for Baltimore and an embarrassment for Dingell. Martin Flax, the chairman of Imanishi-Kari's department at the Tufts medical center, later told a reporter that "Baltimore did brilliantly," adding, "That was Civics 101 for me." O'Toole's introduction of the charge that the subcloning data had not existed in 1986 prompted many scientists, including Herman Eisen, to describe her as a moving target. In a conversation with a reporter, Brigitte Huber tried to explain O'Toole's subcloning claim: "I'm sure that's what she believes. I'm sure she doesn't think she's lying. She must have a different memory."[60]

Along with a number of critics, the biologist Steven Jay Gould faulted Dingell for seeming to fail in the end to understand the difference between error and fraud. Error was in fact a goad to scientific progress, Gould argued in the vein of Baltimore, pointing out that the economist Vilfredo Pareto had once proclaimed: "Give me a fruitful error any time, full of seeds, bursting with its own corrections. You can keep your sterile truth for yourself." The editors of the Washington Post sympathized with Dingell's drive to reform the institutional response to allegations of scientific misconduct, but they warned that his ability to bring change about would "necessarily be impaired" if he appeared "to scientists to be persecuting Dr. Baltimore for no purpose." The Wall Street Journal denounced Dingell for trying to establish a "science police," and the Detroit News, his hometown paper, accused him of mounting another "Galileo Trial."[61]

Whatever his critics in the press and among Republicans might think, however, Dingell controlled his subcommittee and its agenda. He and his staff had long since mastered advanced civics, acquired allies in the press, and become adept at using whatever confidential information it obtained. The subcommittee possessed the correspondence it had demanded from Baltimore the previous summer. Before the hearing, it had evidently leaked the angry letter he had written to Eisen in September 1986 when he momentarily thought Imanishi-Kari had misled him about the reliability of Bet-1. It published the letter in the record of the hearing together with other correspondence that did not appear to do Baltimore credit. The exchanges included his response in the fall of 1987 to the Herzenbergs' suggestion that they invite Margot O'Toole to Stanford for collaborative work; it was reported to reveal "how Mr. Baltimore effectively hindered Ms. O'Toole from 'returning to the laboratory.' "[62]

During a break in the hearing on May 9, O'Toole, in the company of Walter Stewart, had made explicit to reporters what she had implied in her testimony: Imanishi-Kari had produced certain data—presumably O'Toole meant the subcloning data—after the Cell paper had been published. O'Toole said she was sure forensic analysis would demonstrate that she was telling the truth. Stewart added that the Secret Service would test the radiation-counter tapes to see whether they could have been generated before April 1986. At least one reporter asked why this possible charge was bruited to reporters during a break instead of at the hearing. "We haven't brought it out because it's not complete," Stewart said.[63]

Immediately following the hearing, Dingell had extended Stewart and Feder's assignment to the subcommittee. In March, Bruce Singal had certified to Dingell that, even though Imanishi-Kari often entered her data into her notebooks long after she did an experiment, the records of original data—particularly, the radiation-counter tapes—had been generated at the time the experiment was performed. After the hearing, O'Toole urged Dingell to have the tapes dated forensically. In July, Dingell instructed the Secret Service to try to determine whether the dates on the tapes were authentic—that is, whether what Singal said was true. He also told Imanishi-Kari that his staff and several people from the N.I.H. wished to interview her at the end of the month.[64]

Daniel S. Greenberg indicted Baltimore in a newspaper column for achieving his "victory . . . by bare-knuckle, often disingenuous public-relations tactics." Greenberg's *Science and Government Report* declared in its issue for mid-May that the scientific establishment had responded to O'Toole's allegations of error "disgracefully" and in mid-June, SGR, to use the newsletter's acronym, published an interview with John Dingell. Dingell sloughed off the newspaper barrage, telling SGR that the *Detroit News* denounced him several times a month (he might have added that the *Wall Street Journal* regularly attacked him, too).

SGR: What was your assessment of Baltimore's testimony?
DINGELL: Basically an *ad hominen* attack on the Committee.
SGR: Do you think he was responsive to the substance of the inquiry?
DINGELL: No.[65]

By now, David Baltimore's conduct in the dispute had become equally an issue with the veracity of Thereza Imanishi-Kari's research, and not only for Congressman Dingell and his allies in the press but also for some influential scientists. A number of them had participated in the meeting on scientific misconduct that had been held at the Banbury Center in Cold Spring Harbor, New York, in January 1989, where Walter Stewart analogized scientific fraud to the Holocaust. Bruce Chafin and Michael F. Barrett, Jr., the staff director of the Dingell subcommittee, also participated in the discussion, feeling a bit, Chafin recalls, like "the Christians in with the lions." While Stewart discredited himself with the assembled biologists, Chafin and Barrett impressed some of them by the seriousness of their concern with scientific fraud as an issue of public policy and by their resolute eagerness to master the points of dispute in the *Cell* paper.[66]

Walter Gilbert, a professor of biology at Harvard and a Nobel laureate, had been at the Banbury meeting. He found Barrett and Chafin straightforward in their concerns and told a reporter afterward that scientific misconduct was "a small but very real problem." A son-in-law of the liberal investigative journalist I. F. Stone, Gilbert sympathized with the Dingell subcommittee's stated aim of rooting out corruption in science and thought Baltimore wrong in charging that Dingell's probe amounted to an attack on him personally and on American science. Immunology was one of the fields in which Gilbert had worked. He says that he had been unimpressed by Stewart and Feder's analysis of the *Cell* paper but that O'Toole's testimony in 1988 struck him as "sober" when he read it. Although he did not attend the subcommittee hearings in 1989, he heard a lot about it from friends who did and the forensic evidence presented by the Secret Service caught his attention. He says that Imanishi-Kari's testimony raised his suspicions, particularly her claim that she would never commit fraud because her sister had died of the autoimmune disease lupus. Gilbert calls Imanishi-Kari's claim the "typical defense of somebody who is so emotionally caught up in things that they can very easily commit fraud." In October, he told a reporter for the *Detroit News* that the forensic results implied that "those experiments weren't done" when Imanishi-Kari claimed they had been done—"or they weren't done at all."[67]

In September 1989, it became known that the trustees of Rockefeller University in New York City ranked Baltimore high on their list of candidates for the presidency of the institution. More than a third of the senior faculty, including a Nobel laureate named Gerald M. Edelman, openly declared their unhappiness with the choice. The dissidents were upset partly because they had not been fully consulted by the trustees, but what distressed them even more, they said, was the publicly confrontational way that Baltimore had handled the dispute. One member of the professorial opposition—he was Anthony Cerami, a medical biochemist and dean of graduate and postgraduate studies—told a *New York Times* reporter that Baltimore "actually helped precipitate the thing and make it worse for himself and for science in general." Another, the theoretical physicist E. G. D. Cohen, wondered how it would look for Rockefeller if Dingell or the N.I.H. found concrete reason to blame Baltimore, adding, "What would be the public image of us and how would we raise money?"[68]

In the face of the opposition, Baltimore withdrew from consideration, but David Rockefeller, the head of the executive committee of the university's board of trustees, and Richard Furlaud, the chairman of the board, flew to Boston in the hope of persuading him to change his mind. Rockefeller University was organized as an agglomeration of scientific baronies, each under the control of a senior professor—a system that provided few if any junior faculty with their own laboratories or autonomy in research. The university was also running an annual deficit of some $12 million. Rockefeller and Furlaud were convinced that the institution had to be reformed and that David Baltimore possessed the rare combination of scientific authority and administrative talent to accomplish the task. Baltimore says that he conceded to a reinstatement of his candidacy, noting, "David Rockefeller is a very persuasive man."[69]

At the end of September, the Rockefeller faculty was informed that Baltimore had been offered the presidency. David Rockefeller said that the trustees had looked into the controversy "in considerable depth" and had found nothing that "could possibly reflect on the integrity or honesty or honorability of David Baltimore." On October 17, 1989, Baltimore accepted the job. That day, in an address to the faculty, he remarked that in retrospect he might have dealt with certain events differently, declaring, however, "I believe that by confronting Chairman Dingell directly, I was acting in the interests of all scientists. Only time will tell."[70]

Chairman Dingell, on his part, was determined to pursue O'Toole's charges, no matter what Baltimore might have said in defense of the *Cell* paper or what members of the Tufts, MIT, and NIH panels might have testified about the adequacy of their respective investigations. Apparently furious at Baltimore, he told a leading Washington scientist, "I'm going to get that son of a bitch. I'm going to get him and string him up high." Fury aside, the subcommittee's way of doing business virtually demanded the pursuit. Dingell and his staff said that they only wished to ensure that research institutions responded promptly and fairly to allegations of scientific misconduct. But to show that the handling of fraud and misconduct in university science was unsatisfactory, they had to demonstrate that the inquiries at Tufts and M.I.T. into O'Toole's challenges and also the N.I.H. investigation had been inadequate—to prove, in short, that something was seriously amiss with the *Cell* paper and that O'Toole was right. Some time in the months after the hearing, an official at the N.I.H. went to Capitol Hill to talk over the *Cell* paper dispute with Stockton, Chafin, and Stewart. "I felt that if they could have dropped an atom bomb on MIT . . . they would have done it," he recalls. "They were on a mission. They didn't have any perspective. They wanted to get somebody, and they wanted to make a big hit."[71]

NINE

■

Fraud Story

·

THE CONGRESSIONAL cricitism that began in April 1988 had prompted the N.I.H. to initiate reforms of the way it handled allegations of scientific misconduct. In the fall of 1988, the agency announced that it was contemplating the creation of an office of scientific integrity and invited comment on the scope and definition of scientific misbehavior.[1] Contemplation turned into action in the face of moves on Capitol Hill that winter, including a draft bill from John Dingell's subcommittee, to legislate the creation of such an office with powers that would include random audits of the notebooks of N.I.H. grantees. James Wyngaarden worried that congressional action would give control of misconduct cases to an inspector general, a lawyer who knew nothing about science. "We needed a preemptive strike," he said later. On March 8, 1989, the N.I.H. created an Office of Scientific Integrity (O.S.I.) on its own. The O.S.I. would monitor inquiries at its grantee institutions, conduct its own investigations when necessary, and be run by scientists.[2]

Wyngaarden picked Brian Kimes to head the office temporarily, until a regular director could be found. Kimes is a forthright, level-headed, boyishly handsome man, a biochemist by training who loves science and the N.I.H., where he had been a research administrator in the National Cancer Institute for almost fifteen years. He took the job of setting up the O.S.I. reluctantly, telling Wyngaarden that he would stay only until November 1, that "it was not my ambition to send scientists up the river."

What could send scientists up the river was spelled out in the regulations that the Public Health Service, the parent agency of the N.I.H., issued in midsummer. The definition of scientific misconduct included "fabrication, falsification, plagiarism." It also included "other practices that seriously deviate from those that are commonly accepted within the scientific community"—a feature that a number of scientists had objected to, no doubt fearing that it raised "serious risks of undue pressure for scientific conformity," as an official in the Office of Management and Budget had put it when the proposition first came up. Many scientists were no doubt relieved, however, by the Public Health Service's stipulation that misconduct "docs not include honest error or honest differences in interpretations or judgments of data."[3]

Although the O.S.I. was reportedly authorized to have eight full-time staff members, a doubling over the complement of its predecessor, at the beginning its professional crew was tiny, comprising just Kimes, a clinical psychologist named Suzanne Hadley, who was his deputy director, and another on-loan administrator from elsewhere in the N.I.H. Hadley had come to the National Institute of Mental Health in 1977 to continue her research on the effects of psychotherapy but had quickly moved into research policy and management. She is disarmingly soft in manner, energetic, tenacious, and highly articulate. She had acquired increasing administrative responsibilities at the mental health institute, including duties as its misconduct officer. Her first case was Stephen Bruening and his faked data on drug therapies for mentally retarded children. She readily accepted the deputy directorship of the O.S.I., hoping to land the top job after Kimes left in November. She says that she considered the oversight of scientific misconduct to be vitally important, holding that it should be done with close attention to the data and with humane consideration for the accused scientists whose lives were on the line.[4]

The O.S.I. was burdened with a backlog of eighty to one hundred cases. "We were over our heads," Kimes says. "We had to figure out where the political issues were and where the scientific issues were." But they knew that one case dominated all the rest. Congressman Dingell had told Wyngaarden that he regarded the matter of Imanishi-Kari as a "crucial test" of the ability of the N.I.H. "to deal with cases of questioned science." N.I.H. officials were concerned that the agency not appear to be remiss. "We were taking a lot of hits from Dingell on this with the public," Kimes notes. "Wyngaarden was extremely worried about the politics of it because those are things that could impact all of N.I.H. It was a very uncomfortable situation."[5]

In the spring, Wyngaarden reappointed Davie, McDevitt, and Storb to the advisory scientific panel. The N.I.H. asked Dingell's subcommittee for copies of all its Secret Service reports and for a meeting with Walter Stewart about his analysis of Imanishi-Kari's notebooks, particularly a statistical analysis of her data that he was said to have done. Kimes requested that Imanishi-Kari index her notebooks to facilitate the O.S.I's use of them for the forensics-oriented data audit that it was pursuing.[6]

However, Margot O'Toole, having received from the N.I.H. a copy of the data concerning Figure 1, now thought that something more was wrong with it than just the missing support for the fourth point on each of the curves. The figure graphically displayed, among other things, the specificity of Bet-1, the claim that she had disputed from the beginning. In June she wrote to Kimes "that the experimental findings shown in Figure 1 of the *Cell* paper have been falsified." In accord with what she had told the N.I.H. panel about Imanishi-Kari's having not done the June subcloning, she parenthetically ventured the belief that the subcloning data was also fabricated.[7]

On August 10, 1989, Kimes informed Baltimore and Imanishi-Kari that the case was progressing slowly, partly because the Dingell subcommittee had not yet supplied the materials requested from it, partly because Imanishi-Kari had not provided the index asked of her. (The O.S.I. was "extremely disappointed" by her unwillingness to index the notebooks, Kimes wrote.) More important, the investigation would be taking up "new allegations," including the possibility that some of the subcloning data were fraudulent. The case had "become much more complex than originally anticipated." Baltimore rightly surmised that the matter had turned into "much more than a data audit."[8]

To deal with the enlarged allegations, the O.S.I. added two more biologists to the panel—Professor William McClure, of Carnegie Mellon University, in Pittsburgh, a biochemist with expertise in molecular biology, and Professor Stewart Sell, an immunologist at the University of Texas Medical School at Houston. It also enlisted three N.I.H. scientists as internal consultants on the case, each an expert in an area relevant to it. Once the panel was assembled, Kimes and Hadley arranged a conference call with the members. Hardly thirty seconds into it, a participant recalled, one of the panelists exploded that the whole affair was driven by John Dingell and that the O.S.I. were Gestapo.[9]

Bruce Singal, being familiar with the obligations of public agencies to those whom they investigated, told Dingell at the end of July that Imanishi-Kari declined to be interviewed by his staff until she was provided with details of the Secret Service's analyses that he had been asking for, unsuccessfully, since April. He expressed suspicion of the "close collaboration" between the subcommittee and the O.S.I. and doubts about whether the "two overlapping investigations" would be "conducted fairly and in full observance of my client's rights." He protested to Kimes that Imanishi-Kari could not cooperate with the O.S.I. investigation unless she was given not only the forensic evidence but also specific information about the new allegations and their source. Imanishi-Kari, on her part, says that she was reluctant to index her notebooks because supplying additional information might lead to more unwarranted fraud charges, just as the supplemental material she had earlier given to the N.I.H.—notably the June subcloning data—had been turned into a basis for allegations against her.[10]

The O.S.I. had not issued a set of rules and procedures.[11] Kimes says that the staff were too busy that they had to make them up as they went along. Kimes himself wanted the O.S.I. investigations to be open, with everyone having access to all the data and testimony. In August, he assured Baltimore and Imanishi-Kari that once the allegations were completely formulated, they would be made available to "all subjects of the investigation," together with all related evidence. He told Singal that Imanishi-Kari would have "every opportunity to see and comment on all information supplied to the N.I.H. panel of experts" and that she could be confident that the O.S.I.'s inquiry would be "conducted independently" of the subcommittee's. Kimes had given O'Toole to understand, as she noted in a letter to him, that after the N.I.H. had gathered and evaluated the evidence, "all relevant documents will be made available to all involved." No doubt because she had felt blindsided by the unexpected appearance of the June subcloning data in the N.I.H. investigation, O'Toole said that she "strongly advocated that relevant evidence not be kept secret."[12]

In June, O'Toole had told Kimes she was sure she could be of assistance in the reopened investigation, "especially if you let me see the data." Towards the end of the summer, Kimes enlisted her to index the notebooks, since Imanishi-Kari would not, and proposed that she draw up the allegations. Kimes says that they approached O'Toole "as if [she] was telling the truth," but "we did not necessarily believe her," he adds. "Nobody" thought of O'Toole as "an objective evaluator," Kimes recalls. "At the time, she felt she had been messed around with so much there was very little way that you could get a totally rational discussion out of her." He found her aggressively proactive in the investigation, volunteering so often to write him letters on points at issue that he came to discourage her sending them. He nevertheless thought that she would understand the data and that the O.S.I. would not have to be "relying on her comments." Kimes notes that the case "was so political that to try to exclude her from everything would have made us look like bad guys."[13]

By mid-September, O'Toole had been provided with a copy of the notebooks. She told Kimes that she believed Imanishi-Kari had fabricated even more parts of the laboratory record than she had earlier claimed—not only those in support of Figure 1 and the June subcloning but also those undergirding key claims made by Table 3. She said that she wanted the N.I.H. to guarantee her legal representation should she be sued as a result of her cooperation with the agency. In a telephone conversation on October 6, Kimes told O'Toole that the O.S.I. could make her a temporary government employee, which would supply her with both income and legal protection for her services. She would, however, have to work under the same rules as all the staff in the office, which meant that she could not communicate with the Dingell subcommittee on her own. O'Toole apparently did not take an appointment at the O.S.I. and she eventually declined to accept payment for her work, thinking it might be embarrassing, but she did continue to index Imanishi-Kari's notebooks and otherwise assist the office in its investigation. In a letter on October 13, Kimes formally asked O'Toole to draw up a written list of allegations accompanied by an analysis in support of each one. He said that they would be "presented to the coauthors of the *Cell* paper," and that she as well as they would receive copies of the Secret Service reports.[14]

Ten days later, however, O'Toole complained to Kimes that he had assigned her a role that is "totally inappropriate and untenable." She later told a reporter that Kimes's letter was "ludicrous," explaining, "I had agreed to help the panel, and I wanted to see all the evidence. But that was only so that I could document my allegations to N.I.H., not do their work for them." She had also become disturbed by the prospect that her allegations would be given to the coauthors for their response and that the entire package would then to go the advisory panel without her having a chance at rebuttal.[15]

Dingell, made aware of Kimes's letter by his staff, was reportedly very upset by the scenario it proposed for O'Toole. Stockton, Chafin, and Walter Stewart had no more confidence in the O.S.I. than it had in the first N.I.H. investigation. The subcommittee had sent Kimes most of the materials it had promised, but Stewart had still not provided a report of his analysis of Imanishi-Kari's notebooks. Some time in late October, Kimes met with the staff about O'Toole's dissatisfaction and was attacked for his plan to present the allegations to Imanishi-Kari so that she could then have a chance to explain matters. Stewart was especially insistent that the

O.S.I. get to the bottom of the case on its own. Kimes recalls that Stewart "was absolutely obsessed with the case. . . . If you sat across the table from Stewart, his first interpretation was, Where have they cheated?" To Kimes, it was evident that Congressman Dingell "did not want the process to be open; he wanted us to investigate"—meaning, investigate in a way that precluded suspects from covering their tracks.[16]

Kimes discovered that Suzanne Hadley preferred a closed investigation and that so did William Raub, who had taken over the N.I.H. as acting director following Wyngaarden's retirement in July. Kimes says that Hadley, having been a misconduct officer at the agency, had "much more of an investigative mentality" than he did. "She was extremely smart, and she would dig into a case like a detective. I felt that she was obsessed a little bit with it [misconduct] and she was not keeping her balance and perspective at the time." Raub had plenty of balance and perspective on N.I.H. business. He had been with the agency almost thirty years, about half of them spent in administration. He had been deputy director for extramural affairs, the branch of the N.I.H. that gives grants to scientists like Imanishi-Kari, when Stewart and Feder had asked for permission to submit their critique of the *Cell* paper for publication. He says that, like Ed Rall, he held that as members of the scientific community they "had a right and even an obligation to speak out on matters concerning the ethics and mores of the scientific profession." When the O.S.I. was established, Wyngaarden had given him the job of turning it into a working reality. He knew that it would have to satisfy Congressman Dingell, who, Raub says, had wanted the new office established outside of the N.I.H., thinking that if it was inside it would continue the agency's dubious practice of investigating itself. Raub says that the N.I.H. had to show it was "tough enough" to handle its own misconduct problems. That outlook made him leery of risking any compromise of investigations by sharing substantive matters with suspects, a predilection reinforced by the advice of legal counsel that restricted approaches were common in government inquiries.[17]

Towards the end of October, the O.S.I. revised its plans for handling O'Toole's allegations. They would be presented to the O.S.I. and its scientific panel, not to the coauthors. Both would evaluate the evidence that O'Toole provided and then conduct an investigation, talking with the other principals in the dispute. O'Toole understood that during the course of the investigation, as she wrote to Kimes, she would "have access to any information provided by the Secret Service, other technical or forensic experts, and other witnesses."[18] Coupled with Hadley's eagerness for a closed investigation, the new dispensation meant that Imanishi-Kari would be given only indirect access to the allegations and no access at all to the material evidence against her.

All the while, O'Toole had kept working on her allegations. Kimes emphasized to her the need to make her charges strong and precise. On November 6, Margot O'Toole wrote to the O.S.I. that the notebooks "do now contain records which purport to substantiate the claims I challenge, but these records are totally fabricated in some instances and significantly falsified in others." She "therefore made a charge of fraud," submitting eighteen allegations that covered the data in support of Figure 1, Table 2, and Table 3 and summarily concluding that the notebooks contained "no actual experimental findings" in support of the *Cell* paper's central claim.[19]

O'Toole's lengthy conversations with Stewart and her sense of having been "messed around with" had formed the background of her move from accusations of error to allegations of fraud. She believed that her efforts to get the *Cell* paper corrected had earned her only alienation and injury. At the hearing held by Dingell's subcommittee in 1988, she had explained that the dispute had led her to break with "many people I have admired . . . and some who were my friends" and that it had "halted my career, disrupted my social milieu and had a devastating effect on my life." Henry Wortis says that even now O'Toole never comes to social events in the Tufts pathology department, where her husband works.[20] Under ordinary circumstances, the department would be a sustaining part of her professional family, and Wortis, as her doctoral advisor, would be a professional paterfamilias. But the bitter circumstances of the controversy cut her off from what would have been natural scientific relationships and very likely put the dispute at the emotional center of her life.

O'Toole's resentments no longer concerned simply the coauthors' original refusal to correct the misstatements in their paper, or her belief that she had been professionally betrayed by the overstated claims for Bet-1, or what she took to be the animadversions privately cast against her by Baltimore and Eisen in response to Stewart's badgering. They had been substantially expanded by developments since the first Dingell hearing in April 1988, especially by the outcome of the N.I.H panel's inquiry, the testimony of the professors at Tufts and M.I.T. during the second hearings in May 1989, and the documents that the subcommittee had obtained from the principals and released. One of the documents was her letter to Eisen in 1987 asking him to "correct" his report of December 1986 that described her challenge as "allegations of misconduct." O'Toole's irritation that he had failed to respond was compounded by seeing his marginal comments dismissing her request as "imperious." She held that such private attacks against her had now become very public ones, asserting to Kimes that Baltimore and Eisen have been "waging a campaign that has destroyed me professionally."[21]

O'Toole declared repeatedly that her reputation had been broadly "damaged."[22] She testified in 1989 that "my competence and motives have been attacked by scientists from all over the world." She felt that Baltimore had publicly defamed her in his "Dear Colleague" letter of May 1988 by commending her for advancing insightful alternative interpretations of the *Cell* paper's data. She told the subcommittee that she well understood that scientists differ over interpretations and that it had "hurt me greatly" to have had it promulgated to the entire scientific community that she did "not understand this most basic tenet of scientific endeavor." O'Toole said she found it "devastating" that the dean of the Tufts medical school, drawing on a "false report" from the Wortis committee, had suggested by allusion in a magazine interview that she had made a charge of fraud without justification. She queried the subcommittee rhetorically: What laboratory would hire a scientist who had been portrayed as she had been?[23]

At the end of the hearing on May 9, Eisen and the M.I.T. provost, John Deutch, a respected figure among science policymakers in Washington, including Dingell, had learned in a conversation with the chairman that he seemed to be truly distressed by the injuries O'Toole had seemingly suffered. Deutch and Eisen promised that they would attempt to find her a position so that she could resume her scientific career. Deutch soon invited O'Toole to meet with him. To Deutch's surprise, when she arrived at his office on the afternoon of June 5, she was accompanied by her attorney, William P. Homans, Jr. In a memorandum of the encounter that he wrote the next day, Homans said that he had come along because he sensed her expectations of what the meeting might accomplish went beyond obtaining assistance in finding employment. In fact, she indicated to Deutch that her reputation had been seriously damaged, not least by how the controversy had been publicly discussed, and that she wanted the help of M.I.T. in restoring it.[24]

In a letter to Brian Kimes a few days later, O'Toole discussed at least part of what she meant by "help." She wanted to check whether the cancer center housed any gamma-radiation-counter printers that had the same typeface as the one used to produce the June subcloning tapes and that Reis or Imanishi-Kari "could falsely claim to have used." She asked Deutch for permission to take tape samples from any such counters in the building. She wrote, "Not only did he refuse to allow me to do this, he refused to have it done by anyone else. I asked Dr. Deutch to give me assurances that M.I.T. would co-operate fully with the N.I.H. investigation without requiring subpoenas and thus turning the issue away from the facts and into a political question. Dr. Deutch said that he would give me no such assurances because the matter was between him and the N.I.H." Deutch remembers that while he offered to help O'Toole find a position, he told her that he could not otherwise assist her. Some years later, he recalled that O'Toole "made it clear (and was rather emotional about it) that she wanted to pursue the case." She did not approach Eisen at all.[25]

O'Toole interpreted the N.I.H. report as a direct slap, not least because it observed that her own notebooks contained data showing that Bet-1 had the specificity required for the experiment. She held that the coauthors had made "secret accusations against me," that Baltimore had taken "selected pages of my notebook, and used them to convince the panel that the pages cast doubt on my veracity." The panel, she added, "willingly went along with this partisan attack" without even consulting her. Now the report was being cited by "those who are currently attacking me in public." O'Toole feared that the N.I.H. report held her up once again to being "ridiculed by the entire scientific community for making a fuss over nothing."[26]

O'Toole was especially aggrieved by the conviction that her integrity had been deeply impugned. She appeared to think that dissent from her claims on key points in the dispute meant not that others thought she was mistaken or misremembering but that they considered her a liar. It was as though she could no more think that other scientists might judge her guilty of error but not of prevarication than she could now make the distinction in favor of Imanishi-Kari. She took the N.I.H. panel's rejection of her challenges, particularly her insistence on the faultiness of Bet-1, to mean that the senior scientists in the investigation were saying "I was not telling the truth."[27]

Apparently supremely confident of her reliability as a witness, O'Toole was convinced that others should take her word for what she said she knew. She complained to Hadley that the N.I.H. panel should have testified in Congress that "my allegation that the published experiments did not support the claims was absolutely correct." She indicted Baltimore and Weaver for having used the subcloning data in their second letter of correction to Cell even though she had said it was bogus. She told the O.S.I. in a telephone conversation that Moema Reis, who had reiterated to Kimes that she had taken the data for the fourth point in Figure 1, was "in a 'conspiracy' with I[manishi]-K[ari]. At least now." Six of the allegations of fraud that she sent to Kimes were based purely or primarily on what she contended Imanishi-Kari had admitted to in her presence May 1986, and several more depended on those recollections indirectly.[28]

O'Toole drew up and pursued her allegations with purposes that were inextricably linked—both to expose Imanishi-Kari and to vindicate herself. She had wanted to allege a cover-up, including the failure of those in authority to insist on the publication of corrections in 1986 and "the filing of false and secret reports . . . damaging to me." Kimes had managed to persuade her to stick to the science, arguing, as she understood him, that the definition of misconduct did not extend to cover-ups. She nevertheless cited issues such as Wortis's and Eisen's public assertions that she had proposed to explain some of the *Cell* paper data by the formation of *mu-gamma* hybrid molecules. She called the claim "a false representation of my objections," holding that it was "calculated to discredit me among immunologists" and had "convinced scientists I am just plain silly."[29] She embedded her allegations in an account of the dispute that was thoroughly self-justifying. She says that she was fighting to save her reputation and her knowledge of who she was.[30]

At the beginning of November 1989, Brian Kimes returned to the National Cancer Institute as planned, although he remained actively involved in the Imanishi-Kari case as a special consultant. Suzanne Hadley became acting director of the O.S.I., and took charge of the investigation. The O.S.I.'s approach to the case differed sharply from that of its predecessor, the N.I.H. misconduct office. It would consult the scientific panel but the panel members would serve only as individual consultants—not as a group whose collective opinion would be "prevailing," Hadley put it. The O.S.I. staff would run the investigation. They would dig out independent information from the principals in the case, analyze the notebooks themselves, and meld the forensic work of the Secret Service with its own results into a comprehensive assessment. The O.S.I. would control the outcome.[31]

O'Toole had long stressed that an important independent source was Charles Maplethorpe and had told the O.S.I. earlier that he was willing to disclose more about what he knew. In a telephone conversation with the O.S.I. in mid-December, Maplethorpe said that, despite his fear of a lawsuit in 1988, he had given the N.I.H. panel enough evidence to suggest that Imanishi-Kari had committed fraud and that he had been taken aback by the fact that its final report had disregarded his testimony. He said that in the current climate, he thought a lawsuit much less likely. Perhaps in this conversation Hadley first encouraged Maplethorpe, as she would do frequently in succeeding weeks, to recognize that "now is truthtelling time." In an interview at the end of December 1989, Maplethorpe communicated his indictments of Imanishi-Kari to the O.S.I. and its scientific panel, declaring that she was secretive and sometimes lied. He also recounted the several incidents that aroused his suspicions, including Albanese's refusal to show him his molecular data and the conversation between Imanishi-Kari and Weaver, in which he said she admitted that Bet-1 was not specific. Maplethorpe volunteered to go through Albanese's notebooks to find the compromising data he thought they contained. Hadley said she might be able to arrange for him to help out in that way.[32]

Several days earlier, Kimes, Hadley, and their staff had interviewed O'Toole at length, probing her allegations. O'Toole says the questions were tough but that she felt wonderful because they wanted information and truly listened to her even as they tried to punch holes in her story.[33] About this time, Hadley asked O'Toole to analyze the section that Imanishi-Kari had contributed to an omnibus grant application that Herman Eisen had submitted to the N.I.H. for support of research at the M.I.T. cancer center. The section had been written around January 1985 and contained data from experiments with the transgenic mice. The O.S.I. thought that a review of it might shed light on when certain experiments had been performed and with what results.[34] O'Toole, who had created a database of the notebooks in her own computer, was eager to help. On January 10, 1990, in an eighteen-page single-spaced letter to Hadley, O'Toole reported that the grant application upheld four of her allegations concerning the data related to Table 3. "Most importantly, the grant shows that the authentic data was, for the most part, discarded and replaced with fabricated data. It is the fabricated data [that] have been used to refute my assertions."[35]

Drawing on the grant proposal, O'Toole also ventured an account of how Imanishi-Kari had come to fabricate her data. Her story started with Weaver's work. Recall that his molecular analysis of the transgenic hybridomas reported in Table 3 showed that they did not express the transgene. In O'Toole's scenario, these initial molecular results had led Imanishi-Kari to commit herself to what became the central claim of the *Cell* paper—that idiotypically positive antibodies produced by the transgenic mice were generated by genes native to them. However, as O'Toole saw it, in January 1985 Imanishi-Kari found serological evidence that contradicted Weaver's molecular data: Most of the antibodies produced by the transgenic hybridomas of Table 3 could have come from the transgene. In June, Reis obtained results showing that they definitely came from the transgene and soon thereafter got similar yields from the Table 2 hybridomas, recording the data on some of what became the seventeen pages. The data from these experiments was "authentic," O'Toole insisted, and they

SUZANNE HADLEY
Second Acting Director, Then Deputy Director
Office of Scientific Integrity
Photo Credit: © Annie Adjchavanich

"actually *disproved*" the central claim of the *Cell* paper because they showed that the antibodies with the distinctive idiotype did not derive from native genes. However, Imanishi-Kari "kept quiet," O'Toole wrote. "By this time the project was backed by Dr. Baltimore's prestige, and Dr. Imanishi-Kari's career depended on it." At some point, she fabricated the data now in the notebooks that purported to supersede the original, undermining results.[36]

Further, according to O'Toole's scenario, Imanishi-Kari did not tell Weaver that she had detected pervasive expression of the transgene. Weaver had some evidence of transgene expression in his own hybridomas. O'Toole contended that he did not realize the significance of such expression and was in any case led to ignore it "because of his mistaken beliefs" that Imanishi-Kari had not found transgenic antibodies. She also suggested that he may have "concealed" some of his data. The Secret Service had, after all, shown that the transgene smudge in his Figure 4 had been suppressed, and the coauthor's claim that its presence was "scientifically unimportant is most definitely false."[37]

O'Toole continued that Imanishi-Kari may well have realized the implications of her "discordant data" long before the publication of the *Cell* paper. O'Toole recalled that Imanishi-Kari had told her not to discuss scientific issues with the three men in the laboratory, particularly Maplethorpe, because they were " 'poisoning' the atmosphere and causing her great pain." She now knew that, along with the other two men, Maplethorpe had begun piecing together information that "pointed to the truth about the data." He had made Imanishi-Kari aware of his suspicions, and it was obviously for that reason, O'Toole wrote, that Imanishi-Kari had instructed her not to speak with him. O'Toole added parenthetically that what she said about Imanishi-Kari's eagerness to keep her data secret found support in Maplethorpe's testimony that Chris Albanese had "told him that he was not allowed to show Dr. Maplethorpe his data."[38]

O'Toole was willing to give Imanishi-Kari the benefit of some doubt, holding that she may not have fully understood the import of her authentic data until O'Toole challenged the central claim. O'Toole said that Imanishi-Kari "appeared genuinely astonished" when she pointed out the implications of her data. She added that she herself "really believed" Imanishi-Kari was astonished. O'Toole said that now, however, she suspected that Imanishi-Kari's astonishment had been "feigned." Imamishi-Kari had "not yet had the time to fabricate the complex story of superseding data in every instance," O'Toole continued. "She, like me, had no way of knowing in advance that all these people would simply cooperate in an ongoing cover-up. She may therefore have chosen to respond with astonishment in order to convince them that, although she had erred grievously, at least she was not dishonest."[39]

O'Toole held that the indulgence of Baltimore, Weaver, Eisen, Wortis, Huber, and Woodland had permitted Imanishi-Kari to slide through the Tufts and M.I.T. reviews. But when Imanishi-Kari was warned that the N.I.H. would mount "a more serious and independent" inquiry, "she fabricated a large amount of data to make it appear that the data I had found . . . had been superseded by 'good' and compatible data." O'Toole added that Imanishi-Kari "must have had good reason to believe" that Baltimore, Eisen, Wortis, Huber, and Weaver, all of whom "knew the truth," would nevertheless "continue to cover for her." O'Toole exempted Woodland from the group aware of the truth because she had no "personal knowledge of what [he] actually knew." She was in any case certain that Woodland "did not see the data he said he saw" (presumably she meant the June subcloning data). "Huber and Wortis might have told him he saw it, and he might have believed them." However, Baltimore and his colleagues knowingly covered for Imanishi-Kari, O'Toole argued, because they "had vouched for the [*Cell*] paper, and staked their own reputations on its authenticity."[40]

Neither O'Toole's fraud story nor Maplethorpe's contribution to it, however, squared with all the evidence that came to the scientific panel during the winter of 1990. In a letter sent to Ursula Storb shortly after the subcommittee hearings in 1989 that Storb turned over to Hadley in February 1990, Imanishi-Kari's former doctoral student Nicholas Yannoutsos had provided an impression of Imanishi-Kari different from that of O'Toole and Maplethorpe. Yannoutsos declared that what was reported in the *Cell* paper was "not an isolated collection of ambiguous experiments"; it was part of an ongoing research program that included painstaking repetition of the experiments "time and again" by Imanishi-Kari and others, including himself, to "clarify the mechanisms" operating in the mice. All the effort was carried out "in an atmosphere of openness and intellectual integrity," Yannoutsos continued.

> The data . . . were always scrutinized, discussed and analyzed extensively, whether in group and department meetings or even as they were coming out "raw" in the corridor and at the benches of M.I.T. and Tufts. I cannot emphasize enough the genuine attitude with which Dr. Imanishi-Kari conducted her own work and invited other people to participation, criticism and contribution, and her willingness to pursue not a particular theory, but any valid interpretation for her own findings and the findings of other researchers in related work. She was ever present in the laboratory, working long and hard hours and in constant communication with the people in it. She was particularly strict about the technical aspect of the work and demanded that every experiment be well controlled and repeatable.[41]

In an interview with the O.S.I. scientific panel at the beginning of February 1990, Imanishi-Kari's technician Christopher Albanese told how she ran an open lab, how she had never instructed him not to show his data to anyone, and how he, on his own, had declined to reveal it to Maplethorpe because he knew that Maplethorpe's relationship with Imanishi-Kari was acrimonious. Albanese said he was "disturbed" by O'Toole, explaining that she never stated clearly, "I have this big problem with every piece of data in the paper." Her complaints started with Bet-1 and then developed piecemeal. "As things fell apart on one argument, she moved to the next." Albanese declared himself dumbfounded by the course events had taken since 1986: "All of a sudden, it's kind of mushroomed like a nuclear explosion into a full-fledged investigation into our own private lives."[42]

Albanese rejected the scientific implication of the Secret Service finding concerning David Weaver's alleged suppression of the smudge in his radiograph—that is, the smudge indicating that the hybridoma under examination might have expressed the transgene. Since Weaver's radiograph showed clearly that it expressed a native gene, the smudge could be taken to mean that the hybridoma was what O'Toole had claimed—a double producer: In one way or another, it generated some antibodies, or parts of antibodies, characteristic of the transgene and others, or parts of others, characteristic of native genes. Albanese protested, "Any concern that came up about double producers . . . was cross-checked by more than one data point of technique. . . . I think we did a very thorough job of making sure that if there were double producers . . . we would have picked them up." They discussed the issue and they found "no evidence for double producers."[43]

When Maplethorpe talked with the scientific panel on February 11, he thought them indifferent to some of his evidence and suggesting that innocent explanations could be conceived for much of the rest. For example, he had by now scrutinized Albanese's notebooks—Hadley had sent them at the beginning of January, saying they were his to study for almost two months—and discovered that Albanese had done a molecular test on a hybridoma that appeared to express the transgene but that had been reported in the Cell paper not to express it. Storb suggested that the test might have been carried out to check whether the apparent transgene expression was real or not, a view that Maplethorpe considered "preposterous." He complained to Hadley that the panel did not even ask him about the conversation he said he had overheard between Imanishi-Kari and Weaver, an omission that irritated him profoundly.[44]

O'Toole spoke separately with the panel the same day. She was at pains to convince them that Bet-1 had not worked before the publication of the Cell paper as Imanishi-Kari claimed it did. She had been told, she said, that the panel had been shown "a number" of trials of Bet-1 in her notebooks that "showed me to be not truthful" about the reagent. Not "a number," McDevitt said. Just one. O'Toole stressed that she had gotten Bet-1 to work "only once" and that she had been unable to repeat the success even with that same batch of the reagent. She reported that she had assayed perhaps seventy-five other batches but that all of them had failed. Somehow, she added, none of the records of the unsuccessful trials was left in her notebook. She could thus not prove that in her experience Bet-1 usually did not work, but she contended that she could prove that she had to have been doing the trials to make sense out of the cell-transfer experiments she was then attempting.[45]

During a discussion of the June subcloning, O'Toole declared, "I know that the experiment is totally made up." McDevitt asked how she knew it. "Because I know for certain the experiment was not done. I know it for absolute certain," O'Toole answered, invoking her experience at the Tufts inquiry on May 23, 1986. She did not repeat what she had written in her response to the N.I.H. report and in the 1989 subcommittee hearing—namely, that Imanishi-Kari had said explicitly that no subcloning had been done. O'Toole rather indicated she knew it had not been done because Imanishi-Kari had not volunteered that it had been. "I begged to be shown any data on anything done with those Table 2 clones other than what was in the seventeen pages. It was crucial. . . . I wanted very much to believe that there was data."[46]

Eventually, the discussion turned to O'Toole's allegations concerning Table 3, which challenged Imanishi-Kari's claim that she had determined that many of the hybridomas produced idiotypically birthmarked antibodies with *gamma* isotypes. The determination rested partly on a test that biologists call by its acronym, ELISA (for *enzyme-linked immunosorbent assay*). Like a radioimmune assay, this test begins with using a capture reagent to grab antibodies of a certain type in a well; but checking for whether the antibody has been captured depends on the action of an enzyme coupled to it or to another reagent used to detect it. In an ELISA, a colorless chemical solution is added to the well and responds to the enzyme by turning color—yellow in the version Imanishi-Kari used. Imanishi-Kari said that her capture reagent selected antibodies carrying the idiotypic birthmark and that she had then run ELISAs to test them for isotypes other than *mu*.

O'Toole charged that Imanishi-Kari had used a capture reagent that indiscriminately selected mouse antibodies of any kind for isotyping, not just those that carried the idiotypic birthmark. Thus, although many of the Table 3 antibodies were, indeed, *gammas*, they did not necessarily carry the idiotype of the transgene and could not compensate for Imanishi-Kari's failure to have run tests for isotype other than *mu* on the hybridomas reported in Table 2. In her allegations, O'Toole contended that at the meeting of the Wortis committee on May 23, 1986, Imanishi-Kari had admitted the fact, calling it a mistake, and that "all agreed that this was a fatal flaw in the study." Now, in her interview, she invoked in further support of her charge that the data for the Table 3 hybridomas as they were recorded in Imanishi-Kari's notebook and appeared in the grant proposal were discrepant. McDevitt said he could imagine that Imanishi-Kari had done two different assays. O'Toole advanced a plausible supposition about why she must have done only one. McDevitt observed, "But it's still a supposition," whereupon O'Toole, distressed to the point of breaking down by what she took to be hostile badgering, asked Hadley if she could see her outside and left the room. She remonstrated to Hadley that she had been promised that she would not be subjected to such treatment. Davie evidently joined the two women and, with Hadley, pointed out that O'Toole needed to clarify what she was saying.[47]

On returning, O'Toole explained that she was concerned not only with the discrepancy between the grant data and the notebook data but also with the substance of the assay results reported in the grant. In contrast to the *Cell* paper, the grant proposal did not report that the hybridomas tested produced a high frequency of idiotypically birthmarked antibodies. In O'Toole's reasoning, that omission supplied proof of Imanishi-Kari's telling her at the Tufts inquiry that the assay had not been done with a reagent that selected for idiotype. She had to concede, however, that the record of the assay itself was not in Imanishi-Kari's notebook. She had only her "personal knowledge" of what Imanishi-Kari had said on May 23, 1986.[48]

Later in the interview, an O.S.I. staff member asked whether anyone else had been present at the meeting who might confirm her account.

"Yes," O'Toole said. "They're all lying . . . Not only do I know Huber and Wortis are lying, I know that they know they're lying."[49]

A few days after his interview, Maplethorpe declared to Hadley that he had "serious doubts" about the panel's willingness "to conduct a fair and impartial investigation." He insisted to Hadley that Imanishi-Kari "had full knowledge" that Bet-1, as she used it, "could not have provided the data she published in Figure 1" and that the figure "therefore misrepresents the data" she obtained. In a memorandum to the O.S.I. and the N.I.H. panel in March, O'Toole complained that some of the panel appeared willing to "discount my word in instances where my account does not have the support of independent physical evidence." She insisted, "It is time that the investigators in this case gave intrinsic value to my word."[50]

Brian Kimes says that early in the investigation, "it was basically [O'Toole's] word against somebody else's," and the O.S.I. kept encountering testimony that contradicted O'Toole's. In March, William Paul, at the N.I.H., and John T. Kung, his collaborator in the development of

Bet-1 who was now at the University of Texas Health Sciences Center in San Antonio, wrote to Hadley, attesting that, according to their experiments, the reagent could have performed with the "discriminatory power" described in the *Cell* paper.[51] Between late April and mid-May, in telephone interviews from Brazil with the O.S.I. and the scientific panel, Moema Reis reaffirmed that she had made the measurements for the fourth point on Figure 1 and believed that she had entered the values directly on the graphs. When Hadley asked her opinion of Stewart and Feder's analysis of the paper, she addressed the attention they gave to the data concerning Bet-1 in the seventeen pages, pointing out with apparent impatience, "I worked there for a whole year and . . . Dr. Imanishi-Kari has worked longer than that and they have picked one experiment that didn't work. . . . As far as I'm concerned, this paper [by Stewart and Feder] doesn't mean anything."[52]

Hadley raised the issue of the "June subcloning," a term that was by now commonplace among investigators of the notebooks but that Reis knew nothing about. Hadley explained that it meant the subcloning Imanishi-Kari said had been initiated in June 1985 and asked whether Reis "had personal knowledge of these experiments." Reis declared, "Yes, I was there, we did this together. I did most of the [sub]cloning and Dr. Imanishi-Kari did part of the [sub]cloning because it was too much for me to do by myself." She added that they started with three clones identified in the notebook and then [sub]cloned others.[53]

Hadley and Kimes say that, faced with so much conflicting testimony, they made their central concern the data. Thus, for example, the question of whether Imanishi-Kari had admitted to O'Toole in May 1986 that she had not subcloned any of the Table 2 hybridomas was of interest but not high on their list of concerns. The main issue was: Did the evidence in the laboratory notebooks support the claim that the subcloning had been performed beginning in June 1985 and completed before the publication of the paper?[54]

O'Toole held that such questions might be answered by forensic analysis of the radiation-counter tapes in Imanishi-Kari's notebooks. Her request on her visit to John Deutch in early June 1989 for permission to obtain tape samples from the cancer center printers was a preliminary foray into the subject. She had then called Kimes's attention to what she observed in the sections of the notebooks that the N.I.H. had given her in response to her request for the data supporting Figure 1. Even from the copies, it was obvious that the print on the tape that recorded a control test of the reagent AF6-78.25 differed in intensity from the print on the tape that purported to register a test of Bet-1. The two tapes were mounted on the same notebook page and had purportedly been produced on the same day, but O'Toole took the difference in intensities to indicate that they had been generated at different times—indeed, that "neither tape was generated" on the day the notebook said they had been. To her mind, the disparity demonstrated that "the controls claimed were not actually obtained and the experimental results have been falsified."[55] O'Toole argued that the only way Imanishi-Kari could escape that conclusion was to show that the two tapes had been produced by two different printers. She therefore urged the O.S.I. to compare the tapes forensically against those generated by any other printer Imanishi-Kari might claim to have used.[56]

In her allegations in November, O'Toole had extended her ideas about forensic analysis to the June subcloning tapes and now, in February 1990, she urged the O.S.I. to compare the gamma-radiation-counter tapes in Imanishi-Kari's notebook with similar tapes in the notebooks of other scientists who had used the laboratory's counting equipment during the same period. In her interview with the O.S.I. staff and its panel, she averred that she would "be able to prove that many of the tapes could not have come at the time [Imanishi-Kari] said [they were generated]," adding that "the facts will have to fit the truth," meaning her truth that the experiments had not been done. O'Toole asserted that she would "practically bet my life" that the subcloning tapes had been printed before the transgenic hybridoma cells even existed.[57]

The Dingell subcommittee had provided the O.S.I. with what were called "supercopies"—very-high-quality copies—of Imanishi-Kari's notebooks that showed the colors of the different counter tapes pasted in them. O'Toole now asked to see the tapes from the subcloning experiments that Imanishi-Kari said she had begun in June 1985. She pointed out that the color was "sort of a green," different from the colors of the tapes in any of the other notebooks. Such tapes were used in 1982 or 1983, she said but not ever from June 1985 to May 1986, when she was in the laboratory. During that period the tapes were all yellow. O'Toole assured the O.S.I. and the panel that a search for green counter tapes in the notebooks of other scientists would show, from the dates they were obtained, that Imanishi-Kari's subcloning tapes could not have been produced in experiments that occurred between June 1985 and November 1985, when the *Cell* paper was written. It would demonstrate, in short, that "the experiment is wholly fabricated."[58]

Kimes, prompted by O'Toole's expectations, had written in mid-September to John Deutch, the provost at M.I.T., asking for samples of the gamma-counter printouts produced month by month by each of the printers at the cancer research center from January 1984 to June 1986. Deutch soon replied that he would do what he could, pointing out that M.I.T. did not "have access to the universe of past laboratory notebooks" and that it was "essentially impossible to identify tapes on a monthly basis." In late February, after O'Toole in her interview with the panel had stressed the importance of the tape evidence, Hadley reiterated the O.S.I.'s request to Deutch. She also turned to soliciting tape samples directly from scientists who had worked in the laboratory contemporaneously with Imanishi-Kari. In his interview with the O.S.I. in December, Maplethorpe said that he had "an extensive set of countertapes that overlap the entire period." Hadley asked then that he make them available to the O.S.I. and she wrote to him in February that "we are much interested in your reviewing any gamma counter tapes from your own research . . . to determine if you have dated tapes from 1981 through 1983." If he found any, Hadley wanted to be notified "right away" so they could be made available for forensic analysis.[59]

By the beginning of March 1990, Deutch had been able to provide a list of the different radiation counters and printers but only some sample tapes from two machines and a few copies of tapes found in one of the notebooks at the cancer research center. Late that month, Maplethorpe told Hadley that he had given many of his tapes to Walter Stewart, at the subcommittee. The O.S.I. wanted the Secret Service to attempt to determine when Imanishi-Kari's tapes had been produced by comparing them with notebooks such as Maplethorpe's. The subcommittee, however, was reluctant to give them over for the work that the O.S.I. wanted done. It had scheduled another hearing on scientific fraud in May and wanted the Secret Service to complete before then the analysis it had commissioned the previous summer. In mid-April 1990, the inspector general of the Department of Health and Human Services arranged for the Secret Service to analyze the notebooks in the subcommittee's possession to the extent necessary to accommodate the O.S.I. Hadley informed the scientific panel that "the Secret Service has our request for forensic work, and is proceeding to do the work, to the extent this is compatible with work commissioned by the Subcommittee," adding, "The Secret Service reportedly had done *no* analyses of gamma counter tapes until very recently, but now is putting great effort into this work, and may complete it as soon as April 30."[60]

In the meantime, O.S.I. staffers had been studying Imanishi-Kari's notebooks and had concluded that they looked fishy. The subcloning data had come in for special scrutiny. Only part of it comprised radiation-counter tapes pasted onto the notebook pages. The rest consisted of handwritten records of the data—columns of numbers that Imanishi-Kari had purportedly transcribed from counter tapes, which she then discarded. Walter Stewart seems to have suggested that the handwritten counts might be fabricated because they did not display the random distribution of numbers that a radiation counter would likely produce. For that reason, James Mosimann had been made one of the internal N.I.H. consultants to the investigation. A Ph.D. biologist, he held a masters degree in bios-tatistics and had spent most of his many years at N.I.H. applying statistical and computing methods to biological problems.[61]

According to Maplethorpe, McClure remarked to him after his inter-view that he thought the procedures in Imanishi-Kari's laboratory were "crazy." In February, McClure had begun entering various experimental results from Imanishi-Kari's I-1 notebook into his Macintosh computer, using readily available programs to plot the data on graphs and get a feel for the type of experiments she had done. At a meeting of the scientific panel, Mosimann reported that he had been finding what statisticians called "spikiness" in the data—that is, certain digits appeared in the num-bers far more frequently than one would expect in numbers generated at random. After listening to Mosimann, McClure focused his computerized attention on the numbers in several sets of Imanishi-Kari's data, partic-ularly those that purportedly recorded the background radiation she had measured in the June subcloning. The subcloning background counts pro-vided a large sample—270 numerical values—and was marked by obvious rounding to the digit for tens.[62] (*Figure 13*).

In early April, McClure reported his tentative conclusion to the O.S.I. and his fellow panel members: "The data look far too nonrandom for any natural explanation based on experimental variability or even imaginable systematic errors. Rather, it seems to me that these numbers were made up in the mind of T[hereza] I[manishi-]K[ari], who on the day she wrote these numbers into her laboratory book had an unconscious preference for numbers ending in 10, 30, 70, or 80" (*Figure 14*). The background counts, McClure declared, were "fabricated." Mosimann, drawing on his own statistical analysis of the handwritten counts, estimated later in April "that the I-1 notebook contains at least one data set that is not authen-tic."[63]

On May 14, 1990, Dingell opened the scheduled hearing, insisting it was "nonsense" that the subcommittee wished to " 'police' " science. It expected scientists to police themselves, but it had been "severely dis-appointed by the response of the scientific community on a number of occasions," particularly the current instance. In a notable display of anon-ymous accusation, he announced that "a number of prominent scientists,

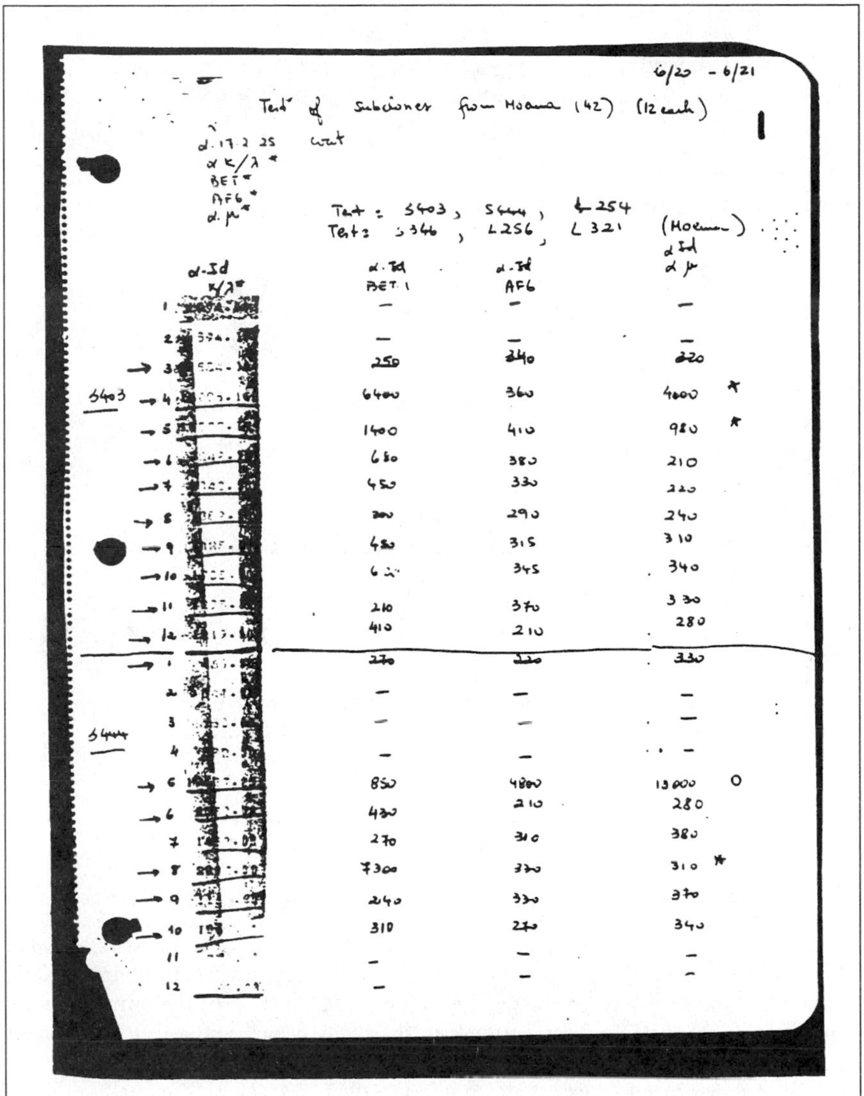

FIGURE 13
June Subcloning Data

Part of the June subcloning data was entered on page 124 from Imanishi-Kari's I-1 notebook, shown above. On the left is a "green" radiation-counter tape; its numbers record how much the antibodies produced by each subclone react with a reagent to detect idiotype. The handwritten numbers record counts taken from radiation-counter tapes indicating the degree of reactivity of the subclone's antibodies with, from left to right, Bet-1, AF6-78.25, and reagent to detect the *mu* isotype. The handwritten numbers below 1,000 constitute the background counts.

Any primer should have a concise summary of the topics covered. Here's my summary of TIK digit preferences:

A. Count up to 10 with TIK!

1 . 3 4 . 7 8 . .
(2) (5 6) (9 0)

The point size of each digit is scaled to reflect the occurrences in TIK's laboratory notebook.

B. Count up to 10 with KERMIT *et al.*

1 2 3 4 5 6 7 8 9 0

For comparison, the expected uniform occurrences are shown above as they are taught on Sesame Street!

FIGURE 14

Statistical Analysis of the Background Count Digits

William McClure's illustration, with a bow to the children's television program *Sesame Street*, of the nonrandomness in the occurrence of the second digits in Imanishi-Kari's handwritten transcriptions of the background counts. The size of the numbers in his summary reflects the frequency with which the digits occur. The row at the bottom expresses the idea that each digit from 1 to 0 occurs with equal frequency. The row at the top shows that Imanishi-Kari's numerical digits occur with unequal frequency, with "1" occurring most often, "3," "7," and "8" occurring next most often, and so on. (From William McClure, "A Primer on the Digit Preferences of Thereza Imanishi-Kari," attached to McClure to Suzanne Hadley, July 13, 1990, Onek Files.)

under a promise of confidentiality, examined the suspect notebook and agreed that it was obviously bogus," adding, "But these same scientists were unwilling to advance their professional opinions in public for fear of the disapproval of their colleagues."[64]

Secret Service agents John Hargett and Larry Stewart addressed the issue that Dingell had posed the previous July: Had the counter tapes in

Imanishi-Kari's notebooks been generated at the times she said she had done the experiments that produced them? To answer that question, Hargett's branch had determined that the notebook tapes could have been produced by only two of the gamma-counter printers in use at the M.I.T. cancer center between 1984 and 1986. These machines printed like a typewriter, with metal letters striking a cloth ribbon, and they registered a counter number—a kind of index number—on the tapes.[65]

On what Hargett called "authentic" tapes, these counter numbers ascended at regular rates—about 11.5 numbers per day—and the intensity of the print faded gradually from one tape to the next. On some of Imanishi-Kari's tapes, however, the register numbers jumped far more from one tape to the next; in one case, the difference between the counter numbers on two tapes dated a day apart was 1,353, a change so large as to indicate they had been produced four to five months apart. On the same tapes, the print intensity also tended to change abruptly rather than smoothly. Hargett testified that, on the basis of this evidence, the Secret Service concluded that twenty to thirty counter tapes were "not authentic with respect to time"—meaning, he explained, that "they were not produced by the gamma counters at the dates indicated on the pages to which they are now attached." Taking into account the dubious pages identified in 1988, the total of "questionable pages" came to forty-four.[66]

Congressman Ron Wyden, evidently unwilling to hazard ambiguity, pressed Larry Stewart: Did he believe "a small or large portion" of Imanishi-Kari's notebooks was "false"?

Agent Stewart responded, "We believe a large portion, we showed one third of the notebook at least was false."[67]

Wyden suggested that the evidence of falsification from both the subcommittee and the Secret Service should be shared with the U.S. attorney for the district of Maryland, the home of the N.I.H. Dingell said that he would direct his staff to do so. Subcommittee members sought a criminal investigation of Imanishi-Kari—for possibly having lied in her congressional testimony the year before, having attempted to obstruct justice, and having submitted false documents to the subcommittee and the N.I.H. Breckenridge C. Wilcox, the U.S. attorney for Maryland at the time, had already successfully prosecuted one scientist—Stephen Breuning—for fraud. Now Wilcox told a reporter, "Anything that John Dingell sends us we'll be glad to look at."[68]

T E N

∎

Burden of Proof

ON JANUARY 10, 1990, Hadley and the N.I.H. legal adviser, Robert Lanman, informed Bruce Singal that the focus of the investigation had shifted from "the accuracy of reporting in the *Cell* paper" to "the authenticity and integrity" of its data. They asked for any and all data that Imanishi-Kari might not yet have provided bearing on Bet-1 and the subcloning, and they reiterated the O.S.I.'s request for her cooperation, particularly that she provide an index of her laboratory notebooks. On February 1, in letters to Baltimore and Imanishi-Kari, Hadley laid out the O.S.I.'s suspicions, framed in what amounted to O'Toole's scenario of fraud: When the *Cell* paper was submitted, substantial portions of it lacked the support of "proper experiments and reliable data"; after it was challenged, support was invented by "systematic fabrication and falsification of data"; and phony data was provided to the N.I.H. and published in the letters of correction. Hadley supplied a two-page list of eighteen "issues"—they corresponded exactly to O'Toole's allegations—with which the O.S.I. was specifically concerned and tied each issue to specific pages in the notebooks of Imanishi-Kari, or Moema Reis, or both. She included copies of the notebooks containing the pages—Imanishi-Kari's I-1 notebook and two of Reis's. She added that as the O.S.I. continued with its work, "others of the coauthors might be formally identified as subject of the investigation."[1]

Imanishi-Kari declined to respond to either communication from the

O.S.I. or to another later in February. She remained fearful that if she provided an index, it would be used to raise new charges against her, and Singal advised her not to respond, continuing to hold that the O.S.I. was inadequately informative about the existing charges. Hadley's list of issues was stenographic, limited to declarations such as that of number 5: "Contrary to what was published, most hybridomas reported in table 3 secreted transgene product," with a list of references to twenty-five pages in the notebooks; or of number 9: "Possible fabrication of June subcloning data . . . ," also referencing several notebook pages. The document said nothing about why, for example, the O.S.I. suspected the subcloning data had been fabricated or what specifically in the referenced notebook pages suggested they were; and so it was for the other seventeen items on the list. Imanishi-Kari recalls that she felt she could not know "what they are talking about if they don't even spell it out."[2]

In January, Hadley informed Imanishi-Kari that she had been entered into the N.I.H. ALERT system, a list of people suspected or found guilty of misconduct that was circulated to officials throughout the agency, in order, Hadley explained, "to facilitate informed decision-making regarding funding decisions." On April 11, 1990, the N.I.H. canceled one of Imanishi-Kari's research grants—a rare event at the agency—without explanation other than an allusion to incriminating evidence that the O.S.I. was finding. Singal telephoned Hadley on April 18 to ask on what grounds the grant had been terminated. She said that evidence had been accumulating against Imanishi-Kari but refused to provide any details.[3]

Writing the next day, Hadley admonished Singal that, despite repeated requests, the O.S.I. had received "no communication whatsoever" from Imanishi-Kari and urged that it was in her interest to respond to its "body of evidence." Singal, in a letter that crossed hers and in another at the beginning of May, reiterated what, as he pointed out, he had repeatedly told Kimes in 1989: Imanishi-Kari was eager to respond to the O.S.I. but could not provide additional information unless and until she was "informed of the charges and evidence against her." That requirement satisfied, she would be "happy to meet with N.I.H.-affiliated scientists to discuss her data."[4]

All the while, Singal kept battling with Congressman Dingell and his subcommittee staff. In mid-December 1989, five months after Singal's request for the full Secret Service analysis of Imanishi-Kari's notebooks, Dingell replied that much of the information he wanted could not be made available since the forensic studies were ongoing and because the Secret Service had not yet prepared a written report of its work. He declared that he found Imanishi-Kari's unwillingness to cooperate with the investigations of both the N.I.H. and the subcommittee "deeply troubling." He again requested that she submit to an interview with his staff. In a telephone conversation early in January, Peter Stockton, Walter Stewart, and the chief of Dingell's staff laid out the general areas that the subcommittee wanted to cover with Imanishi-Kari, including how and when she had constructed her notebooks and performed her experiments. Singal, dissatisfied, subsequently asked Dingell for a detailed statement of allegations. In a letter to Singal at the end of January, Dingell bluntly summarized them—"that she fabricated much of the notebook I[-]1 in an attempt to mislead N.I.H. and the Subcommittee." He added, no doubt reporting Stewart's view and perhaps O'Toole's, "It is also alleged that certain scientists, including Dr. Imanishi-Kari, have conspired to conceal the truth."[5]

Singal, responding, decried the allegations as "stunning" and demanded to know who made them. Who were the scientists involved in the alleged conspiracy? And where was the evidence for such "draconian" accusations?[6]

On May 7, 1990, Dingell supplied his principal evidence—two reports that the subcommittee had received from the Secret Service—and invited Imanishi-Kari to testify at the upcoming hearing, on May 14. The first report, dated April 16, 1990, comprised a three-page summary of the results that agents Hargett and Larry Stewart had presented at the hearings the year before. The second, dated May 4, 1990, was a two-page précis of the new forensic report that the Questioned Documents Branch had been rushing to complete and that Hargett and Stewart would discuss at the hearing. The reports were far less informative than the testimony the two agents provided at the hearing ten days later. For example, the new one said of tapes on various pages of her notebook, including those that contained the subcloning tapes, simply that they were "not consistent" with tapes found in the notebooks of other researchers produced around the dates Imanishi-Kari's had purportedly been generated. It said that tapes on certain other pages were "consistent" with having been produced after February 26, 1986, rather than at the earlier times the dates on them indicated.[7]

In a letter to Dingell on Thursday, May 10, Singal derided the first report as "nothing more than a re-hash" of the Secret Service findings in 1989 and pointed out that Imanishi-Kari had already responded to them in the 1989 hearings. He attacked the second report for its presentation of new findings "in such skeletal and conclusory form," calling it "so devoid of support or documentation" as to be "virtually useless." He complained that it did "not even attempt to explain what its so-called conclusions of "consistent" or "not consistent" mean." Singal advised Dingell that Imanishi-Kari would like to respond to the new Secret Service report but could not do so without a lot more information. He attached a list of twenty-four items that she needed, including identification of the tapes in question, copies of the notebooks of the other researchers, a full explanation of how the conclusions of consistency or inconsistency had been reached, and access to her original notebooks so that she could subject them to an independent forensic analysis.[8]

That same Thursday, at a press conference in Washington, Singal announced that the subcommittee had "struck again" and attacked the new Secret Service report as "nothing less than a sham." He continued:

> I submit to you that it was produced in order to create a pretext or an aura of fairness, when in reality its contents are so vague and so . . . unsupported and so skeletal that they could not possibly furnish a basis for any meaningful response. The report was prepared so that this time the [sub]committee could say, "Well, you wanted a report last time. Okay, we give you a report this time," and then they could . . . say, "Well, Dr. Imanishi-Kari is not responding. She must have something to be afraid of." Well, let me tell you, Thereza Imanishi-Kari has nothing to be afraid of and she has nothing to hide. I have advised Chairman Dingell this afternoon, by way of a letter which is out on the chair by the door, that under these circumstances in which the [sub]committee has once again sought to ambush my client with little notice and even less information, that she is left with no choice but to decline his invitation to testify . . ."[9]

Subcommittee sources disparaged Singal's request for further information, dismissing it, according to Daniel Greenberg's *Science and Government Report*, "as a legalistic wile designed to bog the inquiry in endless disputes about whether all the sought-after information had actually been provided."[10]

What Hadley found most incriminating in the developing evidence was the statistical and, especially, forensic analyses of Imanishi-Kari's notebooks that the O.S.I. now had in hand. Interviewing David Weaver in June, about a month after the hearings, Hadley wondered what his reactions might be to what the Secret Service had revealed about the authenticity of Imanishi-Kari's counter tapes. He responded that he just didn't accept the Secret Service's conclusion that they had been fabricated. Generating a gamma-counter tape that would serve as a suitable replacement for unsuitable real data struck him as demanding "a lot of effort," too much trouble and energy. "If you were going to fabricate data, it seems like it would be easier just to write it out."[11]

"What if you already had a stockpile of tapes from other experiments?" Hadley asked. "That'd be fine," Weaver replied, "but then you would have to find the right stock . . . that had the right order of things . . . You'd be lucky, I guess, if you could find the right accumulation of data." On further probing, Weaver laid out his bedrock view of the matter: "It's not a case of fabrication." He added that if he was to examine the Secret Service report, "I would be reading it from the point of view of trying to discount the allegations. I just don't believe it."[12]

Lanman pressed Weaver whether his position was based on his faith in Imanishi-Kari or his perception of weaknesses in the case for fabrication. Weaver answered, "Some of each."[13]

Hugh McDevitt and Ursula Storb wondered about the Secret Service's results, having not been shown any of the laboratory work the Questioned Documents Branch had performed. They were also skeptical of the statistical analyses that their fellow panel member William McClure and the O.S.I. staff scientist James Mosimann had done of Imanishi-Kari's handwritten entries of the June subcloning data. McDevitt recalls, "I kept saying, 'You know, I don't see how any of this can convince you.'" He could imagine someone's just being "very erratic" in how she rounded numbers, rounding one way "in one phase of the moon" and another in a different phase of it. He told the O.S.I., "Well, you know, it's still just statistical evidence. It's circumstantial. It maybe points to the fact that [Imanishi-Kari did something] that was not quite kosher, but it doesn't establish fraud."[14]

However, McDevitt and Storb formed a minority of two in the O.S.I. investigation. The three other members of the scientific panel disagreed with them and so did Hadley and the O.S.I. scientific staff. Hadley and her fellow investigators may have been naturally disinclined to question the reliability of the forensic and statistical evidence, but by the spring of 1990, any such disinclinations were being strongly reinforced within the O.S.I. by the investigation's direction, configuration, and circumstances.

In January 1990, the N.I.H. appointed a regular director of the O.S.I.—a biologist named Jules Hallum. Recently retired from the chairmanship of his department at Oregon Health Sciences University, he had been busying himself with writing projects and sailing the Columbia River on his small sloop, *The Irregardless*. He says that he took the directorship of the O.S.I. partly in the hope that he could prevent the new agency from growing into a bureaucratic science police. The choice of Hallum dashed

JULES HALLUM
Director, Office of Scientific Integrity
Photo Credit: Marty Katz

Hadley's hopes for the post. Raub says that the N.I.H. wanted someone from the academic world with the kind of scientific credentials that would instill confidence among academic biologists that the O.S.I. was not "a bunch of mindless bureaucrats." Hallum had helped write his professional scientific society's code of ethics and had similarly assisted the National Institute of Medicine in writing a report on the responsible conduct of research. "Ethical behavior in science is not optional," he had declared in 1988 at an Institute-sponsored conference. Hallum says he took the job to ensure that the O.S.I. enforced that dictum in a reasonable way, building up a sensible body of case-by-case precedent.[15]

By most accounts, Hallum is an honest, decent, and well-intentioned human being, but Kimes says that, after interviewing him for the directorship, he told Raub, "Don't hire this guy. He doesn't know what he's doing. He's not going to understand how to deal with this stuff." Hallum acknowledges that it took him six months to learn how to run the O.S.I., partly because he spent only a few days each month in Washington until April 1990, when he was able to move from Oregon and take over full-time. He adopted the management model of an academic department chairman, thinking he should facilitate the work of the staff while offering helpful criticism of it. Hallum gave Kimes the impression of being "very weak." His approach irritated Hadley, who thought he should be investigating. But no matter for the time being: Hallum appointed her to be deputy director of the O.S.I. and designated her to serve as acting director until he moved to Bethesda. Between his extended period of commuting and his hands-off administrative approach, Hadley was left to run the investigation of Imanishi-Kari pretty much as she saw fit.[16]

Kimes says that Hadley was "obsessed" with rooting out scientific misconduct. To her mind, the misstatements in the *Cell* paper concerning the isotyping of the Table 2 mouse hybridomas constituted fraud because the experiments had in fact not been done on those hybridomas. Hadley believed that although creative hypothesizing plays a role in science, "the phenomenon is always right," as she put it later in a lecture on biomedical ethics at the University of California, San Diego. Philosophers of science call such a view "naive inductivism." She held that the selection, reporting, and interpretation of data must be "as free as humanly possible of 'taint' due to the scientist's hopes, beliefs, ambitions, or desires." Thus, for Hadley, out-of-bounds scientific conduct included "data selection, failure to report discrepant data, over-interpretation of data."[17]

Kimes thinks that when he was acting director of the O.S.I., he had provided a counterweight to Hadley's views but that after he left, her obsession increasingly dominated the office. He says that Hadley was capable of "an investigative mentality where you're guilty until proven innocent." Both Storb and McDevitt say that Hadley struck them as fair-minded and competent at the beginning of the O.S.I. investigation into Imanishi-Kari, but that as months went by, she seemed increasingly to be "looking for fraud," to use McDevitt's phrase.[18]

When Moema Reis, in one of her telephone interviews with the O.S.I., repeatedly declared that she had done the subcloning in June 1985 with Imanishi-Kari, Hadley was not concerned with Reis's recollection of what the subcloning showed. She was primarily concerned to obtain from Reis testimony that it had been Imanishi-Kari, and not Reis, who had written the data into the notebooks. Hadley thought David Baltimore had eloquently defended the rights of scientists against the interpositions of Congressman Dingell, but she observed to Maplethorpe that Baltimore's eloquence had not swayed everyone at Rockefeller University and that no matter how much he might insist on "holding down the fort against the goths," his eloquence would have to meet the test of evidence.[19]

At the end of April, the O.S.I. invited Baltimore for an interview, assuring him that he was no longer a target of the investigation and telling him that the staff only wanted to discuss the ways in which their investigative process might be improved. Baltimore's attorney, Normand Smith, recalls that, instead, they got "a lot of questions on the science and Imanishi-Kari."[20]

As the interview wore on, Baltimore grew progressively more irritated, feeling that he had been misled as to its purpose. Hadley asked him whether he had spoken further with Imanishi-Kari about the forensic analyses.

"No," Baltimore said, turning the question back to Hadley. "Why are you asking me about my thoughts and my contacts with Thereza? I don't know what you're trying to get at, but I don't particularly like it."[21]

Hadley asked whether Baltimore had further examined the data sets in the light of the allegations the O.S.I. had sent out in February.

Baltimore replied, "I don't know what you're talking about here, because you say 'possible fabrications of data.' How am I supposed to look at it and see possible falsifications? . . . What was I supposed to do in response to this? . . . Am I supposed to go and say, 'Thereza, did you falsify?' " A few minutes later, Baltimore asked Hadley just what she meant by "possible fabrication." Hadley said she meant experiments not done and data made up, but she declined to specify which data, explaining, "I'm not wanting to be purposely mysterious about this, but neither are we going to disclose things that are not well articulated and well laid out."[22]

After some sparring, Hadley asked Baltimore if he wanted to tell the O.S.I. anything else. Baltimore did, declaring with regard to the June subcloning data that he found it "bizarre" if the O.S.I. was planning "to try to hang Thereza on the basis of data that we were forced to publish by Jim Wyngaarden as a condition of being able to continue in the scientific community." Baltimore went on, "If those data were not real, then she was driven by the process of investigation into an unseemly act," elaborating several moments later, "To my mind you can make up anything that you want in your notebooks, but you can't call it fraud if it wasn't published. Now, you managed to trick . . . Thereza into publishing a few numbers and now you're going back and see if you can produce those as fraud."[23]

Baltimore soon qualified what he had said, asserting that he had "no reason to believe . . . anything in this whole study . . . fraudulent" and that "from everything I know about Thereza Imanishi-Kari, she wouldn't, couldn't [commit fraud]." He says he had never intended to defend the publication of fabricated data, even if the publication was forced, but he later acknowledged that it sounded as though he had done so. Saying he was trying to point out that the O.S.I. appeared to want to "trap" Imanishi-Kari, he reflected, "It was not a good argument on my part and I should never have gotten into it, but you have to realize how high the emotion had become in that room. They were trying to get me to say something stupid, and they succeeded."[24]

Jules Hallum insisted, "We're not here to railroad anyone or to entrap anyone or trick anyone," a declaration that Normand Smith asked to address from his lawyer's perspective. Smith said that if the proceeding was "a criminal trial, if these people were shoplifting," the O.S.I. would be unable to hold back its evidence; "you'd have to tell me now." Smith thought it "remarkable that you're taking world-class scientists and giving them less credibility and ability to defend themselves than you would if we were in a court in downtown D.C."[25]

Hallum insisted that the proceeding was not a trial but an investigation, whereupon Baltimore asked, "So one's rights are severely curtailed because we're not in a trial situation?"

Hadley reminded Baltimore that he was not a subject of the investigation. "We look at you as a source of information, as a colleague who has a common interest in helping us resolve this." Normand Smith says, "In retrospect, I believe that Hadley thought that if David was 'given immunity' he might talk freely, and give them something they might use to get Thereza."[26]

Kimes judges that Hadley, especially after the interview with Baltimore, was "very interested in casting aspersions" on him because, in her view, he had responsibilities as a distinguished senior scientist yet had ignored O'Toole. Then, too, O'Toole's complaint that Imanishi-Kari had used her data with inappropriate selectiveness was inherently compatible with Hadley's commitment to a seemingly indiscriminate empiricism as the basis of scientific practice. Moreover, Hadley "identified with whistle blowers in general," Kimes says, especially those she thought had been mistreated, and she was "very involved with O'Toole," as they all were, "more . . . than we should have been." Hadley spent a lot of time on the telephone with her. She says that O'Toole was vulnerable and suspicious of the N.I.H., and that she had to work to gain her trust.[27]

Withal, while Imanishi-Kari was denied access to information critical to her defense, O'Toole enjoyed what amounted to an insider's role in the investigation. Even when she was drawing up her allegations, Kimes sent certain Secret Service documents that she requested and assured her that she could have access to similar documents in the future. Later, she was told about the results of the Secret Service's ongoing work and was provided specific information about the tapes and the print formats of the radiation counters. She was asked to assess new evidence as the O.S.I.

developed it.[28] She also identified some on her own, using her copies of
the notebooks and knowledge of the forensic analyses to bolster her allegations of fraud. Early in March 1990, speaking with the O.S.I. on the
telephone, she declared that she had discovered " 'an extraordinary piece
of proof'—[a] 'powderkeg' " that she expected would convince the scientific panel of the validity of her charge that the isotyping of the Table
3 hybridomas had not been done on antibodies selected for carrying the
idiotypic birthmark.[29]

In mid-March, O'Toole sent the O.S.I a lengthy memorandum detailing
the "proof"—it amounted to an intricate, convoluted argument—and on
the same day supplied three more intended to clear up what she thought
were misunderstandings on the part of the scientific panel. In June, she
noted to the O.S.I. the "crowning irony . . . that the only evidentiary support still standing . . . [for] the authors' Bet-1 claims comes from an experiment done by me," and went on to indicate why that result should be
heavily discounted. She also directed the O.S.I.'s attention once again to
the matter of the green tapes in Imanishi-Kari's notebook, reiterating her
confidence that they would be seen to match tapes generated much earlier
than June 1985.[30]

All the while, the Dingell subcommittee kept a knowledgeable eye on
the O.S.I. investigation, including O'Toole's involvement in it. In October
1989, Stockton asked her for a letter she had received from Kimes and
for copies of her letters to him. She apparently sent them all, but informed
Kimes "so that you do not think I did anything behind your back."
O'Toole appeared to feel caught between the O.S.I and the subcommittee
and urged Kimes to reach some agreement with Dingell's staff regarding
her role as their witness. In mid-December 1989, William Raub proposed
to Dingell that in the future the subcommittee should seek the O.S.I's
correspondence with O'Toole from the O.S.I. rather than directly from
her. Dingell refused.[31]

Kimes and Hadley wanted to keep the O.S.I. investigation independent,
free of taint by the subcommittee. They interviewed Walter Stewart once
and sought the subcommittee's information, data, and the forensic results
it had been provided by the Secret Service. The O.S.I., however, commissioned its own studies by the Secret Service, and Kimes says that, while
they shared general information about the inquiry with Dingell's staff,
they refrained from providing details. Hadley avers that the O.S.I. never
felt pressured to support Dingell's view of the case.[32]

Kimes nevertheless recalls that Dingell "could subpoena anything; he
could get anything from our office any time he wanted." The subcommittee obtained documents such as all the notes taken by the N.I.H. staff

during the previous investigation and copies of the statistical analyses of Imanishi-Kari's handwritten data as well as copies of the forensic reports that the O.S.I. commissioned. Dingell's staff sometimes visited the O.S.I. to demand materials directly. Hallum, who spent three years in the Marines during World War II, says that they were "the most foul-mouthed staff" he had ever come across. "They were bullies."[33]

Kimes says that under the circumstances the O.S.I. never felt that it could conduct the investigation with "total autonomy," pointing out that, since the subcommittee was conducting a parallel investigation with access to all the data and reports, they felt threatened with being second-guessed. Whenever Hadley had to make an important decision, she and her staff would ask themselves if they could defend it in a hearing. At one point in August 1990, Hadley met with several staff members of the Dingell subcommittee and the next day spoke with Walter Stewart about the Imanishi-Kari case. Stewart's full-time assignment to the subcommit-tee had ended in May, but he remained available to it several days a month. Hadley's notes of the conversation reveal that Stewart, emphasiz-ing some particular feature of the case, conveyed to her that the subcom-mittee considered it "crucial—not to be compromised. If we don't do—they will have to. Word to wise." The note added, "Test of our ability to do these things." Kimes recalls, "We were always under the watch of Dingell"—and under the watch of Dingell's allies in the press. His staff leaked material to journalists, including "very confidential, private infor-mation" about people who, like Imanishi-Kari, were only under investi-gation. His staff "were totally unethical," Kimes says.[34]

The O.S.I. gave the impression of having an unseemly relationship with Dingell's people, one that Bruce Singal certainly considered prejudicial to his client. He holds that the constitutional separation of powers was being violated to Imanishi-Kari's disadvantage. Congress controls the purse strings of the N.I.H. and the committee with authorization power over the agency was publicly claiming that she was guilty without having held a proper trial. "We absolutely felt that pressure was being exerted by staffers on the Dingell committee to get the O.S.I. to arrive at a preor-dained result that would vindicate the committee."[35]

Jules Hallum says he felt that the case was "politicized," mainly by Dingell but also by attitudes within the O.S.I. Although he came even-tually to believe that Imanishi-Kari was guilty, he recalls that when he arrived full-time in Bethesda in the spring of 1990, the office seemed pervaded by a "hanging-judge" atmosphere. About six months after the conclusion of the N.I.H. investigation, Bruce Maurer had returned to his regular post in the agency, but he remained peripherally involved in the case, sitting in on several meetings. He remembers feeling "a kind of anger at that office and at Suzanne [Hadley] and the process" because of "the intensity of their prosecutorial perspective." He says, "It seemed as if they were almost single-minded in trying to show that Thereza was guilty . . . I remember Mosimann . . . talking about some of the statistical analysis. He was really quite excited about some of the things he'd found. It was almost like a feeding frenzy."[36]

Hadley appeared at the Dingell subcommittee hearing on May 14. Congressmen Dingell and Wyden obtained an account of her prolonged and unsuccessful attempts to get Imanishi-Kari's cooperation and they had her enter into the record the O.S.I.'s several months of correspondence with Singal. They queried her about the closing down of Imanishi-Kari research grant. Hadley explained that the O.S.I.'s accumulating evidence against Imanishi-Kari had raised questions about her "fitness" to hold one. She said that the grant had not been canceled on grounds that Imanishi-Kari had falsified her laboratory records but that such grounds had not been ruled out.[37]

"A clear attempt to smear her," Baltimore characterized the hearing to a reporter. Six weeks later, on July 1, 1990, Baltimore became president of Rockefeller University, taking office against the continuing opposition of an estimated third of the senior faculty. Many of them, along with a small but growing cadre of scientists elsewhere, feared that his attacks on John Dingell might jeopardize congressional support for biomedical research. They now faulted him even more for his defense of Imanishi-Kari.[38]

Several days before the hearing, in a letter to Singal, Hadley said that she was glad to learn that Imanishi-Kari would cooperate if the O.S.I. would provide a list of specific allegations. In June, Hadley sent Imanishi-Kari a list of fourteen detailed questions about her data. Taken together, the questions expressed the main issues the O.S.I. investigation had identified, including the allegations concerning the specificity of Bet-1, double producers, O'Toole's "powder keg" proof, the disparity between the grant proposal data and the *Cell* paper data, the June subcloning, and the statistical peculiarity of the handwritten data counts. Hadley advised Imanishi-Kari that she could respond in person, if a meeting could be arranged, or in writing.[39]

A number of the questions turned on the forensic results that the Secret Service had summarized in its two brief reports, especially the alleged mismatch between the dates written on the notebook pages and the period when, according to the Secret Service, the tape on them had likely been produced. The Dingell subcommittee—and now the O.S.I.—had not provided the fuller account of the analyses that Singal had requested and it had not given Imanishi-Kari access to her original notebooks. She complained to the O.S.I. that she did not have all the exhibits to which the Secret Service referred, could not in some cases even identify the exhibits, and was, as a result, unable to answer some of the questions "as completely and accurately as I would like." She said again that she wanted her original notebooks so she could have them subjected to independent forensic analyses—a scientific process that would be properly "accompanied by detailed lab reports" and that "would help to confirm the honesty and reliability of my data."[40]

Imanishi-Kari nevertheless responded to all the O.S.I. questions in a thirty-seven page memorandum in mid-July and submitted in mid-October to an extensive interview with the O.S.I. staff and its advisory scientific panel. Singal explains that he permitted Imanishi-Kari to respond despite the absent information and materials because "it came down to a question of her having to defend her honor," observing, "It's one thing for a lawyer to take a legalistic position, which I think was justified, but what's more important ultimately is her reputation to her."[41]

Writing in July, Imanishi-Kari demonstrated that the "powder keg" proof was no proof of anything and that the discrepancies between the grant data and the *Cell* paper data were only apparent, largely the product of judgment calls based on giving molecular results greater weight than serological ones.[42] She said that she and Weaver believed that hybridomas appearing to be double producers were really double clones—that is, two different hybridomas in the same well, each generating a different type of antibody. They did not discuss double producers in the paper because they had reasonable evidence in support of their belief and none to the contrary. Double producers were "Dr. O'Toole's theory and came up only after publication of the *Cell* paper," Imanishi-Kari noted. She declared that the frequency with which different digits occurred in her handwritten background counts was meaningless. The O.S.I. had no information on how she normally rounded numbers, and it had provided no evidence that the rounding she used was "unusual." The numbers were in the range of several hundred counts per minute. In background counts, only the first digit, representing the number of hundreds, meant something; the second digit, representing the number of tens, meant nothing; and so she had rounded the second digit "casually."[43]

Imanishi-Kari stressed to the O.S.I. that most of the pages questioned by the Secret Service "were not even published in the article and are simply unpublished records in my own notebooks." She thus "had no motive or reason to falsify such information." As Imanishi-Kari understood it, the O.S.I. theorized that she had created the allegedly phony data to refute the challenges to the *Cell* paper that began with O'Toole's in May 1986. For one thing, however, most of the disputed data was irrelevant to the allegations O'Toole made in May 1986, Imanishi-Kari said. For another, according to the Secret Service, most had been produced at the latest several months before that date. Why would she fabricate data when the paper had not yet been attacked?[44]

So far as Imanishi-Kari could tell, the Secret Service had identified only four pages in her notebook that might have been prepared after she first learned of the allegations. To the best of her recollection, the experiments reported in those pages had been performed and the pages written no later than 1985. But even giving the Secret Service some due for the sake of argument, Imanishi-Kari contended that none of the four pages provided evidence of fraud if considered in the context of the science they had been allegedly fabricated to support. One of them had nothing to do with any issue raised in 1986 or since. Another, numbered 113 in the I-1 notebook, displayed data indicating that Bet-1 was not specific. "If I was trying to make up data showing Bet-1 was specific in response to Margot's allegations, would I make up data that would hurt me?" Imanishi-Kari pointedly asked.[45]

One of the remaining two pages—numbered 41 and 43 in the I-1 notebook—contained data showing that an idiotypic birthmark resembling the transgene's was found on antibodies whose isotype, IgG—that is, *gamma*—could only have come from native genes. Imanishi-Kari supposed that "someone could say these were data I would have had a motive to make up." However, she continued, referring to other pages in both her notebook and Moema Reis's, "I did not have to make them up because other data exist which prove the same point." Imanishi-Kari noted that, not having kept a daily diary of her work, she could not accurately explain all the findings of the Secret Service. "What I can say with absolute certainty is that my data were *not* fabricated."[46]

At her interview in October, Imanishi-Kari pressed the issue of motive again, pointing out that she had not been told before she met with Davie, McDevitt, and Storb in June 1988 that the matter of wells versus clones would arise. She asked how she could have made up the June subcloning data before she knew they would get into that issue. She said she had no idea why O'Toole or anyone else would think that she had fabricated data that made Bet-1 appear to be more specific than it really was. There was no more scientific reason to have done so than deliberately to have overstated the specificity of the reagent in the *Cell* paper. The scientific disincentive for such fabrication was identical to what the Wortis committee had recognized in concluding to ignore the overstatement: The lower the discriminatory power of Bet-1, the higher the number of hybridomas appearing to produce transgenic antibodies and the lower the number appearing to produce native ones—and, withal, the weaker the support for the *Cell* paper's central claim.[47]

Before the interview, Hadley had sent Imanishi-Kari two reports that the Secret Service had recently delivered in response to the work the O.S.I. had commissioned. The most significant results involved the comparative analyses that the O.S.I. had requested of tapes in Imanishi-Kari's I-1 notebook, particularly the green tapes, with those of known date in other notebooks produced in the lab during the same period. The comparisons were exemplified by the determinations concerning the green June sub-cloning tapes: According to one of the Secret Service reports, the printer ribbon ink and the tapes differed from the ink and tapes in another notebook of Imanishi-Kari's "even though these experiments were performed using the same counter machine on or around the same dates." These tapes were "most consistent with" tapes found in a notebook of Maple-thorpe's—one of a series designated "C" that the Secret Service used for its comparative purposes—that spanned November 1981 to April 1982, long before June 1985, when Imanishi-Kari claimed the subcloning had been initiated. The Secret Service stated that "no other matches were found to these counter tapes."[48]

At the interview, Bruce Singal held that the two new reports meant no more than the two prior ones, pointing out that the reports did not reveal the basis "for the contention that it was the same counter machine." In her July response, Imanishi-Kari had noted that the Secret Service report of May 1990 spoke of comparing her tapes with "authentic" ones in the notebooks of other researchers. Information about those notebooks was one of the items that Singal had requested from the subcommittee, to no avail. Imanishi-Kari wondered in July how she was "to compare my tapes to these mystery tapes," to explain " 'findings' based on a comparison of my tapes to tapes I cannot see?" Now, in October, Singal insisted that it was difficult to respond to the new comparative analysis without being able to inspect the C notebook and he challenged the leap from "saying something is most consistent with" to "saying it is a match."[49]

Imanishi-Kari suggested that comparisons based on ribbon inks and tapes were misleading. Spare ribbons had been kept on a shelf near the printers and had been changed at will, depending on whether the scientist using the machine was satisfied with the printout. Researchers in the lab often changed the paper rolls, too; if one seemed about finished at the end of the day, someone might install a fresh roll so that an overnight counter run would not be ruined by a lack of printer paper. The remnant rolls were usually left near the printers, where they were used by scientists reluctant to find a new one. When a printer had to be sent out for repair, the tapes were produced by a substitute that likely had a different ribbon and different font.[50]

Singal reminded the O.S.I that the dates in Imanishi-Kari's notebooks "do not reflect the dates of the experiment." Imanishi-Kari herself declared that she had never annotated her pages for "legal" purposes, never jotted anything down "for people to go and check A plus B or C on a particular date." She explained that dates on her notebook pages might refer to when the cell line was established or when it was frozen, or when the supernatants were collected. They did not refer to when the radiation counter was run and the resulting printer tapes were taken.[51] She told the O.S.I. that the June dates marked for the subcloning indicated when the hybridomas had been selected for the procedure, but it took time for cells to grow, and a number of hybridomas had been cloned. Thus, the subcloning had been done in successive parts extending over a number of months, possibly to as late as January 1986, and counter tapes had correspondingly been generated at different times, when each part was ready for assay. Thus, the subcloning tapes, or any of the tapes from similarly extended experiments, would have been done at widely separated times and on dates different from those entered in the notebooks.[52]

On November 26, in a memorandum to Hadley, Singal drew on the recent interview to elaborate on his long-standing dissatisfactions with the O.S.I. investigation. Instead of a coherent set of all allegations, Imanishi-Kari had been "confronted with a free-floating and diverse set of questions, issues, and suggestions which have failed to detail any bill of particulars against her." Furthermore, from the outset of the investigation, a fundamental obligation of American law had been reversed: The burden of proof had "been imposed on Dr. Imanishi-Kari to prove her innocence, rather than on her accusers to prove her guilt." The task of proving a negative was difficult enough, Singal noted, but it had been exacerbated by "the unavailability of the evidence" that the O.S.I. had purportedly developed against her. Just in the last several weeks, in an exchange that was symptomatic of the obstacle, he had asked for but—by instruction of the U.S. attorney's office in Maryland to the O.S.I.—had been denied copies of the C notebooks that the Secret Service had used in its comparative analyses and cited in its September reports. Singal held that, on the face of it, the Secret Service work was "flawed by its overreaching and lack of precision."[53]

Singal summarily contended that the O.S.I. had left Imanishi-Kari with "a hodgepodge of isolated pieces of data allegedly written on different dates than stated or in a manner which deviates from the statistical norm," continuing:

> There is no finding that the particular challenged data are significant with respect to the *Cell* paper or the unpublished data challenged by Dr. O'Toole. There is no showing that these data were written after May 1986, when they first could have been used to respond to such challenges. There is no evidence that these data were so different from other data derived from comparable experiments that they must have been falsified. . . . In short, there is no case.[54]

At the time of Imanishi-Kari's interview, the O.S.I. was already working on a confidential draft report of its finding that would be sent for comment to Imanishi-Kari, Baltimore, Weaver, and O'Toole and then turned into a finished document. While working on the report, the O.S.I. kept accumulating evidence. At the end of November, O'Toole telephoned Hadley to tell her, according to Hadley's record of the conversation, that at a recent social occasion Brigitte Huber's husband said that Henry Wortis had " 'pressured her' " to sign "a document saying she saw [the] subcloning data" in 1986 even though she "didn't remember if [she] really saw it." In mid-January, the Secret Service delivered a final forensic analysis to the O.S.I. It had located green tapes in the notebooks of four scientists besides Imanishi-Kari. Using paper color, ribbon ink, and type font for comparison, it found a "full match" between the tapes on ten of her notebook pages and the other scientists' green tapes, none of which was dated later than January 1984.[55]

In January, too, Hadley queried Weaver about evidence the O.S.I. had obtained that Imanishi-Kari tended to rely on his molecular characterization of the hybridomas when it conflicted with her serological results and that some of the hybridomas might be double producers. He replied at the beginning of February that he had been unaware that Imanishi-Kari considered his molecular studies definitive, and that the combination of his molecular results and what he knew of her serological ones indicated that only a small minority of hybridomas, if any, contained cells that expressed both the transgene and a native antibody gene.[56]

Hadley wrote the draft report, which was completed within a several weeks of Weaver's reply. It presented seven O.S.I. "Findings." They were labeled "A" through "G," in descending order of "definitiveness." Finding A on the list was the June subcloning data. The draft report concluded that Imanishi-Kari had fabricated it and—Finding B—that she had fabricated another set of data termed "the January fusion" that concerned hybridomas from normal mice. The January fusion data, some of which also appeared on green tapes, showed that those hybridomas rarely if ever generated antibodies marked by an idiotypic birthmark resembling that produced by the transgene. The June subcloning determination rested on the Secret Service's green-tape forensics—the report called them "crucial" to the finding—and the statistical analysis of the digits in the handwritten records of the radiation counts. The way that some numbers were avoided and others favored had persuaded the O.S.I. that the counts were "not authentic." The determination for the January fusion was based similarly on green-tape forensics and on another type of statistical analysis of certain features of the numbers that appeared on yellow tapes.[57]

The O.S.I.'s findings on all other issues fell far below those on the first two in definitiveness. The assessment of the data for the specificity of Bet-1, Finding C on the list, opened by declaring that the reagent "played a crucial role" in the experiment. However, it then claimed that inadequate specificity of the reagent would lead Imanishi-Kari to "detecting transgene rather than endogenous [that is, native] genes"—which was the opposite of what it would lead to. The assessment proper comprised an intricate, eighteen-page exegesis of the "discrepancies" in dates, ink and counter-number evidence, and Imanishi-Kari's testimony. It pointed to various "unexplained discrepancies" and proposed that they might be accounted for by "an obvious explanation": that the record of Bet-1 experiments had been "falsified," prepared to "answer the challenge raised by Dr. O'Toole" and the deficiencies in the reagent indicated by the data in the seventeen pages.[58]

Finding D dealt with double producers. No matter that the coauthors had convinced themselves that few if any transgenic hybridomas fell into that category: The O.S.I. substituted its own scientific judgment for theirs, citing data in the notebooks and drawing on points that O'Toole had raised in June 1986 to argue that the possibility of double producers should have been addressed experimentally and discussed in the paper. The draft report used Weaver's recent letter to chastise Imanishi-Kari for failing to keep him adequately informed of contradictions between her serological results and his molecular ones. It asserted that Weaver "knew or had reason to know" that some double producers existed. It pronounced his handling of the *Cell* paper's composite Figure 4 "the most troubling matter," insisting that he should have called attention to the smudge on the radiograph that indicated possible expression of the transgene.[59]

Finding E questioned the legitimacy of several of Imanishi-Kari's notebook pages undergirding the utility of another reagent but could not show they were fraudulent. Finding F concerned the veracity of the *Cell* paper's Figure 1, with the fourth points that Moema Reis said she had entered directly on the graphs. The O.S.I. held that the recording of data on the figure was marked "at best" by "sloppiness and laxness" but acknowledged that it had found "no clear evidence pointing to fabrication or falsifaction."[60]

Finding G comprised what the draft report titled "Commentary on Immunologic Issues." Here the O.S.I. criticized Imanishi-Kari for "poor scientific judgment," pointing to her use of the thousand-count-per-minute cutoff—one of Stewart and Feder's critiques—in deciding the significance of measurements. Here it also held that the *Cell* paper was "not an accurate reflection of the serological results" from Imanishi-Kari's lab, declaring that a good deal of the antibody activity in the hybridomas could be attributed to the transgene rather than to native genes; that Bet-1 was a poor discriminator between native and transgenic antibodies; and that Imanishi-Kari had given inadequate attention to showing that idiotypic birthmarks similar to those from the transgene actually occurred on nontransgenic molecules.[61]

The O.S.I.'s censure of Imanishi-Kari's uses of serology placed what amounted to an imprimatur on O'Toole's original scientific claims against the paper and on the allegations she had later made concerning the fraudulence of the data for Bet-1 and Table 3. The O.S.I. had managed to prove—according to its standards—only two of O'Toole's eighteen charges, which corresponded to Findings A and B. It evidently found no merit in eight others, most of which addressed Moema Reis's notebooks; they were simply ignored in the draft report. It melded the remaining eight into Findings C through G, summarily characterizing its conclusions for them in the draft report's closing section, "Determinations and Commentary": "It is *probable* [emphasis added] that a substantial portion of the I-1 notebook . . . was falsified." Even then, the O.S.I. had to acknowledge that, as Imanishi-Kari had pointed out, much of the allegedly fabricated data had not been published. It nevertheless insisted that "all the data are relevant . . . in determining the reliability and authenticity of the experimental record." The draft report emphasized that Imanishi-Kari herself had said as much, quoting a remark from her interview in October: "The bad data and the good data, everything is relevant for us to know what is going on." She, however, had obviously meant that all the data figured in understanding the mice, not in developing a record for federal investigators to scrutinize.[62]

Elsewhere in the draft report, the O.S.I was similarly selective towards reasons and evidence. In finding that Imanishi-Kari's unpublished data was likely fraudulent, the O.S.I. ignored her argument that she had no motive to fabricate such results. It paid no heed to her contention that she was not likely to have fabricated data before May 1986, as the Secret Service said she seemed to have done, in order to respond to a challenge from O'Toole that had not yet occurred. Instead, the O.S.I. invented vague scenarios of its own, offering one in the draft report to explain the fabrication of the June subcloning data. It was unanchored in time, and it raised questions about the integrity of Moema Reis:

> Dr. Imanishi-Kari used results in the paper from the 17 pages that could not stand up to scrutiny; subsequently she fabricated unusually clean and convincing subcloning data from these wells (the June subcloning) as a way of legitimizing the use of the suspect data. In this scenario it is possible that the subcloning was never done. If so, the protocol for the subcloning on page 35 [in the I-1 notebook] would indicate that Dr. Reis was a party to the deception. However, it is also possible that the sub-cloning was done and gave unconvincing results.[63]

The O.S.I. found it "noteworthy" that the subcloning data had not been submitted as part of the *Cell* paper, omitting to mention Imanishi-Kari's scientific arguments about why those results were inappropriate for the purpose. It sought to discredit the testimony of the Wortis committee members that they had seen the subcloning data in May 1986, invoking Wortis's reply to Dingell's question about it at the 1989 hearings—"I don't know which June subcloning data you are talking about"—as though he would then have known the argot of the investigators. The O.S.I. observed that at the same hearings Joseph Davie had stated that he believed the *Cell* paper's misstatement about isotyping constituted misconduct. It neglected to add that Davie had gone on to qualify his testimony, saying that his N.I.H. panel was unconvinced that Imanishi-Kari had engaged in misconduct if the term meant intent to deceive. The O.S.I. mentioned Singal's assertion that his client had been denied essential investigative information but said nothing about the nature of the information that she wanted or why she needed it.[64]

The draft report said that O'Toole was "dismissed from her position shortly after she challenged the *Cell* paper," implying that she was dismissed because she had challenged it and neglecting to note that her position was scheduled to end within about two weeks anyway. The document marshalled Herman Eisen's memorandum of December 1987, with its heading, "Allegations of Misconduct," against "later assertions by Tufts and MIT officials that Dr. O'Toole's concerns focused on 'differences in scientific interpretation' and not on fraud''; it neglected to mention that the memo had been written after Stewart and Feder had entered the picture and that the summary Eisen had written six months before, the day after the meeting at M.I.T., dwelled entirely on differences in scientific interpretation, praising O'Toole for the serious merit of the scientific questions she had raised. The draft report stated that the morning in Boston that Imanishi-Kari told the N.I.H. panel about the June subcloning data, she "brought [the data] with her to the panel meeting," a version of the event that invited suspicion that she might have fabricated it the night before. The truth, of course, was that when on the morning in question she arrived at the panel, she simply called its attention to the subcloning data in the notebooks that were already in its possession.[65]

The draft report exceeded the scope of the O.S.I.'s authority, going beyond determinations of misconduct to offer appraisals of scientific judgment and character. The tendency to overreach pervaded the text—but nowhere more so than in its closing "Determinations and Commentary." Here the document rendered verdicts not only on Imanishi-Kari but also on the behavior of Weaver, Baltimore, and O'Toole. Although the O.S.I. had discovered "no clear evidence" that Weaver had attempted "to deceptively manipulate" Figure 4, it found that he had not been sufficiently "alert to, nor concerned about," double producers except insofar as they might have affected the paper's central claim. The document added, "This narrow focus was inappropriate for the first author of an important paper, and it seems to have been shared by Dr. Baltimore."[66]

The draft report discredited Baltimore, whose persistent defense of Imanishi-Kari it called "difficult to comprehend" and vindicated O'Toole, whose actions it called "heroic." It quoted Baltimore's angry outbursts at the interview with the O.S.I. the previous April, taking them out of the context in which they occurred, omitting his temperate qualifiers, and glossing them as holding, in effect, that the N.I.H. was "somehow responsible for this act of scientific misconduct." The O.S.I. called Baltimore's statements "extraordinary," adding, "They are all the more startling when one considers that Dr. Baltimore, by virtue of his seniority and standing, might have been instrumental in effecting a resolution of the concerns about the *Cell* paper early on, possibly before Dr. Imanishi-Kari fabricated some of the data" (that is, fabricated it in the judgment of the O.S.I.). The declaration was unaccompanied by any specific suggestion of what Baltimore should have done to resolve the dispute.[67]

The O.S.I. determined that O'Toole had "suffered substantially for the simple act of raising questions about a scientific paper," citing "the loss of her position in Dr. Imanishi-Kari's laboratory . . . [as] only the most visible price exacted of her." It praised her for confidently asserting "the truth as she knew it" and for "willingly and actively" working with the O.S.I. "to sort out the tangled web of data and testimony." The draft report concluded: "She deserves the approbation and gratitude of the scientific community for her courage and her dedication to the belief that truth in science matters."[68]

Hadley kept the five members of the scientific panel apprised of the contents of the draft report as she was writing it and, when it was finished, invited their comments on the completed document. Joseph Davie, William McClure, and Stewart Sell agreed with its conclusions of fraud, but Hugh McDevitt and Ursula Storb still disbelieved the statistical arguments and were highly critical of most of the forensic analyses. "If you take each one of them as a finding," McDevitt says, "the impression analysis . . . the different colors of tapes or the fading of the ink or all of those forensic things, it's not hard to come up with a scenario that explains each one of them alone." The O.S.I.'s reasoning was a "bit like saying, if you have ten quasi-wrong observations and put them all together, they'll make a really [right] one. That's not the way I reason it," he continues. "If they're quasi-wrong, they're quasi-wrong. . . . Ten of them or a hundred of them, they're still questionable. And if you take that point of view, then where's the proof of fraud?"[69]

McDevitt says that he had a "running disagreement with Hadley." She pointed out that a finding of scientific misconduct did not require proof beyond a reasonable doubt but only proof supported by a preponderance of evidence. McDevitt says that he would tell her he wanted "beyond a reasonable doubt" and that she would respond, "But that's a criminal criterion." He continues, "I said, 'Well, if you're going to take somebody's livelihood away from them, that's about as serious as sending the person to jail for a few years.' If I had my choice, I'd rather go to jail for a few years for cheating the phone company and still be allowed to be a scientist rather than have my profession taken away from me." McDevitt adds that in a number of the panel discussions he and Storb "just sat there and said, 'No dice.' "[70]

Storb says that by this time her earlier enthusiasm for how Hadley had been running the investigation had cooled considerably. She recalls, "I became suspicious that she had sort of her own agenda and that it was really to have Margot O'Toole's opinion win, and that she was really not impartial enough." Storb adds that she began to suspect that the presentations given the panel were "selected." She says her impression of selectivity was compounded by the way the O.S.I. relied on the statistical evidence, which she considered meaningless, and the forensic evidence, which both she and McDevitt thought was flimsy.[71]

McDevitt and Storb were particularly skeptical about the Secret Service's finding that the inks on Imanishi-Kari's tapes did not match the inks on any of the tapes in the notebooks of other scientists that were printed around the putative time of the January fusion and the June subcloning. The inks on her tapes matched the inks found only on tapes generated several years earlier. McDevitt insisted that the O.S.I. check that finding against as many other notebooks as possible, which the investigators did, obtaining sixty notebooks for comparison. He remembers that, even then, he said he wanted to see the ink analysis and "they said, 'We can't show you the ink analysis . . . because it's in the hands of the U.S. attorney.'" McDevitt kept insisting, declaring, "I'm not going to buy this. Either I see the goddamn tapes or I'm not going to accept it."[72]

Finally, the O.S.I. arranged for McDevitt to visit the Secret Service with McClure. (Storb was unable to go because of another commitment.) The agents explained how they had done the ink analysis with a technique called *thin-layer chromatography*. The technique uses a glass plate that is covered with a thin layer of a material like cellulose and that, after a sample of ink is applied to it, is dipped into a solvent. McDevitt explains, "Black ink is a mixture of yellow, red, and blue inks. If you put a little dot of it on a thin layer plate, you'll very easily see it resolve into different dots of yellow, red, and blue. To compare inks, you just compare the positions of the dots on the plates, setting the plates side by side."[73]

McDevitt recalls saying on his visit to the Secret Service, "I've looked at a lot of thin-layer plates" and asking to see the ones used for the inks. He says that he was told the plates were in the hands of the U.S. attorney and unavailable—and that, in any case, the Secret Service was expert at its work. He retorted with some annoyance, "But who reviews you. If you're the experts, who decides whether you're right or wrong?" He remembers being told that the Secret Service's "evidence is challenged in court," adding, "That's where we left it. . . . I didn't see the plates."[74]

Joseph Davie recalls that, technical issues aside, the panel had "lots of complaints" about the tenor of the report. Hadley had become so convinced that Imanishi-Kari committed fraud that "she took this on as a cause," making the report "too personal," including material that seemed "inappropriate to a sterile scientific assessment." He adds that, although she modified some of the text, she kept most of it "intact." In the end, Davie, McClure, and Sell agreed to support the document that Hadley produced, but McDevitt and Storb told Hadley that they wished to file a minority opinion. McDevitt says he wanted especially to dissociate himself from the assertions in the final section that "Baltimore is a fraud and O'Toole is wonderful."[75]

McDevitt and Storb's dissent was short, barely more than three pages, but pointed. They noted that the evidence addressed in the draft report findings labeled C through G, which ran from Bet-1 through the "Commentary on Immunologic Issues," might be accounted for by "several alternative explanations." They thus contended that these O.S.I. findings were "not, in our judgment, justified." They said that they also disagreed with the statistical analyses used to help show that the data from the June subcloning and the January fusion—Findings A and B—were fraudulent. They explained that the analysis of digit frequencies in the June subcloning counts was "new and untried, particularly in establishing proof of fraud"; it seemed "interesting, but not compelling." They declared themselves simply unable to accept as proof the statistical evidence derived for the January fusion. They expressed regret that they had not been given access to the thin-layer plates involved in the chromatographic analyses of the inks, observing that "ultimately, the actual chromatograms need to be examined." Subject to that reservation, they concluded that the findings of fraudulence concerning the June subcloning and the January fusion were "likely" true. McDevitt and Storb pointed out, finally, that the conclusions in the closing section of "Determinations and Commentary" were "those of the Office of Scientific Integrity," and did "not accurately reflect" their own.[76]

Hadley did not like the idea of having to incorporate the minority opinion, McDevitt says. It was included in the draft report but not mentioned in the table of contents. On March 14, 1991, the O.S.I. sent copies of the document to Imanishi-Kari, Baltimore, Weaver, O'Toole, and the scientific panel. Someone, somewhere also provided a copy to Congressman Dingell's subcommittee.[77]

■

Bad for Science

THE REPORT, stamped "DRAFT" and "CONFIDENTIAL" on almost every one of its 219 pages of text and appendices, was promptly leaked. On March 20, 1991, it arrived via Federal Express at the *Washington Post*, together with a sheaf of twenty-four photocopied articles that had appeared about the case in the previous two years. A young reporter named Malcolm Gladwell was assigned to write the story. He remembers his editor telling him that the materials came from "Stewart and Feder, of course." In a telephone conversation with one of the pair, Gladwell learned that they had sent the report together with the kit of background articles to a number of press organizations across the country. The next day, it was page-one news that the O.S.I. had found Thereza Imanishi-Kari, the collaborator whom the Nobel Prize winner David Baltimore had defended, guilty of fraud.[1]

Detailed adumbrations of what the O.S.I. would conclude had actually been published in *Nature* six months before and in *Science* several weeks earlier, in articles that said they were based on a variety of sources—all of them sounding like Dingell's staff; they included "congressional investigators," "sources close to the N.I.H. investigation," or "sources briefed by the N.I.H. Office of Scientific Integrity."[2] Likely enough, after the Dingell subcommittee received the draft report, a staff member leaked a copy to Stewart, who in turn promulgated it to the media. Several months later, in connection with an investigation of the leak by the N.I.H., Dingell wrote to Louis Sullivan, the Secretary of Health and Human Services, "We . . . understand that there is no legal constraint on anyone *outside* the N.I.H. from releasing a draft report, and presumably no prohibition on anyone disseminating a draft report after it has been leaked." He called it "totally unacceptable" that anyone should "intimidate and silence critics of the way major research institutions have handled scientific misconduct cases." The review by the N.I.H. determined that the subcommittee had "authorized" Stewart to release the report. Stewart adamantly refuses to discuss how the leak occurred, but he holds that the public needed to know the draft report's contents. He notes that the speedy release of the report dispelled the cloud that he says was hanging over O'Toole's head.[3]

The leak decisively fixed the impression of the case in the public mind. In the several weeks following the initial release of the document, Imanishi-Kari was unquestioningly taken to be guilty in newspapers, on National Public Radio, and, in Boston, on a broadcast of the widely watched "Ten O'Clock News" that featured an interview with Margot O'Toole.[4] The *Wall Street Journal* complained that reporters had wolfed down what had been served them "like a Domino's pizza." Only two newspapers besides the *Journal* itself mentioned Storb and McDevitt's minority dissent and hardly any called attention to the fact that the draft report was a draft, a statement of tentative findings to which Imanishi-Kari had not yet had a chance to respond. Malcolm Gladwell later reflected, "It's just completely inexcusable that we never understood what we had. Perhaps worst of all, the O.S.I. never told us what we had." Hadley had nothing to say about the leak except that it was "regrettable."[5]

The media hailed Margot O'Toole the way the O.S.I. had presented her—as a courageous hero who had suffered grievously for challenging the *Cell* paper. Earlier reports about the price she had paid for her whistle-blowing now turned into a commonplace story—she had been fired, denied recommendations for jobs, and blackballed in the immunology community. The *New York Times* reported that she had also "lost . . . her house," a misimpression that *People* magazine unintentionally reinforced by noting that she had moved her family in with her mother. "She should not have had to endure such pain over trying to be honest," the *Baltimore Sun* editorialized.[6]

O'Toole, saying that she disliked being portrayed as a victim, explains the complexity of the circumstances that led to her housing arrangements. In the late spring of 1987 she and her husband sold a house they owned in Jamaica Plain because they planned to buy another large enough to accommodate themselves as well as her sister and her sister's roommate, who were being evicted from their apartment. They soon purchased such a house, paying about $66,000 more than the price they had obtained on their previous residence, but they could not afford to live in it for a time, partly because she was unemployed. She insists that she was particularly offended by the reports that she had lost her house and the attendant implication that it had somehow been a hardship to move in with her mother. However, she said nothing publicly to correct that misimpression or those concerning her departure from Imanishi-Kari's lab. In the press, she tended to enlarge on her long-standing claim that she had battled for truth in science and been greeted with callous cynicism. "When I first brought the challenge, they considered it improper . . . and I was denigrated for having analyzed data," she said on "The Ten O'Clock News." She told a *Newsweek* reporter that she came "under intense pressure to fudge my own data," adding, "[Imanishi-Kari] would tell me what to do, down to which [data] values to keep and which to throw out."[7]

O'Toole informed other journalists that once she saw the seventeen pages "I knew [the conclusions] had been fudged," and that she had challenged the paper so as to warn others not to try extending its findings "because they don't exist." In the program "All Things Considered," on National Public Radio, she reiterated her story about having been told after the Tufts inquiry "that a retraction of the paper or a correction of the paper would jeopardize funding and that the best thing to do was to not say anything about the paper being wrong." She declared that Baltimore simply did not care to correct the errors in the paper, explaining that his attitude amounted to: "Don't retract stuff—just wait for some poor fool to repeat it and find out it's wrong." She added, "The thing most upsetting to me is the contempt they held for the labor of people trying to repeat the work. I felt it acutely because I had worked for a year trying to do just that."[8]

Virtually no journalist questioned O'Toole's version of events. She advanced her indictments on television as well as radio and presumably to print reporters with soft-spoken, measured, and resolute articulateness. Anthony Gottlieb, the science editor for the *Economist*, told a TV interviewer in Boston that he was "extraordinarily impressed by her honesty," elaborating, "Whenever it was possible to check anything that she said, it was exactly right. . . . She struck me as a person who would very rarely exaggerate and who would attempt to be precise and as honest as possible."[9]

The media paid even more attention to Baltimore's conduct than it did to O'Toole's. David Warsh, of the *Boston Globe*, judged that Baltimore's "big crime is that he stood up to Dingell" and the "thuggish congressional behavior" that harked back to "Senator Joe McCarthy and the House Un-

American Activities Committee." But Warsh was in a tiny minority. Most
columnists and editorialists lambasted Baltimore for wronging O'Toole—
and for a good deal more. Exemplifying the reaction in an essay in *Time*
magazine, the commentator Barbara Ehrenreich held that "Baltimore
pooh-poohed O'Toole's evidence and stood by while she lost her job,"
adding, "What he lost sight of, in the smugness of success, is that truth
is no respecter of hierarchy or fame. It can come out of the mouths of
mere underlings, like the valiant O'Toole."[10] *Science* published extensive
excerpts from the draft report's commentary on Baltimore's conduct, and
several newspapers took unadmiring note of his heat-of-the-moment
observation that scientists "could make up anything they wanted" in their
notebooks. Baltimore was attacked for refusing to acknowledge that a
paper with his name on it might be faulty and for being concerned to
protect himself as much as to defend Weaver and Imanishi-Kari.[11]

At the *New York Times*, Philip Boffey, now an editorial writer for the
paper, appeared to embrace Daniel Greenberg's view of the matter. Several months before the leak, Greenberg's *Science and Government Report*
had declared the Baltimore case "a mini-Watergate" and attacked Baltimore himself for having "orchestrated and led a national campaign to
depict Dingell as an anti-science McCarthyite." Now in an editorial written by Boffey, the *Times* dubbed the scandal "A Scientific Watergate,"
scathing Baltimore for "stonewalling" and for having "orchestrated" an
assault against congressional intervention in American science. It likened
the long affair to a huge cover-up, declaring that "the Baltimore case
started with apparent fraud by a single scientist and soon led to a widespread denial of wrongdoing by almost everyone in a position to right the
wrong."[12]

Greenberg offered kudos to Stewart and Feder for their role in the
dispute, and so did O'Toole, declaring that "without their involvement
. . . the case would have died, and I would have been the scum of the
earth forever." John Maddox agreed to publish Stewart and Feder's analysis of the *Cell* paper in *Nature*.[13] Greenberg also pronounced Dingell
"vindicated," and the *Detroit News* published an editorial grudgingly conceding, "Dingell: He Was Right." Dingell himself castigated academic
science as a fiercely self-protective, "closed world" that treated "intruders
. . . poorly . . . at best." He held on National Public Radio that Tufts and
M.I.T. "were recalcitrant, took punitive action rather against Margot
O'Toole than against the parties who were involved in the actual wrongdoing."[14]

For some months, the Dingell subcommittee had been interested in
whether Tufts and M.I.T. had responded inadequately to O'Toole's chal-

lenge. Early in March 1991, Hadley had said that the O.S.I. would soon initiate a "who-knew-what-when" investigation of the two institutions, and at the end of the month, shortly after the leak of the draft report, Peter Stockton told a reporter that the subcommittee intended to hold hearings on the institutional response, calling Imanishi-Kari and Baltimore to testify.[15] About this time, an audit of the Whitehead Institute by the Office of the Inspector General of the N.I.H. revealed that in 1988 the institution had paid about $100,000 to Akin, Gump, the Washington law firm that had sought to intervene on Baltimore's behalf with the Dingell subcommittee; and that the Institute had improperly charged $68,000 of the total to the federal government, as part of the indirect costs of doing research. Baltimore had in fact specifically instructed Whitehead officials to exclude such legal fees from indirect costs; they had been included by mistake; and the Whitehead promptly reimbursed the government for the amount in full.[16] The episode nevertheless left the impression that Baltimore's administration of the Whitehead had been marked by shady practices, a possibility that the Dingell subcommittee indicated it proposed to pursue.

Stockton told one of Baltimore's attorneys that the subcommittee wanted to charge a cover-up but lacked hard evidence of any kind. The subcommittee seemed bent on a fishing expedition. In a conversation with a reporter about the case, one of Dingell's aides remarked of Imanishi-Kari, "Will you tell her lawyer that if she wants to cooperate with us, we'll do great things for her? She's the poor, pathetic soul who bears the brunt."[17]

The day the newspapers broke the story of the draft report, a copy of the document arrived via Federal Express from Stewart at the Harvard office of Walter Gilbert. Gilbert had sections of the report photocopied and circulated them to colleagues at Harvard. He told a reporter for the *Boston Globe* that the whole affair amounted to "a Greek tragedy," explaining, "It's hubris that brings down the tragic hero. The most charitable thing to say is that [Baltimore] took some arrogant view that 'I can do no wrong,' that 'whatever I write in science is true.' He couldn't admit to a mistake. If he had, this wouldn't have happened."[18]

Baltimore's detractors at Rockefeller University were quick to raise I-told-you-so questions about the meaning for his presidency of what they took to be his stubbornly misguided conduct. One of the opponents of his appointment, a distinguished biologist named Günter Blobel, let a reporter know that he felt "sadly vindicated," continuing, "We'd had concern about Baltimore being involved in an unsettled affair; we felt he hadn't handled it right, and that's why we told the Board of Trustees, don't do it. The board decided to ignore our recommendations, so it will have to live with their actions and the outcome."[19]

To many scientists in and out of Harvard and Rockefeller, the post-leak storm could not have blown in at a worse time. Federal support of basic research and training in biomedical science had boomed in the 1980s, but now, amid the recession of the early 1990s, it was flattening at a level inadequate to fund the research projects of the much-expanded corps of biomedical scientists. In the several years before 1990, the combined number of competitive grants awarded annually by the N.I.H. had fallen by 25 percent, to 4,600, a smaller number than had been funded a decade earlier, and the proportion of meritorious applications that the agency was actually able to fund had plummeted by a greater fraction. Money was so scarce as to prompt a prominent Harvard biologist to speak of a "famine" in the biomedical sciences and an advisor to the N.I.H. to predict that "a few rounds of funding at the present award rates will very quickly result in a reduction in the number of active laboratories to less than half their current number."[20]

Adding to the reverses was the case of Robert Gallo, the biologist at the N.I.H. who in 1984 reported that his lab had identified the virus, HIV, that causes AIDS. A number of AIDS activists suspected him of self-promoting indifference to their plight, and in November 1989, a reporter named John Crewdson, a Pulitzer Prize winner at the *Chicago Tribune*, published a 50,000-word article claiming that Gallo might have stolen the discovery of HIV from a French team at the Pasteur Institute. It was a theft of lucrative proportions because Gallo's laboratory used its knowledge of the virus to develop and patent a test to detect its presence in the blood. Congressman Dingell demanded that the N.I.H. investigate the charges. Early in 1990, at the request of William Raub, the acting director of the N.I.H., the National Academy of Sciences nominated a blue-ribbon panel under the chairmanship of Frederic M. Richards, a senior biologist at Yale, to advise the N.I.H. as individual consultants on the matter. In October 1990, the Richards panel members said that Gallo was innocent of the charges of theft, but they tentatively concluded that his laboratory records seemed inconsistent with his published claims about HIV. By the beginning of 1991, the O.S.I. was embarked on an investigation of Gallo and a collaborator named Mikulas Popovic, a recent immigrant from Czechoslovakia and a coauthor with Gallo on a crucial paper in the identification of the AIDS virus.[21]

There was also the indirect-costs scandal at Stanford University. In the fall of 1990, reports had surfaced that Stanford University had been billing luxury items to federally reimbursable overhead and might have over-charged the government as much as $200 million. Congressman Dingell held a hearing on the issue on March 13, 1991, just before the draft findings on Imanishi-Kari were leaked. The principal witness for the uni-versity was its popular president, Donald Kennedy, a distinguished biol-ogist who had been commissioner of the Food and Drug Administration during the Carter administration. Dingell announced that the public would hear that day a story of "excess" and "arrogance." He confronted Kennedy with a long list of seemingly indefensible charges to federally reimbursable overhead, including expenditures for a luxury yacht and for cedar-lined closets, monthly flowers, and an antique fruitwood commode in the president's house, not to mention a retreat for the university's trustees at a camp on Lake Tahoe. Kennedy responded that some of the charges had been billed to overhead in error but that, while some of the rest might perhaps be inappropriate, all were legal under federal rules. Nevertheless, Congressman John Bryant, a Texas Democrat, felt justified in accusing Stanford of gouging, declaring, "I think the president has to take the responsibility for it—[attempts] to get more money out of the taxpayer than your university deserved." In July, Kennedy resigned from the Stanford presidency.[22]

And now, topping them all, was the scandal of the Baltimore case, inviting more headlines about hubris, including the judgment of *Time* that it symbolized "the fallibility and arrogance of modern science." The outcome seemed to confirm the claims made by William Broad and Nich-olas Wade in their *Betrayers of the Truth* that the self-regulation of science was a sham. It was not peer review that exposed scientific fraud, journalists like Daniel Greenberg pointed out. It was whistle-blowers and Congress and the press. *Newsweek* noted in April 1991 that, according to a con-gressional study of ten recent misconduct cases, "the most trou-bling aspects were the cover-ups, failures to investigate, antipathy to the whistle-blower, and a closing of ranks against an accuser."[23]

The prevailing climate exacerbated Walter Gilbert's dissatisfaction with David Baltimore and prompted extraordinary criticism of him from Gil-bert's friend James D. Watson, the famed co-discoverer of the structure of DNA, who had been a professor at Harvard for twenty-one years and was now director of the Cold Spring Harbor Laboratory, on Long Island. Gilbert, who had started in science as a theoretical physicist, began his career in molecular biology in Watson's lab. It is said that he owed his tenure at Harvard partly to Watson's strong support. Like Baltimore, both Gilbert and Watson had ridden to riches on the wave of the rapidly devel-oping biotechnology industry. Gilbert left Harvard in 1981 to run Biogen, a biotech firm that he had helped found. Gilbert was soft-spoken, but he could be brusque, imperious, and coldly dismissive. The company did not do well under his leadership. Criticized as too academic in managing it, he left Biogen in 1984 and returned to Harvard, which installed him in a distinguished super professorship. Gilbert and Watson early supported the Human Genome Project, a several-billion-dollar federal initiative to identify all the genes in human cells. In October 1988, Watson was appointed director of the project. In 1990, its awards included a $6 million grant to Gilbert's lab.[24]

Baltimore's conduct distressed several other Harvard faculty, notably a molecular biologist named Mark S. Ptashne and a protein chemist named John T. Edsall. People who knew Ptashne in his scientific salad years remember him as brash, combative, self-centered, brilliant, and so ambitious that one scientist recalled, "Anyone who stood between Mark and what he wanted would soon have a Mark-sized hole in him." Harvard embraced him as one of its own after he graduated from Reed College, awarding him a Ph.D. and a coveted term in its Society of Fellows. In 1971, shortly after he scored a triumph by isolating a protein called a *repressor* that was fundamental to gene regulation in bacterial viruses, the university promoted Ptashne, age 31, to a full professorship. Ptashne, whose parents were socialists, displayed an indulgence for radical politics and a liking for underdogs. He went further than most of his Cambridge colleagues in opposing U.S. Cold War policies, traveling to Cuba in the late 1960s and to Vietnam in 1970, protesting the war partly by lecturing on molecular biology in Hanoi. But he parted company with a number of his friends on the left when in the mid-1970s they endorsed the Cambridge City Council's attempt to ban recombinant DNA research within its jurisdiction. Ptashne scoffed at the fears of such research and outspokenly advocated that it go forward in suitably protected facilities, a policy that the city eventually adopted.[25]

Ptashne disapproved of the growing professorial involvements in the biotechnology boom, but he was persuaded by Harvard officials to file for patents on an elegant system that he had devised for high-volume and reliable expression of genes in bacteria. James Watson, one of his graduate mentors, touted Ptashne's work among venture capitalists. In 1981, Ptashne joined with a consortium of investors to found Genetics Institute, a private corporation that would develop and market his system and that before long made him rich, too.[26] He disliked dwelling on his pecuniary good fortune, apparently preferring to be known as a scientist of high culture and contrarian sympathies. He was an accomplished violinist, who liked to trade up to better violins; eventually he bought himself a Stradivarius. Remaining on the Harvard faculty, he told a reporter that his involvement in the biotechnology industry was just a sideshow in his life. His entry in Who's Who for 1990, saying nothing about his role in business, lists his several scientific prizes and his avocations—opera and classical music. He wore Calvin Klein jeans and khaki army shirts. He might dine at a gourmet French restaurant, then repair to a McDonald's for a shake.[27] He was a wealthy scientist yet still left wing in his predilections, someone inclined to listen to a whistle-blower like Margot O'Toole.

Ptashne attended the conference on scientific fraud at the Banbury Center in January 1989 and he says that he was reassured there about the Cell paper by Henry Wortis. Wortis vividly recalls chatting with Ptashne at the front of the conference room near the blackboard, telling him that it didn't matter that the Table 2 hybridomas had not been tested for isotypes other than mu. He remembers explaining, "We can ignore Table 2 because the critical isotype data is shown in Table 3. There you have idiotypically positive antibodies"—meaning antibodies with the distinctive idiotypic birthmark—"that are gamma." Wortis continues, "At this point Walter Stewart walked up, asking, 'Do you mind if I join you. What's the discussion?' I started to say, 'It's about the isotyping of the hybridomas,' whereupon Ptashne interrupted, declaring, 'Let me explain.' He went through the Table 3 data, explaining why the gamma isotyping was important. Stewart responded, 'Yes, if the data are true as reported, then the paper's claims are valid.' " Ptashne says that he concluded the Cell paper was all right and that some time during the conference he denounced Walter Stewart as a fool.[28]

But the conference piqued Ptashne's interest in the dispute. He attended the Dingell subcommittee hearings in May. Stockton recalls, "He was really coming in to watch them shove it up our ass on the science of this thing. He said that he was absolutely shocked that they [Baltimore and Imanishi-Kari] didn't deal with the issues but that they dealt with the smoke and mirrors." Ptashne was particularly struck by Baltimore's testimony that the Cell paper had drawn support from experiments by other scientists. He read the relevant papers, decided that Baltimore had exaggerated, and introduced himself to O'Toole on the day she testified. In the following weeks, he telephoned Stewart, who says he attempted to convey to Ptashne that the whole paper was a lie, discussed the work in snatches of conversations with O'Toole, and ultimately concluded that she was the only person who had the science right.[29]

In one of their conversations, O'Toole asked for help in getting a job. Ptashne said he would recommend her to the Genetics Institute. Some months later, in April 1990, she went to work at the company as a technician. In mid-June, she told a reporter, "I'm really enjoying it. I'd love for it to turn into a full-time position." Shortly thereafter, she left Genetics Institute to have her third child, but in September, the month that *Nature* published the leak of what the O.S.I. investigation would conclude, the company hired her as a research scientist. Upon the leak of the draft report, Ptashne was quick to tell a reporter, "One of the most surprising things to me is the way so many members of the scientific community and the scientific press were ready to denigrate Dr. O'Toole. They were willing to go to battle with absolute certainty, without bothering to read the paper and think about the likelihood that the paper was wrong."[30]

John Edsall, a tall, somewhat stooped man, was eighty-eight years old in 1991 but agile, energetic, and mentally acute. The son of David Edsall, who had been dean of the Harvard Medical School and of the Harvard School of Public Health, he was a kind of academic patrician, an unpretentious champion of principle. In the 1950s, he had attacked federal research agencies that denied support to allegedly subversive scientists and declared that he would not accept grant money from any government agencies that practiced such discrimination. He encouraged Stewart and Feder's attempts to publish their study of Darsee and his coauthors, finding their paper on the subject "a real contribution to scientific ethics." He had long been concerned with the welfare of whistle-blowers. He offered to write on Feder's behalf against the N.I.H.'s adverse evaluation of his job performance in 1986. In testimony that Stewart and Feder arranged for him to give before the Dingell subcommittee hearings in April 1988, he emphasized that a whistle-blower faced considerable "dangers" in raising an issue of fraud, explaining that he "would have serious anxiety about the future of that individual." He said nothing about Margot O'Toole, thinking that she was honest and sincere but that, given the outcome of the inquiries at Tufts and M.I.T., she might not fully comprehend the scientific issues involved in the *Cell* paper. He wrote to John Dingell expressing dismay that the coauthors of the *Cell* paper had not been invited to testify and pointing out that, so far as he could tell from O'Toole's written testimony, she did not seem to have charged that the paper was fraudulent.[31]

Edsall changed his mind, however, after he read O'Toole's far more accusatory testimony before the Dingell subcommittee in May 1989. "I concluded that her words carried greater weight and authenticity than those of her critics," he says. Edsall was in fact mistaken about several key points in the case, including the substance of the objections that O'Toole had originally raised against the paper, but it was clear in his mind that O'Toole had been "badly treated."[32] By 1991, he had decided that the attacks on Dingell had amounted "inevitably, in effect, [to] an attack on Dr. O'Toole." Edsall declared that the O.S.I. draft report "has convinced me that she was telling the truth from the beginning." He held that if O'Toole's original objections had been "thoroughly investigated at the time," everyone might have been spared "the long ordeal through which they have all suffered." He decided that M.I.T. and Tufts, in attempting to respond to O'Toole's criticisms of Imanishi-Kari's data, had clearly been "in a position of conflict of interest." Edsall also wondered whether a cover-up might have occurred. In his judgment, O'Toole's "standards of scientific conduct coincide with those on which I was brought up, and I deeply regret the unnecessary ordeal to which she has been subjected."[33]

Baltimore's phalanx of critics at Harvard did not form primarily by reason of long-standing affinities among all its members. The group was marked by a variety of differences, including age, professional and research interests, leftovers from past scientific rivalries, and degrees of corporate involvement or lack of it. What most united Watson and the Harvard dissidents in the *Cell* paper dispute was their disapproval of David Baltimore's handling of it and what they believed it revealed about his character as a scientist. Gilbert, Ptashne, and Watson were particularly exercised. Gilbert continued to attack Baltimore openly, faulting him for an excess of "pride," as he put it on a Boston television program, "defending a particular piece of scientific work which in certain aspects never should have been defended."[34]

Watson was said to be telling people that Baltimore had stolen the work on reverse transcriptase—and hence, his share in the Nobel Prize—from Howard Temin. In fact, the claim was wrong, as Watson later acknowledged. Baltimore and Temin had not been in touch for many months before early May 1970, when Baltimore began his decisive experiments on the problem of how RNA viruses reproduce, and Baltimore says he had not known that Temin had turned to a biochemical approach to the puzzle. On Monday, June 1, he telephoned Temin to tell him about his detection of reverse transcriptase and learned that Temin had just announced his own discovery of the biochemical at a meeting in Houston, Texas, several days earlier.[35] Baltimore's paper, already completed, arrived at *Nature* on June 2; Temin's, which required some further work, was received on June 15; and the two were published back to back in the issue of the journal for June 27.[36]

Watson says that he had been involved very little in the Baltimore affair. He had attended the meeting on scientific fraud at the Banbury Center in January 1989 and began to fear that the dispute was headed

"It's in a certain sense a Greek trag-
edy, a great scientist sort of brought
down by hubris, by overweening
pride, in which he couldn't let go of
what was happening."
WALTER GILBERT,
Harvard University
Television Interview with Chris
Lydon, Boston, March 27, 1991
Photo Credit: WGBH, Boston

MARK PTASHNE, HARVARD
UNIVERSITY
At the Banbury Center Conference,
January 1989
*Photo Credit: Herb Parsons, Cold Spring
Harbor Laboratory Archives*

"O'Toole has the kind of scientific standard I was brought
up on."
JOHN EDSALL, Harvard University, Meeting at Harvard,
June 4, 1991
Photo Credit: Courtesy John Edsall

for disaster unless someone attempted to stop it. He says that he tried, by visiting Congressman Norman Lent, whose district was on the South Shore, before the Dingell subcommittee's hearings in 1989. He recalls that he told Lent that the dispute really concerned a fight between two women, that it had nothing to do with David Baltimore. Watson adds that after the hearing, he dissociated himself from the controversy. He holds that he doesn't remember what he said in 1991 about Baltimore's Nobel work but that he thinks he was only repeating rumors about the alleged "theft from Temin" that had been around for at least several years. He adds, "Maybe I shouldn't have said such a thing." That April, at the annual meeting of The National Academy of Sciences, the buzz among biomedical scientists was that Watson had suggested that Baltimore should be stripped of his Nobel Prize and kicked out of the academy. Whether Watson went beyond repeating rumor or not, his circulation of the slander about the theft further blackened the cloud of condemnation over Baltimore.[37]

On April 25, 1991, Ptashne telephoned a member of the Rockefeller University Board of Trustees named Roy Vagelos. A biochemist and physician with extensive experience in research, both at the N.I.H. and Washington University, St. Louis, Vagelos was chairman and chief executive officer of Merck and Co., and Ptashne knew him from his own business involvements. Ptashne says that a high-ranking member of the Rockefeller faculty had asked him to call any member of the board whom he might know because his colleagues judged that Baltimore was placing the university in jeopardy but they felt too cut off from the board to convey the danger. According to a contemporary record of what Vagelos told Baltimore later that day, Ptashne said he was convinced that Baltimore had participated in a cover-up of Imanishi-Kari's fraud, that congressional investigations were continuing, and that the university should get Baltimore to resign and establish as much distance from him as possible. Ptashne said that Watson, Gilbert, Edsall, and others at Harvard all agreed with him. Vagelos, shaken, immediately called several of the group and learned that Watson and Gilbert at least shared Ptashne's opinions.[38]

The depth of feeling against Baltimore that Watson, Gilbert, and Ptashne displayed beginning in 1991 baffled many scientists familiar with all the parties then, and continues to baffle them now. Ptashne had seemed warmly disposed towards Baltimore through the years, enjoying his company in the scientific and social life of Cambridge and living with the Baltimores in a house they rented in La Jolla, California, one summer during the late 1970s while Ptashne participated in a course on retroviruses that Baltimore was teaching. Ptashne says that after hearing Stewart's critique of the *Cell* paper at Cold Spring Harbor, he called Baltimore offering to write a defense of it. Gilbert was a good friend of the Baltimores, especially of Alice Huang, who had brought Gilbert into their social circle. The Baltimores and the Gilberts had sailed together in the South Pacific and toasted in several New Years together at intimate black-tie dinners. Baltimore relished talking with Gilbert. "For many years, I considered Wally as smart as anyone else in the field," he says. "He was one of my real heroes." When Baltimore invited a dozen or so close friends and colleagues to his home to celebrate the creation of the Whitehead Institute, Gilbert was on the guest list. On departing from Biogen, Gilbert approached Baltimore about joining the Whitehead and M.I.T., an idea that Baltimore supported (but that he was unable to deliver on because a preliminary sampling of opinion revealed that the M.I.T. biology faculty was unenthusiastic about the appointment). Although Baltimore—along with many other biomedical scientists—had opposed the Human Genome Project as Gilbert and others originally conceived it, he eventually helped reshape the project in ways that defused opposition to it, signed up in support of the new version, and even joined Watson in persuading James Wyngaarden to enlist the N.I.H. on its behalf. When Baltimore was offered the presidency of Rockefeller University, Watson was one of the scientists to whom he turned for advice.[39]

Asked why Gilbert, Watson, and Ptashne turned against Baltimore with such ferocity, friends and acquaintances speculate that the trio were driven by a Harvard-M.I.T. rivalry or by various jealousies or resentments. Cambridge scientists say, for example, that Gilbert was angered by the turndown at Whitehead and M.I.T. Some of the speculations may be right, but a few are offset by facts—such as Gilbert's having landed on his feet with a distinguished professorship at Harvard—and most are simply impossible to verify. However, if personal animosities were at work, they only further energized a view that members of the Harvard group shared and that some expressed to reporters, to other scientists, and to each other, notably at a small meeting that a number of them attended in early June to discuss the case with Herman Eisen: David Baltimore's conduct in the case had made him a liability in the leadership of American science.[40]

All took the conclusions of the draft report as definitive, disregarding that it was only a draft and that Imanishi-Kari had not had a chance to respond to it. They had been prepared to accept it as gospel by the suspicions that each had come to hold, for different reasons, of Baltimore's defense of the paper and Imanishi-Kari. Watson says that he paid little attention to the details of the report itself and was unaware of the questionable fairness of the procedures that had produced it. He says that he was strongly influenced by Edsall's views after the leak. Gilbert, knowing something about immunology, had examined the report. He was convinced that the June subcloning data was fraudulent by reason of the Secret Service evidence, of McDevitt's having signed off on the matter, and of the O.S.I.'s revelation that Baltimore had seemingly declared with a straight face that Imanishi-Kari had been tricked into publishing phony data.[41]

Ptashne found O'Toole credible on virtually all points. For example, he declared that he knew virtually for certain why the N.I.H. had cut off Imanishi-Kari's grant in 1990: He had learned from O'Toole that the data in the application appeared made up. Gilbert considered O'Toole a scientist who displayed a degree of "moral courage" that was "quite unusual." The Stanford biologist Paul Berg recalls that, in a conversation with James Watson, he expressed puzzlement at Watson's disparagement of Baltimore, reminding Watson that Baltimore had helped him gain support for the genome project. He also wondered why Watson accepted at face value what O'Toole said about the inquiry at M.I.T. in June 1986. Could Watson really "imagine David's walking into a room full of people and saying that he thought the data was fudged but that he wasn't going to do anything about it?" Watson replied that he just believed that "good Irish girl." Watson remarked some years later that he had never met O'Toole but had read her written testimony. He noted that his mother was Irish and added, "Contrary to what some people say, I don't believe the Irish are inherently stupid."[42]

Gilbert and Ptashne held that in 1986 O'Toole had given Baltimore and Eisen ample reason to think that she was really challenging phony data. In Gilbert's view, she had exposed a degree of carelessness in Imanishi-Kari's work that suggested misconduct. Both Ptashne and Gilbert contended that Baltimore and Eisen should have recognized what O'Toole was getting at and acted on it. Publicly, Gilbert insisted that Baltimore should have taken the trouble closely to monitor Imanishi-Kari's data throughout the experiment instead of just trusting her. Ptashne faulted him for having not "found out what the hell was going on." Holding to what was commonplace among the Harvard dissidents, Ptashne contended that, in the face of O'Toole's argument that the paper was in error, Baltimore had been obligated to check the results of the experiment, to redo it if necessary as a matter of "scientific honor."[43]

Ptashne emphasized that, by noting Imanishi-Kari's misreport concerning the isotyping of the Table 2 hybridomas, O'Toole had surely hinted at fraud. He seemed to hold it scientifically insignificant that Imanishi-Kari had tested the Table 2 hybridomas for *mu*, with results showing that most produced antibodies that could not have come from the transgene. He also differed from Wortis in his recollection of what Wortis had explained to him about Table 2 at the Banbury Center in January 1989. Ptashne no longer took the lack of isotyping for *gamma* as unimportant; he now understood it as a crucial deficiency in the paper, enough decisively to undermine its central claim. In a letter to *Nature*, he argued, in contrast to Wortis, that Table 3 could not make up for the deficiency. He suggested that it was unclear which, if any, of the hybridomas reported in the table produced antibodies that were not transgenic and also carried the idiotypic birthmark. He added that the O.S.I. had not been able to resolve the ambiguities in the table even though it presumably had access to far more extensive records.[44]

Ptashne, like many other critics of the *Cell* paper, was asking Table 3 to shoulder a burden that it had not been designed to carry. The table had been constructed by David Weaver to show the surprising behavior of the transgene in the hybridomas of the Black/6 mice. Imanishi-Kari observes:

> Weaver's point in making this table was not to show that [the hybridomas were] all idiotype positive. It was to show there are lots of hybridomas that have transgene DNA and that don't express the transgene. And they are producing a lot of isotype. And they are producing some other things. . . . It happened that most of them are idiotype positive. But they were not selected for that. They were selected for growth. So the idea for this table from a molecular point of view is really that if you have transgene DNA, the expectation according to allelic exclusion would have been that they should be expressed. That's what we expected. The O.S.I. people kept saying that this table made no sense unless [the hybridomas] were all idiotype positive, but it doesn't say anywhere that they are.[45]

Yet even though the table did not explicitly demonstrate that the *alpha* and *gamma* antibodies it reported also carried the idiotypic birthmark of the transgene, the unpublished data undergirding it did, a point that Ptashne apparently declined to accept. Wortis says that after the letter appeared, he called Ptashne, asking, "Why did you write this?" According to Wortis, Ptashne replied that he wanted to show that the paper was scientifically worthless independent of O'Toole's charges. He contended that since the Table 2 hybridomas had not been isotyped other than for *mu*, the central claim of the paper rested on Table 3 and that Table 3 wasn't valid. Wortis remembers asking how he knew it wasn't valid and Ptashne's declaring, " 'Because Margot says it's not.' "[46]

Baltimore had not attempted to repeat the *Cell* paper experiment or critical parts of it in his own laboratory. He says that having examined and discussed the data, he was convinced that nothing was wrong. To be sure, he continues, his lab had no good explanations or at best tentative ones for certain aspects of the science—for example, why the transgene affected the antibody repertoire in the Black/6 mice—but he did not think that progress in understanding would come from repeating the same experiment. He says he felt that the matter might be moved forward by other people taking "a different approach to the problem." Baltimore had declared in his "Dear Colleague" letter in May 1988 that scientific disagreements were resolved by "publishing the results and having other laboratories try to repeat and evaluate them," noting that he had been freely supplying the transgenic mice to investigators elsewhere, like the Herzenbergs. At the Dingell hearings in 1989, he held that the work in other labs had so far supported several important observations reported in the paper and none had contradicted its main finding. More needed to be understood about the control of antibody synthesis to explain the odd immune response in the mice. His own lab was "deeply involved" in research on that question, and he hoped that "one day we can return to the transgenic mice and understand the puzzles they present."[47]

The Harvard critics interpreted Baltimore to be saying that the dispute over the substance of the *Cell* paper was not his problem, and they faulted him for it. Ptashne says that he took Baltimore to be dishonoring the ethics of scientific practice, arrogantly refusing to acknowledge that a paper with his name on it might merit reconsideration. Ptashne recalls that the night before the Dingell subcommittee hearings in May 1989, he had urged Baltimore to be conciliatory, to assure Dingell that he would withhold judgment on the dispute until he could get to the bottom of the science. Both Ptashne and Gilbert held that Baltimore had not attacked Dingell to defend science against congressional intrusion or Imanishi-Kari against persecution. They were convinced that he had only been defending himself. The Harvard critics feared that Baltimore was sending the wrong message about science to the lay world. To their minds, his conduct announced in effect that scientists did not scrupulously weigh every bit of contradictory evidence; that they brushed aside serious challenges; that they sought primarily to publish what they could get away with.[48]

The Harvard critics seemed to prefer the contrary message—that scientists returned to the lab to check out whatever objection might be raised against their results. Their preference was in keeping with a mythic image of science, and some invoked gods in support of it—notably Richard P. Feynman, the late Nobel laureate in physics, who had declared in his bestselling *Surely You're Joking, Mr. Feynman!* that an essential feature of genuine science is "a kind of utter honesty—a kind of leaning over backward . . . [to] report everything that you think might make [an experiment] invalid." Feynman was himself actually something of a myth-maker. A theoretical physicist, he rarely faced the task of compelling experimental apparatus to yield articulate data. He spent a year at the beginning of the 1960s trying his hand at the lab bench in molecular biology and discovering that experiments with living organisms were not easily repeatable. Biology was messy. It often involved guesswork, just like some of Feynman's most important physics. Indeed, he acknowledged that doing creative science required more than unalloyed honesty. In a lecture to students in 1967, he declared that admonitions to pay attention to all data missed the importance of "the imagination and the judgment of what to record and what to omit," adding, "You can't look at everything. When you look at everything, you can't see the pattern."[49]

The human and experimental realities of scientific practice discourage treating equally all data or, for that matter, all objections to published claims. What scientists feel obligated to check depends on the significance of the research that may be challenged, the merits of the challenge itself, and the time and resources available for redoing the work. Baltimore says that if someone counters your data with fresh experimental observations, then you should by all means return to the lab and probe further. But O'Toole's critique of the *Cell* paper's central claim did not rest on new data and constituted a challenge whose merits he deeply doubted. He adds that redoing the *Cell* paper experiment would have required an expertise in serology that his laboratory did not possess and that it would have been "daunting" to acquire, not least because making the reagents to detect the idiotypes was exceptionally difficult.[50]

Baltimore's Harvard assailants appeared inattentive to such considerations. In the early 1990s, they counted the image of science critically important. Ptashne allowed at the meeting in June that if the nation's scientists did not show they could root out their own problems, "they'll say they have to have a policeman in every laboratory. They'll destroy us." Ptashne's associates were undoubtedly less extreme in their apprehensions, but all worried that the image they felt David Baltimore was projecting could seriously damage American research.[51]

The bothersome image, of course, was not that of science as a war-making ally of the military-industrial complex that had invited attack a quarter century earlier. It was an image worrisome for the current era and—for much the same reasons Congressman Gore had articulated in 1981—especially for the biomedical sciences, with their increasing powers to intervene in the living world. It was a projection of science as a corrupt enterprise no different from, say, defense contracting or the savings and loan business, and no more deserving of public trust.

Gilbert, Ptashne, and Watson were all concerned in varying degrees that Baltimore's conduct, particularly his confrontation with Dingell, had generated serious congressional distrust of science. Admiring Dingell for going after corruption in science, Gilbert later said that whenever he encountered scientists who called Dingell " 'a bad guy,' I know they are mouthing almost a Republican propaganda." Watson holds, "My belief is that when a policeman stops you, you apologize whether you're right or wrong." He argues that Baltimore, with his commanding reputation, could have forestalled the fiasco of the case but that instead he "mobilized a mob scene" at the May 1989 hearings.[52]

Watson was known to be acutely concerned with the image of American science. (Watson himself had done a good deal to affect that image when, in 1968, he published *The Double Helix*, his frank memoir of the work that led to the discovery of the structure of DNA. The critic John Lear called the book "a bleak recitation of bickering and personal ambition . . . therapy for those who think of science as a realm permeated with unalloyed idealism and of scientists as plumed knights searching always and exclusively for truth.")[53] Members of the biology community had the impression that Watson was disparaging David Baltimore on grounds that scientists had to show that they could take care of their own problems and, by so doing, avert a negative reaction in Congress that would result in losses in the funding of research. In April 1991, in the annual report of the Cold Spring Harbor Laboratory, Watson in fact warned, in apparent reference to cases like the one involving David Baltimore:

> We must never forget that in the past, our nation has treated us very well. It has valued us for our objective analyses of the natural world, not for bending the facts to suit or own personal needs and prejudices. If our charlatans are not treated for what they are our nation's scientists risk being perceived as just another lobby group more interested in their own private gain than in the perpetuation of our nation's greatness. Drawing the wagons around us to protect our sinners from punishment is not the way to ensure that the voices of scientists are heard more, not less, in the Washington corridors of power.[54]

Three days after Ptashne's telephone conversation with Roy Vagelos, Vagelos spoke with a scientist at Harvard Medical School named Bernard Davis about Baltimore's situation at Rockefeller University. According to Davis's notes of the conversation, Ptashne had given Vagelos the impression that he felt the Rockefeller board "must sacr[ifice] David for [the] sake of science." Some years later, Ptashne averred that Davis's report was "completely preposterous," adding, "The 'sake of science'—what's that? I would never urge sacrificing anybody for anything." However, when on April 25, 1991, Vagelos apprised Baltimore of his conversations with Ptashne, Watson, and Gilbert, he urged that Baltimore talk to Ptashne to find out precisely what the cover-up charges he was predicting might be. Baltimore telephoned Ptashne that day. Ptashne seemed to have no information on the rumored cover-up that wasn't already in the newspapers, but he told Baltimore, whose wife, Alice, was listening on an extension and taking notes, what he was exercised about.[55]

The matter was "very painful," Ptashne said. He conceded that, "yes, perhaps I should have called you first," before telephoning Vagelos, but declared that he was Baltimore's "only honest friend" and that he was "doing this for the American scientific community." He insisted that, despite his pleas to take a conciliatory tone at the hearings in May 1989, Baltimore had publicly "insulted" Dingell, made statements to him that were "very damaging to science." He said that Baltimore, given his scientific stature, "should have demanded a clean accounting of what was going on" from Imanishi-Kari. Now "there will be indictments and felony charges." He continued, "American science is put on the line and David [is] casual about it. MAYBE THE PROBLEM IS TOO MUCH RESPECT [*sic*]. . . . This thing is going to twist in the wind for 2 years and it will be terrible for American science."[56]

Several days later, Alice Huang typed up her notes, summarizing Ptashne's main point in a memo to files: "He said that because science is being attacked and will continue to be attacked if David is President [of Rockefeller], that his resignation would save American science."[57]

T W E L V E

■

"Rough Justice"

DAVID BALTIMORE was completely unprepared for the storm. Within hours of the leak of the draft report, he announced that the document raised "very serious questions about the veracity of" Imanishi-Kari's serological data and that he was asking his coauthors to join in retracting the paper "until such time as the questions are resolved." Beyond that tack, he felt conflicted about how to navigate. On the one hand, he continued to believe that Imanishi-Kari was innocent and that Dingell had abused his power. On the other, his lawyers advised that he would have to eat some crow, predicting that otherwise there was no telling what Dingell might try to do to him. He desperately wanted to defuse the situation, not least so that he could remain president of Rockefeller, his graduate alma mater, and press ahead with the reforms that the trustees had hired him to advance.

Change was already underway at Rockefeller when Baltimore had taken office the year before, important parts of it initiated by his friend, mentor, and textbook coauthor, James Darnell, who had come to the university as a professor in 1974. Younger faculty were being hired and given greater status in the institution, a new, twelve-story lab, dubbed "The Tower," was under construction, and cost-cutting measures were under consideration. In his inaugural address, delivered in September 1990, Baltimore made clear that he intended to proceed a good deal further. He called the university "a paternalistic institution" dominated by "older scientists" that in some ways was "out of touch with the times," scientifically and otherwise. He proposed that, to restore Rockefeller to preeminence, "we must modify the academic structure to give young people a more central role, we must renew the strength of the faculty, we must reconstruct older facilities and complete our new one, and we must carefully evaluate our scientific directions in light of emerging currents in biomedical research."[1]

Baltimore encouraged appointments in cutting-edge areas of molecular biology, fostered the hiring or promotion of several dozen junior faculty, enabled a number of them to have their own labs, and initiated reforms in the governance of the university so that they could participate in it. His modifications may have been locally sweeping, but they aimed at little more than bringing Rockefeller into line with other major research universities in the United States. Although the reforms irritated some senior professors, they delighted many junior faculty. One of the new young lab heads, a molecular biologist named Titia Lang, said of the attacks prompted by the leak of the draft report: "I don't know the details about it. I don't care about it. What I do know is that Baltimore is the best thing that's ever happened to this university."[2]

Baltimore tried to explain himself to Rockefeller faculty and reporters, but the attacks kept coming, taking a steady toll on his family. His daughter, Lauren, then 16 and away at school, recalls that she was deeply saddened when on visits home her parents told her that they had lost family friends like Wally Gilbert, whom she had known and cherished since she was small. The public assaults sometimes drove Alice Huang to despair, but she drew strength when she remembered escaping the threat of the Japanese army during the winter of 1945, when she was a small child in China. While her father, an Anglican bishop, remained behind at their home in Kweiyang, she and the rest of the household packed their books and, joining five other families, fled west towards Tibet in a coal-driven steam truck that often had to be pushed up the steep grades of the snow-storming mountains. To her, the tumult arising from the draft report paled in comparison. Baltimore remembers that when the morning paper carried the obituary of someone they knew, she would say, "Well, David, you're far better off than he is."[3]

Baltimore soon drafted a response to the O.S.I. report. "I tried to walk a fine line between saying things I didn't believe and saying things I did believe," he remarks. On April 22, he read the draft to a gathering of the senior faculty at Rockefeller, requesting their comments. Some said it stopped short of admitting blame. At the beginning of May, a week after the conversations with Vagelos and Ptashne, he sent his final response to the O.S.I. and to *Nature* and *Science*, both of which published the state-

"I have to believe the ancient Chinese proverb: True gold fears no fire."
ALICE HUANG, Wife of David Baltimore
Dean for Science at New York University
Photo Credit: Courtesy Alice Huang

ment.[4] In the middle of the month, he formally retracted the *Cell* paper
pending the final outcome of the O.S.I. investigation. He was accompanied by Weaver, Constantini, and Albanese, but Moema Reis refused to
join in the withdrawal, writing Baltimore that she remained "convinced
of the authenticity" of the work, and so did Imanishi-Kari, declaring that
she was the victim of a "witch hunt."[5]

Baltimore sent the O.S.I. a separate, additional response to the draft
report that went unpublished. It strongly protested the animadversions
the O.S.I. had cast against him in the closing section of "Determinations,"
contending that some parts of it were "inflammatory" and inappropriately
"personalized" and that others ripped both his actions and statements
out of context. He sharply criticized the O.S.I. for spotlighting the "ill-
advised comments" about unpublished data that he had advanced during
his interview "in the heat of the moment." His remarks "were not at all

intended to condone fraud or to sanction the submission of misleading materials to N.I.H.," he wrote. "It is unarguable that even one's unpublished experimental record must not contain assertions that the tests were done or results obtained when such assertions are untrue. . . . Clearly, even if the dissemination of the previously unpublished notebooks was driven by the investigation, that circumstance could not excuse or justify deliberate falsification." Baltimore called it "unseemly" for the O.S.I. to use against him isolated comments made in what he had taken to be an intellectual discussion.[6]

In the published section of his response, Baltimore avowed that, if Imanishi-Kari had committed fraud, he had no knowledge of it; otherwise, the rest of his statement amounted to a full-scale *mea culpa*. He said that he had been "too willing to accept Dr. Imanishi-Kari's explanations and to excuse discrepancies as mere sloppiness," that he had done "too little to seek an independent verification of her data and conclusions." He had focused too long and too narrowly on whether the paper could stand up scientifically. He offered an olive branch to Dingell and to the Harvard dissidents, conceding that he should have heeded the "warnings" implicit in the Secret Service report in 1989 and suspended "further comment on the matter until I had a full opportunity to review and digest all of the new information." He found the allegations of Imanishi-Kari's fraud "deeply troubling," not only because of their impact on the *Cell* paper but because such exposures of bad faith "undermine public confidence in the entire scientific community." He declared that he had "come to appreciate the legitimate role of government as the public sponsor of scientific research and to respect its duty to protect the public interest and hold the scientific community accountable for its stewardship of public funds."[7]

Baltimore endorsed the scientific community's obligations to whistleblowers, acknowledging that senior scientists must vigorously pursue questions that scientists at any level might raise about research; that they had to protect them "from retribution or discrimination"; and that they had to "press for a full airing" of the suspicions voiced by junior scientists, who might be "reticent to allege outright misconduct." He also offered kudos to O'Toole, declaring that he had "tremendous respect" for her "personally and as a scientist" and that he had always believed "her analyses were insightful, her expressions of concern . . . proper and appropriate, and her motives . . . pure." He concluded, "I commend Dr. O'Toole for her courage and her determination, and I regret and apologize to her for my failure to act vigorously enough in my investigation of her doubts."[8]

The national press covered Baltimore's apology, and *Nature* editorialized that he "has said enough to restore his own reputation," noting, "To make an error may reflect on a person's judgement, but to confess it in the circumstances in which Baltimore now finds himself is a mark of courage. He deserves a break."[9]

Margot O'Toole promptly told a reporter that, while she was pleased by Baltimore's praise of her, his "apology did not go to the heart of the question." She explained why in her own response to the draft report, which she sent to the O.S.I. in mid-May. Walter Gilbert telephoned John Maddox, the editor of *Nature*, and urged him to print O'Toole's assessment. Maddox ran excerpts from her response in its issue for May 16 under a header announcing that her comments "contradict at several points" Baltimore's recent statement.[10]

O'Toole appeared to crave even fuller vindication than the draft report had awarded her, insisting that many more of her allegations of fraud— for example, those concerning Figure 1—had more merit than the O.S.I. had concluded to be the case. What she meant by "the heart of the question" was revealed by an observation she made concerning the letter to *Cell* that the coauthors had published in the fall of 1988 correcting the misstatements about the specificity of Bet-1 and the isotyping of the Table 2 hybridomas. While the misrepresentation of Bet-1 had been central to her original objections to the paper, she now characterized both misstatements as "fairly insignificant."[11] As she now saw it, her challenge had adduced evidence of fraud, the real issue, from the beginning. She had reconfigured the history and substance of the dispute, refracting it in 1991 not only through the sense of injury that she had advanced to the O.S.I. but also, it seemed, through the data and documents to which she had been given access during the course of the O.S.I. investigation.[12] At the heart of the question now was how the Wortis committee, Eisen, and, especially, Baltimore, had behaved towards the evidence—and her—in the reconfigured story.

Through the excerpts in *Nature*, O'Toole broadcast an account of events that was in some parts vague and unsupported and in others, measured against the documentary record, including her own testimony since 1986, exaggerated, distorted, or selective. For example, she insisted that at both the Tufts and M.I.T. inquiries Imanishi-Kari had "admitted that a large series of the published experiments had not even been performed, and that some that had been performed had not yielded the claimed results." She pointed to nothing specific in support of the claim. Several weeks later, challenged on it, she responded with a kind of circular argument, holding that her "knowledge that certain experiments were not done and that others were misrepresented came in large part from" her participation in the Wortis and Eisen inquiries. "Now that the draft of the O.S.I. report has supported my position, it should be growing clear that the description of how I obtained this detailed and accurate information must also be correct."[13]

O'Toole transformed the suggestion made at the end of the M.I.T. meeting that she and Baltimore might air their scientific differences in an exchange of letters to *Cell* into an assertion that Baltimore warned that "he would personally oppose" any effort she might make to get the paper corrected. She attacked Baltimore for having "submitted the fabricated [June subcloning] data" as part of the coauthors' additional letter of correction in 1989 even though he must have known from her response to the N.I.H. report that Imanishi-Kari had "admitted" that the subcloning experiments "had not been performed." She virtually accused the Wortis committee of lying for having affirmed that they had seen the subcloning data in May 1986 and by clear implication she faulted Baltimore for not having taken her word over theirs.[14]

O'Toole averred that the charge that she had lost her job because she had raised questions was "essentially true," omitting to acknowledge— except for noting that "the facts were a little complicated"—that her fellowship had been due to end on May 31, 1986. She claimed that, more importantly, because of her challenge she had "lost my career." She held that Eisen had failed to defend her against "pressure" from Imanishi-Kari to "misrepresent my own results." The assertion differed dramatically from her first testimony to Dingell, in which she said she had complained to Eisen that Imanishi-Kari had taken her off experiments and assigned her to mouse husbandry. O'Toole added that after the M.I.T. inquiry Eisen had "refused my urgent request for help and redress." The reference alluded to Eisen's irritated refusal to satisfy her demand in 1987 that he recast the memorandum he had written implying that she had wrongly alleged misconduct when she had only charged error. He had sent the document only to several officials at M.I.T., but she was sure it had helped Imanishi-Kari make her unwelcome at Tufts.[15]

She concluded that she "was left unemployed," asserting that she had been "subjected to five years of slander and libel from Drs. Baltimore, Eisen and Imanishi-Kari" and that the "cover up at MIT" enabled "Drs. Huber and Wortis to continue their slander and libel of me." She later explained that her indictment of Eisen, at least, rested on her belief that for five years he had "assured the scientific community that my scientific concerns were trivial." Eisen had in fact said repeatedly in public and private that he considered her challenge to the *Cell* paper cogent and serious, but O'Toole held that his take on her concerns had subjected her to "professional ridicule."[16]

O'Toole's attack provoked Baltimore to override his *mea culpa*. His wife reflected some months later that his people had been running scared before Ptashne, Gilbert, and the *New York Times* "until we all gained our senses and realized that [we] . . . had nothing to be guilty about."[17] Baltimore and Eisen angrily contested O'Toole's construction of events, particularly of their behavior, in letters that *Nature* published at the end of May. Baltimore recalls that his rejoinder was far closer to the center of gravity of his beliefs than his *mea culpa* had been. Imanishi-Kari had just submitted her own response to the draft report to the O.S.I., and the same issue of *Nature* carried a much-abridged version of it that in part addressed O'Toole's claims. Wortis, Huber, and Woodland followed suit in rebuttals that appeared in the journal in mid-June.[18]

All expressed resentment that, as Woodland put it, "anyone who has disagreed with Dr. O'Toole's analysis of the scientific issues has been accused of either incompetence or deceit." Each cited chapter and verse in refutation of her assertions, pointing to the various ways that her account was contradicted by, inconsistent with, or incompletely representative of the accumulated record of the case. They affirmed that at the inquiries in 1986 Imanishi-Kari had not admitted that she had failed to perform a large series of experiments. The Wortis group wondered why, if she had, O'Toole had not mentioned the fact in her memorandum to Eisen in June 1986; Eisen found it difficult to understand why she had not mentioned such a flagrant admission to the Dingell subcommittee in April 1988.[19]

Baltimore, Eisen, and the Wortis group all denied that O'Toole had even hinted at fraud, and certainly not at anything that corresponded to the evidence in the draft report. Eisen observed that if fraud had been on her mind, her memo to him had effectively masked the fact, making no specific reference to the seventeen pages let alone to phony data. It advanced a scientific dissent, arguing points such as the relative sensitivities of reagents and molecular tests and developing the case that the *Cell* paper data could better be explained by double-producing hybridomas and suggesting that some formed antibodies that were heterodimers. Eisen said—independently of Wortis, Huber, Woodland, and Imanishi-Kari— that he had taken O'Toole's mentioning heterodimers to mean the transgenic mouse cells generated *mu-gamma* hybrids and that he found the creation of such molecules "highly implausible." Yet while he then respected her serological critique as "cogent and thoughtful," he now feared that she "confuses disagreement with slander and libel." Baltimore observed that O'Toole had not cited "one example of any comment by me in which I publicly disparaged her or her ideas," and he pointed to various compliments he had paid O'Toole in his written statements and testimony over the years.[20]

Imanishi-Kari acknowledged that O'Toole was right in recalling her having said that no further tests had been done on the Table 2 hybridomas, but she explained that O'Toole was wrong in the inference she had drawn from the statement. What she had meant, Imanishi-Kari said, was that the hybridomas had not been isotyped other than for *mu*, not that they had not been subcloned. In any case, O'Toole had drawn the inference later, when she saw the report of the N.I.H. panel in the fall of 1988. Her memo to Eisen had not mentioned anything about subcloning either. Baltimore noted that his having taken the subcloning data as authentic in early 1989 was surely justified, given that the N.I.H. review had accepted the assurances of the Wortis committee that its members had seen the subcloning data.[21]

Within the severe limits of the space *Nature* allowed, Imanishi-Kari also raised serious questions about the case against her, particularly the reliability of the forensic evidence. She pointed out that she could not yet "disprove any of the Secret Service allegations" because she had been "denied access" to the original laboratory notebooks and tapes. She emphasized that her notebooks contained data unquestioned by the Secret Service that supported the central claim of the *Cell* paper. In addition, she spotlighted the egregious fact that the draft report had backwards the way Bet-1 worked in the experiment, attributing to her a motive for fabricating data about the reagent that simply did not exist. "You have no idea," she noted, "how upsetting it is to find that after five years of controversy and two years of O.S.I. review, an investigation still gets . . . the question of the Bet-1 reagent—all wrong."[22]

Long before the letter volleys, Baltimore had gained an energetic defender in the prominent biologist Bernard Davis. An emeritus professor at Harvard Medical School, Davis had long been engaged in debates on issues of science and society, unabashedly embracing positions—notably on affirmative action in medical school admissions and on genetics and race—that were often unpopular in many quarters of the Cambridge community. Davis had sympathized with Stewart and Feder's exposé of the Darsee affair. He wrote to a reporter in the fall of 1989 that he considered Dingell thoroughly justified in investigating the general issue of fraud in science, partly because the public had a right to expect accountability from scientists who were spending tax dollars and partly because of "the miserable responses of universities, and the N.I.H., in a number of cases."[23] But Davis parted company with Stewart and Feder on the Baltimore case and drew the line at the way Dingell had gone after Baltimore and Imanishi-Kari. After the subcommittee's hearings in May 1989, he told a Boston television newscaster that a congressional hearing was an inappropriate place to assess the merits of a scientific paper and that Dingell's hounding of the principals had led to Baltimore's being smeared in the nationwide press. He added that Baltimore had "showed a lot of courage in standing up to this very powerful congressman."[24]

At the time, Davis believed that the issue in the case was error rather than fraud. He had only a casual professional relationship with Baltimore but ranked him high in science and wondered what society gained "by publicly focusing on occasional errors and cutting down his reputation, rather than leaving the matter to be evaluated by his peers." If the *Cell* paper's science was wrong, Davis expected that the scientific process would plaster Baltimore's face with egg enough. Even in cases of possible fraud, he held, better to rely on the self-policing mechanisms of science to root it out than on intimidating bureaucracies like the O.S.I. He did not condone sloppiness or fraud, but he told the *New York Times* shortly after the leak of the draft report that "it might be better to tolerate a low level of that rather than create an inhibitory atmosphere in science."[25]

The report convinced Davis for the moment that Imanishi-Kari, whom he did not know, was probably guilty of fraud, but he was sure that Baltimore should not be dragged down for having defended her. On April 22, the day that Baltimore read his draft *mea culpa* to the Rockefeller senior faculty, Richard Furlaud, the chairman of the university's board of trustees, assured the *New York Times* that "the trustees have total and complete confidence in [Baltimore]," adding, "We certainly hope he would stay on. It has never been suggested that he wouldn't." Six days later, in his telephone conversation with Roy Vagelos, Davis learned that Mark Ptashne had suggested that Baltimore shouldn't stay on. Vagelos, incredulous at the level of noise about the case, advised that Davis might help matters by generating expressions of support for Baltimore from within the scientific community, a message that Furlaud gave Davis about the same time.[26]

The next day, April 29, at the annual meeting of the National Academy of Sciences, in Washington, D.C., Davis let it be known that a member of the Academy had urged Baltimore's dismissal to the board of Rockefeller University and that statements of support from other scientists might offset the damage. Davis distributed what he called "a minimalist" document that addressed the issues Ptashne had raised for the use of anyone who might wish to write to the Rockefeller board. Its key sentence:

The incidence of fraud among N.I.H. grantees is "way below one percent. I'm not sure that the number of crooks in Congress is less than one percent."
BERNARD DAVIS
Professor Emeritus, Harvard Medical School
At the President's Forum, American Society for Microbiology, May 15,1990
Photo Credit: Courtesy Elizabeth Davis

"Whatever may have been Dr. Baltimore's errors of judgment in dealing with this case at various stages, we have seen no evidence that impugns his integrity."[27]

In a follow-up letter to Furlaud, Davis, noting the sharp divisions in the scientific community, remarked that it was distressing to see "how much the myth of Margot O'Toole's ostracism has generated emotional reactions." To Davis's mind, her public statements at the first Dingell hearing revealed a view of science that was "very sincere but rigid in demanding perfection and security." Mark Ptashne assured him, in a letter on April 10, that O'Toole did not believe that "all instances of error must be ruthlessly excised from the literature." At Ptashne's suggestion, Davis paid O'Toole a visit at Genetics Institute several days later. According to his notes of the conversation, O'Toole insisted that the M.I.T. inquiry in June 1986 had elicited an admission of fraud and that the coauthors had subsequently lied under oath in testifying to the contrary. Davis broke off the meeting when she unequivocally refused to allow even the possibility that she might be mistaken about Imanishi-Kari.[28]

Davis told Furlaud that he was thinking about writing an op-ed piece on O'Toole's "alleged 'firing' "; he did more than think about it after seeing what he deemed O'Toole's "reckless attack" against Baltimore's *mea culpa*. Taking it upon himself to dig out some facts about her professional career, he discovered that she had a rocky record as a postdoctoral fellow; had not lost her house as a result of having challenged the *Cell* paper; and had not, so far as any available evidence showed, actively sought employment as a scientist for several years after 1986, let alone been ostracized by the immunology community. Davis laid out his findings in the *Wall Street Journal* in July, hoping indirectly to assist David Baltimore by showing that O'Toole was not the victim of "outrageous injustice" that she had been portrayed to be.[29]

In late May, weeks before the appearance of the article, Richard Furlaud had assured Davis that the Rockefeller board felt no pressure to dismiss Baltimore for his errors in handling the dispute over the *Cell* paper, not least because the errors were "far more obvious with 20/20 hindsight than . . . at the time they were committed." By then, many scientists had telephoned Furlaud, Vagelos, and David Rockefeller himself to urge that the board retain Baltimore in the presidency. Ed Rall, at the N.I.H., recalls that he wrote Roy Vagelos saying, "David's strong point is not humility. . . . He has an enormous amount of hubris, but I know few people more entitled to it. Why don't you keep him?" In June, Paul Berg, a member of the board, heard that Dingell's further investigation into the case, presumably including the rumored cover-up, had been postponed indefinitely.[30]

Still, at a garden party that month in Princeton, Maxine Singer, the head of the Carnegie Institution of Washington and a devoted friend of Baltimore and his wife, heard a lot of "troubled conversation about David Baltimore," as she wrote to one of her daughters. David Rockefeller was there and so was Furlaud, whom she had not previously met but who had called her several times for advice. Furlaud told her, she said, that the Rockefeller "board has repeatedly confirmed their full backing of David B, but there has been a relentless flow of problems and they have spent endless hours at it," continuing, "And there is enough uncertainty raised that they all look wherever possible for help and support from 'outside.' "[31]

Bernard Davis said he could not get any reporters to pay attention to what he had dug out about the injustices O'Toole had allegedly suffered.[32]

Apart from the scientific press, most of the media ignored all the rebuttals to O'Toole and continued by and large, to take her version of events at face value.

The *New York Times* both reflected and, because of its enormous influence, undoubtedly advanced the tendency in the press, especially through the reporting of Philip J. Hilts, who in 1991 was the principal journalist covering the case for the paper. Hilts had been an award-winning reporter on science and health for the *Washington Post* before joining the *Times*, in 1989. While on the the *Post* staff, he published a book called *Scientific Temperaments* that profiled three scientists, one of whom was Mark Ptashne. Hilts turned to Ptashne for a comment in his first article on the leaked draft report, and Ptashne obliged him with his chastisement of the scientific community for denigrating O'Toole, a critique that Hilts published in two different page-one stories. It is likely that Hilts's views on the Baltimore case were shaped to some degree, perhaps significantly, by Ptashne as well as by Gilbert, who also appeared in the accounts he wrote.

It was Hilts who, in one of the page-one stories, first reported that O'Toole had lost her house. In the same article, he wrote that during the development of the *Cell* paper dispute Baltimore "did not hide his dislike of Dr. O'Toole," pointing to Baltimore's one-time description of her in his angry telephone conversation with Stewart in 1987 as a " 'disgruntled postdoctoral fellow' " and ignoring all else that Baltimore had said about her publicly since. Hugh McDevitt recalls that Hilts called him when the draft report was leaked and that he seemed "convinced that Thereza was guilty and Baltimore was terrible," adding, "He kept focusing on Baltimore. I said, 'Why do you guys want to pillory David?' He said, 'Because's he's so arrogant.' "[33]

On June 4, 1991, several days after Imanishi-Kari's abridged response to the draft report had appeared in *Nature*, Hilts published a feature on her under the lead, " 'I am Innocent,' Embattled Biologist Says." A staff member at the *Times* says it assigned the article to give Imanishi-Kari her day in the press. It was not much of a day, however. Hilts, who spent several hours with her, did report that she was enduring terrible stress, felt like Joseph K. in Kafka's *Trial*, and insisted, "I never made anything up." He said that the "Secret Service evidence has become the heart of the matter," but he gave perfunctory attention to the objections that Imanishi-Kari had raised against it and did not mention that so far she had been denied access to the materials with which she might refute it.[34]

Hilts's description of the scientific issues, confined to just a few paragraphs, was in one part wrong and in the other misleading. He erroneously had Imanishi-Kari using the data from the mistyped mouse in the paper itself, then later being called on it and offering data from a genuinely normal mouse in substitution. He misled by reporting, "The central claim of the paper depends on how many mice showed the unexpected antibody properties. But the statement in the paper that said this work was done, she has admitted, was false. 'We did not do it,' she said'." The reference to work not done could only have meant the non-isotyping of the Table 2 hybridomas for *gamma*—one of the principal substantive reasons Ptashne had found for censuring the paper. But the vague phrase "this work," in the frame that Hilts used it, encouraged the impression that hardly any work had been done in support of the central claim—at least nothing other than "a similar characterization . . . on other mouse samples" that Imanishi-Kari told him about. The reference was to the hybridoma tests summarized in Table 3, of course. Hilts reported that they, too, had been called into question, though he did not note that the draft report had been unable to conclude that the tests were fraudulent.[35]

Imanishi-Kari wrote to the *Times* protesting Hilts's errors, especially his having her admitting as "false" what the *Cell* paper had reported as work done in support of the central claim. She asked the *Times* to publish a correction. She says that an editor at the newspaper got in touch with her, but he seemed interested only in verifying that she had not done the reported *gamma* isotyping. He was completely uninterested in her explanation that it was not crucial in the context of the experiment taken as a whole. In the end, the misrepresentation was left to stand uncorrected.[36]

Unlike the national media, the Harvard dissidents paid close attention to the disputation among the principals in *Nature*. Ptashne disapproved of Baltimore's response to O'Toole. Gilbert openly expressed contempt for the earlier *mea culpa*, telling Hilts, "It reminded me of that moment in the movie Casablanca, where Claude Rains stands in the bar and says, 'There is gambling going on here? I'm shocked! I'm shocked!' "[37] So far as Gilbert was concerned, Baltimore's apology omitted too much. Embracing O'Toole's critique of it, Gilbert publicly held that Baltimore had known since November 1988—because O'Toole had "informed" him of the fact—that the June subcloning data was fraudulent. He faulted Baltimore for nevertheless signing the letter of correction that used the data and attacking the Dingell subcommittee. Baltimore "simply refused to notice what was happening," Gilbert said, adding, "that's the best interpretation you can put on it." Several weeks later, Gilbert made the worse interpretation explicit, writing in *Genetic Engineering News* that "there was most likely a cover-up involving fraud."[38]

On the day that Hilts's profile of Imanishi-Kari appeared in the *Times*,

June 4, 1991, the chairman of the department of biochemistry and molecular biology at Harvard brought together Gilbert, Ptashne, Edsall, and Paul Doty for a discussion with Herman Eisen. Bernard Davis had suggested the meeting in the hope that the encounter with Eisen, who was highly respected and close to the early events, might diminish the barrage against Baltimore. Eisen tried to clear the air about what happened to O'Toole at M.I.T. and afterwards, insisting, "It's just not right to say she was fired and we kept her out of science." He observed that the Harvard group seemed to have made up its collective mind about Imanishi-Kari's guilt even though they had not seen her full response to the draft report. Eisen said that in science, fraud was "the equivalent of murder." "I won't vote for a murder conviction until I see all the evidence," he declared, implying that he wondered how they could.[39]

However, Eisen made no more than a small dent in the convictions of any of the critics. On the contrary, in a remarkable display of censorious second-guessing, Ptashne and Gilbert chastised Eisen for having focused on the science rather than on the possibility of fraud when O'Toole challenged the *Cell* paper in 1986. "You should have said, My job is to look into the data for the paper," Gilbert admonished. Eisen repeated the gist of what he had published in his recent rebuttal to O'Toole—that she had stated she was charging error, not fraud, and that her memorandum to him had directed the inquiry away from fraud and towards the science. Ptashne told him, "It puts the scientific community in an impossible position if you now say to the world the problem is how O'Toole acted in the first place." Gilbert chimed in that Eisen had worsened the problem with Congress by attacking the whistle-blower. Eisen said that he insisted on replying when someone called him dishonest. Ptashne reproved Eisen for claiming in his rebuttal that O'Toole had proposed that *mu-gamma* hybrid molecules formed in the hybridomas, holding that he had made O'Toole "sound stupid" and blackened her reputation.[40]

In late June, in another letter in *Nature*, O'Toole derided Eisen for blaming her, the witness, for his "discredited investigation" and he was similarly rebuked by John Cairns, a prominent epidemiologist of cancer at the Harvard School of Public Health. Cairns had talked with O'Toole, dug into the case, and decided that she was "absolutely right and these guys are absolutely wrong." In a letter to the National Academy of Sciences that *Nature* published in July, Cairns warned that fund-raising for the Academy could become "much harder if Congress is left with the image of the Academy as the organization that sided with Baltimore right or wrong." Invoking the familiar trope of Watergate, he maintained, "I do not see how David Baltimore can escape public censure, at the very least. About the only question remaining is whether anyone will actually go to jail."[41]

In a draft of her letter to *Nature* that O'Toole sent to Hadley, she declared, as she had done in her allegations, that she had "never proposed mu-gamma hybrid molecules" and considered the attribution to her of such a scientifically implausible suggestion "professionally damaging." She held that what she had meant by the idea of "heterodimer formation" in explanation of the data in Figures 1 and 2 of the *Cell* paper was the creation of *mu-mu* hybrids. For some reason, *Nature* did not publish the part of O'Toole's letter that dealt with the heterodimer issue, but in a letter that appeared in the journal two weeks later, Ptashne made the point on her behalf. Ptashne's letter was accompanied by one of concessionary support from Eisen saying that he had misunderstood O'Toole and that her suggestion of *mu-mu* hybrid formation "would thus appear to have been reasonable."[42]

Eisen had written his letter in the interest of defusing the charged situation after Ptashne, in the wake of the discussion at Harvard on June 4, had called his attention to two papers that had been published in 1989 and that seemed to Ptashne to support O'Toole's argument for *mu-mu* hybrids. Eisen says that initially they struck him, too, as bolstering her view. On reflection, however, he concluded that he had been wrong to concede that the two papers provided all that much support for O'Toole. Besides, Eisen continued to be bothered by the many hybridomas in Table 2 that bore the idiotypic birthmark but were not *mu*. Whether *mu-mu* hybrids formed or not was irrelevant to accounting for the kind of antibodies those hybridomas produced. Eisen says that he tried to remove the concession from his letter before it appeared, but although the editors of *Nature* incorporated several other revisions that he submitted, they rejected the withdrawal of his acquiescence in Ptashne's version of what O'Toole had meant. In fact, the editors presented the two letters under the caption that Eisen "concurs" with Ptashne.[43]

The next week, in the issue of their journal for July 18, the editors published a commentary on the case by Paul Doty that indicted Baltimore for what amounted, by Doty's standards, to unethical scientific conduct. In his early seventies, Doty was a senior statesman of science—a distinguished professor emeritus of biochemistry at Harvard who had worked on the atomic-bomb project during World War II and had been involved ever since in issues of arms control and international security, notably as a member of the President's Science Advisory Committee during the Kennedy and Johnson administrations. Having known almost a half century of boom in American science, Doty felt acutely the beleaguerment of research in the early 1990s—its worrisome state of "encirclement," he puts it, that was denying young individuals the wherewithal to pursue the rapidly widening intellectual opportunities in biomedical science. Doty says that his views on the *Cell* paper case were strongly influenced by John Edsall, but he brought to the dispute a sense of scientific probity that was all his own.[44]

At the meeting on June 4, Doty had intervened little except to remark upon the "harm" that he thought the Baltimore case was "doing to the image of science." He explained that he was less concerned with whether Imanishi-Kari had committed fraud than with whether David Baltimore's behavior in the case seemed distant from the canon that science is the search for truth and closer to the rule that you publish whatever you can get away with. He says that he not only considered Baltimore's "conduct wrong" but judged that it set a bad example because "he had so much leverage as a charismatic figure in the scientific world."[45]

Doty wrote his commentary while on vacation in England, where he says he had the leisure to collect his thoughts about the case. He had taken the trouble to study much of the extensive printed record that the *Cell* paper case had generated, in the interest, he said at the opening of his commentary, of assessing the extent to which its principals had "met or compromised" the scientific community's "ethic" of research. He contended that Baltimore's "apology, although welcome, does not erase from the record the behavior that occurred and was defended over five years." He charged that Baltimore had fallen far short of "the traditional standards of science" in several ways—notably by having inadequately scrutinized the *Cell* paper data before it was published and after it was challenged, and by having, instead, "organized an attack on his critics and discouraged publication of their views." Doty found that, in all, Baltimore's "pattern of behavior stands in deep contrast to the traditional view that authors of scientific papers have a special obligation to be responsive to criticism and to test their work from every possible angle—to pursue the truth relentlessly."[46]

Doty's letter, quoting Feynman, applied to Baltimore the mythical standard of scientific practice that his fellow Harvard critics had been parading—that scientists are ethically obligated to respond to every challenge by returning to the lab to check their work. In a recent conversation, Doty concedes that the standard is "an ideal" and that redoing the disputed parts of the *Cell* paper would have required "a prohibitive amount" of effort. What really bothered him, he says, was that Baltimore seemed to hold "that someone with his status and reputation" did not need to revisit challenged data in the laboratory.[47]

Like his fellow Harvard critics, however, Doty reached conclusions about the merits of the controversy in ways that seemingly fell short of the scientist's obligation to pursue truth relentlessly. Although he had read the published record of the case, he says that he had probably not sufficiently understood the tentative nature of the draft report. He had not discussed anything at issue directly with Baltimore himself, with whom he was socially as well as professionally acquainted, or, for that matter, with Imanishi-Kari. Nor did he give attention to the follow-up to the *Cell* paper research that the Herzenbergs had performed with Baltimore's mice or to a letter that Imanishi-Kari's former student Nicholas Yannoutsos had published in *Nature* the week before Doty's appeared: Yannoutsos avowed that he and Imanishi-Kari and others in her lab "have painstakingly repeated time and again the work reported in the *Cell* paper" and were doing new experiments with the aim of trying to clarify the mechanisms at work in the transgenic mice.[48]

Doty gave no indication that he was familiar with the details of the research reported in the *Cell* paper, particularly its disputed parts, and he appeared to have misunderstood the origins of O'Toole's challenge to the experiment, mistakenly thinking that it was rooted in her failure to replicate it. His letter was also inherently contradictory. On the one hand, it acknowledged that the conclusions of the draft report were only tentative; on the other, it invoked several of them—for example, the censure of the coauthors for letting the smudge in Figure 4 disappear without comment—as evidence against Baltimore as though the conclusions were unquestionable.[49]

Baltimore was angered by Doty's letter, particularly its charge that he had failed traditional standards of science. He had refrained from responding to most attacks, and people at Rockefeller University urged him not to answer Doty's, but "that one," he says, "went over the threshhold." In a "Dear Paul" response that appeared in *Nature* for September 5, he chastised Doty for arriving at "definitive judgments" by drawing on "a leaked, confidential draft of a government report" and on incomplete evidence, especially "the unsubstantiated, and often refuted, allegations of one participant in events five years old." He explained why Doty's version of the fate of the smudged band in Figure 4 and its scientific significance was "unworthy of you" and urged that he "go back and study this again." He also insisted that his science was "done with rigor and criticality," including the science in the *Cell* paper. Its "data have proved more durable than the data in most papers," he asserted, saying that he knew of "no [published] experiments . . . that contradict" them. In fact, he declared, citing six scientific articles, "there is much published evidence and more coming that support the paper's results in remarkable detail."[50]

Doty lashed back at Baltimore in a letter that was published in *Nature* for October 10 and that was accompanied by a letter from Ptashne deriding the *Cell* paper for the misrepresentation of the data in Table 2 and a lack of clarity and reliability of the data reported in Table 3. Baltimore's claim that the paper had been produced with rigor and criticality, Doty argued, was contradicted by his retraction of it, not to mention by Ptashne's "demonstration . . . that essential data in [it] is either nonexistent or undecipherable."[51]

An editorial in the same issue called an end to the letter war, noting that readers were likely to be "confused by ever more detailed statements," but it declared that one issue raised by Doty remained to be decided: "What are the responsibilities of the authors of a published research report?" The editorial continued, accepting at face value the Harvard dissidents' gloss on what Baltimore had said since the dispute had burst into public notice: "Dr. David Baltimore . . . has from the outset taken the view that it is for the scientific community at large, and for others working in the field concerned, eventually to demonstrate the validity or otherwise of the disputed data, and the conclusions drawn from them. It is a point of view, but hardly a defensible one, especially when the authenticity of the data on which the disputed paper's conclusions were supposedly based has been sharply questioned."

A number of faculty at Rockefeller University had already been influenced against Baltimore by what scientists such as Gilbert and Ptashne were saying. Many now considered his reply to Doty outrageous, particularly his emphasis on the durability of the data in the *Cell* paper. The Rockefeller professors said that Baltimore appeared to be retracting his retraction. They wondered how the data could be durable if much of it had been found to be fraudulent. They were irritated by his refusal to admit that he had made a mistake and pledge that he would have the experiment redone.[52]

Baltimore says that the retraction had been pro forma, a withdrawal of the paper made under pressure pending the final resolution of what was a tentative finding of fraud. He still believed strongly that the data were not made up. It was on that ground that he judged the observations reported in the paper so durable. Supporters, including even Richard Furlaud, urged that he explain himself to the faculty, make himself more available to them. Baltimore says that he was willing to talk to anyone who wanted to listen but that hardly anyone expressed interest. By now, the controversy had been boiling for almost five years. Friends and foes alike at Rockefeller got the impression that Baltimore was adopting a bunker mentality. Reporters who talked with him about the *Cell* paper controversy sometimes found him subdued. Baltimore told one of them in late September, "This is something I need to deal with. This is not something that reflects on Rockefeller and not something that affects the running of this institution."[53]

The mounting uproar nevertheless did affect the internal politics of the institution. Younger faculty continued to "adore" Baltimore, Roy Vagelos recalls, and a substantial fraction of the senior faculty recognized the merits of his reforms, albeit with degrees of support that ranged from unmitigated to grudging. The trustees had reason to be pleased with him, too. During the first year of his administration pledges of gifts to the university had risen 20 percent and expenditures had come in almost a half million dollars under budget. To cut costs, Baltimore had frozen senior salaries, reduced support staff, and told lab heads, some of whom were accustomed to having roughly a third of their research and administrative costs paid from endowment income, that they could count on only half that fraction in the future; they would have to try to raise the rest from outside grants and contracts.[54]

Some senior faculty remained unreconciled to Baltimore's moves to rejuvenate the institution and resented his approach to deficit reduction. When he arrived, he took over half of an entire floor in a campus building for his lab, refurbished the president's house, and was given the use of a new Lexus and drivers from available university guards when he wanted them. Baltimore says that the trustees had written these arrangements into his recruitment package. They had told him he would need the car for time and convenience. He required the lab to continue in science. He used the house, which badly needed renovation, for official entertaining and fund-raising. Nevertheless, Baltimore struck the critical senior faculty as imperious. While they were being forced to tighten their belts, he seemed to be exempting himself from budgetary pain. During the summer one senior professor among the opposition resigned, and in the early fall so did the Nobel laureate Gerald Edelman.[55]

The recalcitrants wanted Baltimore out of the presidency, and they saw the *Cell* paper sensation as a weapon to get rid of him. In the fall of 1991, they were joined by an increasing number of his original supporters who were convinced that his response to his critics, but especially his duel with Doty, very much affected the welfare of the university.[56] The swing group felt that he had discredited himself as a responsible scientist, stained the reputation of the institution, and unnecessarily antagonized Congress. They feared that he had jeopardized the university's ability to raise money. Professor Norton Zinder remarked, "People just see the headlines [and say] 'You work at that place that has those crooks?'" Zinder himself, an initial supporter of Baltimore's, concluded that the best interests of the institution required that he quit the presidency.[57]

Furlaud says that the trustees continued to have high confidence in Baltimore. The worst he thought was that Baltimore had been overly loyal to Imanishi-Kari and had not treated Dingell with proper reverence. He never had any question about Baltimore's integrity or administrative ability. He told a reporter that he saw the dispute over the paper as "an unfortunate distraction for Mr. Baltimore rather than an incident that has damaged the university." Privately, however, Furlaud and his fellow trustees were, in fact, worried about Baltimore's eroding faculty support. In early October, while Baltimore was out of the country, Furlaud and Vagelos asked Torsten Wiesel, a Nobel laureate on the faculty who was not identified with any faction, to conduct a straw-poll of his senior colleagues. On Baltimore's return, according to notes taken by his wife, Wiesel told him that he had the strong backing of about half the faculty, but about a third wanted him out of the presidency.[58]

At a meeting of the board of trustees on October 17, David Rockefeller announced that he was giving the university $20 million, the largest gift in its history. He had been planning the gift for some time, but he chose to announce it then, declaring that it reflected his "absolute confidence" in Baltimore. "The trustees, after very careful analysis of all the facts, are convinced that in no way had Dr. Baltimore done anything improper, scientifically, legally, or morally," Rockefeller said. Alice Huang reflected soon afterwards that people felt the meeting was "a turning point," in light of Baltimore's own fund-raising record since coming to the university, David Rockefeller's huge gift, and his "clear cut confidence" in Baltimore. She also noted, however, that "it is hard for us to feel that it is all over, because of the long and difficult trip so far."[59]

In fact, Huang had reasons to think the difficulties would continue. According to her record of events at the time, Furlaud had been "shocked" by the results of Wiesel's poll because it showed that Baltimore had not managed to increase the support among the faculty that he enjoyed when he first arrived at the university. He was no doubt further disturbed when James Darnell, whom Baltimore had appointed to be vice president for academic affairs, allowed at dinner with Furlaud that Baltimore had few genuine backers among the senior faculty and that philanthropic foundations would now be reluctant to grant funds to the university. The trustees decided that they had better talk with some of the Rockefeller faculty directly.[60]

Wiesel put together a group that reportedly comprised eleven senior professors and two junior ones and that was said to be representative of the faculty as a whole. They met with several trustees, including Furlaud and Vagelos as well as Marnie Pillsbury, David Rockefeller's assistant for affairs at the university, the same day that Rockefeller announced his gift. The faculty first convened as a group and then met with the trustees one-on-one so that each could speak frankly. They delivered almost, though not quite, a unanimous vote of no confidence in Baltimore, arguing that his continuation in office would make life much more difficult for the university with all the institution's constituencies, including Congress, the N.I.H., and donors. "His handling of the situation encouraged them to think that they had been right and the trustees had been wrong," Vagelos recalls.[61]

On Thursday, November 7, Baltimore flew back from London on the Concorde so that he could rendezvous that afternoon with Furlaud and Rockefeller. They told him that they had been shaken by their discussions with the faculty, believed that Baltimore's support would only continue to erode, and were apprehensive that the next meeting of the trustees' executive committee, scheduled for early December, might result in a request for his resignation. Furlaud and Rockefeller stressed that they themselves were pleased with all he had accomplished but were worried about appearances, particularly the impact of appearances on the university's ability to raise money. Rockefeller said that he had refrained from making introductions to potential donors for fear of what they might think. Furlaud and Rockefeller suggested that Baltimore might want to think seriously of resigning then, when things were quiet. Baltimore got the impression that they were virtually asking him to resign.[62]

"For whatever reason, it all seems to be unraveling," Huang noted. A flurry of activity followed during the next several weeks, including a strong show of support from junior faculty and senior faculty at the Rockefeller hospital at a special one-day meeting of the board of trustees on November 21. Baltimore got the impression that the board would continue to back him so long as he had the cooperation of certain key senior faculty, including Darnell. Intimates in and out of the university had been telling Baltimore over the preceding month or so that Darnell was behaving like a Brutus, but Baltimore had refused to hear of it. Baltimore took Darnell, who had helped recruit him to the Rockefeller presidency, to be his staunchest backer. He doggedly resisted believing that the man who had been his mentor, his coauthor, and, for so many years, his friend would turn against him. At dinner with Baltimore the day of his meeting with Furlaud and Rockefeller, Darnell professed that he was indeed "a true friend" and pledged that he would work on Baltimore's behalf.[63]

However, on Monday, November 25, Darnell told Baltimore that he felt painfully conflicted and expected to quit his vice presidency. Baltimore found it hard to believe that Darnell would thus vote no confidence with his feet but took the announcement as another, though not decisive, indication that his presidency was no longer tenable. In fact on Tuesday, Darnell told Furlaud that he thought the university would be best served by Baltimore's resignation as president. On Wednesday, in a letter to Furlaud, Darnell formally submitted his own resignation as vice president for the consideration of the board. He wrote that he strongly disagreed with those trustees who thought the campus would be worse off without Baltimore as president. He said that Torsten Wiesel would make a calming interim president and that, if the board wished, he would continue as vice president for academic affairs. He was confident that under such an arrangement the anxiety pervading the campus would disappear overnight.[64]

On Wednesday, while driving to Woods Hole, Massachusetts, for Thanksgiving, Baltimore called Furlaud on the car phone and said that he realized the situation for what it was, an opening that Furlaud took to be an offer of resignation. In Woods Hole later that day, Darnell telephoned to say that he was in fact resigning his administrative post and also, probably, that he had urged Baltimore's resignation to Furlaud. Shortly before dinner, Furlaud called, telling Baltimore that he and David Rockefeller wanted to straighten out the university themselves and felt they could better accomplish the task if Baltimore were not saddled with it. On December 2, Baltimore formally resigned in a letter to Furlaud and Rockefeller, explaining that the *Cell* paper controversy "created a climate of unhappiness among some in the University that could not be dispelled" and that trying to govern the institution "under these conditions had taken a personal toll on me and my family which I can no longer tolerate." He said that he would remain at Rockefeller as a professor, reviving the research program in AIDS that he had terminated on becoming president.[65]

Furlaud, no doubt attempting to put the best face on Baltimore's resignation, told Philip Hilts that he had "tried to talk him out of it, but he had decided." He and Rockefeller both publicly praised Baltimore as a great administrator but said that the controversy over the *Cell* paper had impaired his ability to lead the university. In late October, Herman Eisen had written to David Rockefeller extolling his $20 million endorsement of Baltimore's presidency. On December 5, Rockefeller replied, "Alas, David's enemies, including Congressman Dingell, have now been successful."[66]

The *Wall Street Journal* also blamed Dingell—"Dingell Gets Baltimore," it editorialized—and so did the head of the American Association of Universities and untold numbers of his scientific constituents.[67] But the balance of editorial reaction was with the *New York Times*, which concluded that Baltimore had been made appropriately "accountable," that a "rough justice" had been meted out to him for his "sins." Daniel Greenberg's *Science and Government Report* summarily declared that "in reality . . . , Baltimore got Baltimore" by his display of "arrogance and intransigence." The "mystery" was, Greenberg added, why he kept "defending his colleague long after she was exposed as indefensible."[68]

THIRTEEN

■

Dr. Healy's Mantra

BALTIMORE'S DEFENSE of Imanishi-Kari was no mystery to her good friend Joan Press, who was an immunologist at Brandeis University. When on March 21, 1991, Press first read about the leaked draft report in the *Boston Globe*, she was outraged. She considered Imanishi-Kari completely scrupulous, dedicated, and trustworthy. Press knew that she cherished two things in life—her daughter and her science—and found it inconceivable that Imanishi-Kari would have fabricated data. That day, Press sent the editor of the *Globe* an angry letter protesting its reporter's uncritical acceptance of the judgment against Imanishi-Kari, pointing out that, among other omissions, he had neglected to mention that she had been only tentatively convicted, and in a procedure that denied her some of the fundamental protections of due process of law.[1]

The O.S.I. had come under criticism for its procedures since its establishment, not only from Bruce Singal, Imanishi-Kari's attorney, but also from other lawyers familiar with the office. They embraced the "truism in the law that 'due process' consists of whatever process is due in a given situation," as Barbara Mishkin, a respected Washington lawyer who specializes in misconduct cases, had earlier put it, adding, "The greater the potential effect on an individual's reputation, freedom, or livelihood, the greater must be the due process afforded."[2] Robert P. Charrow knew a great deal about the O.S.I. In 1989, at the end of a stint as deputy general counsel of the Department of Health and Human Services, Charrow had been involved in the deliberations that led to the creation of the O.S.I., and had fought to place it on a fair due-process footing within a formal legal structure. He continued the battle after he left the government, joining the Washington law firm of Crowell and Moring, where his clients included scientists accused of misconduct, and skirmishing against the O.S.I. in several trenchant articles that were published in the *Journal of NIH Research*. In June 1990, he faulted the N.I.H. for its "crisis-driven and case-specific" response to misconduct, noting that it was operating without a formal, published set of rules and procedures. "It would appear," he wrote, "that the entire ad hoc 'process' has been devised with at least one purpose in mind—protecting N.I.H. from congressional criticism." The absence of formal procedures, he told a reporter for *Science*, "allows people to swing in the wind for years."[3]

At the beginning of 1990, a scientist at the University of Wisconsin named James H. Abbs had found himself under investigation for misconduct by the O.S.I., and in June, like Imanishi-Kari, he was refused access to the evidence in his case as well as the right to examine witnesses against him. He filed suit in the federal district court in Madison to stop the O.S.I. investigation on grounds that the agency had no published procedures and denied him due process of law. He was joined in his complaint by the University of Wisconsin, which argued that the O.S.I. was in violation of the federal Administrative Procedure Act. Contrary to the requirements of the act, it had not published its rules, such as they were, for public comment before adopting them.[4]

During a pretrial hearing in August 1990, Judge Barbara B. Crabb, who heard the case, said that she was "appalled" by the "fluid," "discretionary," and "unspecified" nature of the O.S.I.'s procedures, calling them "embarrassing," "the work of amateurs." She said she was "shocked" that a powerful agency of the U.S. government would be operating under them. On December 31, 1990, Judge Crabb ruled that she could not grant relief to Abbs for violation of his due-process rights for the technical reason that he had failed to show that he had yet suffered damage to his liberty or property. However, although she did not grant his constitutional claim, she did hold that the O.S.I. had violated the Administrative Procedures Act when it adopted its rules.[5]

At the end of August 1990, perhaps in response to the Abbs case, the O.S.I. had drawn up a statement of policies and procedures applicable to itself as well as to misconduct investigations carried out at N.I.H. grantee institutions. The statement, however, left the O.S.I. with a good deal of discretion. For example, the rules regarding what evidence respondents could or could not obtain access to were altogether vague, stipulating that

they could "review and comment on important investigatory documents
. . . unless such disclosure would violate individual confidentiality or significantly impair the investigation." Moreover, the statement permitted
the O.S.I. to "deviate from any particular policy or procedure where it is
determined to be in the best interests of the United States."[6] Now, after
the Abbs ruling, the O.S.I. was compelled to prepare its policies for public
comment, a task that it soon began. Jules Hallum and Suzanne Hadley
contended that the O.S.I. already provided adequate due process, but the
ruling in the Abbs case brought critical attention to the agency, stimulated
the filing of several more due-process law suits against it, and helped
inspire Joan Press to write her letter to the *Globe*.[7]

In the face of the criticism, the O.S.I., Orwellian in name, justified its
practices and procedures with an Orwellian logic. Its leaders contended
that it was engaged in a scientific rather than a legal investigation. They
were scientists and most of their investigative staff were scientists who
had been given some training in case building from government lawyers.
The agency relied primarily on what Hallum and Hadley called "a scientific dialogue."[8] Hallum said that the O.S.I. did not "even use words like
guilt or innocence"—it called the scientists whom it accused "respondents" rather than "defendants"—that it was only "trying to find out the
scientific truth." He said that getting at the truth in a fraud case meant
that respondents could not be given access to their data; they might alter
the notebooks. It also meant that they could not confront witnesses
against them; the prospect of confrontation would discourage whistle-
blowers from blowing their whistles.[9]

Officials at the O.S.I. held that conventional due process was unnecessary because what was at issue was not the credibility of the witnesses
but the veracity of the science. They were only asking respondents for
something similar to what scientific journals required of them—to rebut
doubts about the reliability of their data. They said that respondents knew
from the outset what topics were being addressed; that, although the
faults alleged in the work changed during the course of an investigation,
respondents had plenty of opportunity to cross-examine the scientific
issues when they received a draft report; and that, all things considered,
the rights of accused scientists were protected well enough.[10] Hallum
warned that full due-process hearings would be "very dangerous to science" because they would be "very expensive," and the money would have
to come from the research budget, leaving untold research grants
"unfunded."[11]

However, to critics like Charrow, it was the O.S.I. that seemed dangerous. To many scientists, the office resembled the "Keystone Cops," Char-

row said, handling cases slowly and inefficiently. Respondents could in fact be left in limbo for many months, in some cases a year or more. To a number of scientists and lawyers alike, the O.S.I. seemed to turn American principles of justice upside down, placing the burden of proof on the accused rather than on the state. The "science police" could be "brutal," *Newsweek* noted, pointing to the kind of limitations Bruce Singal had experienced in attempting to defend his client—among them, "no right to know who has made the accusations or what they are, no access to documents or records." During an investigation charges "are revealed to you as time goes on," Barbara Mishkin remarked. "It is as if there is an indictment you can't see." In all, the O.S.I. combined the duties of investigator, prosecutor, judge, and jury and appeared to many scientists to pursue them in the manner of the Star Chamber.[12]

In human affairs, as in science, truth is inseparable from the process employed in determining it. The flaws in the O.S.I. procedures raised sharp questions about the merits of the agency's findings concerning Thereza Imanishi-Kari. The implication was apparent to Bruce Singal now just as it had been since she had become his client. He regarded the leak of the draft report as yet another denial of her due-process rights, and an egregiously outrageous one. In submitting her response to the report, he reiterated that the O.S.I. had repeatedly denied Imanishi-Kari requests for information necessary to her defense, particularly the data from the forensic tests done by the Secret Service on which much of the finding of fraud rested; and he charged that Imanishi-Kari could not respond effectively when much of the evidence was "concealed" from her.[13]

What the due-process deficiencies meant for Imanishi-Kari was evident not only to Joan Press but also to several other immunologists in the greater Boston area with whom she met regularly to talk about developments in their field and who in the spring of 1991 spent some of their time together discussing the case against Imanishi-Kari. Another member of the group was David Parker, who was on the faculty of the University of Massachusetts at Worcester and who had spent a year on sabbatical with David Baltimore. He was as distressed as Press by the unfairness of the O.S.I.'s treatment of Imanishi-Kari. Together, Press and Parker organized a campaign to send an open letter to the O.S.I. protesting its handling of the case, particularly, Press says, its "star-chamber" procedures. Press notes that the aim of the letter, which they drafted themselves, "was to get scientists to say, We're not making a decision about whether she's guilty or not. We don't know enough to decide that. We want to ask, Have the procedures been fair enough to make a determination." She adds that she and Parker thought that, confronted with that question, "most scientists would say, Hey, you're right, We don't want to be treated like this. Criminals get better due process than she's getting."[14]

Press and Parker stuffed copies of their letter into envelopes for a full day, mailed them out, and eventually obtained the signatures of 143 immunologists, including the Herzenbergs, Klaus Rajewsky, O'Toole's postdoctoral mentor Donald Mosier, Maplethorpe's earlier doctoral supervisor Michael Bevan, and other friends, students, and colleagues of Baltimore and Imanishi-Kari. The letter, sent to the directors of the N.I.H. and the O.S.I. on June 14, was promptly reported in *Science* and published in full in *Nature*. It declared that the signatories were reserving judgment on Imanishi-Kari's guilt or innocence, pointing out that the O.S.I. had produced its report in "a politically charged atmosphere under intense pressure from Congress" and that as a leaked draft to which Imanishi-Kari had not yet had a chance to respond, the report formed a poor basis for deciding the case on its merits.[15]

The letter won virtually no attention in the general national media and it probably changed few, if any, minds about Imanishi-Kari within the scientific community. It certainly did not affect thinking at the O.S.I. Jules Hallum responded to Parker and Press that his office had "*not* leaked" the draft report, that it was "outraged" by the act as well as by the media's conviction of Imanishi-Kari, but that the agency itself was treating her "fairly." However, the Press-and-Parker letter added fuel to the increasing dissatisfaction with the O.S.I. among biomedical scientists. So also did the publication in the *Federal Register* for June 13, 1991—by the U.S. Public Health Service, the parent of N.I.H., in response to Judge Crabb's ruling in the Abbs case—of the procedures by which the O.S.I. proposed to continue dealing with misconduct.[16]

In the main, the proposal simply codified the accumulated operating guidelines of the O.S.I.—which meant that it set down in cold print the procedures it had employed in its cases to date, including Imanishi-Kari's case. It took no apparent account of any of the criticism that had been mounted against the way the office operated. Indeed, the Public Health Service published the procedures as a simple "notice," rather than as a "notice of proposed rulemaking," which would have required approval by the Office of Management and Budget. The proposal also revealed that the agency intended in the future to revise its policies and procedures without notice and comment whenever it felt that such action conformed to the public interest.[17]

In late July 1991, Robert J. Cousins, the president of the Federation of American Societies for Experimental Biology, urged its members to "inundate" N.I.H. policymakers with letters protesting the procedures. He

DAVID PARKER
University of Massachusetts Medical School, Worcester
Co-organizer, Letter Campaigns on Behalf of Imanishi-Kari
Photo Credit: Courtesy David Parker

JOAN PRESS
Brandeis University
Co-organizer, Letter Campaigns on Behalf of Imanishi-Kari
Photo Credit: Courtesy Joan Press

warned that the proposal spelled "a disaster for biomedical research" and that *"it must be withdrawn."* The Federation, comprising seven major scientific societies, represented thousands of biomedical scientists from the collective fields of physiology, biochemistry, molecular biology, immunology, and cell biology, among several others. Cousins included a sample letter that specified the procedural deficiencies that he no doubt thought would especially exercise the Federation membership. He included the control of misconduct investigations by "bureaucrats" in Washington, listing in the ALERT system the names of scientists who had merely been accused of misconduct, the lack of due-process protections for such scientists, the absence of sanctions against whistle-blowers who brought "malicious or groundless charges."[18]

By late September, the N.I.H. had received more than 2,000 letters about the misconduct proposal, only five of them in favor of it, the overwhelming majority of the rest attacking it with arguments drawn from the Federation's draft. An independent protest was mounted by the nation's most powerful academic associations, an assemblage representing the leading private universities, state universities and land-grant colleges, and medical schools. Relying on a twenty-five-page legal analysis by Robert Charrow, the associations sent a joint letter to misconduct policymakers calling for reforms in the way the O.S.I. worked, including the institutional separation of the functions of investigation and adjudication. Bernard Davis and a Boston lawyer named Louis M. Guenin submitted a collaborative comment to the government arguing from the vantage point of both science and law:

> There is no reason why the same procedural fairness should not be the rule for N.I.H. as it is for the Securities and Exchange Commission in its investigation of securities fraud, the Federal Trade Commission in its investigation of unfair and deceptive practices, the Department of Justice in general, and numerous agencies and inspectors general who conduct investigations of proscribed activities. Certainly professional scientists conducting research for the advancement of knowledge and the improvement of human welfare are entitled to at least the same procedural protections as those whom federal agencies investigate for crimes of personal gain and injury to the person and property.[19]

The protest against the O.S.I. exercised the new director of the N.I.H., Bernadine Healy. A Harvard Medical School graduate, she had distinguished herself in cardiology research while on the faculty of the Johns Hopkins University and had also earned high marks for the acumen and energy that she brought to administration and policymaking. She spent two years as deputy science adviser in the Reagan White House and a term as president of the American Heart Association. Since 1986, she had been director of the research institute of the Cleveland Clinic Foundation in Cleveland, Ohio, which she called home and to which she returned every weekend to be with her husband and two daughters. She was a Republican who said that she was a feminist in heart and soul. Early in her administration at the N.I.H. she established a long-term research initiative in women's health and initiated a study of teenage sexual behavior.[20]

Reporters tended to describe Healy as a striking woman, very blond, perfectly made up, and bold in the bright colors of the clothes she wore. She also had lots of brains and had been encouraged to use them by her father, a perfume maker who housed his family in an apartment above his mom-and-pop factory in a middle-class neighborhood in Long Island City, Queens. She was headstrong, outspoken, sometimes blunt. She was also intellectually engaged, keeping with her what she learned from the teachings of her Irish-Catholic girlhood, the discussions in her after-hours philosophy club at Hunter High School in New York City, and her books and professors at Vassar College. Healy understood the realities of scientific research, telling a reporter in 1991: "Science is not accounting, it's not pedestrian. Science is actually rather treacherous. It takes a lot of courage and commitment, it takes a lot of ego to be able to take an observation or hypothesis that challenges the rest and move it along."[21]

Healy had firsthand experience with the issue of scientific misconduct. In April 1990, a whistle-blowing junior scientist at the Cleveland Clinic research institute had accused his boss, a scientist named Rameshwar K. Sharma, of misrepresenting data on a grant application. Healy, as director of the institute, presided over an inquiry that, after deliberating for a few hours, reprimanded Sharma for sloppiness but concluded, on the basis of the limited evidence it saw, that his misrepresentation was unintentional. Healy says that for weeks afterward she fretted that the proceeding had been perfunctory. A few months later, after additional documents became available from the chairman of Sharma's department, she convened a second inquiry that came to suspect that misconduct might have occurred. Healy asked for a full investigation of the matter at a higher level in the clinic. On her own hook, she also froze the considerable amount of money that the grant had brought in, a safeguard against the possibility that the grant funds had been obtained under false pretenses. The high-level investigation resulted in a finding of no demonstrable misconduct. The records of all the inquiries were forward to the N.I.H. for review. Healy subse-

BERNADINE HEALY
Director, N.I.H., 1991–1993
Photo Credit: Howie Sacks

quently learned that in late 1990 the O.S.I. elected to open its own investigation of Sharma's grant application.[22]

Healy says that she arrived in Washington with a few notions about handling matters of scientific misconduct, including "that following proper procedures is essential to handling cases that threaten the lives and reputations of all parties—accused and accusers alike." On becoming a candidate for the job at the N.I.H., she had recused herself from the Cleveland Clinic case, and when she took office, in April 1991, she recused herself from anything at all that might involve the clinic itself.[23] Healy says that, at the time, she embraced the common wisdom that Imanishi-Kari was guilty but had no firm opinion about the case. She declined to read the draft report, judging it inappropriate for her to become involved at that stage in a determination concerning a scientist who was not part of the N.I.H.[24]

Healy saw her responsibilities differently in the case of Robert Gallo and Mikulas Popovic, who was the lead author on one of the fundamental papers from the Gallo lab reporting that AIDS was caused by a specific virus. Because the research in question had been performed when both were staff scientists at the National Cancer Institute, the case fell directly within Healy's purview. The N.I.H. had agreed that, before the draft report on the Gallo case was released to Gallo and Popovic, it would be shown to the Richards committee, the panel that William Raub had appointed in 1990, on nomination by the National Academy of Sciences, for advice on how to proceed with the case. In mid-May 1991, Gallo's lawyer, an attorney named Joseph Onek, objected to the arrangement, calling it illegitimate in principle and warning that it would heighten the risk that the report would be leaked. Healy reviewed the report. She also took home the O.S.I.'s internal guidelines for conducting investigations and concluded that giving the Richards panel access to the draft report would violate them. The guidelines had it that allegations and information developed in the course of an investigation were to be made available only to people directly involved in the case, which the Richards panel was not. They also called for honoring the provisions of the federal Privacy Act, which had been enacted in 1974 to protect individuals against a government agency's improperly disseminating information that it had obtained about them.[25]

Jules Hallum told Healy in a meeting also attended by Suzanne Hadley and William Raub, who was now Healy's special assistant, that she had been given the wrong set of procedures. Another set existed, and a discussion ensued among Hallum, Hadley, and Raub about which set applied. Healy says that she was astonished by this "confusion." She decided to withhold the draft Gallo report from the Richards panel until the principals had seen it. She later reflected that "these events opened a Pandora's box" of unsettling disclosures about the O.S.I. In fact, the more she learned about the office in the spring of 1991, the more she was appalled—by the lack of due-process protection in its procedures, the seeming deficiency in objectivity of some of its judgments, the smearing of reputations arising from leaks of investigative information, and the unfairness inherent in what she called the "circus-like" headlining of the draft report on Imanishi-Kari, as though it was a final document. She remembers that she re-read *The Merchant of Venice*, looking for Shylock's line, "You take my life/When you do take the means whereby I live."[26]

Healy wondered whether even the authorized distribution of draft reports to whistle-blowers like O'Toole, let alone to an outside review body like the Academy panel, might violate the accused scientists' right to privacy under federal law. She says that one night she took home the federal Privacy Act, read it through, and discovered that it permitted Congress access to private information that was otherwise protected. She recalls that she studied a text on constitutional law by Laurence H. Tribe, underlining passages on due process, and then called Tribe at Harvard Law

School, where he was a professor, to ask whether the O.S.I. and the Dingell subcommittee could legally do what they were doing. Tribe responded with dismayed surprise when she told him what was going on, assured her that she was right to be distressed about the due-process issue but that limitations on congressional access to information were airy with huge loopholes.[27]

Healy held that the draft report on Gallo would not do the O.S.I. any good. The case turned in considerable part on whether Gallo, Popovic, or both had knowingly misrepresented that their findings had depended in part on samples of the AIDS virus that had been provided by French scientists or whether some of the stock in the Gallo laboratory had been accidentally contaminated with the French viral cultures. Healy says, "It's important to know if someone's sin was mortal or venial. If it was mortal, you should be drummed out of the corps. In the Gallo case, the O.S.I. did not seem to distinguish between the two—that is, between the mortal sin of theft of the virus or the venial one of error in handling it." Healy thought that the tone of the report was unnecessarily inflammatory, with far too much editorializing. She says that she consulted an official in charge of investigations in The Department of Health and Human Services and that the official agreed. Healy told Hadley that the report "read like a novel" and asked her to recast it so that it began with a clear statement of the charges, proceeded through a development of the evidence, and concluded with a finding on each allegation. Hadley refused, declaring, with Hallum's support, that such a revision would necessarily alter the thrust of the report.[28]

Healy was concerned about Hadley's handling of several other misconduct matters as well. At odds with Jules Hallum, Hadley had left the O.S.I. in March 1991, shortly before Healy arrived; but with the approval of William Raub, still then the acting director of the N.I.H., she continued as chief investigator on the Gallo and Imanishi-Kari cases on the understanding that she would bring them to their conclusions. Raub says that he authorized Hadley only to complete the writing of the reports on the two cases. Hadley, however, interpreted her task to include such additional investigation as might be necessary—for example, a probe of the institutional responses of Tufts and M.I.T. to O'Toole's complaints. Hadley was assigned to the Office of Science Policy and Legislation in the central administration building at the N.I.H., just upstairs from Healy but in a different building from the one in which the O.S.I. was housed. She says that she kept Hallum, Raub, and Robert Lanman, the N.I.H. legal adviser, informed of what she was doing, but Hallum felt that he had little, if any, day-to-day supervision over her ongoing inquiries.[29]

Hadley initiated a probe of the institutional responses of M.I.T. and Tufts to O'Toole's original challenge and learned that Ursula Storb had written a letter of recommendation for Imanishi-Kari in 1986, when she was under consideration for her appointment at Tufts. The letter was impersonal, based on Storb's knowledge of Imanishi-Kari's research, including her collaborations with Baltimore, and presentations at immunology meetings. Storb says that she had not disclosed having written the letter when she was originally asked to join the scientific panel because she had forgotten about it. In mid-May, Hadley, supported by Hallum and Raub, nevertheless asked Storb to resign from the panel on grounds of conflict of interest. Storb refused to leave. She says it was "clear to me, if not to reporters, that Hadley used this to kick me off the scientific panel because I was a critical voice."[30]

Healy strongly objected to getting rid of Storb, realizing that she was being subjected to conflict-of-interest rules that were inconsistent with those in force everywhere else at the N.I.H. and, for that matter, throughout the government. She was also concerned that forcing her to step down now would throw into doubt her previous input to the investigation and possibly compel redoing it. In the presence of Raub and Lanman at a meeting on May 23, she dressed down Hadley, pointing out that academics wrote hundreds of such letters, invoking Charrow's Keystone Cops characterization of the O.S.I., and accusing her of submitting to pressure from Dingell.[31]

During the next two weeks, at Healy's urging, Hadley obtained further information on Storb, which showed that Storb had no professional or personal relationship with Imanishi-Kari. Storb's fellow panel members also intervened on her behalf. Hadley nevertheless remained adamant that Storb's involvement in the investigation should terminate, and on June 6, Healy, unwilling to intervene directly in the matter, authorized Hadley to proceed. However, an article by Philip Hilts in the New York Times for June 14 revealed the tempest over Storb, including Storb's refusal to resign and her observation that "the whole thing is ridiculous." According to Hilts, Dingell demanded that she quit the panel forthwith, but at a regular senior staff meeting that morning Lanman noted that a decision to remove Storb from the panel might be difficult to defend. "There was general agreement that O.S.I. had created a mess," Healy recalls. Later that day, Hallum learned that Assistant Secretary for Health James O. Mason was irate that Storb had been asked to resign. Several days later, despite Hadley's ongoing opposition, Hallum told Storb by telephone and letter that the O.S.I. wanted her to remain on the panel.[32]

By now, Healy judged that Raub had committed a serious managerial error in permitting Hadley, as it appeared he had, to keep charge of the Imanishi-Kari and Gallo investigations after she had left the O.S.I. As she saw it, Hadley had essentially established herself in competition with Hallum's office. Worse, Healy says, Hadley was regarded at the O.S.I. as an agent of Dingell, which made it much harder for Hallum to do his job. Healy herself was no fan of Dingell, not least because she knew assistants and secretaries at the N.I.H. who lived in terror of the brutal manner and four-letter words that came with telephone calls from his staff. She says that during some of her darkest moments, Hadley's ongoing involvement in misconduct probes struck her as "almost like having a member of the Secret Police walking around with the authority of the Gestapo."[33]

Healy increasingly had the impression that Hadley's was a "rogue" activity, operating without adequate supervision or attention to the proper behavior of an investigative office. She recalls that one of the secretaries mentioned that O'Toole called Hadley all the time. Healy wondered to what end the conversations were taking place, since the investigation of Imanishi-Kari had been completed. She says that three or four members of her office staff told her that Hadley had gotten too "close" to O'Toole, which seemed plausible to her since she knew from the newspapers that the draft report had called O'Toole heroic. Healy approved a suggestion from Lanman that he review Hadley's telephone logs, if only to check that the O.S.I. could defend itself against someone's charging in the future that she had been less than objective in the investigation of Imanishi-Kari.[34]

On June 7, Lanman visited Hadley, asking her for the notes. Hadley had a staffer collect them from the O.S.I. files and give them to Hallum. Hadley later said that at no time did she refuse to provide her notes, but according to Healy, Hadley also wrote a memorandum protesting that her integrity was being questioned and that she would not open her notes unless she was charged with some kind of wrongdoing. Hallum, in any case, told Lanman that he would not relinquish the telephone logs. At a meeting on June 21, Healy learned that Hadley was conducting a follow-up investigation of the Tufts and M.I.T. responses to O'Toole's challenge and that the probe might explain the ongoing telephone conversations with O'Toole. Healy says that the discussion nevertheless revealed "that Dr. Hadley was *not* properly keeping logs of all phone calls," adding, "She would not be precise on when or why she did or did not keep official phone records in accordance with OSI policy."[35]

Several days later, Healy learned that someone in Hadley's office had inadvertently sent Gallo's collaborator, Popovic, an audiotape of the full final meeting of the investigative panel in his case, thus exposing the pro and con opinions of the members to Popovic's lawyer, who was Barbara Mishkin. Mishkin had returned the tape but kept a copy, no doubt thinking it would be useful in defending her client. Healy also discovered that the draft report, which had now been sent to the respondents, disclosed that the O.S.I. was opening an investigation of one of the witnesses, who had not yet been notified that he was the target of a forthcoming inquiry or why. The new investigation was to be conducted by Hadley. Healy, furious, told Hallum that, as she later put it, "the unusual arrangement that had been approved for Dr. Hadley was failing and was unacceptable."[36]

In a telephone call to Hadley on June 27, Hallum said that Healy had ordered him to "rein you in." He said that he was shutting down her satellite operation and instructed her to return all her files on both the Gallo and Imanishi-Kari cases to the O.S.I. office. He says that he told her she could have space there for any work she wanted to complete; Hadley says she was told she could not be further involved in any of the follow-up investigations. Whatever the truth, on July 1, 1991, Hadley resigned her responsibility in the two cases and took a leave of absence.[37]

Healy later remarked to a reporter, "Everywhere you turned, it was a mess." Further to clean up the mess, she ordered a review of the O.S.I.'s operations and procedures. She insisted that legal steps be taken to ensure that the Privacy Act would apply to records and information concerning people involved in misconduct investigations. On July 15, in a talk to an N.I.H. advisory group on scientific misconduct, she noted, "Breaches of confidentiality might be viewed as among the biggest failings of the OSI's 'system of justice.' . . . Breach of confidentiality might be viewed as misconduct in the investigative process, every bit as much as misconduct is at issue for the accused." She went on to say that the obligation to confidentiality "may only be exceeded by the challenge of assuring *due process* in these proceedings." She elaborated that the questions the review had to address included whether an accused scientist could cross-examine witnesses and see the evidence; whether the accuser should be closely involved with the investigation of a case; and whether the functions of investigation and adjudication ought to be institutionally separated. In what might have been an allusion to Hadley, Healy declared: "The same individual in OSI should probably not be the initial contact, the investigator, the report writer, serve on the jury and be the judge."[38]

Dingell called a hearing of his subcommittee on the recent events at the N.I.H. and, in preparation for it, several of his staff, including Peter Stockton and Bruce Chafin, went to the campus in Bethesda on July 19 to interrogate the principals. Healy says they behaved like "thugs, absolute thugs." She later told a reporter that they laced their talk with "four-letter words, yelling, screaming, insults" and that one of them implied to Raub that she "had a personal relationship" with David Baltimore. "Now, how slimy is that one?" She says that Stockton wanted to know about the house she occupied on the N.I.H. campus: Did she pay rent? Did she have expensive commodes? (She paid $2,700 a month and used her own furniture.) She recalls that they told her that they were in regular contact with the U.S. attorney on the Imanishi-Kari case and expected that Baltimore and Imanishi-Kari would be indicted shortly. Healy adds that they "demeaned the N.I.H. leaders (we were lap dogs, not scientific watchdogs, they said), and they gloated about having taken down two of the biggest names in science—Dr. Baltimore and Dr. Gallo." She reminded them of the lack of due process. "The staffers made it clear that they thought [Imanishi-Kari] was guilty, so who cared about the rest?"[39]

The subcommittee staffers expressed suspicion that Healy's recent actions might be connected to the Cleveland Clinic case, on which Hadley had been the chief O.S.I. investigator. Depending how the case turned out, they said, she theoretically might become the target of an O.S.I. investigation for mishandling it. In a memorandum that same day to Assistant Secretary of Health James O. Mason, Healy reported that Dingell's staff suggested that her crackdown on Hadley and the O.S.I. might be interpreted "as an attempt to influence that theoretical possibility." Healy told Mason that the "insinuation is patently untrue" (she called it "preposterous" when she testified before the subcommittee). However, she voluntarily recused herself from all dealings with the O.S.I. until it closed the Cleveland Clinic case, believing the action "simple and prudent" under the circumstances. Before the hearing, members of Dingell's staff nevertheless let reporters know that Hadley was no longer on the Baltimore and Gallo cases and that Healy might have shown bias toward her because she was looking into the Cleveland Clinic.[40]

At the hearing on August 1, Dingell announced that Healy had "virtually obliterated" the progress that the N.I.H. had made in dealing with scientific misconduct, not least by depriving the O.S.I. of Hadley. He supported the claim with a "chronology" of events since mid-May that was entered into the record of the hearing and was relentlessly unfavorable to Healy. The content of the chronology, which was unidentified as to source or compiler, suggests that it was based on Hadley's notes of events. Dingell provided a summary account of the Cleveland Clinic case, said he understood that the O.S.I.'s draft report on it was "sharply critical" of Healy, and declared that the next step was to look into "the potential cover up." He asserted that the issue of Healy's behavior was bipartisan. Congressman Norman F. Lent, the Long Island Republican who had previously displayed considerable sympathy for scientists called before the subcommittee, indicated that he, too, was displeased with Healy—a director of N.I.H. who would compare her investigative team with "Keystone Cops" and who would approve a survey of teenage "sexual proclivities and frequency."[41]

Dingell said that he wanted to learn more about what had happened to the O.S.I. but he directed his questions on that score almost entirely to Hadley, who responded easily to Dingell's friendly questions, and to Lanman and Raub, who appeared with Hadley and tried to defend Healy. Healy, in testimony and a lengthy written statement, dwelled on the misconduct issues that preoccupied her, quoting the jurist Felix Frankfurter on the merits of procedural justice: "The validity and moral authority of a conclusion largely depends on the mode by which it was reached." She dealt with events at the Cleveland Clinic briefly, but Dingell, Lent, and Chafin grilled her relentlessly on that issue, asking her virtually nothing about her views or actions concerning policies or procedures at the O.S.I. She says that she was unable to prepare for the Cleveland Clinic line of questioning: She had learned only a few days earlier that the issue would be raised and could not be briefed about it because she had recused herself. She was nevertheless not without her resources. When Lent badgered her about why she had signed the grant application even though it might have been faulty, she explained that her signature not only endorsed the merits of the proposed research so far as she could know them but also provided assurance of institutional support for it. "Scientists don't do research in their backyards in a tent," she snapped. "Scientists do research in institutions, and that means a major commitment of space, of resources, of equipment." Lent complained to Dingell, "Mr. Chairman, I can't handle this witness. I am getting unresponsive replies to my questions."[42]

When Dingell suggested that Healy herself might become a target of the Cleveland Clinic investigation, Healy retorted, "It is extremely curious to me, Mr. Dingell, that they [O.S.I.] have had this case open for about 8 months and it is only about 2 weeks ago that this theoretical concern about the investigation of the Cleveland Clinic or my handling this case has arisen." Assistant Secretary Mason testified that the first he had heard that Healy might be a target was when Dingell's staff had visited him in mid-July. Mason, a career officer in the Public Health Service, was a Mormon with a strict code of values. He added, referring to the innuendoes raised against Healy for her handling of the Cleveland Clinic affair, "I really resent the assumptions that have been brought forth in this hearing and by the inquisitors that called upon me."[43]

Healy told a reporter several weeks after the hearing that if she exercised her decision-making authority on behalf of the N.I.H. and the scientific community, "then I'm going to get hit." It was Healy who took the risk of trying to track down the leak of the draft report on Imanishi-Kari, an investigation that she had several times publicly vowed to mount and had ultimately ordered. A clue came a week after the hearing, in the form of an anonymous letter to Hallum from a self-described "Friend of Science and NIH." It urged asking Walter Gilbert how he came to receive a copy of the report "within minutes of its availability and why he chose to circulate copies of that report to many members of the Harvard faculty." The letter added, "Since Dr. Gilbert has been widely quoted as being outraged by any breach of ethics or appearance of impropriety in science, he will undoubtedly be anxious to cooperate with you in this investigation."[44]

On August 26, 1991, the Office of Inspector General of the Department of Health and Human Services sent a special agent to visit Gilbert. He asked where Gilbert had gotten his copy of the report. Gilbert at first balked at the question but several days later, having obtained Walter Stewart's permission, revealed that it had come from him. He also said that he recalled having seen references to the report in the press prior to receiving it. Stewart told Daniel Greenberg's *Science and Government Report* that he had sent the document to Gilbert "about a week or two after the report was leaked and widely available." Some time early in September, the N.I.H. itself discovered that Stewart had in fact sent the document to Gilbert by Federal Express on March 20, the day before the news stories about the report appeared. However, the inspector general's office had already resolved, on August 28, to drop the investigation. Healy recalls that a high official in The Department of Health and Human Services telephoned to explain why and, trying to be helpful, advised, "The only way to deal with Dingell is to get on your knees. That way you won't be kicked in the shins." Healy says she responded, "That's also the way to get your head cut off."[45]

Healy's head was in no danger. Her testimony at the hearings on August 1 had blunted the subcommittee's attempt to besmirch her handling of the Cleveland Clinic case. Her exposition of the necessity for assuring confidentiality and due process to scientists under investigation for misconduct attracted attention to the issue, and influential endorsements of her position. In an editorial a week after the hearing, *Nature* observed that Healy's "instincts are correct," continuing, "The constitution of OSI as it stands is thoroughly unsatisfactory. . . . There is ample reason to believe that people against whom misconduct is alleged may be unjustly pilloried."[46]

In a letter to Healy on October 3, 1991, Assistant Secretary Mason ended Healy's self-imposed recusal from all matters pertaining to the O.S.I. The Cleveland Clinic case had not yet been concluded, but care of it had been transferred to the O.S.I.'s oversight agency, the Office of Scientific Integrity Review, which was under Mason's rather than Healy's control. Mason explained to Healy that the investigation would not be over for another six to seven months, too long for her to be away from the revamping of the guidelines for dealing with scientific misconduct that she had initiated.[47]

" 'We've got to get due-process guidelines'—it was sort of my mantra," Healy recalls. She relied heavily on two aides—Jay Moskowitz and Leslie A. Platt—to see to devising them. "I was policy and Platt was law," Moskowitz recalls, though he adds that he "became a constitutional scholar for several weeks." Moskowitz says that he and Platt sought to refashion the misconduct investigative process in the direction of greater fairness and due-process protection. The work went on during the fall and winter, assisted by Mason as well as study groups in the N.I.H. and the Public Health Service.[48]

Healy hoped that the definition of misconduct might be reconsidered, particularly the vexing phrase that brought within its scope "serious deviations" from "accepted" practices. Healy had instructed the Dingell subcommittee that misconduct policy had to "be able to distinguish error from fraud, unintentional and even careless mistakes from intentional misconduct, and misstatements from deceptive misrepresentation," and that it had to refrain from extending the "serious deviations" clause to cover "bold leaps of imagination, clever tinkering and unorthodox methods." Healy was also eager to get the misconduct office out of the N.I.H. She says that the agency was a "contaminated" environment in which to deal with cases, like Gallo's, that arose from within its own laboratories. If the scientist was exonerated, then the N.I.H. was vulnerable to charges of softness. If the scientist was found guilty; it could be criticized for submitting to congressional pressure. By early March 1992, a comprehensive proposal for the reorganization of the way the Public Health Service dealt with scientific misconduct had been sent to Secretary of Health and Human Services Louis Sullivan.[49]

The proposal moved the entire operation out of the N.I.H, and it provided any scientist tentatively found guilty of misconduct with the right to appeal the finding at a hearing that would be governed by all the rules of conventional due process. Under the existing regulations, only scientists who faced debarment from grant eligibility enjoyed such a right. That policy was consistent with the precedents of federal administrative law, but it ignored the consequences to a scientist of a putative conclusion of guilt by an investigative body like the O.S.I. Barbara Mishkin noted, "They think if they haven't debarred someone, it's not a serious action. It's not the sanction but the label that damages someone's reputation."[50]

Various features of the proposed reorganization were controversial among policymakers and advisers, particularly those arising from the due-process issue. Questions were raised whether whistle-blowers would enjoy equal rights with the accused or whether the hearings should be public, which might risk the reputations of both the whistle-blowers and the accused. However, the disputes did not impair the proposal for an appeal hearing, where the accuser could be cross-examined. Indeed, enthusiasm for instituting the appeal was widespread. Even Hallum, who vowed that he would "strive mightily" to keep the existing misconduct system essentially intact, said that he thought an appeal hearing would be a "wonderful improvement," since it "will make it easier for us to be perceived as an investigative office." The appeal, in fact, comprised the principal step in the proposed reorganization to separate the investigators and prosecutors from the judge and jury.[51]

The reorganization, modified in unessential details from the proposal unveiled in March, was officially promulgated in a notice in the *Federal Register* for June 8, 1992. The O.S.I. was taken away from the N.I.H. and reconstituted as the Office of Research Integrity (O.R.I.) within the office of the assistant secretary for health. The notice called on the director of the new O.R.I. to devise policies that would "ensure that subjects of investigations are treated fairly," where fairness included "clear specification of what constitutes misconduct, a fair hearing process, appropriate time limits on pursuing allegations, and guidelines to discourage malicious allegations of misconduct."[52]

The staff of the new office was slated to increase more than threefold, from around sixteen to close to sixty, including an increment of half a dozen lawyers. The enlargement was no doubt intended to speed up the pace with which the office dealt with allegations of misconduct in conformity with the requirements of fairness. On August 15, Hallum resigned, blasting the O.R.I. for growing "too big" and becoming "driven by lawyers, not scientists." The lawyers, however, were aware of the O.S.I.'s weaknesses. They set out to fix some of the most egregious. They put together a work group that included staff from the office of the inspector general and that drew up what their predecessor office had lacked: a full set of written, self-consistent internal procedures to guide investigations.[53]

In late October 1992, Public Health Service officials let it be known that they hoped in succeeding months to lay out further reforms for the O.R.I., including a more precise redefinition of scientific misconduct and a revision of the rules governing the ALERT system, notorious among biomedical scientists because it flagged scientists who had been only accused of misconduct. A notice in the *Federal Register* for November 6, 1992, formally announced that any respondents found tentatively guilty in a misconduct investigation could, whether they faced loss of grant eligibility or not, henceforth take their cases to the Health and Human Services Departmental Appeals Board, which would afford them all the rights of due process, including the right to question all witnesses and evidence against them and present their own in rebuttal.[54]

All the same, the notice of November 6 left procedural matters considerably beclouded, especially during the investigative phase. While the O.R.I. busied itself with devising final rules of fairness, the rules that had been promulgated in June 1991—precisely the rules that had drawn so many protests—would remain in force. The November notice elaborated that, of course, the rules promulgated in 1992 made "portions" of the old ones "no longer germane" and the old rules remained in force only "to the extent that they remain relevant." But which rules remained relevant and which portions were no longer germane was unspecified. The vagueness was compounded by the declaration in the notice that "deviations from the policies and procedures may be necessary where [the] P[ublic] H[ealth] S[ervice] determines it is in the best interest of the Government to do so in an individual case."[55] The O.R.I. thus retained a great deal of flexibility in how it chose to proceed.

It opted, in fact, to continue conducting its investigations largely under the scientific-dialogue rules that had been drawn up by Hallum and Hadley, including the rule that respondents would engage in dialogue primarily with O.R.I. investigators rather than with their accusers. Barbara C. Hansen, a professor at the University of Maryland medical school, considered the establishment of the universal right to a due-process-protected appeal a good step. But she pointed out that, made available only after a finding by the O.R.I. investigation, it might come too late in the process, "only after there is a strong perception of guilt." It was a point that Thereza Imanishi-Kari well understood.[56]

F O U R T E E N

■

Justice Delayed

THE LEAK of the draft report shocked Bruce Singal. He had seen plenty of corruption and chicanery in his days as a U.S. attorney, but he says that he believed that the system would run its course until Imanishi-Kari had been afforded all her rights. In an indignant response, he blasted the O.S.I. for the persistent breaches of confidentiality that had marked the investigation and that now, with the leak of the report, "have once again caused Dr. Imanishi-Kari to be pilloried nationally." He complained that the report "refers to literally dozens of documents and other items which we have never seen or had the opportunity to see." The pertinent items went far beyond the materials related to the Secret Service's forensic findings that he had requested in the spring of 1990 and had not yet received. They included transcripts and summaries of interviews with O'Toole, Maplethorpe, and Walter Stewart plus O'Toole's assertions that the June subcloning had not been done and her comments regarding the green tapes. Singal asked for all such documents so that he and his client could prepare an appropriately informed response to the draft report.[1]

Although Hadley and Hallum had publicly assured critics of the O.S.I. that respondents were given access to all the evidence in due course, Imanishi-Kari was denied almost all of the additional documents she had requested. It was not O.S.I. policy to provide the type of materials she wanted, Hadley informed her.[2] On May 23, 1991, Singal sent Hadley Imanishi-Kari's reply, pointing out in a letter of transmittal that she had been asked to respond to a finding that was "sweeping in conclusions but cursory in supporting evidence" and to participate in a process that "has kept much of that data in the dark, concealed from the accused." Imanishi-Kari herself noted that she had learned more about the materials undergirding the findings from the leak-based article published early in March in *Science* than she had from the O.S.I.. The withholding of the materials meant, she protested, that she was "denied the means to establish innocence."[3]

Imanishi-Kari nevertheless did what she could to rebut the charges. She emphasized the selectivity that characterized the draft report's use of evidence: Data questioned in one part of her notebooks was often accompanied by data in other parts that supported the same scientific point but that the O.S.I. ignored. Virtually all the testimony on her behalf was dismissed as though "all of these scientists are part of some giant conspiracy" while only the testimony of Margot O'Toole was accorded credibility. Imanishi-Kari provided more reasons than she had included in her statement in *Nature* as to why the credibility was undeserved. Particularly important was a claim in the draft report that one of O'Toole's "earliest and strongest allegations" concerned the procedure used to determine that the hybridomas reported in Table 3 produced antibodies with idiotypic birthmarks similar to that of the transgene. O'Toole held that the procedure had not been done, invoking in support of her charge six pages in the notebook that she said she had seen, among a few others, at the Tufts inquiry. Imanishi-Kari pointed out that O'Toole had declared at the Dingell hearings in 1989 that she had seen "only two sheets of paper" at the Tufts inquiry, testimony that accorded with her own memory. Neither sheet, Imanishi-Kari added, was directly relevant to the procedure in question.[4]

Imanishi-Kari pointed to flaws in the O.S.I.'s implied scenarios of fabrication: The draft report suggested that she had made up the so-called January fusion data on the normal mice to satisfy the N.I.H. inquiry in 1988. The report was as mistaken on that score as on the claim that she had brought the June subcloning data to the N.I.H. panel after they challenged her on Table 2. She had not been asked to supply the normal mouse data, only the data showing that the mistyped mouse was truly mistyped. She had supplied the data on the normal mice voluntarily. Besides, she had no reason to concoct it. Plenty of data were available showing that normal Black/6 mice don't produce antibodies with the idiotypic birthmark characteristic of the transgene. Imanishi-Kari also noted that the draft report not only got the Bet-1 issue backward but also displayed misunderstanding of other critical elements in the science, particularly how she determined whether antibodies with the telltale birthmark were produced by the transgene or genes native to the mice.[5]

From what Imanishi-Kari now knew of the draft report's use of the forensic evidence, all of it appeared highly questionable. Recall that the forensic strategy rested on dating Imanishi-Kari's radiation-counter tapes by searching for matches between her tapes and tapes of known dates in the notebooks of other scientists. The Secret Service had reported, for example, that the green tapes of the June subcloning did not match tapes that other scientists had obtained during the summer of 1985, the period when Imanishi-Kari said those experiments had been done; but a simple calculation revealed that the agency's experts had examined only about 1 percent of the tapes produced on the first floor of the M.I.T. cancer center in those months. The Secret Service held that it found a "full match" between the June subcloning tapes and the green tapes in Maplethorpe's notebooks, which had been generated much earlier than the summer of 1985. Imanishi-Kari's tapes had, however, been generated by a gamma-radiation counter while Maplethorpe's had been produced by a beta-radiation counter. The beta machine printed in a different format, thus confusing the meaning of "full match." The draft report furthermore did not specify what the phrase meant—a full match in the composition or intensity of the inks, the color of the paper or the printer font, or some combination of all these factors.[6]

Imanishi-Kari contended that the draft report suffered "from its overly ambitious effort to transform possibility into proof," adding, "No reputable scientific journal would accept a paper with conclusions based on such flimsy evidence. No journal would accept the Secret Service reports without seeing the evidence purporting to support the results." Singal called the failure to provide the forensic data especially "crippling" to Imanishi-Kari's attempt to frame a response. The denial of access to her original notebooks alone made it impossible for her to evaluate assessments that turned on the color of the tapes and inks. Singal reiterated his request for the materials he had requested in March and urged that Imanishi-Kari be allowed to submit a supplemental response once she had seen them.[7]

The O.S.I. was actually willing to provide Imanishi-Kari her original notebooks and related forensic materials, but they were in the hands of the U.S. attorney in Baltimore, to whom they had been sent in connection with the criminal investigation that the Dingell subcommittee had called for. Robert Lanman had asked the attorney for access to the notebooks but had been refused. At a meeting on June 6, 1991, two weeks after

Imanishi-Kari filed her response, Healy agreed to "hit the pause button" on the process, as Raub put it, authorizing the O.S.I. to defer in preparing its final report in the case until Imanishi-Kari got access to the documents necessary to respond fully to the draft report. On July 2, Hallum formally advised her of the decision, telling her that once her supplemental response was received, the O.S.I. would move promptly to complete its work in the matter.[8]

The O.S.I., however, did not provide Imanishi-Kari with any of the considerable body of materials Singal had requested that the draft report referred to but that were not in the possession of the U.S. attorney. Nor did the N.I.H. concede that she might therefore have a legitimate complaint about how she was being treated. On the contrary, in early October, while Healy was still self-recused from all matters concerning the O.S.I., John W. Diggs, the deputy director for extramural research, wrote to David Parker, Joan Press's collaborator in the due-process protest the preceding June, that the only maltreatment of Imanishi-Kari had been the leak of the draft report. "In no other sense has she been treated unfairly. The respondent was given, and will continue to be given, access to the physical evidence; has the opportunity to cross-examine the evidence as part of the scientific dialogue model we use."[9]

The denials and the delay left Imanishi-Kari and Baltimore vulnerable to the animadversions that continued to be cast against them. In November 1991, Daniel Greenberg's *Science and Government Report* insisted that Imanishi-Kari's "blatant commission of scientific fraud" had been documented by the O.S.I., "fairly or not," and that "the friends of David Baltimore" were only spreading "heavy smoke around" her sin by "pounding on the due process issue." In an exchange with Joan Press and David Parker in the *Journal of NIH Research*, Greenberg declared that the due-process "gambits" left untouched "a large body of unrefuted evidence that fully supports O'Toole's charges." It did not seem to matter to Greenberg that Imanishi-Kari could not refute evidence to which she had no access, nor did it appear to matter to Congressman Dingell and his staff either. On November 4, Hallum attended a meeting of the staff, including Stockton and Chafin, to discuss, among other matters, the investigations of Imanishi-Kari, Tufts, and M.I.T. According to Hallum's notes of the encounter, Stockton opened the meeting by remarking that the O.S.I. was "our child," saying the subcommittee wanted to protect it. The staff inquired whether Healy intended to issue an interim report on Imanishi-Kari in the absence of her full response to the draft report. Hallum said that so far as he knew, Healy did not intend any such action.[10]

To Hallum's apparent surprise, Suzanne Hadley was at the subcommittee that day; she had begun helping with its work after she returned from her leave of absence to another job at the N.I.H. Four months later, during the weekend of March 7, 1992, an informant told Hallum that Hadley was obtaining confidential O.S.I. documents from two members of the O.S.I. support staff. Hallum promptly informed Healy, who in turn immediately asked the inspector general's office to investigate the charge. Healy says that an official in the office determined that such actions might violate federal criminal law and referred her to the Federal Bureau of Investigation (F.B.I.). The locks on Hadley's office were changed so that she could enter it only under supervision. One of Dingell's aides remarked, "This is the craziest thing I've ever seen. Leaking documents is clearly not a federal crime."[11]

On March 11, an F.B.I. agent named Alan Carroll spoke with Hadley, saying that he had already talked with the two O.S.I. support staff and they had admitted giving her documents that she was not authorized to have. A journalist from *Science* interviewed her at the time and reported, "Hadley declined to tell *Science* whether she had received such documents, and she says she also refused to tell Carroll." She declared to Philip Hilts, of the *New York Times*, that the investigation was part of "a pattern of unfounded allegations and false statements against me for months."[12] Healy, in published remarks that did not name Hadley, but that appeared to allude to her, later said that the F.B.I. subsequently brought the case to the office of the U.S. attorney in Maryland "for possible indictment" ·of "a former 'chief investigator'" from the O.S.I. Healy continued that, in the end, he declined to prosecute, explaining why in a letter to her, a copy of which he sent to Dingell. As she summarized part of the reason, "the documents were being given to congressional staff members who could have obtained them legitimately anyway." Healy added, "As a result of this investigation, it was also determined that the former O.S.I. employee had a unique relationship with Chairman Dingell's House Subcommittee on Oversight and Investigations, which involved extensive clandestine activity on the subcommittee's behalf." Hadley later said that the charge that she improperly obtained O.S.I. documents and passed them to Dingell was "absolutely untrue."[13]

In a letter to Healy in March 1992, Dingell protested these "apparent acts of harassment and intimidation aimed at courageous, public spirited whistleblowers," and in the annual Shattuck Lecture, delivered in Boston on May 9, 1992, he chose to discuss misconduct in medical research and to attack Healy for mounting "an unjustified attempt to initiate an investigation of one of the [O.S.I.'s] chief investigators." Dingell took the opportunity of the occasion to defend the subcommittee's actions on the misconduct issue, pointing out that it had "an obligation to the American taxpayer" and chastising the scientific community for its seeming unwillingness to face up to the realities of misbehavior among its members. He castigated Healy for her handling of the allegations against Sharma at the Cleveland Clinic, suggested that Robert Gallo hid behind obfuscations, and slashed at David Baltimore for defending Imanishi-Kari. He asserted that Margot O'Toole "was vilified and effectively driven from her profession after she revealed that a paper in *Cell* . . . relied in large part on data that were falsified."[14]

In May, too, Philip Hilts published a cover story in the *New Republic* on the Baltimore case that read like an elaboration on Dingell's theme. Titled "The Science Mob," the piece called the case "the exemplar of what's wrong with the defensive and self-regulating structure of the American scientific establishment." The article, however, was riddled with errors—wrongly claiming, for example, that O'Toole "could not duplicate . . . important experiments" in the *Cell* paper; that Eisen wrote a report on the dispute on the basis of a quick read of O'Toole's memo and a discussion of the matter "with the Tufts scientists"; and that Baltimore had grandstanded against the subcommittee in May 1989 despite having been told beforehand about all the forensic evidence, including the green tapes (much of which, of course, the Secret Service did not develop until a year or more later). Hilts's errors tended to reinforce his principal claim: "David Baltimore clearly failed as a scientist—through his carelessness, his willful oversight, and his extraordinary attempts to protect his own reputation at the expense of a conscientious young colleague."[15]

On June 19, 1992, National Public Radio devoted its "Science Friday" to scientific misconduct and the host, Ira Flatow, invited O'Toole to help kick off the discussion by briefly recounting the story of the dispute over the *Cell* paper. She told the audience that when Stewart and Feder approached Baltimore with their critique of the paper, "he did a very extraordinary thing," continuing, "He claimed the freedom to ignore this prima facie evidence that his own paper was at best erroneous and at worst fraudulent." She added that Baltimore has "belligerently attacked anybody who stood up for me." At a ceremony in Washington, D.C., six days later, O'Toole received the Cavallo Award, a $10,000 prize established by a Cambridge businessman named Michael Cavallo to recognize people who take risks for the public interest; she was cited for her willingness "to speak out when it would have been far easier to remain silent." O'Toole accepted the award with thanks to a long list of supporters, including Mark Ptashne, John Edsall, Stewart and Feder, Congressman Dingell, his aides, and Suzanne Hadley. In December and, again, in January, Dingell had written to Healy asking that Hadley be assigned to assist his subcommittee as it might need her. Healy said no, but in the late spring Dingell appealed the matter to Secretary of Health and Human Services Louis Sullivan, who authorized the move and in July 1992 Hadley went to work for the subcommittee on a six-month, full-time detail.[16]

All the while, Baltimore found himself dishonored as a public figure, something of a pariah in the world of affairs. Larry Kramer, the playwright and an outspoken AIDS activist, recalls that for several years he had been looking for another Robert Oppenheimer to head a latter-day Manhattan Project against the disease and that Baltimore's name "kept coming up no matter where you looked," adding, "Nobody had anything but extreme praise and admiration for his brains and his ability to get things done." But he was considered too much a political liability to be pushed as a candidate for AIDS czar. Baltimore retreated to his research community and his laboratory. "I enjoy doing science, and I'm going to try very hard to prevent people from taking science from me," he said at the time.[17]

Biologists say that he continued to do brilliant research, and he remained in high demand in the scientific world. He was recruited for a professorship at the Memorial Sloan-Kettering Cancer Center, across the street from Rockefeller University, a prospect that attracted him because his wife had moved to New York and was now the dean for science at New York University. His candidacy was reportedly "sensitive" among some of the Sloan Kettering trustees; the job fell through. The day his resignation from the Rockefeller presidency was publicly announced, Charles Vest, the president of M.I.T., had telephoned to say that his professorship there remained open to him. In May 1992, he announced that he would rejoin the M.I.T. faculty in 1994. His friend and colleague at M.I.T., Phillip Sharp, told a reporter, "He feels comfortable here. He can carry on conversations with friends without having any hesitations as to his history or future."[18]

Imanishi-Kari says that the accusations against her felt like a constant weight bearing down on the back of her head. She says she lost her zest for company and parties and found solace in pottery and gardening. She was angry but kept her anger to herself. She struck friends as stoic. She says, "I could not sit around idle and be mumbo jumbo because I have a daughter. That's my first responsibility. So to collapse is out of the question." She says she thought the scandal hurt her daughter, who was then entering her senior year in high school, "very much, especially every time the newspapers talked about her mother." Imanishi-Kari continued to care deeply about her science, too. "It's all I really do," she says, "the science and my daughter."[19]

When she was not distracted by having to deal with O.S.I. matters, she busied herself in the laboratory. She was strongly supported by her colleagues in the pathology department and by the dean of the Tufts medical faculty, Louis Lasagna. Because of the six years she had spent at M.I.T., she had come to Tufts with the understanding that she would be reviewed for tenure in two years; but mindful of the mounting controversy over the *Cell* paper, the university kept postponing her tenure decision. In December 1991, Tufts entered into a special agreement concerning her employment. Eighteen pages long and complicated in its details, the agreement basically stopped the tenure clock until the charges against her were resolved or until she requested a tenure review, a right that she could exercise at any time through June 1995. Everyone assumed that by then her case would surely be concluded. She could keep her assistant professorship until then, but if in the interim she was found guilty, she would have to leave the university. Although Imanishi-Kari remained without grant money from the N.I.H., she obtained funding from the American Cancer Society. By the spring of 1992, she had results that strongly supported the observations reported in the *Cell* paper and had submitted two papers reporting them to a leading journal of immunology.[20]

In late April 1992, the U.S. attorney for the district of Maryland, Richard D. Bennett, at last sent Singal the original notebooks that Imanishi-Kari needed to respond to the draft report, particularly the findings of the Secret Service. Singal retained a forensic expert named Albert H. Lyter to analyze them. Lyter was a former member of the technical staff of the Federal Bureau of Alcohol, Tobacco and Firearms and now an independent consultant. He had taught forensic science at a number of federal agencies, including the F.B.I. Academy and the Secret Service, was qualified in the civil and military courts in more than twenty states, and was expert in the type of techniques that the Secret Service had used to analyze Imanishi-Kari's notebooks.

Lyter reviewed the forensic materials, including the Secret Service reports, the impression analyses the agents had conducted, the thin-layer chromatograms they had used, Imanishi-Kari's original notebooks and tapes, and those that had been produced by other scientists. In an affidavit of June 17, 1992, he concluded that the Secret Service findings based on the thin-layer chromatograms were "erroneous" and that it was "not possible to draw any conclusions from the T[hin] L[ayer] C[hromatography] analysis as to the dates of Dr. Imanishi-Kari's tapes." In a second affidavit, of August 13, he reported that the impression analysis "alone furnishes no forensic basis to question the genuineness of any page in the I-1 note-book, the date on which it was prepared, or the sequence in which it was prepared." Lyter provided evidence that reinforced Imanishi-Kari's own observation that her tapes and those in Maplethorpe's notebook had been generated by two different printers. He also made concrete her supposi-tion that the Secret Service had simply not looked far enough for tapes produced in 1985 that hers might match: The number of tapes available from the gamma-counter printer comprised "a small percentage of the total number of tapes" that had been generated both in 1985 and in the earlier period, he observed, with the consequence that the sample was "far too small to draw any conclusions" as to whether Imanishi-Kari had produced her tapes in 1985 or not. Lyter also examined the thin-layer chromatograms made to compare the June subcloning tapes with Maple-thorpe's green ones. He found that the inks on each differed and that, therefore, "the Secret Service conclusion that they 'match' (or even that they are 'most consistent with' one another) is erroneous."[21]

On July 13, 1992, Bennett, having seen just Lyter's first affidavit, pub-licly announced that he would not seek an indictment, declaring that "no matter how a charge against Dr. Imanishi-Kari were framed, the central issue in a criminal trial would remain the fundamental validity of her scientific work." Geoffrey Garinther, one of the assistant attorneys in Ben-nett's office, said while trying to explain the decision to reporters that "we were very confident about the Secret Service findings." He proposed that if any of Imanishi-Kari's "results were true, a juror might have had trouble finding beyond a reasonable doubt that she had corrupt intent—that she had intended to mislead with experiments that achieved results that could not actually be achieved." In a letter to Dingell and the regional inspector general of the Department of Health and Human Serv-ices, Bennett elaborated on the problem of proving fraud beyond a rea-sonable doubt: In a criminal trial the statistical evidence on which the N.I.H. had relied would be inadmissible and the Secret Service's forensic analysis would have been contradicted by Lyter's.[22] Singal says that Lyter's analysis was "significant but not decisive" in the outcome. It merely rein-forced the numerous "benign explanations" of Imanishi-Kari's notebooks that were manifestly compelling in their own right against the allegations of fraud.[23] It nevertheless demonstrated, as though the point needed dem-onstration, that allowing Imanishi-Kari access to the evidence against her could make a huge difference.

Baltimore promptly announced that he was canceling his retraction of the *Cell* paper, explaining that Lyter's analysis "is a confirmation of my feeling all along that Imanishi-Kari had not attempted to deceive anybody." He called on government officials to "apologize for putting her through six years of hell . . . give her grants back and let her go on being the good scientist she is." In 1989, scientific advisers to the N.I.H. had recommended funding a request for a competitive grant renewal from Imanishi-Kari, but the award of funds had been frozen. The N.I.H. had instead extended support from a prior grant into 1990, which was the funding that, as Hadley had then reported to Dingell, was cut off. Now, in the wake of the decision not to prosecute her, Imanishi-Kari declared that she would ask the N.I.H. to unfreeze the competitive renewal money that had been denied her in 1989. "I feel that, after all, there is some justice. I was doubting before," she said.[24]

In a supplemental response to the draft report filed on August 18, 1992, Singal cited Lyter's analysis and asked the new O.R.I. to dismiss the case. It should be clear from the new forensic evidence that the O.R.I. "has no basis for continuing this investigation and that Dr. Imanishi-Kari should have to endure no further personal suffering and professional harm from this ordeal," Singal wrote. He now knew that the Secret Service, in its search for matching tapes, had "never examined" the notebooks of at least thirty other scientists who worked on the first floor of the cancer center between 1981 and 1986 and that what it had scrutinized comprised only 3 to 7 percent, if that much, of the thousands of pages generated by the gamma counter that Imanishi-Kari had used. The percentage estimate was probably high because some scientists—for example, Margot O'Toole—completely transcribed the numbers on their counter tapes, then discarded them. There were months in every year from 1981 through 1985 for which no tapes were available. It should not be surprising, Singal argued, that green tapes were not found in the 1985 notebooks of other investigators.[25]

Singal added that if the O.R.I. found it necessary to pursue the matter further, it should begin anew, discarding the draft report which the new forensic evidence has "largely discredited." In that event, the O.R.I. should not follow its predecessor's practice "of sharing information with the accuser but not the accused." In a clear allusion to Dingell and, perhaps, to Hadley, he urged that it should also resist efforts by "various interested personnel, some of whom occupy positions of authority within the federal Government, to influence the outcome of this proceeding so as to ensure a finding of scientific misconduct against Dr. Imanishi-Kari." Singal admonished the O.R.I. to "display the political courage and scientific integrity to rule that Dr. Imanishi-Kari has not engaged in scientific misconduct."[26]

However, neither Lyter's forensic conclusions, the U.S. Attorney's decision, nor Singal's plea swayed the O.R.I. In mid-September, John Dingell made clear that he remained decidedly interested in the Imanishi-Kari matter, sending the O.R.I. a letter inquiring about its plans for completing the case and about the status of its investigation into how Tufts and M.I.T. had responded to O'Toole's allegations. Dingell also asked for copies of all the documents and materials that the O.R.I. had gathered in its probe of the two institutions. In late September, Robert Lanman wrote Singal that the O.R.I. was completing its investigation of Imanishi-Kari and proceeding with the preparation of a final report. In a letter a month earlier, the O.R.I. had informed Imanishi-Kari that the Public Health Service had decided, as an interim policy, to enter into the ALERT list only those scientists who had been found guilty of misconduct, not those who had only been accused of it. Her name was thus being removed from the list. Nevertheless, in September the N.I.H. told Imanishi-Kari that her competitive renewal grant could not be unfrozen: Her application for it had expired.[27]

Joan Press and David Parker urged colleagues around the country to protest to Anthony Fauci, the scientist who headed the health institute to which Imanishi-Kari had applied for the renewal. "This withholding of grant support pending a finding in a misconduct investigation is a terrible precedent," they noted. Press and Parker's appeal produced letters on Imanishi-Kari's behalf from many scientists. Fauci explained in reply that, since the renewal grant had never been funded, it had been inactivated at the end of the 1989 fiscal year with all the rest of the year's unfunded grants. It thus could not be restored, but Imanishi-Kari could apply for a new grant.[28]

In a letter to Fauci, a copy of which he sent to Healy, David Parker angrily challenged the catch-22 character of Fauci's response. Parker found it hard to see how Imanishi-Kari could expect a new application to be successful—given that "the N.I.H., under intense political pressure, had denied her grant support before a finding in this case." Could Fauci assure Parker that if she submitted an application now "it will be evaluated independently of this botched, endless investigation, and *paid* if it competes favorably with other applications?" Early in 1993, one of Fauci's subordinates apprised Parker that the institute could fund a meritorious application from Imanishi-Kari while the O.R.I. final report remained pending.[29]

In April 1993, the two papers that Imanishi-Kari had submitted the previous May appeared in print, impressing immunologists that the observations reported in the *Cell* paper were, as one of them said, "reproducible." In May, the journal *International Immunology* accepted for publication a paper by Leonore Herzenberg of Stanford and Alan Stall and a student at Columbia University that independently confirmed the *Cell* paper's observation that the transgenic mouse cells produced a high level of native antibodies. Neither Imanishi-Kari nor the Stanford-Columbia team attributed the phenomenon to any kind of idiotypic mimicry—new evidence had made that interpretation questionable—and they differed over the mechanisms that were at work in the cells. But Herzenberg told a reporter that the original article by Weaver *et al.* "showed that the antibodies these mice made were abnormal, and our current paper confirms that absolutely."[30]

In June, the *New England Journal of Medicine* published the Shattuck Lecture that Congressman Dingell had delivered in Boston the previous May, with its censures of Gallo, Baltimore, Healy, and Imanishi-Kari. Dingell noted in an addendum that the U.S. attorney's decision not to indict Imanishi-Kari was "not an exoneration of David Baltimore or Thereza Imanishi-Kari, nor did it reflect doubt on the part of the prosecutor's office that the data had been falsified." He also observed that late in 1992, the O.R.I. had concluded in one final report that Rameshwar Sharma had committed scientific misconduct at the Cleveland Clinic and, in another, that Robert Gallo had fraudulently misappropriated the French AIDS virus.[31]

Replying in the September issue of the journal, Baltimore called attention to the "numerous papers" published since 1986 that have "supported and extended" the findings of Weaver *et al.*, including the recent results of Imanishi-Kari herself as well as Stall and Herzenberg. Adding in Lyter's forensic report, he declared, "If the science has stood the test of reproducibility and the evidence of fraudulent data production does not hold up, there is simply no case against Dr. Imanishi-Kari." Healy contributed her own rebuttal to Dingell in the same issue, pointing out that in August the departmental board to which adverse findings of the O.R.I. could now be appealed—it was called the Research Integrity Adjudications Panel—had overturned the verdict against Sharma. It concluded that the O.R.I. had failed to prove intentional deception on Sharma's part.[32]

On November 3, 1993, Gallo's coauthor Mikulas Popovic, whom the O.R.I. had also found guilty of misconduct, was completely exonerated by the appeals board, too. In its decision, the board noted that it had been compelled to parse an enormous record of documents, briefs, and testimony, "all focused essentially on the meaning which we should give a handful of words and notations contained in one heavily-edited paper written by a scientist with limited English skills." The board continued, "One might anticipate that from all this evidence, after all the sound and fury, there would be at least a residue of palpable wrongdoing. This is not the case." The board found that the O.R.I. had consistently read language that was merely "ambiguous" in only one way, meaning consistently against Popovic, and "faulted him simply for doing things differently from how they would have done them."[33] Gallo had filed an appeal in his own defense. On November 12, 1993, three days before his hearing at the board was to begin, the O.R.I. dropped its case against him.[34]

Five weeks later, on December 17, Bruce Singal filed a motion with the O.R.I. calling either for a finding that Imanishi-Kari had not committed scientific misconduct or for a simple dismissal of the proceedings against her. In a lengthy memorandum in support of the motion, Singal reiterated the numerous reasons—notably, the lack of confidentiality, the inaccessibility of much of the evidence, Lyter's forensic report—that he had previously advanced to the O.R.I. and added several telling new ones that had developed since August 1992, when he had submitted the supplementary response to the draft report. He provided letters from two scientists at two different universities that reinforced Lyter's analysis that the forensic report of the Secret Service was badly flawed. One of them, a microbiologist at the University of Pennsylvania named Fred Karush, who had more than four decades of experience with thin-layer chromatography, advised that the technique did not show "significant differences among the counter tapes prepared at different times." He concluded that it was wrong to infer "that the tapes in [Imanishi-Kari's] notebook (I-1) were not generated in 1985."[35]

Singal also pointed to the recent papers by Imanishi-Kari and by Herzenberg and Stall, noting that they reinforced the observations of Weaver *et al.*, and he spotlighted the outcomes in the Gallo and Popovic cases. He argued that because conviction required proof of intentional fraud, the O.R.I. would face "insuperable obstacles" in trying to demonstrate its case against Imanishi-Kari now that the Secret Service results had been refuted. He added that the draft report on Imanishi-Kari displayed one of the features for which the appeals board had slammed the Popovic report—a propensity to interpret ambiguous language in only one way, and consistently against the respondent.[36]

Singal came down hard on the O.R.I. for the vagueness of its announced procedures and for so far failing—now after some two and a half years—to promulgate them in their final form. "A government agency could hardly have concocted a more confounding set of 'rules' if it had intentionally set out do so," he remonstrated. He stressed that the O.R.I. was in gross violation of one of its few explicit rules—that, in general, cases were to be completed within 120 days of their initiation. Depending on where one started, the investigation of Imanishi-Kari "may set new records for violating these important timeframes," Singal pointed out. The O.S.I. had gotten involved in the case about May 31, 1989—940 days earlier. It was now 712 days past May 23, 1991, the date that Imanishi-Kari had submitted her initial response to the draft report and 485 days past August 18, 1992, the date that she had filed her supplemental response. Singal, citing letters from the O.R.I. and its predecessor, noted that Imanishi-Kari had repeatedly been told that the office intended to move promptly to a final report once it had received her final response. Yet according to his recent telephone conversations with O.R.I. officials, the office was not even close to issuing its final report in the case.[37]

Singal put it to the O.R.I. that the consequences were unconscionable.

This investigation lingers interminably, continuing to taint and damage Dr. Imanishi-Kari and her career and reputation irreparably. . . . These seven and a half years of investigative scrutiny . . . have exacted a draconian toll on her life and career. During this time, she has been subjected to unfair publicity on a national and international scale, the unauthorized leak of the "confidential" O.S.I. draft report, non-renewal of her grants, inability to gain tenure at Tufts, loss of funding for her students and research, and incalculable and irretrievable loss of her professional and personal reputation.

Singal stressed that, in all, Imanishi-Kari "has already received a penalty comparable to that imposed on scientists who have actually been found guilty of scientific misconduct."[38]

Chris Pascal, who was then chief of the Division of Legal Counsel in the O.R.I., attributes the delay partly to the fact that in the latter half of 1992 the new office was busy physically relocating out of the N.I.H., hiring new staff, including several additional attorneys, and drawing up its internal procedures. When it became active again in late 1992, it had a backlog of some seventy cases to deal with, and new ones were arriving continuously. In 1993, the staff also had to cope with seven appeals to the departmental board in addition to those of Sharma and Popovic. Pascal says that, as he best remembers, the office did not turn seriously to working on the final report in the Imanishi-Kari case until the fall of 1993. However, an investigative team was created within the O.R.I. in July 1992, and a draft of the final report was being circulated in the summer of 1993. What Pascal perhaps remembers is that the serious involvement of the O.R.I. legal staff began in later 1993, after the appeals board had sounded several sharp wake-up calls.[39]

By then, the O.R.I. had a year's experience contesting—and mostly losing—cases before the appeals board. The rulings had indicated that the board took seriously the stated definitions of misconduct, particularly their patent implication that it included intent. It was not misconduct if a scientist inadvertently published a false result; it was misconduct if the result was the product of "falsification," a knowing and deliberate act of deception. Pascal recalls that, given the board's rulings, the O.R.I. was "not prepared to rubber-stamp" the draft report on Imanishi-Kari. He continues, "I wouldn't want to say the document was inadequate, but some of the early reports did not explain their cases in a way that would be useful to the appeals board. We decided that we had to take a fresh look at all the old reports from the O.S.I. and ask, Do we agree with them or not?" It seems evident from Pascal's circumlocutions that the O.R.I. feared that rubber-stamping the draft report greatly risked losing its case against Thereza Imanishi-Kari on appeal.

In July 1992, the O.R.I. had assigned the primary responsibility for investigating the case to a microbiologist named John D. Dahlberg, a staff scientist at the N.I.H. through most of his career whose research included work with the kind of immunological tests—for example, radioimmune assays—that Imanishi-Kari had used in her experiments. A later O.R.I. account of the investigation noted that Dahlberg "did not review the OSI draft report in any detail; rather, he tried to take a fresh look at the evidence."[40]

After the fall of 1993, the O.R.I. invested an additional six to nine months of work on the final report, completing the penultimate draft and sending it to Imanishi-Kari for comment in the summer of 1994. Far more substantial than the draft version produced by the O.S.I. in March 1991, the final report totaled 231 dense pages of text plus several appendices. It did not read like a novel or a moral tract; it contained no assessments of scientific character like those that Hadley had proffered of Baltimore and Weaver. It opened with a summary of findings, discussed the *Cell* paper in its scientific context, and proceeded, finding by finding, to a presentation and analysis of the evidence for each. It revealed that the O.R.I. staff had gone to considerable lengths not only to strengthen the findings of the draft report but also to extend them.[41]

Like the draft report, the final report found that Imanishi-Kari had fabricated much of the normal mouse data and the June subcloning data, but it also concluded that she was guilty on a number of other counts, including several on which the draft report had been able to determine only probable culpability. The document declared that she had fabricated the data showing the specificity of Bet-1 and misrepresented its specificity by failing to acknowledge its problematic qualities. It also found her guilty of counts on which, in 1989, the N.I.H. had found her innocent, holding that she had been "deliberately misleading" in discussing the wells in Table 2 as though they were clones and that she had indulged in "deliberate misrepresentation of data" in choosing the thousand-counts-per-minute cutoff. It claimed that she had not tested most of the hybridomas in Table 3 for the idiotypic birthmark of the transgene and that she had "concealed the presence" in them of double producers.[42]

The final report made some claims of fabricated data do double duty, using several in support of more than one adverse finding. It said that the letter of correction to *Cell* in November 1988 constituted scientific misconduct because it contained the fabricated Bet-1 and June subcloning sets of data. It declared that at least 20 percent of the I-1 notebook was falsified and that the notebook itself thus amounted to "an independent act of deliberate fabrication and falsification in the reporting of research." It charged Imanishi-Kari with misconduct in submitting grant applications, in 1984 and 1985, that drew on her falsified data. In a summary statement, the final report asserted:

> In effect, Dr. Imanishi-Kari first established the findings and then crafted the data and experimental results to support these findings. When her claimed results were questioned, she attempted to cover up her initial fabrications and falsifications by fabricating additional data purporting to support her claims.[43]

For all its sobriety, the final report resembled much more a prosecutor's brief than the product of a "scientific dialogue." It amounted for the most part to a detailed endorsement of Margot O'Toole's main allegations, with at least one from Stewart and Feder—the cutoff issue—added in. It appeared to raise as many charges as possible in the hope of making at least one stick. It let Imanishi-Kari off on only one count—the fourth points in Figure 1—but grudgingly. It claimed they were fabricated but said that the evidence was "insufficient" to attribute the fabrication to Imanishi-Kari (and that the O.R.I. intended to pursue the matter with Moema Reis). Although it discussed the science, it omitted mention of the papers published in 1993 by Imanishi-Kari and by Herzenberg and Stall. The final report declared it "most significant that *all the authors with the exception of Dr Imanishi-Kari*" had retracted the *Cell* paper, inexplicably ignoring Reis's original refusal to do so and Baltimore's nullification of his own retraction.[44]

The document took note of arguments that Imanishi-Kari had submitted in her defense, but only in the main to refute them. It ignored the challenges to the O.R.I.'s statistical questioning of Imanishi-Kari's transcriptions of counter-tape data. The penultimate draft of the final report did not address Lyter's rebuttal of the forensic evidence, even though a great deal of the findings of fabrication rested on the Secret Service's reports. In February 1992, the Secret Service had sent the O.R.I. a memorandum that responded to Lyter's critical analysis of its forensic work. On the whole, it was more assertive than substantive. On several key points, it was remarkably imprecise—declaring, for example, that the different radiation counters were "often attached to the same printer," or that a review of the chromatograms "clearly shows that no erroneous matches were made." The O.R.I. appeared to take the Secret Service's response as sufficient, but Ursula Storb, to whom the draft was sent as a member of the scientific panel, did not. She contended that the O.R.I. needed to resolve the forensic dispute. The final report showed no evidence that Storb's admonition was followed. It merely mentioned Lyter's affidavits and included them in appendices, along with the response the Secret Service had prepared against them in February 1992.[45]

The O.R.I. had given Imanishi-Kari ninety days to respond to the final report. When she first received it, and again in mid-October 1994, Bruce Singal protested to its director, Lyle Bivens, that it was "unconscionable" to expect Imanishi-Kari to respond in such a short time to a lengthy and complicated report that the O.R.I., with all its staff and resources, had taken more than two years to prepare, and to respond without the documentation that she needed to defend herself.[46]

The O.R.I. sent Imanishi-Kari its final report on October 26, 1994, more than eight years after she had first been brought under suspicion by O'Toole. In an accompanying joint letter from Lyle Bivens, who headed the O.R.I., and a deputy secretary of health and human services, she was told that her motion of the previous December for dismissal of her case was denied on the summary grounds that her arguments were "without merit." The letter informed her that the Public Health Service proposed to bar her from eligibility for N.I.H. grants for ten years. Imanishi-Kari promptly announced that she was innocent and would appeal.[47]

The day after the report was sent to Imanishi-Kari, Congressman Dingell requested a copy. The O.R.I. sent him one, asking that it be kept confidential until such times as it could be made public, which under O.R.I. policy included the point at which an appeal was filed. On November 25, two days after Imanishi-Kari appealed the finding against her, the O.R.I. released the report. Tufts responded to the report by suspending her from the tenure-track faculty, although it permitted her to remain in her laboratory while the appeal proceeded. The university was "running scared of Dingell," its provost frankly recalls, and people were writing letters complaining about "our having a crook on the faculty." Imanishi-Kari focused on her appeal. "For the first time in all these years I can really see all the so-called evidence," she said.[48]

Bivens told *Science* magazine, "We're geared up for this one."[49]

F I F T E E N

■

Matters of Judgment

THE DEPARTMENTAL Appeals Board appointed a three-member panel to hear and judge the case; the panel comprised two of its own regular members and a scientist named Julius S. Youngner from the University of Pittsburgh. Youngner, a specialist in molecular biology and immunology, was officially retired as a professor but he remained highly active in research. He had been earlier enlisted by the O.R.I. to help hear the appeal of Robert Gallo, a task cut short when the case was dropped. Youngner says that the high-profile quality of disputes like Imanishi-Kari's heightened a sense of responsibility that he felt to science. It derived from his experience as a coworker with Jonas Salk, who had recruited him to Pittsburgh in 1949 to work on the polio vaccine. "I'd seen from the inside what the press did with scientific affairs," he recalls, characterizing the headlines about the vaccine as "hyperbole, mythmaking." He had read about the Imanishi-Kari case in the newspapers. He says that he was willing to take on the time-consuming task of serving on the panel because "I wanted to get to the truth of the matter. I didn't want to take as true what was in the *Chicago Tribune* or even the *New York Times*."[1]

The two board members on the panel were Cecilia Sparks Ford and Judith A. Ballard, both lawyers with, between them, some twenty-five years experience handling departmental appeals, including recent duty in the proceeding that exonerated Robert Gallo's collaborator Mikulas Popovic. Youngner says they helped bring him "into the rhythm of what a hearing is" and kept him "from making terrible gaffes," like discussing scientific issues with the witnesses during breaks. He adds that they also "were good on the science, and they learned fast." In preparation for the hearing, the panel worked through the vast body of evidence on the case with the help of two young staff lawyers who say they became consumingly absorbed with the issues and thought hard about them day and night. Usually only one staff lawyer is assigned to a case. The head of the board says the Imanishi-Kari case took more of its time and resources than any other case that had come before it.[2]

The O.R.I.'s findings against Imanishi-Kari had been transformed in a letter that accompanied the final report into nineteen counts of fraud and misconduct, but they constituted an indictment rather than a verdict. The panel was well aware that because of the lack of due process in cases coming from the O.R.I., the evidence against respondents like Imanishi-Kari had not been tested. The panel was interested in scrutinizing all of it rigorously, including how it bore on the credibility of witnesses. In the comparative informality of an administrative trial, it could permit a degree of questioning by lawyers for both sides as well as by the panel members themselves that might be disallowed in the more formal setting of a federal court. Before the hearing, the panel established a detailed process by which both sides would exchange summaries of what their respective experts would say. By the right of pretrial discovery, Imanishi-Kari also obtained access to the O.R.I.'s evidence, including the testimony, memoranda, and allegations of Margot O'Toole, and the research notes, work sheets, correspondence, and reports of the Secret Service. Imanishi-Kari's appeal thus constituted a full administrative trial in which the O.R.I. had to prove its case to the panel *de novo* against Imanishi-Kari's due-process informed defense.[3]

The O.R.I. had no power to issue subpoenas, which meant that witnesses testified voluntarily and the hearing had to be scheduled to accommodate them. To that end, the O.R.I. and Imanishi-Kari agreed to divide the hearing into two phases: A first round, lasting the three weeks to the end of June, would in the main explore the scientific and statistical issues in the case; a second, beginning on August 21, would probe the forensic evidence, examine Imanishi-Kari, and entertain whatever other testimony might be necessary to complete the proceeding.[4]

At the opening session in June, Marcus H. Christ, Jr., from the general counsel's office of the Department of Health and Human Services, introduced the O.R.I.'s allegations with the contention that Imanishi-Kari had produced "one of the more egregious cases of scientific misconduct that has ever befallen" the department. Christ, turning forty, had been a staff attorney with the department for ten years, the majority of which he had spent litigating financial matters in medical administration. His cocounsel was Stephen M. Godek, in his early thirties, a trial lawyer with a reputation for combativeness who had been in private practice before joining the general counsel's office. Neither Christ nor Godek had dealt with cases of a scientific nature before they had been assigned to the O.R.I. several years before, but they were rapidly coming to specialize in scientific fraud. Christ says that he doesn't have any problem dealing with technical cases and doubts that Godek does either. Christ says that both he and Godek thought the case against Imanishi-Kari was strong and that they were happy to pursue it.[5]

In nine days of opening testimony, Christ and Godek developed the scientific and statistical evidence in support of the O.R.I. indictment with testimony from Margot O'Toole; the O.R.I. staff investigators John D. Dahlberg, who was the principal author of the final report on Imanishi-Kari, and James Mosimann, who was now an adjunct professor at American University; and the scientific panel members Joseph Davie and William McClure.[6] Christ and Godek rounded out this phase of their case with a statistician from outside the N.I.H. and with Walter Gilbert, who testified that he had followed the science of the *Cell* paper but also reviewed "a small amount of the laboratory notebooks in this matter." Gilbert insisted that what counted for misconduct was not whether the science was right but whether the paper faithfully reported the experimental data. He made clear that, so far as he was concerned, it did not. "My opinion is that Dr. Imanishi-Kari did falsify the laboratory notebooks, and falsified the data listed in the paper. That opinion is based on my review of the forensic information developed by the Secret Service, [and] by the statistical argument." He added that he regarded the June subcloning data as a "key fraudulent element" and "some of the Bet-1 data as clearly fraudulent."[7]

On June 23, the hearing turned to the presentation of Imanishi-Kari's defense by her principal appellate attorney, Joseph N. Onek. Onek was in his mid-fifties, a product of Harvard University, the London School of Economics, Yale Law School, and coveted clerkships with Judge David Bazelon at the U.S. Court of Appeals in Washington, D.C., and Supreme Court Justice William Brennan. He had spent much of his career working at the intersection of law, ideas, and public policy, serving on the White House staff of President Jimmy Carter, where he had responsibilities in both domestic and foreign affairs. He says that after Carter's defeat, in 1980, he opened a boutique law firm with several other former Supreme Court clerks that specialized in Supreme Court–related issues, particularly law and health care. Judge Bazelon first stimulated his interest in science, and then so did his wife, a psychiatrist who is a daughter of Emanuel Piore, for many years chief scientist at IBM and a high-ranking member of the national science-policy establishment.[8]

Onek's scientific interests were pretty much confined to psychiatry and the law until one day around the end of 1989, an attorney in a firm a few floors below his called to say that Robert Gallo needed a lawyer. Onek says that he took on the case after a three-hour talk with Gallo convinced him that the man was innocent. Onek was by then with the prominent Washington law firm of Crowell & Moring, of which Robert Charrow was also a member. He was offended by what he soon learned about the government's handling of the Gallo case, including the high pressure from Dingell's subcommittee, the lack of due-process protections, and the exercise of Suzanne Hadley's questionable judgments. Onek says Hadley "knows all the facts and dates but doesn't know what they mean," elaborating that her opinions on what they mean remind him of the admonition from one of his law professors: "Mr. Onek, your answer is like the 13th chime of the clock. It's not only wrong in and of itself but it casts doubt on all else."[9]

Onek says that he agreed to defend Imanishi-Kari after talking with her for several hours and concluding that, like Gallo, she was innocent. He took the case largely on a pro bono basis, believing that the stakes in it went beyond Imanishi-Kari's plight. Now, on June 23, he outlined Imanishi-Kari's defense, ending with the blunt assertion: "Prodded in part by a congressional sub-committee, O.S.I. and subsequently O.R.I. have relied on forensic and statistical analysis to assert that there was misconduct. But the evidence will show that these analyses amount to one-sided searches for any data that could be viewed as supporting O.R.I.'s conclusions. . . . This case has been a nightmare for Dr. Imanishi-Kari for almost a decade. During this same decade, a number of other scientists have been falsely accused of misconduct. This panel's decision will not only vindicate Dr. Imanishi-Kari, but will bring to an end an ignoble chapter in the history of American science."[10]

Onek, who says that he had not tried a case in some twenty years, was joined in defending Imanishi-Kari by an attorney named Tom Watson, who was then a senior partner in Crowell & Moring and a seasoned, highly regarded courtroom lawyer. Watson and Onek together developed the testimony of a parade of witnesses who appeared on Imanishi-Kari's behalf, some testifying in person, others by telephone or video conference call. It included not only the familiar figures—David Baltimore, Moema

CECILIA SPARKS FORD and JUDITH BALLARD
Departmental Appeals Board
Lawyers on the Imanishi-Kari Panel
Photo Credit: D. J. Kevles

JULIUS S. YOUNGNER
University of Pittsburgh
Biologist on the Imanishi-Kari Panel
Photo Credit: Courtesy Julius S. Youngner

JOSEPH ONEK
Imanishi-Kari's Appeal Lawyer
Photo Credit: Courtesy Joseph Onek

Reis, David Weaver, and Chris Albanese; Henry Wortis, Brigitte Huber, Robert Woodland, and Herman Eisen—but a dozen other friends and colleagues. The lineup for Imanishi-Kari totaled twenty-three people, roughly twice the number that the O.R.I. deployed. By the time the hearing ended on September 15, it had accumulated twenty-eight days of testimony and generated 6,500 pages of transcript.[11]

The hearing record revealed again and again how the rules of due process—the right to see the evidence, present and examine witnesses, cross-examine the O.R.I.'s experts, and be heard before a neutral panel of judges—permitted Imanishi-Kari to defend herself with an effectiveness that had been impossible under the policies of Suzanne Hadley's O.S.I. and its successor. Her appellate lawyers consulted with Bruce Singal and used many of the arguments that he had advanced, but they were far better armed than he had been to contest each and every element of the O.R.I.'s case. They struck at its foundational presumptions that Imanishi-Kari was dishonest while Maplethorpe and O'Toole were completely credible. They rebutted the charges that she had misrepresented, fabricated, or falsified data bearing on Bet-1, Table 2, and Table 3, partly by challenging the O.R.I.'s standards of scientific conduct and its readings of the I-1 notebook, partly by attacking the statistical and forensic analyses on which it relied.

Immunologists from Harvard Medical School, the University of California at San Diego, and the Scripps Research Institution, none of whom had any previous connection with the case, testified to Imanishi-Kari's scientific character and reputation. They affirmed that she was professionally esteemed, that she generously shared mice, cell lines, and reagents, helped even competitors in using them, and that the reagents—including Bet-1—behaved as she described them.[12] Marcus Christ spent most of one morning in September trying to impugn Imanishi-Kari's veracity, suggesting, among other things, that she had lied about earning the equivalent of a master's degree at Kyoto University and misrepresented her faculty rank at Tufts on grant applications. Onek demonstrated that the suggestion had no merit, and before the lunch break Panel Member Ford admonished Christ, declaring, "We didn't really see, necessarily, the point of a lot of what we heard this morning." The panel grew impatient more than once with the government's lawyers, at one point rebuking Christ for persistently referring to a reagent of Imanishi-Kari's with a sneer and at another chiding Godek for cross-examining one of her witnesses in an inappropriate tone and manner.[13]

Onek permitted Charles Maplethorpe to raise suspicions about the reliability of his own testimony. Vivien Igras, the cancer center technician, testified how Maplethorpe had told her that he intended to "get" Imanishi-Kari. Although Maplethorpe categorically denied having said any such thing, Onek brought out other evidence for doubting his perceptions of Imanishi-Kari and her work. Maplethorpe acknowledged that Imanishi-Kari had declined to make him coauthor on the paper with Boersch-Supan, that he had composed a private account of the dispute—"one never knows when one needs to remember facts," he said—and had later given the document to Walter Stewart. He conceded that Imanishi-Kari had refused to write a reference for him. He admitted that he had surreptitiously taped the conversation between Imanishi-Kari and Weaver in June or July 1985, saying that he had turned on the tape recorder because he expected the discussion "to be very interesting" and since he already had "many suspicions about the results she was reporting to us in the lab and had seen evidence that many things that she was saying were not true."[14]

Onek asked Maplethorpe how he had been able to provide the N.I.H. investigators with a copy of a grant application that Imanishi-Kari sent to the American Cancer Society. He explained that he suspected she might be "committing fraud"; he found the text on the word processor on which he was writing his thesis in August 1985; so he printed it out. "You didn't ask permission . . . to access her grant from the computer?" Onek inquired. "Of course not," Maplethorpe answered. Imanishi-Kari, invited by Onek to comment on the matter, noted that the word processor in the lab at the time was a simple Digital Equipment computer with a system disk and a floppy drive for a user's disk. Her floppy disks were not shared with students, she recalled; they were in the possession of her secretary, implying that Maplethorpe had not stumbled across her grant file but had taken her disk to snoop in it. "The fact that he took . . . something without asking, it's very disturbing," she said.[15]

Moema Reis, testifying by telephone from Brazil, called her relationship with Margot O'Toole "very good." Her laboratory bench was right in front of O'Toole's; they used to lunch together, often at Reis's apartment because O'Toole's boy was in day care in her building. Yet for all their conversations, Reis said that she did not remember ever telling O'Toole that Bet-1 did not work well or O'Toole's ever complaining that Bet-1 lacked specificity. Onek asked Imanishi-Kari whether, as O'Toole claimed, she had ever told her just to ignore unfavorable results that had come from two mice in her cell-transfer experiment while giving credence to favorable ones that had arisen from three others. Imanishi-Kari responded that it was too long ago to remember what she had said exactly, but she observed that the transfer of cells was accomplished by injecting them into a vein in the mouse tail with a very thin needle. It is "a very hard procedure and sometimes you may miss," she noted, with the result that the mouse would receive too few cells or none at all. "So in a set of five mice, you may have results that are very clear on some individuals, but not clear in other individuals. And certainly I raised that possibility with Dr. O'Toole, if that could be an explanation for these experiments."[16]

Onek, cross-examining O'Toole herself, exposed crucial weaknesses in the charges concerning Table 3 that formed part of the allegations she had filled with the O.S.I. in November 1989. Since the document had become available to Imanishi-Kari as a result of her appeal, she and her lawyers now knew what O'Toole had asserted—that during her meeting with the Wortis committee on May 23, 1986, Imanishi-Kari had brought out data intended to show that many of the hybridomas reported in the table produced antibodies that were idiotypically birthmarked yet had the *gamma* isotype of native genes; that the data showed the ELISA tests, which were recorded on six pages, now page numbers 83 to 88, in the I-1 notebook; but that the antibodies that had been isotyped had not been selected for idiotype; that Imanishi-Kari had admitted the fact, calling it a mistake; and that "all agreed that this was a fatal flaw in the study." However, O'Toole had not mentioned the flaw in the memorandum that she sent to Herman Eisen on June 6 detailing her quarrels with the *Cell* paper. On that occasion, her discussion of Table 3 dealt almost entirely with the analytical methods used to characterize the hybridomas reported there and was informed by page 41 in the I-1 notebook, which she had in fact seen on May 23. The page showed the results of tests done on eight of the Table 3 hybridomas after the antibodies had been purified (that is, separated from everything else in the supernatants): All eight produced antibodies displaying the idiotypic birthmark and *gamma* isotypes. Furthermore, in her interview with the N.I.H. in 1988, O'Toole had said nothing about having seen the six ELISA pages. On the contrary, it was in reference to a loose-leaf sheet of data concerning tests of similarly purified antibodies for *mu*, which was page 42, and another, which was page 41, that she had said, "The only data that I saw [at Tufts on May 23] were these two sheets."[17]

Onek called O'Toole's attention to the contradiction in her statements about what data she had seen that day. She indicated that she had not told the N.I.H. staff about the ELISA pages because they were "not relevant to the central thesis." Onek asked why O'Toole had not mentioned the unanimous agreement about the fatal flaw in her memorandum to Eisen, even though the document dwelled at length on Table 3. O'Toole responded that she had written her memo towards "dealing with the science." The ELISA data were irrelevant to her desire for evidence that idiotypically birthmarked antibodies displayed isotypes characteristic of native genes. "Therefore, I did not have to speak to it." She "generously went with the page 41 data," even though she was told it had "just been generated."[18]

In November 1988, in her response to the draft N.I.H. report, O'Toole had contended for the first time that *only* eight or nine of the thirty-four hybridomas reported in Table 3 had been tested for both idiotype and *gamma*. Onek wanted to know why she had done so then but not before— for example, in her memo to Eisen. She responded that the draft report was the first time she had seen a claim that more than eight had been thus tested. That report, Onek observed, had rejected the formation of antibodies that were *mu-gamma* hybrids as a plausible explanation of Imanishi-Kari's Table 3 data. He asked O'Toole whether she might possibly have first developed the fatal-mistake argument in her response because she now recognized that her *mu-gamma* hybrid theory "was wrong" (and because, he implied, if the ELISA data were reliable, the central claim of the *Cell* paper, contrary to her challenge, was right). "I never had a *mu-gamma* theory," O'Toole replied. "That was something concocted to discredit me."[19]

Subsequently in the hearing, Brigitte Huber and Henry Wortis both unequivocally denied that the meeting on May 23 had produced any agreement that the paper was marked by a fatal mistake. Both affirmed that at the meeting O'Toole had proposed what amounted to the formation of *mu-gamma* hybrids. "We discussed this over and over," Huber testified. Onek had Herman Eisen read into the record of the hearing a paragraph in O'Toole's June 6 memorandum to him, the gist of which was that "low level expression of the transgene in hybridomas can explain the high frequency of ID-plus [that is, idiotypically birthmarked] hybridomas that also express IgG [that is, antibodies with *gamma* isotypes]." Asked by Onek how he had interpreted the paragraph in 1986, Eisen replied, "Well, the fact that she's referring to these expressions of IgG, and the fact that she uses the word heterodimer [the technical term for a two-part hybrid]—she's an experienced immunologist—meant to me that she's talking about . . . a *mu-gamma* molecule." Much later, Panel Member Ford asked Imanishi-Kari whether she could explain O'Toole's "emphatic understanding" that the ELISA had not tested for idiotype but only for isotype. Imanishi-Kari replied that she was unable to comment on the reason for O'Toole's belief. She could only say that she had "never" told O'Toole that the ELISA tests had not been run for idiotype and that the six pages of ELISA data had not been shown to her at the meeting.[20]

Panel Member Ballard asked Imanishi-Kari to respond to O'Toole's "similarly emphatic" contention that some of the data she saw on May 23 had been obtained in the days just prior to the meeting—that is, after the publication of the *Cell* paper. Imanishi-Kari responded that the data had not been generated in 1986 but, as best she could recall, in the fall of 1984. Had she put the data into a different form before the meeting? Imanishi-Kari doubted it. She didn't think she had much more time than to "grab everything that was in the lab or in my office to show during investigations." Onek quickly followed up, asking Imanishi-Kari when she had learned that O'Toole wanted to see data showing idiotypically birth-marked antibodies with *gamma* isotypes. In the meeting on May 23, she replied. So such data hadn't been requested of her before the meeting? "No," Imanishi-Kari said, her answer implying that she had not been given any reason to generate the particular data that she had shown O'Toole.[21]

The hearing revealed that parts of the O.R.I.'s case turned on matters of judgment concerning the substance and practice of the science in question—and that the O.R.I. often insisted on substituting its own judgments for Imanishi-Kari's. John Dahlberg conceded in testimony that Bet-1 could discriminate between the *mu-a* and *mu-b* allotypes, but he held that the notebook data on the reagent indicated that it did so neither reliably nor very well. When it worked, it failed to distinguish sharply between the two allotypes and it did not work all that often. So far as he was concerned, Figure 1 in the *Cell* paper overstated its specificity and misleadingly suggested that high specificity was typical of its behavior. On questioning from Panel Member Youngner, Dahlberg declared that such overstatement was equivalent to falsification. Panel Member Ballard, puzzled, asked: Did the scientific community have "some standard . . . that if you are reporting on the specificity of a reagent," you have to use a graph showing "how it acts representatively as opposed to how it reacts in a particular experiment?" Dahlberg replied, "I believe that is exactly . . . the case."[22]

Dahlberg's views were countered by several independent authorities, including John Kearney, a professor of microbiology at the University of Alabama and the coauthor of a basic paper on allotype-detecting reagents, including Bet-1. He affirmed what his paper said: In the ELISA assay he used, Bet-1 "was specific only" for *mu-a*. At Onek's request, J. Donald Capra, of the University of Texas testified that, from the data he had seen in the notebooks, under suitable conditions Bet-1 distinguished unambiguously between *mu-a* and *mu-b*.[23]

Phillipa Marrack, an immunologist at the University of Colorado and a member of the National Academy of Sciences, weighed in against Dahlberg's standard of representativeness. She maintained that it was "not customary" to report on reagent failures, not unless the paper was "absolutely about that particular phenomenon." Imanishi-Kari had cited Kearney's paper in the *Cell* publication as well as the fundamental paper on Bet-1 that had come from William Paul's laboratory at N.I.H. Eisen averred that, having referred to them, she had "no obligation" to report that the reagent was at times unspecific. "Absolutely" none, he emphasized, pointing out that he now knew enough of the notebooks from the O.R.I. report to conclude that Bet-1 showed "perfectly good discrimination" most of the time. Eisen added that it was "extremely uncommon" to call attention to the fact "that some reagent or some procedure doesn't work all the time." Troubles with reagents were "almost an everyday experience . . . in a laboratory."[24]

Capra, testifying to the O.R.I.'s notion of a proper cutoff, declared that he didn't see a serious problem with Imanishi-Kari's having chosen to set hers at 1,000 counts per minute. "Cutoff points are often arbitrarily chosen," he explained, "and judgment must be used, and the scientist doing the work is generally in the best . . . place to make that judgment." He found "nothing" in the data he was asked to review "that convinces me that poor judgment was used."[25]

Marrack, who said she had done numerous hybridoma fusions, addressed the issue of the *Cell* paper's terminology, allowing that any cellular immunologist would understand what was reported in Table 2. The term "hybridomas" meant growths in wells and did not necessarily connote single clones. It was also more appropriate than saying "wells": Wells did not produce antibodies; the hybridomas growing in them did. To be sure, it would have been preferable if the coauthors had not said "clones" in the text of the *Cell* paper when they meant "hybridomas," but they had not called the hybridomas "clones" in the table itself. Besides, the table reported on several hundred hybridomas. "You could never clone that many different hybridomas," Marrack held. "Unless you had an absolute armada of people working with you, it would be technically impossible to do. It would cost a tremendous amount of money and take a tremendous amount of time. . . . Anybody with any common sense would have known that you couldn't possibly have cloned that many."[26]

Onek solicited Marrack's opinion on a key element in the O.R.I.'s charges concerning Table 2—that, by counting wells instead of hybridomas, Imanishi-Kari had deliberately misrepresented the frequency with which the hybridomas reported in the table produced idiotypically birthmarked antibodies of native origin. Marrack responded that it would have been better to have reported the real frequencies, but even the corrected frequencies solidly supported the *Cell* paper's central claim. When Imanishi-Kari herself testified, Onek invited her to address the charge that she had exaggerated the frequencies intentionally. She insisted that she believed the published frequencies were correct when the paper appeared. "I still believe the same thing . . . ," she added, indicating that so far as she was concerned, "correct" did not necessarily mean "exact," at least not in this branch of serology. "To me, 68 or 50 percent, it doesn't make much difference. It's still a high percentage."[27]

The O.R.I. offered several types of evidence in support of Margot O'Toole's insistence that key parts of the data reported in Table 3 were false. One of them was the testimony of Walter Gilbert. Despite his declaration that what counted in determining misconduct was the data but not the science, Gilbert brought his understanding of the science to bear on his determination of the fidelity of the table to the underlying data. He endorsed the O.R.I.'s charge against Imanishi-Kari on the issue of double producers, holding that the significance of the paper depended "critically" on showing that a hybridoma producing antibodies with *gamma* isotypes did not exhibit the idiotypic birthmark because it also generated a small amount of transgenic antibodies whose isotype was *mu*. Imanishi-Kari's notebooks contained data indicating that some of the hybridomas she had reported in Table 3 as *gamma*-producing were also *mu* producers. Gilbert testified that the *mu* evidence unquestionably undercut the paper's central claim, and Imanishi-Kari had been obligated to report it.[28]

Gilbert further contended that Table 3 was scientifically insignificant without evidence that the native antibodies analyzed in it carried the idiotypic birthmark of the transgene. Imanishi-Kari insisted, of course, that the ELISA tests recorded on pages 83 to 88 of her notebook showed evidence of the idiotype on thirty of the hybridomas and that so did radioimmune assays that she had done on a smaller number. Gilbert held that such a demonstration would have amounted to a "killer experiment," one whose results she surely would have fully reported in the paper. Since she had reported the idiotypic character of only some of the hybridomas, he reasoned that she must not have actually tested all of them for idiotype and therefore must have misrepresented the ELISA data.[29]

Dahlberg bolstered the Table 3 charge by invoking, as O'Toole had done in the O.S.I. investigation, Imanishi-Kari's assessment of the table's hybridomas in the section she had contributed to Herman Eisen's omnibus grant application for the cancer center. He pointed out that she had used the ELISA data there as evidence of isotype, not of idiotype. For idiotype, she had relied on the radioimmune assays. They demonstrated that 40 percent of the hybridomas used in Table 3 produced idiotypically birthmarked antibodies, in contrast with the ELISA data, which showed that 80 percent of them did. Dahlberg found it "inconceivable" that Imanishi-Kari would not have "been anxious to use that exciting data of very high level idiotype positivity in a grant application." To his mind, the fact that she did not use it indicated that the ELISA tests had not been run on idiotype in the first place.[30]

Gilbert's understanding of what Table 3 was intended to convey was disputable, since it had been constructed primarily to show that the molecular and isotype data obtained by Weaver and Imanishi-Kari challenged the theory of allelic exclusion. In the judgment of the coauthors, its scientific significance did not depend on demonstrating that all or even most of the hybridomas produced idiotypically birthmarked antibodies, and the paper nowhere said that they did.[31] Although in the hearing Onek did not cover the explanation of the original intent of Table 3, he pointed to control data in the ELISA pages that would have been positive if the test had been run for isotype but that were negative. He walked Imanishi-Kari through the pages in her notebooks in which different Table 3 hybridomas revealed on radioimmune assays that they produced idiotypically birthmarked antibodies. And he gave Imanishi-Kari an opportunity to explain why she had not used the ELISA data in the Eisen grant application. She testified that her section of the grant proposed a study of the nature of idiotypic birthmarks at the molecular level. The subject had little to do with the frequency with which antibodies carrying the idiotype were generated and had a lot to do with the characteristics of the antibodies. Determining the outcome of ELISA tests depended on assessing the degree to which the supernatant turned yellow. A machine could measure the change with quantitative exactitude, but such machines were expensive and she didn't have one. She said that she could have used the ELISA results but that she was "much more comfortable" with a radioimmune assay. It produced numbers from a counter, while the ELISA depended on judging a color shift "by naked eye examination."[32]

Onek, cross-examining Dahlberg, brought out that he was less than precise in what he meant by "double producer." Although the term conventionally meant a single clonal cell line that produced two different antibody genes, Dahlberg used it both conventionally and—in the main— unconventionally to describe hybridomas containing two such cell lines, each of which produced a different antibody. The unconventional meaning had no bearing on whether Imanishi-Kari had misrepresented her data. Pressed by Onek, Dahlberg conceded that the existence of such double producers was not germane to the central claim of the paper. He admitted that if Imanishi-Kari had observed at least one of two such hybridoma cell lines producing idiotypically birthmarked antibodies with *gamma* isotypes, "it would still be a strong finding."[33]

To be sure, Imanishi-Kari's notebooks recorded that some of the Table 3 hybridomas appeared to be *gamma* on some isotype tests and *mu* on others. Onek put it to Dahlberg that someone had "to make a judgment" about the conflicting data, taking all of it into account, in order to prepare a table, adding that Imanishi-Kari's judgments, which designated several of the hybridomas as *mu* producers, did not noticeably favor the thesis of the paper. On questioning by Onek, Imanishi-Kari declared that she thought the double-producer question was irrelevant to the central claim of the experiment. What counted was the assay of the antibodies captured with a reagent sensitive to idiotype. If the assay detected a *gamma* isotype, then the idiotype and the *gamma* constant region had to be on the same molecule—or at least they did so long as *mu-gamma* hybrids did not form (and they did not, witnesses on both sides agreed).[34]

Imanishi-Kari said she had nevertheless addressed the issue of double producers "indirectly," discussing it with Weaver, Reis, and Albanese. They knew of "almost absolutely" no support for double producers in the molecular data. The exception was a hybridoma that Albanese had examined. Albanese, who was now a technical supervisor at Northwestern University Medical Center, said by telephone that he remembered a seemingly rogue hybridoma that on further tests by Weaver proved not to be a double producer. Dahlberg had testified to another possible exception—the hybridoma with the faint radiographic smudge that might have indicated the presence of the transgene but that had disappeared during Weaver's preparation of the *Cell* paper's Figure 4. Questioned by Onek, Imanishi-Kari reported that at the time her lab had found no serological evidence of *mu* antibodies in that hybridoma. She added that, in 1993, she had tested a sister clone of it and determined that it did not express the transgene.[35]

The O.R.I. expected Imanishi-Kari to have published evidence suggesting her hybridomas might be double producers, even though she regarded the evidence as unimportant. However, in advancing the charges against her, it neglected to give explicit attention to evidence that was exculpatory. Cross-examining Dahlberg one day, Onek pointed, for example, to Moema Reis's having said in her interview with the O.S.I. that she had suspected that the batch of Bet-1 that performed poorly as recorded in the seventeen pages was contaminated; that she had seen actual data for the normal mice that the O.R.I. said was fabricated; and that her notebooks contained evidence that she had done the June subcloning with

Imanishi-Kari. Onek observed that Reis's information was "clearly favorable" to Imanishi-Kari's arguments, yet none of it was "mentioned even in passing, even in a footnote, even to be dismissed in the O.R.I. report. Is that correct?" Dahlberg replied, "Yes. Apparently it is."[36]

When the O.R.I. did treat exculpatory evidence, it strained logic attempting to refute it or to transform it into support for the government's case. Godek had an audiotape of an open gathering in the Sackler Auditorium at the Tufts Medical School on May 9, 1988, where, a few weeks after the first hearing of the Dingell subcommittee, the Wortis committee discussed the *Cell* paper dispute. All three members of the committee may have testified to having seen subcloning data at its meeting with Imanishi-Kari on May 16, 1986; Godek elicited from Brigitte Huber and Robert Woodland testimony that during the colloquy at the Sackler Auditorium they had not mentioned having seen it. Woodland understood the clear implication of Godek's line of questioning (which had actually originated with O'Toole, in her memoranda to the O.S.I.): If they had not mentioned the subcloning data in May 1988, they must not have seen it in May 1986. Woodland retorted, "You seem to misunderstand. This was not a defense. This was a scientific discussion of what the issues about the *Cell* paper were."[37]

Dahlberg said he suspected the June subcloning data in part because of what it reported about the behavior of five or six of the sets of the subclones. "It's as if we're starting off with a culture of cells that are secreting one of two types of antibodies when they are subcloned, and they're suddenly secreting four types of antibodies. I don't think that's scientifically credible." Dahlberg reasoned similarly about data on pages 102 to 104 in the I-1 notebook that recorded the results of tests on transgenic hybridomas. Imanishi-Kari had not relied on these data in the *Cell* paper. Dahlberg insisted that they were integral to the O.R.I.'s case anyway; they were part of a series that included tests on hybridomas from normal mice, the results of which were recorded on pages 106 and 107. The normal data had been used in the paper and the O.R.I. considered them fraudulent.[38]

Imanishi-Kari had submitted the normal mouse data on pages 106 and 107 at the request of the N.I.H. in March 1988, at the beginning of the Davie panel's investigation. The N.I.H. had not asked her for the transgenic mouse data on pages 102 through 104. She had sent them voluntarily, along with page 105—which was now also held to be fraudulent—even though they were not part of the *Cell* paper because the experiments corresponding to the results they recorded had been done about the same time as the experiment with the normal mouse. "So I just put all the experiments together," Imanishi-Kari explained. The O.R.I. nevertheless held that Imanishi-Kari had fabricated the transgenic data to make the fraudulent normal data on pages 106 and 107 look legitimate.[39]

Part of what struck Dahlberg as fishy about the transgenic data on pages 102 to 104 was that Mosimann, analyzing it statistically, had found that, going from well to well, it exhibited a cyclical pattern that trended from very low values to very high ones every twelve wells. It thus formed what the O.R.I. investigators referred to as a "cycle of twelve" when it should have displayed no pattern at all. William McClure, of the O.S.I./O.R.I. scientific panel, explained, "The cells are randomly distributed like rain-drops on a windshield; there is no pattern . . . To see the pattern that Dr. Mosimann found is like seeing your name spelled out on a windshield with raindrops. It just stands out, structure where there should be no structure. Makes no sense." Dahlberg said he could only presume that the data came "from a different kind of experiment." He observed that it was "not fabricated because there is tape data" but contended that it was falsified because it was not from the experiment that Imanishi-Kari claimed she had performed.[40]

What to Dahlberg was evidence of fraud in the June subcloning data was to Imanishi-Kari an indication of a scientific puzzle. On examination by Onek, she testified that she thought the behavior of the several sub-clones that proliferated with four different antibodies over time was "very bizarre." Nothing in the scientific literature of the period explained it because research on immunologically transgenic mice was so new. She had tried to figure out the puzzle since then and had arrived at a plausible explanation that she published in the *Journal of Immunology*. (Immedi-ately after subcloning, the cell would produce native antibodies, but as it proceeded to divide, it could lose the genes responsible for them, per-mitting the transgene to produce transgenic antibodies.) Imanishi-Kari testified that, after hearing about the cycle of twelve, she had thought hard about what in her experimental procedure might account for it but had come up blank. She said she was unable to think of any experiments done by other people that might generate such a cycle either.[41]

Imanishi-Kari's bizarre data was exculpatory, too: Why would anyone fake results that invited suspicion? The O.R.I., however, did not think or choose to ask that question. It conceded that the notebooks contained exculpatory data that, for example, supported the specificity of Bet-1 or revealed the presence of idiotypic birthmarks on native antibodies, but it contended that the Secret Service analyses had demonstrated they were also faked. Whatever the type of exculpatory evidence, the O.R.I. repeat-edly fell back to the argument that the data was "questioned" on statis-tical grounds, forensic grounds, or both.[42] The testimony bearing on these issues grew increasingly pivotal as the hearing proceeded.

SIXTEEN

■

Crossing the Experts

IN THE end, the statistical and forensic assessments formed the make-or-break core of the O.R.I.'s case, accounting for almost half the testimony. Statistical analysis figured in the cycle-of-twelve determination. James Mosimann, the O.R.I. investigator and biostatistican, had also used it to quantify the probability of the occurrence of another feature of the transgenic data on pages 102 to 104—that, according to data from one set of lymph-node fusions, hybridomas grew in fifteen wells that came one after the other on one of Imanishi-Kari's plates. Hybridomas that had been pipetted into a well might or might not grow. The wells in which they did tended to be scattered randomly over a plate. Mosimann had calculated that the occurrence of growth in fifteen consecutive wells was unlikely, and the O.R.I. believed that the low probability of the event added weight to the case for fraud.[1]

The O.R.I.'s contention that the June subcloning data was false also rested on statistics, of course—the results of several tests that Mosimann had devised for the purpose. One of them mixed together different versions of the Poisson distribution that described many random phenomena—for example, the rate at which radioactive sources emitted gamma rays. Mosimann explained in testimony that finding such "a Poisson mixture model that fits the data . . . removes the data from suspicion," while not finding one "leaves the data under suspicion." He had been unsuccessful in finding a suitable Poisson mixture model for the June subcloning data, which he took to mean that the handwritten numbers did not express real radiation counts. A related test of Mosimann's yielded a numerical measure showing that the handwritten record was highly "spiky." It was a mathematical way of summarizing what William McClure had illustrated with his Sesame Street graph (see *Figure 14*)— that certain digits occurred in it with a greater frequency than a radiation counter would ordinarily have produced. Mosimann had used the high level of "spikiness" in his arguments that the digits must have come from Imanishi-Kari's head rather than from an experiment.[2]

The O.R.I. brought in the statistician from outside the N.I.H.—he was an adjunct professorial colleague of Mosimann's at American University named Austin M. Barron—to testify in part about Mosimann's work. Godek invited Barron's opinion of the statistical techniques that Mosimann had employed in analyzing Imanishi-Kari's data, including the spikiness test. Barron agreed with Godek that it could be likened to a baseball player's batting average, a single-statistic measure of the player's overall ability. Godek asked whether he believed that Mosimann's measure of spikiness was "based on valid scientific reasoning in the field of statistics," to which Barron responded, "Yes. I think it's an interesting statistic, and it certainly makes some sense."[3]

Joseph Onek's cocounsel Tom Watson handled the examination of the statistical and forensic experts who appeared on behalf of Imanishi-Kari plus the cross-examinations of Mosimann and the Secret Service agents when they testified later in the summer. Watson says that before turning to law he studied a lot of mathematics, chemistry, and biology as a premedical student and that sheer interest has kept him reading in science, medicine, and statistics ever since. He observes that, like most trial lawyers, he is not a specialist, adding that nevertheless about half his cases have tended to center on scientific and technical issues. Watson might approach an expert witness, saying, "Let me see if I can get this straight," to bring out a point for the panel's benefit, when he undoubtedly understood the witness perfectly well.[4]

Watson had tried and won Rameshwar K. Sharma's case against the O.R.I. He says that he regarded Godek and Christ as competent but that, now in his fifties, he had the advantage of being senior to them in trial experience by many years. During the hearing, Godek sometimes slipped on basic science, asking an expert witness if white wasn't the absence of all color and being told that, no, it was the opposite, the sum of the primary colors red, blue, and green. Watson counted Godek's gaffe on the composition of white light as a typical misstep of relative inexperience. "If you're going to handle trial material, especially of a technical nature, you've got to be totally compulsive about mastering every detail." Watson says that cross-examining expert witnesses can be the most crucial part of a trial. "The ability to do that in the Imanishi-Kari case is what provided the opportunity to put the truth on the table." He pursued the O.R.I.'s expert witnesses relentlessly, peeling their arguments down point by point to the bedrock premises. At one point Imanishi-Kari's statistical expert, Terence Speed, declared that he would not conclude from the distribution of digits in her handwritten counts that she had fabricated the June subcloning data. Godek objected that Speed was not an expert in deciding on allegations of scientific misconduct, but Watson beat back the objection, snapping that, if what Godek claimed was allowed, "You could have only one witness in these cases, and that's the O.R.I. witness."[5]

Terence Speed had been in charge of all the mathematicians and statisticians in the principal governmental organization of scientific research in his native Australia and was now a professor at the University of California at Berkeley. He had collaborated extensively with biologists, including those who ran assays like ELISAs. Speed says that when Watson, whom he had not previously known, first called him, he was aware that everyone thought Imanishi-Kari was guilty. The case intrigued him partly because he tends to be skeptical of conventional wisdom and partly because Watson told him that the evidence was flimsy. On reading the statistical appendix in the O.R.I. report, he concluded that the analysis was "pretty amazing—a parody of statistics." Speed agreed to appear on Imanishi-Kari's behalf without charge. He says that he wanted the option to bail out if the matter proved too time-consuming and that he did not want to risk being attacked as a hired gun. Watson pointed out to the panel the *pro bono* character of Speed's work for the defense, emphasizing, that, unlike the O.R.I.'s full-time staff, he had been compelled to prepare his testimony between other commitments. In fact, Speed finished his recalculations of the O.R.I.'s statistical analysis in the small hours of the morning of the day before he first testified.[6]

Speed attacked several features of the O.R.I. statistical work. He pointed out that Mosimann had used an unquestioned data set of David Weaver's in reaching his conclusion about the cycle of twelve in Imanishi-Kari's. The two data sets differed in character, Speed observed, arguing that the difference put the comparison "in the apples and oranges category." Speed contended that the reliability of Mosimann's Poisson mixture model was, to say the least, questionable. For one thing, statisticians did not commonly use Poisson mixture models the way the O.R.I. did. Speed testified that he had never seen the model deployed with more than two components, certainly not with nine. For another, Poisson mix-

TOM WATSON
Co-counsel for Imanishi-Kari
Photo Credit: Courtesy Tom Watson

ture models were "fairly flexible," the product of considerable "subjectivity" in determining what weight to give to their various components. Adjust the parameters in the right way and you could make it fit the data under scrutiny—or not fit it.[7]

Speed saw no reason why anyone should pay attention to the spikiness index. It was not a standard statistical test—Austin Barron acknowledged that he had never heard of spikiness before reading the O.R.I. report—and Mosimann had not yet published it in a peer-reviewed scientific journal.[8] More importantly, Speed argued that the spikiness test merely added a scientific aura to what anyone could see on casual inspection—that Imanishi-Kari's hand-recorded numbers were manifestly spiky. The numbers were obviously not direct tape counts, Speed noted. "They are clearly the result of some sort of human intervention." The question was whether the intervention involved the fabrication of phony numbers or the oddball recording of real ones. Answering that question statistically required extensive information on how Imanishi-Kari took down uncontested real data. The O.R.I. had not obtained any such information. Imanishi-Kari said that the only rule she followed in writing down data was "to smoke." She added, however, that, roughly speaking, if the numbers were low and not significant, she didn't pay "too much attention to the smaller digits."[9]

Speed quarreled with Mosimann's overall approach to the statistical analysis of Imanishi-Kari's data, testifying that it bore "rather marked resemblance" to what people in the statistics business call "data snooping." The phrase did not connote anything "unsavory" or "surreptitious," he explained, making clear—in response to a heated objection from Godek—that he was not impugning Mosimann's integrity. It was a term of art that meant a tendency, often unconscious, to find something suspicious and then subject it to statistical analysis, not to determine whether the suspicions are merited but to reinforce them. Speed says that "when you're asked to look at some group of data, you're already encouraged to be somewhat biased," holding, "You could look at a table of random numbers and find something wrong with it if someone asked you to."[10] In statistics, confirming suspicions might mean showing that the probability of an event's occurring by chance was very small—which is what the O.R.I. had concluded about the feature of Imanishi-Kari's data indicating that she had found hybridomas growing in fifteen consecutive wells on a plate. Since such a series of growths on a plate was highly improbable, the O.R.I. reasoned that the data must have come out of her head rather than out of real hybridoma wells.

Speed found the O.R.I.'s approach to the run of fifteen growths exemplary of its propensity for data snooping. During the course of the experiment Imanishi-Kari had grown hybridomas on scores of plates. Speed argued that Mosimann had in effect asked, "What's the chance of a run in exactly that place?"—that is, on the particular plate on which it occurred according to Imanishi-Kari's data. Speed pointed out that Mosimann's determination of the probability of the run would have been much larger—and thus would have made the likelihood of its occurrence much greater—if he had asked instead, "What's the chance of a run happening anywhere?" That is, on any of her numerous plates. To illustrate the point, Speed supposed that someone tossed a coin 100 times, with the last 20 tosses producing heads. If you asked, What is the probability of a run of 20 heads occurring beginning with the eighty-first toss? the answer would be that the probability is very low—one in two raised to the power of twenty. But if you asked, What is the probability of a run of 20 heads occurring beginning with any toss from the first to the eighty-first? the answer would be that it is much larger, because you would have eighty-one chances to accomplish the run rather than just one. Speed drew the take-home lesson: "If you're going to assign probabilities to patterns that you see, you have to correct for your search procedure as far as you can. . . . What would be regarded as unusual has to be built into your calculations." So far as Speed could tell, the O.R.I. statistical analyses had violated that rule not only in probing its suspicion of the run of fifteen wells but also in virtually every one of its statistical analyses.[11]

Agents John Hargett and Larry Stewart presented the Secret Service's forensic case at the appeal, just as they had before Congressman Dingell's subcommittee. Now, however, they not only reported the results of their tests but also described how they had gone about their task. They said that at the outset of their investigation they adopted the cardinal rule of forensics, which was to have authentic documents against which to assess suspected ones. They did not have authentic scientific notebooks readily available the way they ordinarily had access to, for example, authentic currency or passports. They had no idea how the notebooks of a laboratory scientist should look. With the aim of both informing themselves and obtaining a comparative norm, they got the subcommittee to obtain sample notebooks from M.I.T. that had been created by scientists working on the first floor of the cancer center during Imanishi-Kari's tenure. The subcommittee produced the twenty-six notebooks that the Secret Service examined in detail; it later provided forty-four more, which the Secret Service had asked for to check the conclusions it had reached from scrutiny of the first batch.[12]

Stewart supplied a definition of what the Secret Service meant by a comparative norm: in general, "the normalcy that you are trying to find, so that you will know when a deviation is important." In the case of the notebooks, "the normal procedure, or the usual procedure that others follow." Stewart said that he and Hargett concluded that, in comparison with the sample, Imanishi-Kari's notebooks seemed "unusual." Hargett cited the alterations and obliterations that marked them, including whiteouts on both sides of the page. Stewart called some of them "natural"—the result of, say, an incorrect date having been entered, but said that others struck him as "unnatural," changes that, like the double whiteouts, "required a bit of effort." He also said that some of the radiation-counter tapes seemed anomalous. Several had been cut into strips and pasted together; in various instances, two on the same page differed in print intensity or in color, with a tape on one side of the page being tinted yellow and a tape on the other side being tinted greenish yellow. Stewart said he had found it difficult to understand how tapes generated contemporaneously with each other could appear so disparate.[13]

Stewart and Hargett had the impression that Imanishi-Kari's notebook was organized in roughly chronological order. Despite all that Imanishi-Kari had said to the contrary, they held to the understanding that the data on any of her notebook pages had been produced around the time of the date on the page and entered on it within a day or two. They declared that their understanding rested on Bruce Singal's letter of March 1989, responding to Congressman Dingell's assumption that Imanishi-Kari had created her notebooks "contemporaneously" with the performance of her experiments. (Singal had written that, in the strict sense of the word, "only" the radiation-counter tapes were created contemporaneously with the experiments. In the broad sense of it, the data were "assembled contemporaneously with the . . . experiments and placed in the Notebooks in a timely fashion.") The Secret Service evidently melded the two meanings: It proceeded to determine whether Imanishi-Kari had generated the tapes around the times the pages on which they appeared were dated. Stewart said that three technicians had been assigned to the task over the years, and that he and Hargett had spent hundreds of hours on it. The effort involved the multiple forensic analyses—the work on the inks, page impressions, and tapes—that centered on comparing the data records in Imanishi-Kari's I-1 notebook with the notebooks that the subcommittee had obtained. The results had been presented in the subcommittee hearings, figured in the reports of the O.R.I. and its predecessor, and formed the basis of the Secret Service's ultimate judgment, which Stewart now repeated for the benefit of the panel: Imanishi-Kari had "falsified" roughly a third of the 128 pages in the I-1 notebook.[14]

Imanishi-Kari had long contended, of course, that the Secret Service had not looked hard enough for data records produced at the time she said hers had been generated. Now Tom Watson cross-examined Hargett and Stewart on the adequacy and reliability of the sixty notebooks they had used as a comparative norm. They said that the M.I.T. provost, John Deutch, had told them that the notebooks given the Secret Service comprised a "good representation," as Stewart put it, of the notebooks that had been created in the cancer center between 1979 and 1986. In response to Watson's probing, they conceded, however, that they didn't know Deutch's criterion of "good." They acknowledged ignorance of how many researchers had used the radiation-counting equipment during the period; how many notebooks they had generated; how many had been discarded or taken from M.I.T.; or what fraction of all the notebooks, including those produced at the time of Imanishi-Kari said she had done her experiments, their sample comprised.[15]

Hargett and Stewart also had no idea whether, by the time they received Imanishi-Kari's notebooks or the sample of sixty, they were unchanged from what had been sent to the subcommittee. It was the panel that originally raised the question whether a chain of custody had been established for the notebooks, but Watson pounced on it. Just because the government thought of it and we didn't, he noted, doesn't make it "a dumb idea for us to raise it. . . . [The point] goes to the very integrity of the evidence in this case."[16] Agent Larry Stewart said that Walter Stewart, at the subcommittee, had photocopied Imanishi-Kari's notebooks so that their order could be preserved and that he had provided an inventory of them when he turned them over to the Secret Service. Larry Stewart declared he believed that Walter Stewart was honest.[17]

Hargett had already acknowledged, however, that the Secret Service had no record supporting the fidelity to the originals of Walter Stewart's inventory. It had no inventory or photocopies whatsoever of the sixty control notebooks at the time the subcommittee had gotten them. Its investigators simply did not know whether any materials had been removed from those notebooks, Hargett acknowledged. A number were highly vulnerable to tampering. While some of the sample "notebooks" were spiral bound, others were loose-leaf binders, and still others were merely file folders containing one or two sheets of paper. (Panel Member Ford described Charles Maplethorpe's C-2 notebooks as an ordinary manila file folder "full of a crumpled and folded selection of documents.") Panel Member Youngner asked Hargett whether it wasn't unusual for the Secret Service not to have a chain of custody for documents whose genuineness it proposed to test. Hargett replied, "This whole case was unusual. . . . It was a whole different thing for us. And . . . yes, it could have been done better, no question about it."[18]

Watson inquired of Stewart and Hargett how they decided what sections of the notebooks to investigate if they had no inkling about the science contained in them. Had they been directed to focus on certain parts of the data by, say, the subcommittee, particularly Walter Stewart? Hargett testified that the O.S.I. and the O.R.I. sometimes asked for analyses of specific pages, but he couldn't recall whether requests had come from the subcommittee. Hargett added that Walter Stewart had "showed up regularly," though he was unsure whether Stewart wanted information about certain pages or simply wanted an update on what the Secret Service was finding. But Hargett told Watson "point blank" that they tried to put out of their minds whatever Walter Stewart requested, explaining,

"We did not like to work with Walter Stewart.... We viewed Walter Stewart as a loose cannon." Larry Stewart, prompted by Watson, allowed that he meant Stewart might act irresponsibly. Hargett said that he certainly struck him as "very much an advocate" of the position that Imanishi-Kari was guilty of scientific misconduct. Larry Stewart emphasized that the Secret Service had not taken any direction from Walter Stewart. It had followed its own guidelines of correct forensic practice.[19]

Tom Watson wondered aloud about the merits of some of the forensic standards that agents Hargett and Stewart had adopted in the case, particularly those that led them to conclude that Imanishi-Kari's notebooks were especially odd. He asked Hargett how notebooks that were particular to the experiments and record-keeping of individual scientists provided a comparative norm. Hargett explained that the Secret Service believed the sample notebooks were of unquestioned legitimacy, revealing that he tacitly equated the veracity of data with the way it was recorded. Agents Hargett and Stewart may not have paid attention to Walter Stewart's requests, but they had decided, on advice from the subcommittee, to exclude one of Moema Reis's notebooks from their assemblage of normal notebooks. Stewart testified that its contents looked a lot like Imanishi-Kari's so they "were unwilling to use that as an example of what a normal notebook or a usual notebook would be." They thus made Imanishi-Kari's notebook abnormal by definition. Was it not true, Watson asked them, that the notebooks allowed into the comparative norm also contained date changes, alterations, obliterations, scribblings in different inks on the same page, and, in some cases, small strips of counter tapes? Yes, it was true, Hargett and Stewart conceded.[20]

Watson asked Stewart whether he could distinguish between what he had called a "natural" and an "unnatural" change of dates. "That's part of what they pay me to do, yes," Stewart replied. Stewart took semantic cover when Watson called him on his insistence that it was "noteworthy" that Imanishi-Kari's notebooks differed in the appearance of specific features from most of the samples. Watson inquired whether it was forensically significant that she had used little strips of tape. No, it was not, Stewart responded. What are you saying then? Watson shot back. Stewart replied that it was "significant enough to be noteworthy," adding that he had presented the information to the Dingell subcommittee. Watson pushed harder: Was it important? Stewart: "I think noteworthy and important are self-evident. They're both terms that mean about the same to me." He elaborated that the oddities in the I-1 notebook were important enough to have contributed to his professional opinion of its authenticity.[21]

Watson posed to Stewart that it might be essential that notebook comparisons take into account the kind of immunological assays that the pages recorded, especially if the experiments differed from each other. Stewart replied that such considerations "had nothing to do with the forensic findings." Watson observed that the pair of tapes—the one yellowish, the other greenish—on page 103 of Imanishi-Kari's I-1 notebook that Stewart found suspicious recorded two separate tests of a supernatant. Each used a different reagent and each likely had been done at a different time. Stewart admitted that a researcher could put two differently colored tapes from distinct experiments on the same page, but he still insisted that the disparateness of the two tapes was forensically significant. They were on a page that was part of a dated sequence and thus must have been part of the same experimental run. He had "absolutely no idea what the numbers mean, other than the dates." He had "no idea" about tests with two different reagents.[22]

Watson's principal forensic expert was Gerald B. Richards, an independent consultant recently retired after twenty-three years with the F.B.I. Having spent a number of them as head of all the bureau's document research, Richards had dealt with bad checks, phony food stamps, and investigations—for example, espionage cases that often involved voluminous materials. (He said that the F.B.I. relied on him for sensitive work and that he had recently been called out of retirement by the Justice Department to help out with the case of Aldrich H. Ames, who had been accused of compromising the security of the C.I.A. Richards possessed a lithe and incisive intellect that he deployed with a light touch. Invited by Watson to introduce himself to the panel, he noted, "My opinion of the Secret Service Laboratory is they are probably one of the finest labs in the United States and . . . the world. They . . . probably are only second to the F.B.I."[23]

Richard's testimony bolstered most of what Watson had brought to light in his cross-examination of Hargett and Stewart. It also joined with the testimony of Imanishi-Kari and witnesses who appeared on her behalf to form a powerful rebuttal to each and every element in the Secret Service's forensic conclusions about the I-1 notebook: the alterations as well as the cut and pasted tapes; the sometimes huge one-day increase in the register numbers on her counter tapes; the impression analyses purporting to show that her notebook pages had not been produced in the order of the dates written on them; the differences in ink intensity in putatively contemporary tapes; and the comparisons of ink composition and paper to show that her counter tapes, especially the greenish ones, matched those from the notebooks of other scientists—notably Maplethorpe's—which had been printed long before she even had the transgenic mice for the experiment.

Richards instructed the panel that forensic science could reveal a great deal about how documents were assembled or prepared but that only rarely could it "tell us the intent of the person" behind them. While the signature on a check might be forged, the forger might have had authorization to sign the account holder's name. Richards observed that Imanishi-Kari's date changes seemed par for the course, pointing out that one of Maplethorpe's notebooks had a good half dozen similar alterations. Besides, Imanishi-Kari's emendations were "obvious," Richards noted. "In my experience, when someone is conducting fraud or trying surreptitiously to change a document—let's say, a draft card back in the sixties—they don't make it obvious. . . . They try to change it so you can't tell or eliminate it completely, if that isn't possible. Normally, it isn't done with scratch-outs and write-overs." Had Imanishi-Kari cut and pasted tapes? Moema Reis said that she cut and pasted tapes "many times," explaining that in certain experiments the printout would include large spaces of blank paper between the ranks of numbers and that she would excise the blank areas so that the ones with numbers would fit on her notebook page. Other researchers from Imanishi-Kari's days in the lab testified that, if another scientist had an urgent need to use the radiation counter, counting would be interrupted, then resumed later, a process that also produced separate printouts for the same experiment.[24]

Stewart testified, as he had told the Dingell subcommittee, that one of Imanishi-Kari's tape register numbers—the page number that the radiation counter put there—showed a change of 1,353 between two consecutive days in March 1985, which was huge compared with the average increase of only 11.5 per day in the comparative-norm notebooks. However, David Weaver's notebooks showed an average daily change in register number of 270, a fact that Stewart could not explain when Watson confronted him with it. Richards said that the Secret Service had not established an accurate basis for estimating the rate at which register numbers increased. "They did not use all the tapes that were at hand in the notebooks. There are many, many other tapes with register numbers in there that didn't seem to be considered in their analysis."[25]

Other testimony brought out that jumps in tape-register numbers were produced by interruptions and by a feature of the only radiation counter in the lab at the time that printed register numbers on the tapes. Manufactured by Beckman Instruments Corporation, the counter was known in the lab as the "paper eater" because it double-spaced its lines of numbers. More importantly, it knew to stop counting and printing when it detected a red-colored regulator that had been inserted into one of the tubes on the counting rack. If by mistake or forgetfulness, the wrong regulator or no regulator was inserted, the counter—especially if left over-night—would print out for hours, not only eating up paper but raising the register number enormously. Richards told the panel, "Sometimes they could go through literally hundreds if not thousands of counter num-bers within a very short period of time."[26]

The Secret Service had concluded from its impression analysis—to which everyone referred by its technical acronym, ESDA—that the pages in the I-1 notebook were out of sequence with regard to when information had been written on them: Page 41, for example, showed impressions of page 43's having been written on top of it, and thus earlier, which the Secret Service took as evidence of fabrication. However, Imanishi-Kari testified that when she had organized the notebook, she had grouped the pages roughly by subject; she had not arranged them in the order in which they had been produced. She also recalled that the pages had undergone a good deal of shuffling during the inquiry by the N.I.H. panel in Boston, in June 1988, and that she had then sent the notebook to the N.I.H. as it was. She had not numbered the pages. Richards pointed out that her notebooks were given page numbers after they had been turned over to the subcommittee and that the sequence of numbers bore no necessary relation to when the pages had been prepared. "The problem we have with this notebook," he emphasized, "is that we have locked ourselves into cement so far as the page numbers go." The panel should not be so cemented, he argued. Simply reverse pages 41 and 43, for example, and the indentations would occur in perfect "logical sequence." Watson had earlier made the same observation to agent Hargett, eventually asking him how much he relied on the ESDAs in reaching his overall conclusion. Hargett replied, "The ESDA by itself, standing alone, I think if that was the situation, we wouldn't be here today."[27]

The Secret Service's suspicion of abrupt changes in print intensity on Imanishi-Kari's tapes rested on the assumption that the ribbon in the printer remained continually in use, gradually fading until it was replaced. Testimony revealed, however, that the assumption was mistaken. The rib-bons were like those in mechanical typewriters, and so was the ribbon mechanism in the teletype printers. If a new ribbon was needed but una-vailable, researchers in the lab would turn the ribbon, looking for a dark stretch, or rewind it entirely, or toggle it up or down the way you could change a mechanical typewriter ribbon from black to red or vice versa. Any of these manipulations could produce a noticeable change in print intensity. On cross-examination, Stewart said that he knew nothing about whether ribbons were switched between machines, how many the lab possessed, or in what order they were replaced.[28]

Watson told the panel that Imanishi-Kari could not afford to challenge the Secret Service's ink analysis by doing her own; it was too expensive. Instead, she had engaged an independent expert to review the thin-layer chromatography results—the thin-layer chromatograms that the Secret Service had used to compare the inks of Imanishi-Kari's tapes with those such as Maplethorpe's. It was much the same kind of work that Albert Lyter had done in 1992, once the U.S. attorney for Maryland had released her notebooks. Imanishi-Kari's expert in 1995 was Robert L. Kuranz, who had worked for the Parker Pen Company for thirty-six years helping to control the quality of the company's ink production and who was now a full-time forensic ink analyst. Kuranz, responding to a probe by Stephen Godek, assured the panel that he had done his analysis independently of Lyter, but what he had to say was even more discrediting than Lyter's assessment to the Secret Service's efforts. Its lab work sheets were filled out inadequately and were somewhat obscured by cross-outs and obliterations. It had not controlled sufficiently in each of its thin-layer chromatography runs for the kind of paper on which the inks printed. Kuranz noted, "We're looking for subtle differences in many cases. And we have to be as sure as we can of the contribution of the paper to any of those subtle differences." Kuranz held that, in any case, for several technical reasons thin-layer chromatography was an unreliable method of analyzing printer ribbon inks. Watson asked him the bottom-line question: Does the Secret Service's thin-layer chromatography analysis help justify the conclusion that the documents in this case were fabricated or falsified? Kuranz replied, "No, I don't believe it does."[29]

Richards suggested that the green-tape evidence might have been compromised by the breaks in the chain of custody. "Unless you have some type of method for logging and accounting the materials, there is the possibility that materials can be taken out. They can be placed in. They can be lost. They can be stolen, what have you."[30] Alluding to the incompleteness in the sample of comparative-norm notebooks, he declared that it was very difficult to conclude that fraud had been committed because contemporary matching materials were absent. If the conclusion is "based on the absence of materials, you pretty much have to have them all."[31]

Stewart, in testimony, had recounted the development of his knowledge of the tapes that the Secret Service did get, including the printers on which the tapes had been produced. He had started at M.I.T., in 1989.

The equipment room on the first floor of the M.I.T. cancer center had housed three radiation counters—the one manufactured by Beckman Instruments and two produced by the Packard Company. The teletype-writer to which each counter was hooked was a dumb terminal. The counters determined the formatting of the numbers on the printouts, and the Beckman format differed from the format of the Packard machines. Stewart concluded that some of Imanishi-Kari's tapes had been produced by one of the printers connected to the Packards. The zeros on those tapes were slashed, and the printer connected to one of the Packards was the only model that hammered out slashed zeros. The formats on all the rest of Imanishi-Kari's tapes, comprising the large majority of them, indicated to Stewart that they had been generated by the printer to which the Beckman counter was connected.[32]

Stewart said he noticed that the printer attached to the Beckman was on a cart, which suggested to him that it might be movable from one counter to another. He wanted to be sure that the tapes in the I-1 note-book had not been produced by "multiple printers that are appearing as one to me," so he contacted Beckman Instruments. A scientist in mar-keting and technical support at Beckman—her name was Leiko Dahl-gren—told him, he recalled, that the printers could "be attached to any of these types of machines and used with very little effort." She also said the printers that the company shipped with its radiation counters were manufactured by Western Union.[33]

Stewart said that he then visited the Western Union office in Wash-ington, D.C., where he talked with a veteran of some dozen years at the company about both the printers and the paper they used. He recalled having learned that the company did manufacture teletypewriters and that they were indeed interchangeable between systems. He said he was told that the printer paper was fundamentally "very yellow." None of it was green, but it came in different shades of yellow, including "a yellowish green." Stewart testified that he tried to turn the very yellow paper to green by subjecting it to ultraviolet light, heat, and various chemicals. Although the color darkened, the shift was in the direction of brown, not green. Stewart apparently concluded that the greenish tapes in the note-books of Imanishi-Kari, Maplethorpe, and several other scientists must have been the greenish-tinted yellow rolls of paper that Western Union provided with its printers to Beckman Instruments.[34]

Stewart was not altogether clear about the forensic strategy his inves-tigation had then prompted him to adopt, and not even persistent ques-tioning by Watson could elicit a sharp definition of it. He was right to have worried initially about multiple printers whose output looked the same. A comparison of tapes from different printers would reveal nothing about whether one had been generated at the same time as the other. Two printers equipped with the same font, paper, and ribbons might well be used at widely separated times for any number of reasons, including, for example, one's being kept in storage as a backup for the other. Yet though Stewart had discovered that the printers could be interchanged, he seems to have decided that such possibilities did not matter. He said that he could solve the problem of not knowing from which machine a tape came by the use of forensics. He somehow held that comparisons of ink, font, and paper could determine whether Imanishi-Kari's tapes and those from the sample notebooks had been generated about the same time and by the same printer.[35]

Richards had cast his field-investigative net far more widely than Stewart, talking with all the relevant manufacturers—for example, Packard as well as Beckman. According to his testimony, the original manufacturer of the printers was not Western Union but the Teletype Corporation, which built the machines for Western Union as well as for Packard and Beckman. The two Packard radiation counters and the Beckman counter each had a dedicated Teletype printer, all of the same model. Information about the printers could not be obtained from the Teletype company, since it had gone out of business in the early 1980s. However, unlike Stewart, Richards had spoken with the two employees at the M.I.T. cancer center who had been responsible for the upkeep of the counters and their printers. One was Vivien Igras, who had seen to all the experimental apparatus on the first floor of the cancer center when she worked there; the other, Eliot W. DeHaro, the center's billing manager for twenty-two years, had regularly moved equipment around. Both told Richards—and ultimately testified at the hearing—that the printers often broke down and were replaced by substitute Teletype printers that DeHaro obtained from inactive counters or the basement. However, both testified they understood that Packard printers were incompatible with Beckman counters and vice versa. They had only switched Packards with Packards and Beckmans with Beckmans. Richards explained the incompatibility: Although the printers were the same model, they were wired slightly differently to interface with the respective radiation counters.[36]

The evidence of incompatibility weighed heavily against the O.R.I. case. Charles Maplethorpe's tapes had been used as the sample of comparison with key tapes from Imanishi-Kari, including those for the June subcloning. But Maplethorpe's tapes had been produced by one of the Packard printers. Matching his tapes with Imanishi-Kari's was "like comparing apples and oranges," Richards testified. Godek tried to poke a hole in

Richards's rejection of Maplethorpe's notebook as forensically defensible, asking him whether having "the same ink, same type font and same paper" wasn't consistent with the Packard and Beckman counters "being used with only one printer." Richards replied, "No, sir."[37]

In later testimony, a longtime Beckman salesman named William Fitzgerald, whose rounds had included numerous regular visits to the cancer center, confirmed Richards's conclusions, declaring it was unlikely that Beckman printers could be switched to another manufacturer's radiation counter. Godek, declining to give up, said that Leiko Dahlgren had told him that the Packard and Beckman printers were compatible with the counters made by either manufacturer. Fitzgerald remarked, "If that's true, then Laiko Dahlgren must have information that I don't have."[38]

Both sides had been eager to hear what Dahlgren had to say. The day before Fitzgerald's appearance, she had testified by telephone from California as a witness for the O.R.I. Under relentless cross-examination by Watson, Dahlgren admitted that she did not know whether Western Union ever supplied printers to Beckman Instruments: She had an "understanding" that the printers that came with the Packard and Beckman counters were interchangeable. Her understanding was based on having "an acquaintance that knows a Packard employee." She had no direct knowledge that the signals put out by the counters were compatible with both printers. She did not know what model printer was used with the Packard counter in the cancer center, whether it was compatible with the Beckman counter, or whether the plugs that hooked the counters into the printers were the same. Dahlgren conceded that she had not visited the center between 1979 and 1985, whereupon Watson asked her summarily: "So, you wouldn't know, of your own personal knowledge, whether any printer that was in the lab, M.I.T. labs in 1979 to 1985, was able to be connected to a different counter machine?" Dahlgren replied, "No, I do not know."[39]

Richards's field investigation challenged the O.R.I.'s tape evidence even more forcefully. He testified: Neither Beckman nor Packard supplied green tape. Packard sold canary-yellow tape rolls, while Beckman Instruments had sold no tape to M.I.T. Beckman had provided a single 300-foot-long roll of beige tape with its counter and printer when M.I.T. had bought the machines. The original manufacturer of the tape rolls was Cutting USA, which produced rolls of printer paper in four different colors: a white bond, a three-ply pink, a canary yellow of poor-quality paper, and an off-white or "beige" of equally low paper quality.[40]

Richards had no doubt heard from Vivien Igras what she declared at the hearing. She had never purchased green tapes. The only tapes she ever ordered were canary-yellow ones, and those came from Packard. She did recall seeing "newsprint-colored tape," rolls that might have come with the Beckman when it was purchased. William Fitzgerald corroborated Igras's recollections, declaring that he had never seen his company's counters and printers accompanied by green or greenish-tinted paper; only by paper rolls that were either yellowish or grayish to beige. Richards concluded that "we have two colors of paper . . . a yellow and . . . a beige and both of them fade to different colors." Stewart was right that the yellow did not fade to green; it shifted to dark brown. But the beige did seem to fade to green.[41]

In response to a skeptical question from Godek, Richards declared, "if you look at all the green tapes, particularly in C-2, you will see a span. It isn't all green. It's a span of color, and that color goes backward very distinctly almost to the beige." Under the scotch tape used to attach the counter tapes to the page, Richards pointed out, the yellow shifts to dark brown and the green shifts to a stronger green (see the darkened border strips in the counter tape shown in *Figure 13*). The fact that only one roll of beige paper came with each Beckman counter was consistent with the sporadic appearance of the greenish tapes. They were "interspersed throughout the notebooks over periods of time, depending on . . . when that particular roll happened to get onto a machine for one reason or another."[42]

Richards's detective work was freighted with a crucial forensic point: In its search for matches of Imanishi-Kari's greenish tapes that were contemporary with their dates—say, the summer of 1985 for the June subcloning tapes—the Secret Service had looked only for other green or green-tinted tapes. It had paid no attention to the much larger universe of whatever beige tapes the sixty notebooks might have contained. "So when we talk about the green paper," Richards argued, "we also have to take all the beige paper that can be found into consideration, because it's all one and the same."[43]

Stewart, noting that white is a combination of colors, said that he couldn't see how sunlight would remove one of them to turn beige-colored paper green.[44] Richards countered that a variety of storage conditions—exposure to heat, ultraviolet light, or chemicals—might turn the beige paper to green, and he bolstered his theory with an experiment, obtaining a roll of beige paper from Beckman Instruments and subjecting it to ultraviolet light for several hours. He showed the panel the results: The paper had acquired a greenish tint like that in the C-2 notebooks, and he observed that the roll's original beige was similar to the color on the edges of the C-2 tapes. Maplethorpe's tapes and the new Beckman roll were "truly not a green paper." They were "just a transition of the beige over a period of time."[45]

Richards acknowledged that he did not know for a fact that the new beige roll from Beckman was the same as the beige paper available in the 1980s, but he said he had no reason to think it was any different. He also emphasized that he could not say for a fact that the green tapes in the notebook were color-shifted beige tapes. He could only say that his theory was "the most likely scenario" consistent with the information he had gathered. Those points acknowledged, he contended in summary that the sample of notebooks used to date Imanishi-Kari's tapes was inadequate, by his estimate amounting to only about a third of the research notebooks on the first floor of the cancer center at the time, and that the inattention given beige tapes was a forensic failure. The result was that the attempt to date Imanishi-Kari's green tapes by comparing them with green tapes found in the sample notebooks such as Maplethorpe's had "no real validity."[46]

On the last day of the hearings, September 15, Panel Member Ford addressed John Dahlberg, the O.R.I.'s chief scientist, noting that the case was marked by "a great deal of complexity"—scientific, forensic, and statistical—and that combining these findings required "a lot of attention to detail." Stepping back from the detail, she wanted to know: "If you were going to set out to fabricate data of this sort, why would you fabricate data that can be described as not the best data to support what you are about? . . . Why wouldn't you fabricate the best possible set of data, all of which was related to what you purported in the paper?" Dahlberg responded, "That's a hard one."[47] In mid-October, the panel, giving instructions for the preparation of their briefs to the respective counsels, insisted in a telephone conference that the O.R.I. answer the hard one. The panel's staff lawyers summarized that part of the conversation: "The O.R.I.'s brief should clarify for the Panel what the O.R.I. views as the theory of its case, i.e., the scenario of what Dr. Imanishi-Kari did, when and with what intent, and a common sense understanding of what is alleged to have occurred."[48]

In a post-hearing brief that it filed with the Appeals Board on December 22, 1995, the O.R.I. provided such a scenario, arguing that Imanishi-Kari had "motive, means, and opportunity" to cheat. The Cell paper loomed large for Imanishi-Kari's reputation, and her prospects at Tufts were threatened by O'Toole's challenge. Imanishi-Kari's contributions to the paper were undergirded by some real data, but she knew the data "was simply not strong enough" to satisfy either O'Toole or her friends and coauthors. The O.R.I. rejected Imanishi-Kari's insistent claim that she could not have fabricated data before the meetings of the Wortis committee to counter objections of which she was as yet unaware. The O.R.I. found it hard to believe that in the week before the first meeting of the committee her friends Wortis and Huber had not at least hinted at the issues the inquiry would pursue.[49] More important to the O.R.I., when Wortis approached Imanishi-Kari on May 9, 1986, about reviewing the data, she realized that the questions about it must have originated with O'Toole and she probably made "a reasonably educated guess as to what Dr. O'Toole's questions would be." She did not have adequate data to show and she had ample time before the Wortis committee meeting on May 16 to concoct it.[50]

Beyond that point, the O.R.I. scenario grew murky about the actions that Imanishi-Kari took following her remarkable guesses. It held that at the meeting of the Wortis committee on May 23, Imanishi-Kari did not show O'Toole all the data that might have satisfied her—because she did not have the genuine variety and had possibly not yet created the phony substitute. It did not attempt to explain why she hadn't taken her opportunity to forge and exhibit all the data she needed to kill suspicion. It simply noted that in the succeeding months Imanishi-Kari felt increasing pressure arising from the dispute about the *Cell* paper, declaring— wrongly—that the controversy kept her appointment at Tufts in a tentative state at least through May 1988, when the N.I.H. investigation was beginning. Withal, she must have seen "little to risk in doctoring the record—changing dates, changing headings, whiting out data, and even creating pages of tape or transcribed data." She thought that she could safely submit to the N.I.H. what she had fabricated, but she was mistaken.[51]

In the remainder of the brief, the O.R.I. pressed its charges with arguments that at times seemed disconnected from the evidence. It reversed the point of Phillipa Marrack's discussion of the language of hybridomas and wells, exploiting it to support the charge that Imanishi-Kari had deliberately misrepresented her results in Table 2. It turned Gerald Richards's arguments about color changes in the tapes upside down, holding that he had shown that green tapes turned to beige. It cited as suspicious Imanishi-Kari's open acknowledgments that she could not remember what she had done for this or that nine years earlier and could not now explain certain odd features in her data. It took Margot O'Toole's account of events as reliable, ignoring all the inconsistencies in the accumulated record and discounting the considerable testimony against her recollections. It faulted Imanishi-Kari for "blackening the reputation of the paper's challenger, Dr. O'Toole."[52]

The brief held that some of the subcloning data was not scientifically credible because it was bizarre, despite Imanishi-Kari's having testified that the pecular behavior of the hybridomas was a scientific puzzle that she had since resolved. It dismissed Terence Speed's statistical criticisms, calling them "interesting" but "generally academic." It contended, for example, that his attack against Mosimann's Poisson mixture analysis was disconnected from the science, that Speed knew little about how radioactivity could be statistically modeled. For the most part, it ignored the fact that Speed's quarrel was not with using statistics to model nature but with the flexibility, complexity, and untested character of Mosimann's particular statistical model.[53]

To some degree, the O.R.I. brief attempted to get around Gerald Richards's sharp critique of the Secret Service's forensic conclusions. The O.R.I. now argued, backing off from a previous claim, that matching Imanishi-Kari's tapes to earlier ones was not "particularly critical." The "critical fact" was that her June subcloning tapes did not match other tapes generated in 1985 from the same counter. Otherwise, the brief held tenaciously to the Secret Service's forensic methods and conclusions. The sample notebook size was sufficiently large (a point that the brief attempted to support with a specious calculation); the information about paper and printers that Stewart got from Western Union was authoritative; and matching tapes on the basis of font, ink, and paper was appropriate, regardless of whether the tapes had been produced by different printers. In any case, the brief held that the Secret Service's determination of the compatibility of the Packard and the Beckman printers was supported by the "vast weight of the evidence"—including the dubious information that Larry Stewart had obtained from Leiko Dahlgren.[54]

Onek and Watson organized their brief on behalf of Imanishi-Kari—it was filed on February 9, 1996—by striking at what they called "cross-cutting" flaws that ran throughout the O.R.I.'s case, then contesting each charge in its particulars while showing how each was undermined by one or another of the flaws.[55] To their minds, the fundamental evidentiary flaw comprised the statistics and the forensics. Watson says that the test for spikiness, being new and not yet even submitted to peer review by statisticians, would not likely be admitted in a court of law. He and Onek held it in any case to reveal nothing about whether Imanishi-Kari had fabricated the June subcloning data, summarizing in their brief its inherent tautologousness: "Spiky data turns out to be spiky in tests designed to highlight spikiness." They noted that Austin Barron, the O.R.I.'s own witness, had conceded that the statistical evidence counted only in context, that the evidence by itself was not "enough for me to say it's fabrication."[56]

Onek and Watson, calling the Secret Service's field investigations "extraordinarily sloppy," leveled all of Richards's and Kuranz's critiques against the O.R.I.'s forensic case. They also extrapolated from them to make a tellingly exemplary point against its green-tape analysis, pointing out that the green tapes present in Maplethorpe's notebooks dated December 1983 and January 1984 appeared "out of nowhere chronologically" with a series of identically colored tapes neither preceding them in the same notebook nor existing in the contemporary notebooks of other scientists. Using its own reasoning, the O.R.I. might as well have accused Maplethorpe of having fabricated the dates on his tapes. The O.R.I. was "once again" relying on the Secret Service's "peculiarly one-sided analysis of the evidence." Onek and Watson contended that the O.R.I. seemed "to believe that if it proffers enough arguments, however, weak, they will somehow add up to a strong case," continuing pointedly, "But a succession of weak links does not create a strong chain."[57]

In a generous moment after the hearing, Onek remarked, "You can't blame the Secret Service. *Mu, gamma*—it was all Greek to them. They understood nothing about the science." In their brief, Watson and Onek did blame the Secret Service for close-mindedly drawing on Bruce Singal's 1989 statement about contemporaneity to insist that Imanishi-Kari had entered data into her notebook about the same time she did an experiment. They stressed that researchers might well want to compare assays done at different times by pasting data strips on the same page. They proposed that radiation-counter tapes could differ from each other in color and ink because at some point Imanishi-Kari sat down "with a few pads of paper and legitimately organized (primarily through cutting and pasting) counter tapes that had been generated days, week[s], or months before." The O.R.I. presumably did understand the science, but it "refuses to discuss" any such alternative scenario to fabrication.[58]

Onek and Watson did not dwell on O'Toole save to impeach her reliability, pointing out that even though she might believe what she said, her testimony was contradicted by numerous witnesses and was likely distorted by her conviction, evident in her testimony, that "her competence and character would be forever impugned unless Dr. Imanishi-Kari were found guilty of misconduct."[59] They did dwell on indications that the O.R.I. had pursued Imanishi-Kari far beyond fairness. For one thing, the O.R.I. violated the rules of the hearing by adding three new charges after its end. For another, it had a difficult time defending the underpinnings of its charge that Imanishi-Kari had committed scientific misconduct by not reporting her discrepant results from tests of Bet-1. Everyone knows that reagents like BET-1 are temperamental, several witnesses had testified, and no one is obligated to report discrepant results that lack serious consequence. Onek and Watson declared, "With this charge [the] O.R.I. has gone beyond manufacturing evidence of a crime. [The] O.R.I. has manufactured the crime itself."[60]

The O.R.I. may have taken a stab at answering Panel Member Ford's hard question about Imanishi-Kari's motives for misconduct, but Onek and Watson pointed out that it had yet to explain why she had fabricated particular data it alleged to be phony. They emphasized that she had no reason to falsify the Bet-1 data or the large fraction of questioned notebook pages—they amounted to almost half the total—that had not appeared in the *Cell* paper or been cited in the corrections. They also argued that if she had fabricated the data set indicating growth in fifteen consecutive wells, she obviously could have written down numbers exhibiting a normal growth pattern rather than an improbable one. One of the pages she had showed O'Toole on May 23, 1986, included sticky data for Bet-1, and it was hard to imagine that, to save herself, she would have created subcloning data that the O.R.I. found scientifically incredible. If Imanishi-Kari had wanted to fabricate data, she had time and brains enough to do a far better job, Onek and Watson held. Her challenged notebook "contains many examples of data which undermine [her] scientific position or fail to support it adequately or which are bizarre or seemingly impossible." The O.R.I. "fails to explain why [Imanishi-Kari] would falsify or fabricate such data or why she would have included such data in the materials she provided to N.I.H."[61]

Onek and Watson held that Imanishi-Kari's notebook data expressed "all the inconsistencies and surprises that one would expect in the laboratory notebook of a real bench scientist engaged in cutting edge research"—sometimes supporting the *Cell* paper and sometimes not, sometimes demonstrating that BET-1 was working well and sometimes not. It was one of their principal contentions that making sense of such data often required calls of interpretive judgment. The O.R.I. was entitled to second-guess Imanishi-Kari's decision to give more weight to one set of, say, idiotype data than to another, but it was "not entitled to call it scientific fraud." Imanishi-Kari similarly had the right to make a judgment about the cutoff value without risking "misconduct charges by ORI's science police."[62]

On March 1, 1996, the O.R.I. filed a brief in reply to Onek and Watson's, deploying counterarguments that were often intricate but sometimes simple—for example, some of the allegedly phony data was bizarre because Imanishi-Kari had fabricated it in haste. In the final oral arguments on the case, held on March 19, Christ contended that, despite the anomalies in her made-up data, Imanishi-Kari had "fooled the N.I.H., and she fooled the original panel of her friends at M.I.T., at Tufts . . . that looked at the data. But, [she] didn't fool the O.R.I., and [she] should not fool this Board." Onek went rapidly over the key points in Imanishi-Kari's defense, trying to make them as sharp as possible in the short time allotted for the final session. Having fifteen wells positive for growth in a row was not statistically impossible, he insisted, alluding to testimony from Phillipa Marrack that the clumping of cells in the source container might lead to more of them being distributed into the later wells. The greater the number of cells in a well, the greater the probability that growth would occur in it. "It's statistically impossible if you would choose randomness, but with clumpy and lumpy, it is not random," he said. "If basketball games were random, the Chicago Bulls couldn't win fifteen games in a row."[63]

Overlaying the final sparring, however, was the blunt point with which Onek and Watson had closed their brief: It had been almost ten years since Margot O'Toole had stumbled on the seventeen pages in Moema Reis's notebook. "The ensuing decade of staged hearings, leaked documents, ever-changing charges and inaccurate reports shattered [Imanishi-Kari's] scientific career." Although the panel could not restore the time, it could at least "provide a modicum of justice."[64]

S E V E N T E E N

■

Final Verdicts

JOSEPH ONEK was out of town on the afternoon of Friday, June 21, 1996, when his office was informed that the decision of the Appeals Board panel would be released at 4 P.M. It would be available in hard copy and also on a Web page of the Department of Health and Human Services. His secretary called Imanishi-Kari, who had been told some days earlier that a decision was imminent and was spending the day at home in nervous anticipation that it might come down this last day of the work week. She telephoned the news to Baltimore's office at M.I.T. Baltimore, at a meeting in New York City, got her message when he happened to check with his office shortly before four o'clock. Anxious and drained, they both waited while, in Washington, Onek's secretary obtained the decision by a messenger sent to the Appeals Board. About a half hour later, she read Imanishi-Kari its front page over the telephone and then faxed her the first two pages, which summarized the overall finding—exoneration on all nineteen of the O.R.I.'s charges.[1]

Both Imanishi-Kari and Baltimore, who got the word from Onek's secretary about the same time, soon downloaded the full decision from the Web and stayed up late reading it—Imanishi-Kari until 4 A.M. In an introduction, the panel ventured a series of hammer-blow "General Findings" about the O.R.I.'s case. Among them: A good deal of the evidence "corroborated [Imanishi-Kari's] statements and directly contradicted representations made by O.R.I." In some respects the O.R.I.'s attempts to establish a motive for falsification or fabrication required "very convoluted reasoning." The questioned data were in many cases "conflicting or bizarre in ways more likely to raise than resolve scientific questions." While the forensic and statistical analyses "identified some anomalies," they "provided no independent or convincing evidence that the data or documents were not authentic or could not have been produced during the time in question." "Many of the most troubling forensic attacks are mounted against the most trivial or peripheral data." If the O.R.I. were right, Imanishi-Kari's coauthors and the Tufts and M.I.T. scientists who investigated her would have had to engage in a remarkable cover-up of her misconduct—a view to which O'Toole appeared to subscribe but that the panel called "implausible," explaining that "there is no basis in the record to suggest that all of these scientists would jeopardize their own careers by lying to save hers."[2]

In its detailed review of the nineteen charges against Imanishi-Kari, the panel found the O.R.I. often simply mystifying or inept. The O.R.I. employed a definition of double producers that was "erroneous," one "not confirmed by any of the witnesses, even O.R.I.'s own independent experts." It aimed "one of the most troublesome" forensic attacks at the data from the transgenic hybridomas recorded on pages 102 to 104 in the I-1 notebook, which included the run of fifteen wells positive for growth. The analysis "fails to consider" the likelihood—what Phillipa Marrack had noted—that the clumping of cells could contribute to such a run. In any case, these data "were not published anywhere" and if Imanishi-Kari fabricated the numbers, "she also fabricated the totals [of hybridomas] and had no need to create a run of 15 positives." In presenting Imanishi-Kari's data for the specificity of Bet-1, the O.R.I. scientist John Dahlberg "may have engaged in data selection and interpretation not unlike that for which he criticized" her that permitted him to exaggerate the reagent's nonspecific characteristics. "During and after the hearing, O.R.I. sought to expand its charges beyond those stated in the charge letter." The attempt "raised questions of fundamental fairness"; Imanishi-Kari "could not reasonably have been expected to respond to charges not previously made, concerning conduct not mentioned in the O.R.I. report."[3]

The panel considered it highly unlikely that Imanishi-Kari would fabricate data that was included in neither the *Cell* paper nor the published corrections, and it judged much else in the O.R.I.'s case comparably improbable. Notebook emendations revealed nothing about intent. "It is not self-evident that a change made at a later time or using a different pen is less likely to be an honest or accurate correction than one made immediately or in the same ink." If Imanishi-Kari had fabricated or misrepresented her data, "it seems hardly likely she would hire Dr. O'Toole to extend the work reported in the paper, have her review a draft of the paper, and give her access to a notebook containing much of the problematic data." O'Toole's account of the meeting at Tufts on May 23, 1986, was not only "uncorroborated" but unconvincing. "We find it implausible, were Dr. Imanishi-Kari guilty of having fabricated these pages immediately before the meeting in order to satisfy Dr. O'Toole, that she would promptly complain to Dr. O'Toole of the trouble it involved to fabricate the data."[4]

Besides, one of the pages shown O'Toole revealed Bet-1 behaving badly, which made believing that it was fabricated difficult. Indeed, the data on many of the questioned pages were "not helpful" to the *Cell* paper, which ran contrary to Walter Gilbert's testimony that when people fake experiments they tend to construct data that is "generally too accurate." The O.R.I. may have attributed the bizarre features of Imanishi-Kari's data partly to her haste in fabricating them, but in the panel's view, "Haste does not explain creating new scientific questions." Dahlberg had dismissed her later scientific understanding of the strange results in her June subcloning data because, the panel suggested, he misunderstood it.[5]

The panel itself had examined the notebooks and discovered key flaws in the O.R.I.'s reading of Imanishi-Kari's data. Its "own visual observation and examination" showed that the allegedly cycling data displayed "a significant number of interruptions" in the general trend of rising and falling values. In short, the cycle of twelve did not cycle. Casual inspection of the run-of-fifteen data revealed that the hybridomas probably had not grown in the fifth well: The run of fifteen was only a run of ten. Imanishi-Kari was well within defensible scientific bounds to have set her cutoff at roughly two and half times the background level of radiation, and the data indicated that she had consistently abided by the standard, even when it worked against the *Cell* paper's claims. The panel observed that she had no obligation to report discrepant results concerning Bet-1 and no reason to overstate its specificity. In any event, from its inspection of the data it did not think that her results with the reagent were "seriously out-of-line" with the laboratory's experience with the reagent or that Imanishi-Kari had disregarded "contrary results . . . without reason."[6]

To the panel, several of the O.R.I.'s charges against Imanishi-Kari boiled down to differences of scientific interpretation. The panel was mindful of Imanishi-Kari's contention that it was inessential to the *Cell* paper that every hybridoma in Table 3 generate idiotypically birthmarked antibodies. It held that whether or not the table implied that they did was "of very limited importance." Imanishi-Kari had several "scientifically reasonable" assays to use on the table's hybridomas; selecting which to employ involved "scientific judgment." The panel found "ample basis" in the data for her determination that the birthmarked antibodies were produced by particular hybridomas. It also held that the dispute over whether the Table 3 hybridomas were double producers was clouded by O'Toole's "lack of clarity"—whether she was proposing *mu-mu* hybrids or *mu-gamma* ones. Whatever she may have had in mind, Imanishi-Kari had checked out the evidence that the Table 3 hybridomas might be double producers and made "honest judgments" that they were not.[7]

The panel found it hard to believe that when O'Toole attended the meeting of the Wortis committee on May 23, 1986, Imanishi-Kari showed her the ELISA data for the Table 3 hybridomas and admitted that they did not test for idiotype. The panel's reasons were Onek's: O'Toole's recollections were uncorroborated. She did not mention the alleged admission in her memorandum to Herman Eisen, and she told the N.I.H. in 1988 that she had seen only two pages of data that day at Tufts. The panel, having heard the testimony and examined "all" of O'Toole's statements over the years, was moved to "question the accuracy of Dr. O'Toole's memory and her increasing commitment to a partisan stand."[8]

The panel rejected the O.R.I.'s statistical and forensic analyses for reasons of its own as well as those advanced by Terence Speed and Gerald Richards. It observed that Mosimann's spikiness test revealed nothing about Imanishi-Kari's intent in writing down the June subcloning data; that he had offered no proof of the validity of his Poisson mixture model as it was applied in the case; and that Walter Gilbert's testimony in support of Mosimann "did not clearly set out either an empirical or theoretical basis for his opinion, that directly relates to the issues here."[9]

The Secret Service's conclusions were unpersuasive in part because "they were reached in a vacuum of information about the kind of experiments done in Dr Imanishi-Kari's laboratory and the substantive nature of her scientific work which influenced the way her data were organized and presented." Unlike the Secret Service, the panel insisted on distinguishing in time between "the creation of *pages* and the creation of *data*," which nullified the discrepancies in Imanishi-Kari's dates as evidence of fabrication. It agreed with Gerald Richards that absent proof that two counter tapes were produced by the same printer, it would be "irrelevant whether they were produced by ribbons using the same formula of ink." The panel held that whether or not the printers were switchable, Igras and DeHaro believed they were not. They thus probably did not switch them, which meant the Maplethorpe's tapes and Imanishi-Kari's had been produced by different printers. The panel observed: Not only was the comparative-norm sample of other scientists' notebooks incomplete; the dates in some of them appeared to be entered as cavalierly as those in Imanishi-Kari's notebook. The O.R.I. had introduced no testimony or affidavits as to what the dates in the sample notebooks meant. The panel thus felt it had "no basis to presume" that the samples "more accurately represented dates" on which their experiments had been conducted—and to which Imanishi-Kari's experimental records could therefore be temporally tied.[10]

The panel made clear its overall view of the case:

> Much of what O.R.I. presented was irrelevant, had limited probative value, was internally inconsistent, lacked reliability or foundation, was not credible or not corroborated, or was based on unwarranted assumptions.[11]

"A goddamn sad day for science," Peter Stockton said. He decried the decision as "a stunning repudiation of the truth." Congressman Dingell applied a logic all his own to the ruling, writing that it "confirmed that Dr. O'Toole was right to have serious concerns about the validity of significant parts of" the *Cell* paper and that it "did not rebut much of the Secret Service's testimony that portions of Dr. Imanishi-Kari's laboratory records had apparently been altered or falsified." Marcus Christ says that the O.R.I. lawyers were extremely disappointed by the outcome. Walter Stewart declared that the decision "only proves what people already know: the system does not work." To Margot O'Toole, the decision was a travesty. After spending many years going over the case in detail, "scientists" had reluctantly but unanimously concluded that fraud had been committed, she said to a reporter for the *Washington Post*. "Then all this is put in front of a legal panel of judges, and this panel thumbs their noses at the scientists." Suzanne Hadley insisted that "the findings of the investigation that I led were absolutely correct" and that the panel's conclusions "had virtually nothing to do with scientific facts and everything to do with legalities." She pointedly added, "Analogies with the O. J. Simpson case are not out of line here," seemingly alluding to the fact that despite many months of testimony the jury had acquitted Simpson after deliberating less than four hours.[12]

But the decision was soon being downloaded all over the country, and its thoroughness gave readers reason to think that analogies like Hadley's were far out of line. The panel's judgments rested on meticulous probing of the huge corpus of accumulated evidence, including the hearing testimony, the extensive exhibits of memoranda, interviews, and Secret Service documents, and even the laboratory notebooks that the Secret Service had gathered to establish a comparative norm. Its exhaustive examination of these materials was manifest in the decision itself, a book-length analysis that ran to 183 single-spaced pages and was studded with 235 elaborative notes.

Imanishi-Kari and Baltimore found themselves bombarded by telephone calls from friends, colleagues, well-wishers, and journalists. "It's a great day for Dr. Imanishi-Kari and a great day for science," Onek proclaimed. "The era of open season on scientists has come to an end." Ursula Storb privately welcomed the decision and Hugh McDevitt publicly celebrated it, observing, "Ten coincidences, even if you add, them together, don't prove guilt." The *New York Times* conceded that its harsh editorials in 1991 now seemed "a rush to judgment." Both Imanishi-Kari and Baltimore told reporters that it felt wonderful to be vindicated, but Imanishi-Kari could not help but mention her distress at having had to endure ten years of slander. Baltimore called the victory "bittersweet," noting that he felt "a sense of relief but not accomplishment." Thereza had been "unfairly prosecuted," he said. "A lot of people owe her a serious apology."[13]

Several brief postmortems of the case quickly appeared. They variously called it "an American tragedy," a "travesty of justice," and, with an eye to preventing a recurrence, dwelled on the public contributors to it, especially Congressman Dingell and the O.R.I./O.S.I.[14] Neither the postmortems nor the media in general gave much attention to Margot O'Toole. She had been attractive to celebrate as a post-Watergate saint of science ("We both could have made the choice to keep quiet for the sake of our own careers," she said in 1990, comparing herself with David Weaver. "He did and I didn't. . . . The science was more important to me than the career.")[15] But O'Toole evidently did not compel media interest in the wake of a ruling that openly expressed the judges' doubts about her credibility and impression of her increasing partisanship.

O'Toole not only initiated the case; she helped to sustain it by the certainty and resourcefulness of the convictions she expressed. A friend from her postdoctoral days at Fox Chase notes, "Margot doesn't easily see gray scales. She sees the world very starkly in black and white." Herman Eisen detected in her a "kind of rigid mind," suggesting that she might say, "If the reagent is wrong, then it doesn't matter which way it's wrong. Everything it does is wrong." Brigitte Huber adds that O'Toole tended to heed what supported her position and to be deaf to what didn't. She converted Imanishi-Kari's acknowledgment that no isotyping tests other than for *mu* had been performed on the Table 2 hybridomas into the

Thereza Imanishi-Kari and David Baltimore
June 23, 1996
Photo Credit: Bettyann Holtzmann Kevles

claim that the hybridomas had not been subcloned either. When in November 1988 she saw the N.I.H. panel's report, which raised the issue of wells having been called clones in the *Cell* paper, she added to her original complaints that Imanishi-Kari "had repeatedly told me to report results from multiclonal wells in a similar manner." She said that she had repeatedly refused and that it was for this reason that Imanishi-Kari had told her to stop doing research, a claim that contradicted her testimony to the Dingell subcommittee in 1988. The appeals panel noted that she did not mention seeing the ELISA results at the Wortis committee meeting "until after she likely had heard about the ESDA results, but attempted to explain this by saying that questions about the ELISA simply did not arise until later."[16]

O'Toole seemed capable of hearing silence speak. Wortis, Huber, and Woodland may have attested to having seen the June subcloning data on May 16, 1986; O'Toole told Kimes and Hadley that she knew of "a tape recording of their assertions to the contrary." No such tape was ever introduced into evidence despite the considerable doubt it would have cast on the Wortis committee's veracity. What O'Toole likely meant was a tape of the public forum on the *Cell* paper dispute held at the Sackler Auditorium in May 1988, right after the first Dingell hearing. Wortis, Huber, and Woodland had in fact asserted nothing to the contrary there. They had simply not mentioned the subcloning in their discussion of the scientific issues.[17]

Imanishi-Kari says, "I could see that Margot was telling half truths—or half lies—because, yes, there were things that happened a certain way, but it wasn't 100 percent the way she would tell it. It was half right. She also says things very vaguely, so you can put many twists to it. . . . I kept thinking . . . Did I really do that? . . . It was very disturbing, because very often I wondered, maybe I'm crazy." Huber says she doesn't "think [O'Toole] lies." She calls her "a very honest person . . . a very diligent person." She adds, "I think she's just a plain fanatic."[18]

Leonore Herzenberg, speculating why O'Toole and Imanishi-Kari "locked horns so hard over" Bet-1, noted, "Thereza had a small lab. She had a certain amount of money." She really didn't consider it important to go back and check out Bet-1. "In science, you don't ever really do something by repeating it over and over again." You "validate what came before" by exploring phenomena that the original result predicts. "Thereza's better idea was, We'll do some cell transfers and we'll see if the same thing happens." Robert Woodland says that "Thereza is extremely tough on people because she's a very demanding person," and "Margot is tough on herself." Thereza was basically telling Margot that she was "incompetent because she couldn't repeat things" in the lab. Margot was driven by the "anger . . . of humiliation . . . to . . . prove Thereza's wrong." She would "just keep generating new" points of criticism until somebody would say, "You're right. She's wrong."[19]

Bruce Maurer calls the course of events a "true tragedy" for O'Toole. "Margot is a bright person, and things might have turned out differently for her, too." So far as she was apparently concerned, her career had been thrown off track by her difficulties with Bet-1. Her frustrations with the reagent still rankled in 1991, when shortly after the leak of the draft report she explained to a TV interviewer in Boston, "Science is a jigsaw puzzle. Each piece that's right fits with each other piece. You can waste years of your life trying to do something that's based on something that isn't right. It's like throwing a piece of a jigsaw puzzle in from another picture." Leonore Herzenberg reflects that not inviting O'Toole to Stanford in 1987 to run down the double-producer issue was the "biggest mistake of my life." She could have done then what she accomplished later with Alan Stall and saved an enormous amount of "aggravation." Indeed, such a chance at Stanford might have restarted O'Toole's career in research and, perhaps more important, given her the legitimation as a capable scientist that she evidently craved and that she felt she was increasingly denied as the case moved from the private to the public arena. She declared in the Boston television interview, "The hardest part was being portrayed, believe it or not, as someone who did not understand science, when I felt I was the only one who stood up for science."[20]

For all her resentful eagerness to call Imanishi-Kari to account, O'Toole still brought her complaint to the public arena reluctantly, being forced into it by Walter Stewart and Ned Feder, who had been surreptitiously assisted by Charles Maplethorpe. Anthony Lewis, in his *New York Times* column a few days after the exoneration, derided Stewart and Feder as "the Savanorolas of the N.I.H." Stewart holds that whether their critics consider them "vindictive, nasty, mean, self-interested, glory-seeking, or whatever" is besides the point. "Eventually people have to address the *merits* of our argument." Their arguments may have been impressive in the Darsee case, but the Appeals Board's decision in the Imanishi-Kari case strongly implied that their arguments in that one had no merit. Many scientists believe with Herman Eisen that Stewart and Feder had gotten "carried away" by the heady turn that their careers took when they went into fraudbusting. A few weeks after the leak of the draft report, Feder, then 63, told a reporter, "I never would have imagined that at my age I'd be doing something so exciting, and more than that, pioneering work."[21]

In February 1993, Stewart and Feder sent a fourteen-hundred-page document to the American Historical Association reporting that Stewart's plagiarism machine had revealed that a historian named Stephen Oates, the author of biographies of Abraham Lincoln, Martin Luther King, and William Faulkner, had lifted many passages in these works from other sources without attribution. The American Historical Association had already cleared Oates of charges of plagiarism in the Lincoln book; he was a private citizen; biographies of Lincoln, King, and Faulkner had nothing to do with science; and on examination, a number of the passages fingered by the computer program that Stewart had devised suggested to many observers that while the pair had detected some that seemed problematic, they appeared to know little, if anything, about what constituted plagiarism. In April, the N.I.H. abolished their laboratory, prohibited them from working on scientific misconduct, and assigned them each to separate administrative jobs. Dingell's office said it had no objection. Stewart promptly went on a hunger strike.[22]

Robert Charrow remarked of the O.R.I., "We created a monster, and now it's time to bury it." Several years earlier, he had contended, with the authority of having been in on the event, that the N.I.H. established the

O.S.I. mainly to protect itself from Congress, especially John Dingell.[23] Once in place, it had a mandate but little design about how to accomplish its purpose. Jules Hallum recalled that it was beset internally by constant arguments over whether it should be run like a court of law or like the editorial office of a scientific journal. An O.S.I. official of the early period recalled thinking "that one of the worst things that could happen to this enterprise would be for it to become the science cops—for it to be taken over by the office of inspector general and managed and implemented by a bunch of gumshoes who had no experience or understanding or love of science."[24]

The O.S.I. managed to keep control of its operations in scientific hands, but it failed to exercise any critical oversight of the work done by its principal technical collaborator, the Secret Service. Julius Younger says that agents Hargett and Stewart were given a "virtually impossible task" when they were asked to determine the authenticity of Imanishi-Kari's notebooks. They might have been saved some of their errors if the O.S.I. had not encouraged them to ignore completely the science that she was attempting to accomplish and had thought through the pitfalls in O'Toole's strategy of checking Imanishi-Kari's tapes by scrutinizing the printers and the tapes in the notebooks of other scientists. During the appeal hearing, Onek offhandedly called O'Toole "The real prosecutor in this case." The appeals panel suggested that the O.S.I. and the O.R.I. had failed to maintain sufficient distance from Margot O'Toole, observing in one of its notes, "While we share others' concern that a 'whistleblower' be protected from adverse consequences, we are also concerned about the implications of involving a whistleblower too heavily in an investigation. Such involvement can compromise both the ability of the investigators to maintain objectivity and the ability of the whistleblower to avoid becoming too vested in the outcome. We think that happened here."[25]

Joseph Onek says that the principal trouble was not so much that O'Toole was asked to draw up the allegations against Imanishi-Kari; it was in the institutional arrangement—that the O.S.I. acted as prosecutor, judge, and jury without granting Imanishi-Kari due process.[26] A ranking legal officer at the N.I.H. says that when the O.S.I. laid down its investigative policies, it made two "miscalculations." One was rooted in the belief that, since a verdict for debarment from grants gave an accused scientist the right to a full hearing, the O.S.I. provided sufficient due process to respondents. The miscalculation was to rely on the legal underpinnings of that view, which were strong, and to ignore the political support for it, which was weak. The second miscalculation—an astonishing blindness, the legal officer might have said—was not to have realized that a finding of misconduct alone, even without debarrment, could damage a person's reputation.[27] As it was, the O.S.I. had no regular internal mechanism for critically assessing O'Toole's allegations. Its refusal to give Imanishi-Kari due-process access to key materials during the investigation denied her the opportunity to contest them effectively, and it placed the burden of proof on Imanishi-Kari's shoulders rather than on its own.

The structural flaws in the O.S.I. were compounded by the attitudes of Suzanne Hadley, the chief investigator on the Imanishi-Kari case. Brian Kimes says that Hadley's having gone to work for Dingell just reinforced his view that "she was obsessed with misconduct," noting, "She knew as well as anybody that Dingell and his staff were unethical and did not care about fairness or objectivity in any way." While working for Dingell, Hadley evidently prepared a report for the subcommittee on Robert Gallo, producing a draft reviving the charges that Gallo's claims to the discovery that HIV causes AIDS were fraudulent and contending that the United States Government had engaged in an extensive cover-up of the fraud to protect its patent rights in the blood test arising from the discovery. "While there will always be individual scientists who go wrong, what marks this is the way much of the U.S. Government got sucked in, perpetrating and promoting a big lie," Hadley said. In January 1995, Stewart promulgated the report on his Web page and sent a copy to, among others, the head of the N.I.H., Harold Varmus. Dingell himself promptly repudiated the document, writing to Varmus that he could not vouch for its "authenticity or accuracy," that it had not been "reviewed, much less evaluated," by the subcommittee, and that the subcommittee's staff director had "rejected" an early draft several months earlier.[28]

Anthony Lewis identified John Dingell as the "principal villain" in the "ugly tale of persecution" that was the Baltimore case. "He went after Professor Baltimore, and through him, M.I.T., with a viciousness that suggested a desire to humiliate the academic establishment." Dingell persistently denied that either he or his staff had any interest in attacking universities or destroying American science. He reminded people that his brother was a scientist at the N.I.H. and that his wife donated both time and influence to the Children's Inn on the N.I.H. campus. He said that he sponsored hearings into scientific misconduct so as to shine a light on the issue. His belief that it needed further illumination was, in truth, well warranted; since the Gore hearings in 1981, the nation's research institutions, including the N.I.H., had acted slowly, and often in a desultory manner, to establish or bolster their machinery for dealing with allegations of scientific misconduct. The N.I.H. had been slow to respond to the Imanishi-Kari case and naive, to say the least, in selecting two of Baltimore's associates to help judge it. Dingell saw no reason to apologize for the special brand of ferocity with which he pursued the matter. "We do not wear lace on our drawers as we conduct our investigations," he told a reporter. "I'm not paid to be a nice guy. I'm paid to look after the public interest. Our purpose is simply to compel universities and scientists to clean up their act and to see to it that public money is properly spent."[29]

William Raub, experienced in dealing with Congress, observed at one point in the controversy, "What angers Mr. Dingell and other members of Congress on occasion is the scientific community seeming to set itself apart, putting itself on a higher plane or otherwise expecting some special recognition and privilege. I think he would say in some areas special treatment is deserved. . . . He's also prepared to say where 'I don't think you're doing well and I'm going to hold your feet to the fire.'" Dingell publicly called the charge that his subcommittee engaged in McCarthyism "most ironic, and moronic," noting that "all witnesses received a full opportunity to state their cases." Chafin, in private, says angrily that "it's bullshit that we're like McCarthy." He stresses that, unlike McCarthy, the subcommittee disclosed evidence to witnesses ahead of their testimony and did not trap them into pleading the privileges of the Fifth Amendment. Stockton notes that Dingell had told them not to treat scientists like defense contractors.[30]

However, the subcommittee emulated Senator Joe McCarthy in ways that it perhaps did not realize and that people like Raub overlooked. It sought to intimidate the O.S.I., bullying the agency's staff and demanding information about the probe of Imanishi-Kari's ongoing case. It leaked privileged investigative information damaging to her to favored members of the press. In order to hold the scientific community's feet to the fire, it was willing to sully the reputations of innocents. It turned its hearing room into a forum for smear on April 12, 1988, limiting the proceedings concerning the *Cell* paper dispute to the one-sided testimony of Stewart, Feder, Maplethorpe, and O'Toole and making no effort beforehand to check the reliability of their respective accounts. The House has rules for the protection of witnesses' reputations, but the subcommittee had no rules for fair protection of the reputations of people whom the witnesses might attack. Stockton and Chafin say that such people are given a fair chance to respond, if not in the same hearing then in one soon thereafter. Baltimore and Imanishi-Kari had to wait a full year for their chance.[31]

By turning a laser beam on Imanishi-Kari to spotlight the issue of scientific misconduct, the Dingell subcommittee, with its deployment of Stewart, Feder, and the Secret Service, in effect prosecuted as well as persecuted her. It pursued her, an individual, as though she was a defense contractor, with a defense contractor's means and resources. She had neither. By the end of her appeal, she had been provided with legal services valued at roughly $1 million. David Baltimore defrayed $100,000 of the costs, and friends like Joan Press and David Parker raised some $30,000 more. All the rest was offset by the willingness of her lawyers and expert witnesses in immunology and statistics to work for her *pro bono*. The subcommittee's prosecution was worse than the O.S.I.'s. It not only denied her the elemental protections of a defendant; it went after her publicly. When she tried to defend herself knowledgeably, the subcommittee did not cooperate, ignoring, for example, her attorney Bruce Singal's repeated requests to examine the Secret Service reports on her notebooks. In the comment of Vice Chancellor C. Kumar N. Patel, at UCLA, "What Dingell carried out was a public lynching."[32]

Donald Kennedy, the former president of Stanford University and object of the subcommittee's attention, reflected on why Dingell had succeeded in his campaign of intimidation. Ordinarily, the main lines of defense would have formed in the media and the scientific community, he said. "Where the hell was everybody?"[33]

By and large, the national press had been with the courageous martyrdom of O'Toole and the drama of the confrontation between Baltimore and Dingell. The scientific media—notably, *Science* and *Nature*—did deal with the technical dispute and attend to Imanishi-Kari. The conservative *Wall Street Journal*, no friend of Dingell's in general, attacked him for abusing his investigative power in the matter. The general press, however, tended to relegate Imanishi-Kari to a minor role in the drama and cover perfunctorily—often, like Philip Hilts, getting it wrong—the bearing of the technical issues on the question of misconduct. Kennedy singled out Hilts and the *New York Times* for the "relentlessly negative" tone of their coverage, stressing that "at no time" did they "call attention to the failures of due process." Neither did most of the rest of the mainstream press, probably because, Kennedy held, a number of reporters were being fed leaks by the Dingell subcommittee. Not wanting to jeopardize their sources, they tended to write about the controversy from one side—Dingell's. More important, the flavor of the subcommittee's news no doubt appealed to much of the media. In the post-Watergate environment of American political culture, many reporters were likely inclined to think Dingell, O'Toole, and Stewart and Feder's take on the dispute—that fraud had occurred and been covered up—was right.[34]

Kennedy reckoned that the scientific community remained largely silent in the affair because "too many of us were content to presume that Imanishi-Kari was guilty." However, that presumption did not energize the angry attacks against Baltimore. His Harvard critics and his opponents at Rockefeller were not alone in believing that he had been wrong or foolish or both to confront Dingell. Even many of his admirers held that by taking on the congressman he had brought science into disrepute and jeopardized the funding of biomedical research. Bernardine Healy says that a number of scientists warned that her testimony before the subcommittee in 1991 might prompt Dingell to punish the N.I.H. and its grantees. The physicist Freeman Dyson, appalled at how people like David Baltimore were treated by their scientific colleagues, deplored the tendency among academics "to be more zealous than Dingell in their extirpation of deviant science." Dyson added, "The forces now driving academic institutions to join the bandwagon of moral rectitude are the same forces that drove academic institutions in the 1950s to join the bandwagon of spy mania. These forces are now, as then, cowardice and venality." One did not have to embrace Dyson's likening of scientific fraud now to loyalty and security then to recognize that the country's biomedical scientists had become wards of the state and the way that many of them responded to the issues in the case was shaped by fear of antagonizing their most munificent patron.[35]

The Republican House that the 1994 congressional elections swept into office relegated Representative Dingell to ranking minority member on his committee and somewhat reduced the committee's jurisdictional range. Dingell's power was thus diminished, at least temporarily. Stockton and Chafin soon left for private business. Yet Dingell's relentless pursuit of Baltimore and Imanishi-Kari had, indeed, gotten the scientific community's attention. It was what he said he had set out to do, and he left an indelible mark on the research enterprise.

A number of scientists still doubted the necessity of elaborate fraud machinery. Brian Kimes says that most of the cases that came to the O.S.I. on his watch were "trivial"—postdocs in close quarters, at personal odds and engaged in sibling-like quarrels. The dissidents continued to hold that the serious form of misconduct—fabrication and falsification of research— is rare. There was as yet no reliable data on its incidence to gainsay them, and such data as did exist provided support for their position. In 1996, for example, the O.R.I. closed forty-nine cases, seventeen of which resulted in a finding of misconduct, and that year it conducted twenty-five investigations in response to new allegations. Bernard Davis, the Harvard biologist and critic of federal misconduct investigations, had observed several years earlier that, compared with the thousands of grants sponsored annually by the N.I.H., the number of allegations then before the

O.S.I., fewer than 100, "represent a level of cheating to which Congress—and even the clergy—might aspire."[36]

Kimes, like so many others, says that the fraud-policing enterprise is "a waste of resources, because science basically corrects itself." He adds that scientists might be able to get away with faking data in an "esoteric field that no one cares about," but that they will be discovered if they publish at the cutting edge in a field about which people do care. In a talk on the eve of the Dingell subcommittee hearing in May 1989, Baltimore held that if nothing else prevented scientists from cheating, it was that they "risk their reputations" when they published their work. Just about that time, the self-correcting mechanism of science worked with crushing effectiveness against the claims recently made by two scientists in Utah that nuclear energy could be released through the process of cold fusion. Their wrongheadedness was exposed by other scientists who could not repeat their experiments or account for the alleged phenomenon with theoretical calculations. The effort saved the people of the United States the $25 million that the Utah scientists—and their university—urged the federal government to give them and that some in Congress seemed eager to provide.[37]

However, after more than fifteen years of congressional inquiries and media reportage, it was clear that the mechanisms of self-policing did not suffice to expose scientific fraud and relying on them alone would not satisfy the public patrons of science. The scientific community may have previously handled misconduct with a slap on the wrist or a discreet word to colleagues, resorting in extreme cases to excommunication, but that kind of clubby protectiveness—represented by David Baltimore's initial response to Herman Eisen's mistaken news about Imanishi-Kari's knowledge of Bet-1—was no longer tolerable. Now the public consensus in science was: All wrongdoing deserved to be exposed; public investment merited protection; and phony data needed to be flagged, not least to prevent other scientists from wasting time and money following it up. Daniel Koshland, the editor of *Science*, noted in 1989, "Scientists have lost the ability to run their affairs as a cozy collegial group that rewards the good guys and agrees to throw out the bad guys with a minimum of formality."[38]

Faced with Dingell's scrutiny, universities strengthened their misconduct policies, sharpened the procedures for dealing with allegations, and encouraged their faculty to report quickly any suspicions of fraud or fabrication in the labs. At M.I.T. in 1991, a review of the response to O'Toole's complaints concluded that the university had done "a reasonable job according to the rules and mores of the time," but added that

"M.I.T. would clearly do a much better job under the guidelines in force now." Professor Gerald Rubin, the head of the genetics division of the University of California, Berkeley, remarked, "I think a lot of people have learned from the pain and suffering David Baltimore went through."[39]

Yet Imanishi-Kari's exoneration demonstrated to many that a good deal of work still needed to be done. Partisans on both sides of the case—from John Dingell to Joseph Onek—agreed with the *New York Times* that the O.R.I. had "no doubt . . . turned in an awful performance" and that "a clean-up of the whole investigative process should be a top priority." A few people continued to defend the agency's original notion of approaching its task primarily through a scientific dialogue that placed the burden of proof on the respondent. Among them was O'Toole, who found it justifiable that Imanishi-Kari had been denied eligibility for grants before the O.S.I. had found her guilty of misconduct. Pointing to Imanishi-Kari's initial refusal to respond to the O.S.I.'s queries, O'Toole declared heatedly, "It's your right under the constitution to avoid self-incrimination, but don't tell me that they violated her rights as a scientist when they took away her money after that. Don't tell me that, because it makes me angry." She said that as she sees it, "scientists are stampeding at the bidding of the lawyers," continuing, "They're ignoring the principle of accountability in science. They're stampeding to trash their own profession. They're trashing it by saying, 'Innocent until proven guilty. She hasn't had a chance to respond.' "[40]

However, resistance to strong legal involvement in misconduct cases was no longer compelling now that the O.R.I. had won only one of the four appeals against its findings—an uncommonly low batting average in federal procedures of this type, Robert Charrow pointed out. The O.R.I.'s own staff was now heavy with lawyers. Scientists in and out of Washington could agree with Robert Gallo: "If it wasn't for the lawyers, we'd all be dead. Where would we be if we left it to the politicians and, if you'll pardon me, the media, and if I'll pardon myself, the scientists among us?" Nevertheless, *de facto* the O.R.I. still joined the duties of prosecutor, judge, and jury. Onek, among others, insisted that it had to be reconstituted so that accused scientists were given a "fair set of rights." He added, "You shouldn't have to wait nine years for an opportunity for cross-examination."[41]

Still, like Congressman Dingell, many partisans of policing science remained concerned with protecting accusers at least as much as they were with ensuring rights to the accused. Margot O'Toole was often cited as the paradigm example of the fate that awaited whistle-blowers, despite the fact that she had not been fired for challenging the *Cell* paper.

O'Toole's difficulties in finding a new job in science—once she started looking again—seem to have been real, but they seem to have stemmed in part from the character of her charges against Imanishi-Kari, the high visibility they achieved, and the pursuit of them by John Dingell. A Boston area immunologist recalls, "Many of us were very upset, and we thought what was going on was unreasonable. But nobody said, We're going to get O'Toole because of this. On the other hand, frankly speaking, there was no way any of us would have taken her into the lab. . . . If this is what's going to happen when there's a conflict between a mentor and a junior scientist, who needs it?" Yet if O'Toole was atypical, ordinary whistle-blowers were certainly vulnerable to a variety of retaliations, including termination of employment and blackballing in the competition for jobs and grants. According to a survey of whistle-blowers conducted by the Department of Health and Human Services, a majority said they suffered some form of adverse response and about 12 percent reported that they were fired in the wake of filing their complaints.[42]

In late 1995, a federal commission on research integrity headed by Kenneth J. Ryan, a physician and professor emeritus at Harvard Medical School, had recommended thirty-three measures for dealing with misconduct in science in a report to Secretary of Health and Human Services Donna Shalala. The Ryan Commission had been authorized by Congress in 1993 "in reaction," the report noted, "to continuing misconduct in research and retaliation against whistleblowers." The recommendations, revealing a sensitivity to the charges that the O.R.I. was a Star Chamber, called for separating the functions of judge and jury from those of investigator and prosecutor. They urged federal adoption of a detailed "bill of rights" for "responsible" whistle-blowers, including protection against retaliation, guarantees of impartial assessment of charges, and provision for vindication upon the substantiation of allegations. However, the commission suggested some cures that to many scientists seemed worse than the disease. It proposed a kind of sunshine rule for misconduct cases, a requirement that outcomes should be publicly disseminated, whether or not the accused scientist was found guilty. Responding to part of its formal charge, the Ryan Commission argued for replacing the existing criteria of misconduct—fabrication, falsification, plagiarism, and practices that deviated from the norm—with a broader definition of the term: "misappropriation," which would include stealing words and misusing ideas and information; "interference," which would cover, for example, unauthorized removal of data from a laboratory; and "misrepresentation," which aimed to incorporate in misconduct such transgressions as omitting information from a paper that might cast its claims into doubt.[43]

In the months following the release of the Ryan Commission's report, it was blasted by the National Academy of Sciences and denounced by the Coalition of Biological Scientists, a conglomeration of representatives from fifty professional societies that together listed 285,000 members. Spokesmen for the coalition or its constituents called the report "so seriously flawed that it is useless as a basis for policy making and should be disavowed" by the Department of Health and Human Services. The new definition of misconduct threatened scientists with "unpredictable and ill-defined charges" and was thus "an open invitation to litigation." ("The very principle of due process is to define a crime before you convict people of doing it," a scientist involved in policy making for misconduct remarked.) Publicizing cases no matter their outcome would profoundly damage the careers of innocent researchers, "since in science even the faintest taint of misconduct associated with one's work can be professionally fatal." The protections for whistle-blowers were "extraordinary" and unwarranted. They ignored "the possibility that accusations may be ill-founded, malevolent, or simply wrong." They made no provision "for violations of confidentiality, false statements, or other unlawful behavior on the part of accusers." Worse to the dissidents, they awarded whistle-blowers rights in determining the composition of investigative panels and obtaining investigative information—provisions that would "make the accuser part of the investigating team and create an assymetric relationship with the accused."[44]

On receiving the Ryan Commission's recommendations. Secretary Shalala appointed an interdepartmental group under the chairmanship of William Raub, the former acting director of the N.I.H., to advise her whether and how to put them into practice. The group, which reported to Shalala in the late spring of 1996, supported roughly two thirds of the Commission's proposals but shied from endorsing those that had been attacked. It found the proposed protection of whistle-blowers "more attentive" to their rights than to the "rights of other parties" and urged that the issue be approached with greater balance. It argued for postponing action on the redefinition of misconduct, holding that it required far more public comment. It rejected the airing of investigative outcomes. The group, which included Chris Pascal, the acting director of the O.R.I., also declined to recommend separation of the investigation of misconduct from its adjudication in cases that the O.R.I. handled. Secretary Shalala took the group's suggestions under advisement. In March 1998, a year and a half later, they were still under advisement while, following one of its recommendations, the presidential Office of Science and Technology Policy attempted to finish devising a definition of scientific misconduct that would be acceptable throughout the government.[45]

Whatever the differences between the Ryan Commission and its critics, they shared the crucial belief that, as the Coalition of Biological Scientists put it to Raub, "Falsification, fabrication and plagiarism are so detrimental to the conduct of science that government action is appropriate when [other] institutions fail to provide proper oversight of federally-funded research." Raub says that Ryan was "surprised," even "shocked" that the scientific community regarded his commission's proposals as draconian. Ryan himself held that breakdowns of integrity in American science were "pervasive" and that, for example, the government had been justified in pursuing Imanishi-Kari so vigorously. Her case "didn't rise to the certainty of fraud," he told a reporter, "but the way the notebooks were kept left something to be desired. We're talking about whether the government is interested in the quality of the work when they fund it. It goes beyond just, 'Am I lying.' "[46] While Ryan went further than most biomedical scientists in his embrace of federal oversight, most had come to hold a different set of tenets from those that undergirded the antiregulatory posture that their leaders had adopted at the Gore hearings in 1981. Signaling the shift, the Ryan Commission urged that institutions receiving federal monies for biomedical science be required to sensitize "**all individuals**" involved in the enterprise "to the ethical issues inherent in research." It was one of the proposals that Raub's group solidly endorsed.[47]

In his inaugural at Rockefeller University, David Baltimore had declared: "Biomedical science has become big business in America, attended by keen competition for funds and ideas; managers and politics intrude. No longer ignored because of its impracticality, biomedical science today attracts armies of patent lawyers, corporate funders, auditors, personnel managers, Congressional investigators, and peer review panels."[48] By the later nineties, biomedical science was bigger business still, with the annual budget of the N.I.H. amounting to more than $13 billion, biotechnology companies continuing to grow and proliferate, and an avalanche of new life-related knowledge and technologies pouring out of the labs. Under the circumstances, the federal government was in the labs to stay, regulating research in multiple ways, including the handling of misconduct. The submission of the scientific priesthood to lay oversight of its integrity was widely understood to be necessary and justified.

In the storm that followed the leak of the draft report, David Baltimore had been in a sense defrocked. While his fellow scientists continued to value highly his opinions on scientific issues, his views on public affairs were suddenly no longer of interest to the media and he was shut out of science policymaking circles. His friend Maxine Singer observed, "The university world, sadly in need of visionary and outspoken leadership, lost

an articulate voice when Baltimore himself left Rockefeller University."
David Baltimore missed the town square, but he was reluctant to speak
out, fearing that he might once again become a target for attack.[49]

Several days after Imanishi-Kari's exoneration, the *New York Times*
reporter Gina Kolata wrote that the outcome "vindicates the long and
eventually lonely campaign [that Baltimore] waged in her defense." In
mid-December 1996, William Paul, in whose laboratory Bet-1 had been
devised, and who was now head of the Office of AIDS Research at the
N.I.H., announced the appointment of Baltimore as head of a new AIDS
vaccine advisory panel, part of a stepped-up effort to develop a vaccine
against the disease. Anthony Fauci, the director of the National Institute
of Allergy and Infectious Diseases, publicly applauded his appointment,
citing Baltimore's "breadth of experience and extraordinary creativity and
intelligence." On May 13, 1997, it was announced that Baltimore had
been named president of the California Institute of Technology, in Pasadena, California. The chair of the presidential search committee was
asked by a reporter how Baltimore's entanglement with the fraud controversy and the Rockefeller faculty had affected the choice. He responded
that those issues "were simply not relevant—except the extent to which
the experience strengthened David's character."[50]

Imanishi-Kari's exoneration changed no minds among Baltimore's Harvard critics. Mark Ptashne and Paul Doty say that he has suffered enough,
and Ptashne volunteers that he wishes him well in his return to public
life. But none of them appears to regret the virulence of the criticism
leveled against Baltimore at the height of the controversy. Indeed,
Ptashne's judgment of Baltimore's handling of the dispute over the *Cell*
paper remains as negative as it was in 1991. Doty insists that Baltimore's
defense of the *Cell* paper had done "lots of long-term harm" to scientific
practice and that he had irreversibly relinquished the moral authority to
speak for his—that is, Doty's—brand of research standards. The appeal
panel's decision made no difference to Walter Gilbert's belief that Imanishi-Kari had committed fraud. He continues to fault Baltimore most,
he told a reporter, for "trying to mobilize scientists against the Congress,"
particularly Congressman Dingell. "I thought then and do now that David
Baltimore damaged the future of American science at that time."[51]

It is not difficult in retrospect to think of actions that Baltimore might
have taken at different stages of the dispute to defuse it. Instead of simply
suggesting that O'Toole publish her critique of the experiment in a letter
to *Cell*, he might have urged her to participate in an exchange of letters
and tried to persuade her that she would be taken seriously. He might
have called on his coauthors to join him in submitting a letter to the

journal in 1986 correcting the overstatement about Bet-1 and the mis-statement about Table 2, corrections that they voluntarily published in 1988, albeit under different circumstances. Once Stewart and Feder began raising questions, he might have asked for a more thorough investigation at M.I.T., as Healy would do at the Cleveland Clinic. He might have done more to explain himself to other scientists, including the Rockefeller faculty, particularly with regard to why he believed in the validity of the experiment, why it was impractical for his lab to repeat it, and why he felt that O'Toole's critique did not obligate him to redo it. James Watson says that he blames Baltimore partly because "he never talked with anyone. We asked him to come to the Banbury meeting and he didn't."[52] He might have turned the other cheek to John Dingell.

But the might-have-dones entail a good deal of second-guessing and, more importantly, a variety of improbabilities. It is not obvious that many of the actions he might have taken would have significantly altered the course of events. If he did not call for a more thorough inquiry at M.I.T., he did request one by N.I.H., but neither O'Toole nor Stewart and Feder accepted the results of the first N.I.H. investigation. It is difficult to believe that a renewed investigation at M.I.T. would have persuaded them to halt their complaints or satisfied the O.S.I., which found reasons not to accept the conclusion of Sharma's innocence from the second inquiry at the Cleveland Clinic. It is equally implausible that a reopening of the inquiry or offering more explanations to colleagues, let alone a redoing of the experiment, would have appeased Dingell and his staff or mitigated the negative coverage in the media.

In the spring of 1997, a reporter asked Baltimore what he had learned from the affair. He responded, "I learned a lot of things . . . , one of which is not to give up when you believe in something." His temperament undoubtedly contributed to his tenacity. A less self-confident person might not have persisted against the combined pressure and power of Congressman Dingell, the O.S.I., and the press. Baltimore's friend and fellow Nobelist Paul Berg notes that he can be "very obstinate."[53] But if Baltimore was unyielding, he was stubbon on issues worth being stubborn about.

Baltimore unabashedly defended the common-sensical legitimacy of collaborators' taking the results of participants in other specialties to a considerable degree on trust. He insisted on the personal, contingent nature of science, including having inevitably to make judgments about data. In his testimony to the Dingell subcommittee, he explained that errors cropped up in most biological research. They arose from the variability of biological materials, from "analyzing only some fixed amount of data," from "the finite sensitivity of the measuring systems or by not having made a measurement that later turns out to be crucial." He continued:

> No study is ever complete. . . . Deciding when to write up a study is an arbitrary and personal decision. A paper is written when an investigator decides that a story can be told that hangs together, that makes sense and that others will want to read and build on. The scientific literature is a conversation among scientists. . . . It is crucial to remember, and often forgotten, that a paper does not claim to be an absolute assurance of truth, only a moment's best guess by one group of investigators. Because all of these judgments are less than wholly objective, another investigator might have come to a different conclusion using the same data. In a real sense, a scientific paper is a subjective product.[54]

Baltimore, like Eisen, was right to contend that O'Toole's critique of the *Cell* paper could be resolved not by argument but only by further research. The issue of double producers was greatly elucidated by the joint work of Alan Stall and the Herzenbergs. Additional work in the United States and Europe has strongly indicated that the introduction of the transgene does not activate an idiotypic network in the recipient mice, but it has also led to the conclusion that at least part of the core of the central claim of the *Cell* paper was correct. As J. Donald Capra testified at the appeal hearing, it "is now an accepted paradigm in the field that anytime you introduce . . . rearranged *mu* [DNA] into a transgenic mouse, you profoundly alter the repertoire" of its antibody response.[55]

Baltimore may not have appreciated that John Dingell did not intend to attack American science as such; he was justified in thinking that the way Dingell pursued the misconduct issue in effect threatened the flexibility and freedom with which scientists approached the understanding of their data. Dingell did not directly force the *Cell* paper's coauthors to supplement Table 2 with the June subcloning data, but his intimidating presence must have figured in shaping the charge to the N.I.H. panel and in James Wyngaarden's ultimate insistence that the coauthors publish the correction to the *Cell* paper containing the supplement.

It was a chilling precedent. Joseph Davie conceded that the panel's disagreement with the coauthors concerned science, not truthfulness. The coauthors submitted to Wyngaarden for political reasons. Bernard Davis discerningly recognized the import of the coercion to publish. "It deprives a distinguished scientist of his right to expert judgment about what is a significant error, and to have the outcome of the dispute determined in the traditional way by further work, rather than by legal pressure."[56]

No one was put in greater jeopardy by Dingell and his staff than Imanishi-Kari. She was gravely disadvantaged in the contest by her weakness in English, by her lack of resources, and by Margot O'Toole's overpowering articulation of events. If Baltimore had abandoned her, Bernard Davis wrote to a reporter, "what would we then think of him?" Freeman Dyson judged that Baltimore showed "great courage" in his "refusal to sacrifice" his friend on the altar of scientific rectitude, despite "knowing that the personal cost" to himself "would be heavy." It might be added that he was equally estimable in attacking the Dingell subcommittee for using its power to hound and prosecute her.[57]

After the exoneration, Imanishi-Kari was promptly reinstated as an assistant professor at Tufts. The N.I.H. awarded her a grant designed to assist women who were re-entering science after having taken time off. She considered it a kind of revival of the grant that she had been deprived of during the O.S.I. investigation. In the fall of 1996, she was asked whether she intended to sue the government for damages, as Gallo's collaborator Mikulas Popovic was doing, claiming $500,000 for damages that the O.S.I. had done to his career and reputation. She said she did not, explaining that no amount of money could repay her for what she had suffered, the time she had lost, and the harm done her career. She also did not want to revisit her ten years of pain. She preferred to get on with her life.[58]

In June 1997, Imanishi-Kari received official notification that Tufts University had promoted her to associate professor with tenure. She said at the time, "It is my understanding that the Tenure Committee recommended unanimously for promotion. Well, a new chapter in my life starts. I am very busy trying to bring my lab to full speed of research, and am working hard as I used to. . . . It is wonderful to be the same as everyone else again. I can also hug my friends freely when I see them. It's my hope that I still have a couple of decades to do what I have to do."[59]

Glossary of Technical Terms

17.2.25 Shorthand for the transgene that was inserted into Black/6 mice. It produces antibodies against the chemical **NP** that display the distinctive idiotypic birthmark and have a **heavy-chain** constant region whose allotype is *mu-a*.

AF6-78.25 A biologically produced **reagent** used to test for the presence of the allotype *mu-b*.

allotype A defining feature of an antibody's **heavy-chain** constant region that distinguishes antibodies of the same **isotype** from each other.

antibody(ies) A protein produced by a **B cell** that attacks foreign agents entering the body. It is constructed from discretely identifiable elements and structurally resembles the letter Y(See *Figure 1*). One element is a short chain of amino acids termed the **light chain**. Another is a long chain of amino acids termed, in comparison, the **heavy chain**. Identical light chains form part of each arm of the Y. A second part comprises identical heavy chains running down alongside the light chains to a joint, where they turn to parallel each other, becoming a double strand that forms the trunk of the Y. The **variable regions** of the two chains combine to give the antibody its specificity. The **constant regions** define the antibody's **isotype**.

B cells A type of cell in the immune system that produces antibodies. Each B cell normally generates only one specific type of **antibody**.

Bet-1 A **reagent** manufactured by injecting rats with an antibody drawn from BALB/c mice, from which the gene used as the transgene was derived. Under suitable conditions, it was far more likely to latch on to *mu-a*, the transgenic antibody, than to *mu-b*, the native one.

Black/6 Short for C57BL/6, the inbred strain of mice into which the **17.2.25** transgene was inserted.

clones A population of cells that derive from a single cell. Two or more such populations of cells may happen to grow together—as in the case of a **well** that contains two or more different **hybridomas**. In that case, single cells may be isolated from the mixed population to form separate new cells lines. These lines are called *subclones* and the process of obtaining them is called *subcloning*.

constant region See **light chain, heavy chain.**

DNA Deoxyribonucleic acid. The molecule is configured as a double helix whose two strands are joined at regular intervals by one of two pairs of chemicals, called *base pairs*. The sequence of base pairs determines the genetic information that the DNA carries. The expression of the DNA involves transcription of the information into a molecule called **RNA**. Thus, checking, as Weaver and Albanese did, for the presence of the kind of RNA found in a cell reveals the identity of the DNA that is expressed.

ELISA An acronym for *enyzyme-linked immunosorbent assay*: an assay that uses an enzyme to detect how much **antibody** of a particular type there may be in a **well**. The well is coated with a **reagent** that will grab antibodies of that type. The enzyme is chemically coupled to the antibodies, and a colorless chemical solution that interacts with it is added to the well. If the enzyme is present, the solution will become colored. The intensity of the color is proportional to the amount of enzyme contained in the well, which is in turn proportional to the amount of antibody that the reagent has captured. Imanishi-Kari used an ELISA that turned yellow to test for the presence of antibodies carrying the **idiotype** of the **transgene.**

gamma See **isotype.**

heavy chain. A long chain of amino acids that comprise one of the two building blocks of an **antibody.** It includes a constant region, which defines the antibody's isotype, and a variable region, which contributes to the idiotype.

heterodimer A molecule composed of two different chains. O'Toole used the term to mean an antibody constructed of two different **heavy-chain** constant regions.

hybridoma A hybrid cell that is a fusion of a normal **B cell** with a cancerous myeloma cell. It grows indefinitely and produces an abundance of antibodies of the single type characteristic of the B cell.

idiotype A distinctive feature of the variable region of an antibody that is genetically determined and is akin to a birthmark. It can be used to tell one antibody from another and to study the inheritance of genes that produce the variable region.

IgG, IgM See **isotype.**

immunoglobulin A general term for the different classes of **antibodies,** such as **IgG** or **IgM** generated by the immune system.

isotype The class of immunoglobulin to which an **antibody** belongs—for example, IgG or IgM. It is defined by the chemical composition of the **heavy-chain** constant region—for example *gamma* or *mu*.

light chain A short chain of amino acids that comprises one of the two building blocks of an **antibody.** It includes a variable region and a constant region.

lymph node A small organ in the immune system that supports the proliferation of lymphocytes. Lymph nodes were used as a source of cells for the hybridomas in the experiment.

lymphocytes Cells of the immune system derived from precursors in the bone marrow, for example, **B cells.**

mu See **isotype.**

mu-a The allotype of an **antibody** produced by the transgene.

mu-b The allotype of an **antibody** that has a *mu* isotype and that is produced by a gene that is native to the **Black/6** mouse.

NP Pronounced "nip"; one of a class of small organic chemicals called *haptens* that, when combined with a protein, will stimulate the generation of an antibody against itself.

radiograph A picture of a gel through which radioactive molecules have migrated. Comparing the distance of travel of unknown RNA with how far RNA whose originating DNA is known migrates indicates whether the RNA derives from the known DNA or some other DNA. Radiographs were used in the experiment to determine whether RNA from the **hybridomas** that was treated to become radioactive derived from, for example, the DNA of 17.2.25 or of something else.

radioimmune assay The assay determines whether an **antibody** is present and in what quantity by measuring the intensity of the radiation given off by a radioactive **reagent** that has bound to it. Imanishi-Kari's lab used the assay extensively, relying, for example, on reagents such as **Bet-1** that had been combined with a radioactive element such as **radioiodine**.

radioiodine Radioactive iodine. It can be combined with a reagent such as **Bet-1** so that the presence of the reagent can be detected in a radiation counter.

reagent A chemical or biological substance such as **Bet-1** used in an experiment.

rearrangement The process in which segments of raw immune DNA combine as a **B cell** matures to produce an antibody. The huge number of possible combinations accounts for the vast number of different antibodies that the immune system generates.

RNA Ribonucleic acid. The molecule is constructed in the cell as a vehicle for the expression of one of its genes. Its constituent parts complement the base-pair sequence in the DNA that is expressed. It thus provides a proxy for the DNA that can be identified in a **radiograph**.

sequence See **DNA**.

serology The study of the properties or contents of organic fluids such as blood sera, which includes immune-response products such as **B cells**, **T cells**, and antibodies.

spleen A glandular organ in the immune system that figures in the production of lymphocytes. The spleen was used as a source of cells for the hybridomas in the experiment.

subclones See **clones**.

supernatant The fluid in a well that contains the antibodies that the B cells in it produce, as well as the nutrients on which the cells grow.

T cells A type of **lymphocyte** that is produced in the thymus and that figures in the immune response, notably in priming and regulating **B-cell** action.

transgene In the general parlance of molecular biology, a gene that is taken from one animal and inserted into the nucleus of the newly fertilized egg of another. The egg is then inserted into the uterus of a surrogate mother, where it develops into an animal that contains the transgene in each of its cells. In the experiment reported in the *Cell* paper, the transgene, 17.2.25, was isolated from an inbred strain of mice called BALB/c and inserted into the newly fertilized eggs of **Black/6** mice so that Black/6 mice with the transgene in all their cells, including those that generate antibodies, were obtained.

transgenic mice Mice that contain a gene—the **transgene**—from another organism.

variable region See **light chain**, **heavy chain**.

wells Hemispheric indentations that are about one-quarter inch in diameter and occur at regular intervals on a plastic plate. Wells were used in the *Cell* paper experiment as containers for hybridomas and to characterize the kind of antibodies they produced.

Glossary of Source Abbreviations

AE Appeal Exhibit

Appeal Decision Department of Health and Human Services, Departmental Appeals Board, Research Integrity Adjudications Panel, Subject: Thereza Imanishi-Kari, Ph.D. Docket No. A-95-3, Decision No. 1582, June 21, 1996.

Appeal Proceedings "Transcript of Proceedings before the United States Department of Health and Human Services," In the Matter of: Thereza Imanishi-Kari, Ph.D., Board Docket No.: A-95-33, Case 072, 1995

Baltimore Files David Baltimore Files

Davis Files Bernard D. Davis Files

Doty *et al.* tape Tape recording of a discussion among Paul Doty, John Edsall, Walter Gilbert, Herman Eisen, Stephen Harrison, and Mark Ptashne, June 4, 1991

Eisen Files Herman Eisen Files

Friedly Files Jock Friedly Files

Hearing, Scientific Fraud and Misconduct and the Federal Response U.S. Congress, House, Subcommittee [on Human Resources and Intergovernmental Relations] of the Committee on Government Operations, *Scientific Fraud and Misconduct and the Federal Response*, 100th Cong., 2nd Sess., April 11, 1988

Hearing, Fraud in N.I.H. Grant Programs U.S. Congress, House, Subcommittee on Oversight and Investigation of the Committee on Energy and Commerce, *Hearing, Fraud in N.I.H. Grant Programs*, 100th Cong., 2nd Sess., April 12, 1988.

Hearings, Scientific Fraud, **May 4 or 9, 1989** U.S. Congress, House, Subcommittee on Oversight and Investigation of the Committee on Energy and Commerce, *Hearings, Scientific Fraud*, 101st Cong., 1st Sess., May 4 and May 9, 1989

Hearings, Scientific Fraud (Part II) U.S. Congress, House, Subcommittee on Oversight and Investigation of the Committee on Energy and Commerce, *Hearings, Scientific Fraud (Part II)*, 101st Cong., 2nd Sess, May 14, 1990

Hearings, Scientific Fraud, **Aug. 1, 1991** U.S. Congress, House, Subcommittee on

Oversight and Investigations of the Committee on Energy and Commerce, *Hearings, Scientific Fraud*, 102nd Cong., 1st Sess., March 6 and August 1, 1991

Huang Files Alice Huang Files

N.I.H. National Institutes of Health

Onek Files N.I.H. and O.S.I. Files on the Imanishi-Kari Case Obtained on Discovery by Joseph Onek

O.R.I. Office of Research Integrity

O.R.I. Files Office of Research Integrity Files, Case 072, Thereza Imanishi-Kari.

O.S.I. Office of Scientific Integrity

Parker and Press Files David Parker and Joan Press Files

Singal Files Bruce Singal's Imanishi-Kari Case File

Singer Files Maxine Singer Files

Stewart and Feder Files Walter Stewart and Ned Feder Files

Endnotes

NOTES FOR CHAPTER ONE

1. Author's interview with Peter Stockton and Bruce Chafin, March 24, 1993.
2. Author's interviews with Margot O'Toole, Nov. 8, 1992; with Shirley Tilghman, March 22, 1996; James O'Toole, *Man Alive: A Comedy in Three Acts* (Dublin: Allen Figgis, 1962), pp. 25, 42, 67, 64, 86–87; Judy Sarasohn, *Science on Trial: The Whistle Blower, the Accused, and the Nobel Laureate* (New York: St. Martin's Press, 1993), p. 2.
3. Author's interview with Margot O'Toole, Oct. 5, 1992; Sarasohn, *Science on Trial*, p. 2.
4. Bernard Davis, notes on a telephone conversation with Thomas Wegmann, May 12, 1993; Wegmann to Davis, May 18, 1993, Bernard Davis Files on the Baltimore case, in the possession of Elizabeth Davis (hereafter, Davis Files).
5. Author's interviews with Henry Wortis, Oct. 2, 1992, and, by telephone, June 12, 1997; N.I.H. Staff, Notes of Interview with Henry Wortis, May 20, 1988, copy in N.I.H. and O.S.I. Files Obtained on Discovery by Joseph Onek (hereafter, Onek Files).
6. Author's interview with Donald Mosier, by telephone, Oct. 8, 1996; Donald E. Mosier to Bernard Davis, July 16, 1991, Davis Files.
7. Author's interviews with Mosier, by telephone, Oct. 8, 1996; with O'Toole, Oct. 5, 1992; author's conversation with Thomas Vogt, Feb. 7, 1996. Mosier says that O'Toole's charge was conveyed to him from several sources, including a *New York Times* reporter who asked him about it.
8. Donald E. Mosier to Bernard Davis, July 16, 1991; Bernard Davis, note of telephone conversation with Melvin Bosma, April 8, 1993, Davis Files; author's interviews with Melvin Bosma, June 22, 1993; with Henry Wortis, Oct. 2, 1992; with Donald Mosier, Oct. 8, 1996; with Vogt, April 13, 1994.

9. Bernard Davis, notes of telephone conversations with Pat Harsche and Jay McKay, March 18, 1993, March 26, 1993; Mosier to Davis, July 16, 1991, Davis Files; author's interview with Shirley Tilghman, March 22, 1996; author's conversation with Thomas Vogt, Feb. 7, 1996; author's interview, by telephone, with Jay McKay and with Patricia Harsche, Oct. 7 and 8, 1996.

10. Bernard Davis, notes of telephone conversations with Pat Harsche and Jay McKay, March 18, 1993, March 26, 1993, Davis Files; author's interviews with McKay, by telephone, Oct. 7 and 8, 1996.

11. Author's interview with Martin Flax, Oct. 7, 1992.

12. *Ibid.*; author's interviews with O'Toole, Oct. 5, 1992, Nov. 8, 1992; Martin Flax, Testimony; Henry Wortis, Testimony, U.S. Congress, House, Subcommittee on Oversight and Investigations of the Committee on Energy and Commerce, *Hearings on Scientific Fraud*, 101st Cong., 1st Sess., May 4 and 9, 1989, pp. 224, 229 (hereafter, *Hearings, Scientific Fraud*, May 4 or 9, 1989).

13. Author's interview with Imanishi-Kari, March 20, 1996.

14. *Ibid.*

15. *Ibid.*; Margaret Pantridge, "In the Eye of the Storm," *Boston Magazine*, July 1991, p. 41; Judy Foreman, "Fraud Charge Leaves a Career in Shambles," *Boston Globe*, May 6, 1991, p. 29; Malcolm Gefter to Thereza Imanishi-Kari, Oct. 8, 1979; Herman N. Eisen to Thereza Imanishi-Kari, Oct. 26, 1979; Herman N. Eisen *et al.* to Gene Brown, Dec. 3, 1979, Herman Eisen Files, in possession of Eisen (hereafter, Eisen Files).

16. Foreman, "Fraud Charge Leaves a Career in Shambles," *Boston Globe*, May 6, 1991, p. 29; Imanishi-Kari interview, March 20, 1996; N.I.H. Staff, Margot O'Toole interview, May 17, 1988, Transcript, Appeal Exhibit H-109, p. 10 (hereafter, AE and exhibit number); Walter Stewart, Notes of Telephone Conversation with Margot O'Toole, Oct. 14, 1986, N.I.H. and O.S.I. Files on the Imanishi-Kari Case Obtained on Discovery by Joseph Onek (hereafter Onek Files).

17. O.S.I., "Meeting of a Special Scientific Panel in re M.I.T. [with Moema Reis]," April 23, 1990, Transcript, AE H-104, pp. 4–5, 7, 8, 11–12, 22–23; Moema Reis, Testimony, "Transcript of Proceedings before the United States Department of Health and Human Services," In the Matter of: Thereza Imanishi-Kari, Ph.D., Board Docket No.: A-95-33, Case 072, (hereafter, *Appeal Proceedings*), June 29, 1995, p. 2503. Albanese recalled regular laboratory meetings and a generally open, data-sharing atmosphere. O.S.I., C. Albanese interview, Feb. 2, 1990, Transcript, AE H-118, pp. 11–13, 20–21.

18. N.I.H. Staff, O'Toole interview, May 17, 1988, Transcript, AE H-109, p. 10; Philip Weiss, "Conduct Becoming?" *New York Times Magazine*, Oct. 21, 1989, p. 68; O'Toole Interview, Oct. 5, 1992; Pantridge, "In the Eye of the Storm," pp. 41, 104–105.

19. Author's interviews with Herman Eisen, Oct. 2, 1992; with Imanishi-Kari, March 20, 1996; Foreman, "Fraud Charge Leaves a Career in Shambles," *Boston Globe*, May 6, 1991, p. 29.

20. David Baltimore to Martin H. Flax, May 23, 1986, AE H-229; author's interview with Imanishi-Kari, March 20, 1996.

21. Baltimore to Flax, May 23, 1986, AE H-229; Henry Wortis to Dean Banks, June 17, 1987, copy in David Baltimore Papers (hereafter, Baltimore Files), item 44.

22. N. K. Jerne, "Towards a Network Theory of the Immune System," *Annals of Immu-*

nology (Inst. Pasteur), 125C(1974), 377–378. On the immune system, see Bruce Alberts *et al.*, *The Molecular Biology of the Cell* (2nd ed.; New York: Garland Publishing, 1989), pp. 1002ff.

23. Author's interview with David Baltimore, March 20, 1996.

24. In specific chemical terms, NP is (4-hydroxy-3-nitrophenyl)acetyl. It belongs to the class of chemicals that the great immunological chemist Karl Landsteiner recognized in the 1920s as having special immunological properties. Landsteiner named the chemicals *haptens*, after the Greek word for "attach," a reference to how they work immunologically. Haptens have proved to be advantageous tools in experimental immunology. Unlike microorganisms, they are chemically well defined, readily synthesized in the laboratory, and easily manipulated to suit particular experimental purposes. Anne Marie Moulin, *Le dernier langage de la médecine: Histoire de l'immunologie de Pasteur au Sida* (Paris: Presses Universitaire de France, 1991), pp. 264–266.

25. During her first year in graduate school in Helsinki, Imanishi-Kari noticed a peculiarity in the immune response that NP provoked in a certain strain of inbred mice. When immunized against NP, the mice produced antibodies that had a much higher affinity for a hapten related to NP than they did for NP itself. Imanishi-Kari recalls that she mentioned her "funny" results to a scientist named Geoffrey Haughton from the University of North Carolina Medical School in Chapel Hill, who was spending a sabbatical year in the lab. He thought that the antibody's strikingly higher affinity for the related hapten rather than for the one that had provoked the immune response was "amazing," Imanishi-Kari says. To Haughton, it appeared that the antibody's variable region was generated from a seemingly fixed ensemble of DNA segments. Such a combination would be passed down from one generation of mice to the next, so Haughton suggested to Imanishi-Kari that she check to determine whether the antibody's funniness was inheritable. Imanishi-Kari bred the mice and tested whether the peculiar immunological response to NP showed up in the offspring. She developed methods of distinguishing one such response from another, compared the antibody responses of hybridomas from related and unrelated mice, and demonstrated that the funniness of parental responses was in many cases transmitted to offspring in accord with Mendel's laws of inheritance. Author's interviews with Thereza Imanishi-Kari, March 20, 1996, Oct. 28, 1996; Thereza Imanishi and O. Mäkelä, "Inheritance of Antibody Specificity: I. Anti-(4-hydroxy-3-nitrophenyl)acetyl of the Mouse Primary Response," *Journal of Experimental Medicine*, 140(1974), 1498–1510.

26. Idiotypes had been recognized in the mid-1950s. The chemical structure of the NP-stimulated idiotype was unknown, but the idiotype was recognizable by its multiple affinities for related haptens as well as by several other features of the antibody. Moulin, *Le dernier langage de la médecine*, pp. 327–333; Mary E. White-Scharf and Thereza Imanishi-Kari, "Characterization of the NP[a] Idiotype Through the Analysis of Monoclonal BALB/c Anti-(4-hydroxy-3-nitrophenyl)acetyl (NP) Antibodies," *European Journal of Immunology*, 11(1981), 903.

27. John E. Hopper and Alfred Nisonoff, "Individual Antigenic Specificity of Immunoglobulins," in F. J. Dixon, Jr., and Henry G. Kunkel, eds., *Advances in Immunology*, (Vol. 13; New York: Academic Press, 1971), pp. 58, 59, 60; Thereza Imanishi and O. Mäkelä, "Strain Differences in the Fine Specificity of Mouse Anti-Hapten Antibodies," *European Journal of Immunology*, 3(1973), 323–330; Klaus Eichmann, "Genetic Control of Antibody Specificity in the Mouse," *Immu-*

nogenetics, 2(1975), 491–506; John Maddox, "Can a Greek Tragedy Be Avoided?" *Nature*, 333(June 30, 1988), 795–796; Leroy Hood *et al.*, *Immunology* (2nd ed.; Ballinger, 1984), pp. 371–372.

28. Baltimore proposed that his laboratory try to isolate the gene. All three scientists thought it would be "fantastic" to have the gene to work with, Imanishi-Kari remembers. Rajewsky had Imanishi-Kari send Baltimore some hybridomas that contained it. Baltimore's group successfully isolated and characterized the gene, publishing a paper on it in 1981 with Imanishi-Kari as a coauthor in recognition of her contribution of the cell line. N.I.H. Staff, David Baltimore interview, May 18, 1988, Transcript, AE H-107, p. 2; author's interview with Imanishi-Kari, March 20, 1996; Alfred M. Bothwell *et al.*, "Heavy Chain Variable Region Contribution to the NP[b] Family of Antibodies: Somatic Mutation Evident in a γ2A Variable Region," *Cell*, 24(June 1981), 625–637.

29. Mary E. White-Scharf and Thereza Imanishi-Kari, "Characterization of the NP[a] Idiotype . . . ," *European Journal of Immunology*, 11(1981), 897–904; Dennis Y. Loh *et al.*, "Molecular Basis of a Mouse Strain-Specific Anti-Hapten Response," *Cell*, 33(May 1993), 85–93.

30. N.I.H. Staff, Baltimore interview, May 18, 1988, Transcript, AE H-107, pp. 1–4. The process of inhibiting rearrangement is called *allelic exclusion*.

31. *Ibid.*; O.S.I., Interview with David Baltimore, April 30, 1990, Transcript, AE H-108, pp. 6–8.

32. Author's interview with Baltimore, March 20, 1996.

33. *Ibid.*

34. O.S.I., Interview of Thereza Imanishi-Kari, October 13, 1990, Transcript, AE H-103, p. 170.

35. Rudolf Grosschedl, David Weaver, David Baltimore, and Frank Constantini, "Introduction of a μ Immunoglobulin Gene into the Mouse Germ Line: Specific Expression in Lymphoid Cells and Synthesis of Functional Antibody," *Cell*, 38(October 1984), 647–658; O.S.I., Interview with David Weaver, June 14, 1990, Transcript, AE H-116, pp. 31–32.

36. Author's interview with David Weaver, Dec. 11, 1992.

37. N.I.H. Staff, Interview with David Weaver, May 18, 1988, Transcript, AE H-115, pp. 3–4.

38. Weaver *et al.*, "Altered Repertoire of Endogenous Immunoglobulin Gene Expression in Transgenic Mice Containing a Rearranged Mu Heavy Chain Gene," *Cell*, 45(April 25, 1986), 247–259; N.I.H. Staff, Interview with David Weaver, May 18, 1988, Transcript, AE H-115, pp. 3–4; Imanishi-Kari, Testimony, *Appeal Proceedings*, Aug. 29, 1995, pp. 4705–4713.

39. O.S.I., Baltimore interview, April 30, 1990, Transcript, AE H-108, pp. 6–7. For an account of much of Imanishi-Kari's part of the experiment, see her letter to Katherine L. Bick, March 28, 1988, AE H-245.

40. For how the nomenclature of immunoglobulin classes came about, see Howard C. Goodman, "Immunodiplomacy: The Story of the World Health Organization's Immunology Research Programme, 1961–1975," in Pauline M. H. Mazumdar, ed., *Immunology, 1930–1980* (Toronto: Wall and Thompson, 1989), pp. 254–256.

41. O.S.I., Interview with Imanishi-Kari, Oct. 13, 1990, Transcript, AE H-103, pp. 177–178. Clues to what would become the central phenomena reported in the *Cell* paper had cropped up in the Baltimore group's previous transgenic experiments. When Grosschedl and Weaver earlier looked at the antibodies in the

bloodstream of their transgenic mice, they found antibodies of the IgG class with a *gamma* isotype. O.S.I., Weaver interview, June 14, 1990, Transcript, AE H-116, pp. 45–47. In a paper that they completed in the spring of 1985 and that was published in August, Baltimore, Weaver, and Imanishi-Kari said that in many cells of the mice the presence of the transgene did not inhibit rearrangement—that, indeed, new research on their hybridomas suggested that many of the NP-sensitive antibodies appeared to have been produced by genes native to the mice. David Weaver, Frank Constantini, Thereza Imanishi-Kari, and David Baltimore, "A Transgenic Immunoglobulin Mu Gene Prevents Rearrangement of Endogenous Genes," *Cell*, 42(Aug. 1985), 124–125. At the beginning of the new experiment, Imanishi-Kari's observations on the transgenic hybridomas revealed that, contrary to expectation, the idiotype on the antibodies from them was not quite the same as the idiotype on the antibodies produced by the transgene, an early indication that something odd was occurring. Author's interview with Imanishi-Kari, Oct. 28, 1996.

42. Author's interview with Imanishi-Kari, Oct. 29, 1996.
43. *Ibid.* Bet-1 was manufactured by injecting rats with an immunoglobulin that was drawn from BALB/c mice, the mice from which the gene used as the transgene was taken. To the rats, the immunoglobulin was just another foreign agent and they protected themselves by producing antibodies against it, including against the *mu-a* region that it contained. Bet-1 was thus an antibody against antibodies generated by the transgene from the BALB/c mice. Imanishi-Kari first got just batches of Bet-1 but then she was given rat-cell cultures in which it was being produced and from which she could harvest it. She says that she had problems getting the cultures to manufacture sufficient quantities of Bet-1. One of the first things Moema Reis did when she arrived in February was to manipulate the cell cultures for Bet-1—and also those for AF6, which was generated in a similar fashion—so that they produced the reagents in greater abundance, several liters at a time. Reis's preparations were used throughout the rest of the experiment. John T. Kung *et al.*, "A Mouse IgM Allotypic Determinant (Igh-6.5) Recognized by a Monoclonal Rat Antibody," *The Journal of Immunology*, 127(Sept. 1981), 873–876; Meeting of N.I.H. Panel and Authors, May 3, 1989, Transcript, AE H-101, p. 49; O.S.I., Meeting of a Special Scientific Panel in re M.I.T. [with Moema Reis], April 23, 1990, Transcript, AE H-104, pp. 14–15; author's interview with Imanishi-Kari, March 20, 1996.
44. O.S.I., Special Scientific Panel, Interview with Charles Maplethorpe, Feb. 11, 1990, Transcript, AE H-114, pp. 20–21; O.S.I., Interview of Thereza Imanishi-Kari, Oct. 13, 1990, Transcript, AE H-103, pp. 14–15; Meeting of N.I.H. Panel and Authors, May 3, 1989, Transcript, AE H-101, pp. 50–51.
45. Author's interview with Imanishi-Kari, Oct. 29, 1996; O.S.I., Meeting of a Special Scientific Panel in re M.I.T. [with Moema Reis], April 23, 1990, Transcript, AE H-104, pp. 9–10.
46. Weaver *et al.*, "Altered Repertoire . . . ," *Cell*, 45(April 25, 1986), 250, Table 2.
47. N.I.H. Staff, Interview with Baltimore, May 18, 1988, p. 3; O.S.I., Baltimore interview, April 30, 1990, Transcript, AE H-108, pp. 7–9; O.S.I., Weaver interview, June 14, 1990, Transcript, AE H-116, pp. 12–14, 28; author's interview with David Weaver, Dec. 11, 1992.
48. O.S.I., Weaver interview, June 14, 1990, AE H-116, pp. 18–19, 29–30; O.S.I., Albanese Interview, Feb. 2, 1990, AE H-118, pp. 172–173; Weaver *et al.*, "Altered Repertoire . . . ," *Cell*, 45(April 25, 1986), 247, 252.

49. O.S.I., Weaver interview, June 14, 1990, AE H-116, pp. 18–19; O.S.I., Baltimore interview, April 30, 1990, AE H-108, pp. 14–15.

50. N.I.H. Staff, Baltimore interview, May 18, 1988, Transcript, AE H-107, pp. 4–5.

51. O.S.I., Baltimore interview, April 30, 1990, AE H-108, pp. 9–12; Weaver *et al.*, "Altered Repertoire . . . ," *Cell,* 45(April 25, 1986), 250, Table 2.

52. David Baltimore, Statement, *Hearings, Scientific Fraud,* May 4, 1989, p. 95; author's interview with Margot O'Toole, Oct. 5, 1992; copy of annotated draft of the *Cell* paper, n.d., AE H-227; N.I.H. Staff, O'Toole interview, May 17, 1988, Transcript, AE H-109, pp. 16–17, 19.

53. N.I.H. Staff, Interview with Baltimore, May 18, 1988, Transcript, AE H-107, p. 14; O.S.I., Baltimore interview, April 30, 1990, AE H-108, pp. 12–13; Weaver *et al.*, "Altered Repertoire . . . ," *Cell,* 45(April 25, 1986), 247–259; referees' reports, attached to Baltimore to Brian Kimes, Sept. 26, 1989, AE H-273. Joseph Davie, a prominent scientist who was later involved in the federal investigations of the *Cell* paper, called it "really quite remarkable" and "very important," "in many ways a seminal change in the way people approached this whole area" of immune response. Davie, Testimony, *Appeal Proceedings,* June 20, 1995, pp. 1269–1270.

54. Idiotypic mimicry featured in a network theory of the immune system that Niels Jerne advanced at a conference on immunology in Paris in 1974 on the occasion of Pasteur's birthday. Jerne was concerned to provide an explanation of the immune system's ability to regulate itself. According to his theory, the formation of antibodies with a given idiotype—call them A—would generate antibodies against themselves—anti-A—which in turn would generate more A antibodies. The A antibodies would promote the body's immune response; the anti-A antibodies would suppress it. The immune system was thus a network in equilibrium that could be perturbed into the promotion of an antibody response by an infectious agent but would then restore itself by generating defenders against the idiotypic antibodies it produced. Jerne's theory was weakened by what Jerne himself conceded was "its lack of precision" and it rapidly fell into disrepute. Jerne, "Towards a Network Theory of the Immune System," *Annals of Immunology (Inst. Pasteur),* 125C(1974), 373–389; Moulin, *Le dernier langage de la médecine,* pp. 335–339; Sarasohn, *Science on Trial,* pp. 20–21.

55. Author's interviews with Imanishi-Kari, Oct. 6, 1992; with David Baltimore, Jan. 23, 1992; Baltimore, Statement, *Hearings, Scientific Fraud,* May 4, 1989, p. 104. For Weaver's contributions to the paper, see N.I.H. Staff, Interview with Weaver, May 18, 1988, AE H-115, pp. 1–2.

NOTES FOR CHAPTER TWO

1. N.I.H. Staff, O'Toole interview, May 17, 1988, Transcript, AE H-109, p. 10; Henry Wortis to Dean Banks, June 17, 1987, copy in U.S. Congress, House, Subcommittee on Oversight and Investigations of the Committee on Energy and Commerce, *Hearing on Fraud in N.I.H. Grant Programs,* 100th Cong., 2nd Sess., April 12, 1988, p. 209 (hereafter, *Hearing, Fraud in N.I.H. Grant Programs,* April 12, 1988); Martina Boersch-Supan, Testimony, *Appeal Proceedings,* June 23, 1995, pp. 1851–1852.

2. Author's interview with Margot O'Toole, Oct. 5, 1992.

3. Author's interview with Imanishi-Kari, Oct. 6, 1992; N.I.H. Staff, O'Toole interview, May 17, 1988, Transcript, AE H-109, p. 24.

4. Author's interview with O'Toole, Nov. 8, 1992; Gary McMillan, "A Third Witness Testifies," *Boston Globe*, July 16, 1985, p. 24; Judy Foreman, "Researcher Blew the Whistle Once Before," *Boston Globe*, May 10, 1989; Philip Hilts, "Hero in Exposing Science Hoax Paid Dearly," *New York Times*, March 21, 1991, pp. 1, 11.

5. *Boston Globe*, May 1, 1985, pp. 25, 31; May 4, 1985, p. 22; July 16, 1985, p. 24.

6. Mary White-Scharf, telephone conversation with an O.S.I. investigator, Oct. 22, 1990, summarized in "Memorandum to the Record," Nov. 7, 1990, Re: Conversations with Dr. Mary White-Scharf, Onek Files; Maggie Hassan to File, Re: Subcommittee Interviews of Dr. Herman Eisen and Dean Gene Brown, May 11, 14, 1988, p. 3, Eisen Files; Sarasohn, *Science on Trial*, p. 91.

7. Author's interviews with Brigitte Huber, Oct. 7, 1992; with Henry Wortis, Oct. 2, 1992, March 20, 1996; with Joan Press, June 16, 1996.

8. Author's interview with O'Toole, Oct. 5, 1992.

9. N.I.H. Staff, O'Toole interview, May 17, 1988, Transcript, AE H-109, p. 10; Walter W. Stewart, Notes of Telephone Conversation with Margot O'Toole, Aug. 25, 1986, Onek File.

10. Author's interviews with O'Toole, Oct. 5, 1992, Nov. 8, 1992; N.I.H. Staff, Interview with Thereza Imanishi-Kari, May 19, 1988, AE H-100, p. 4.

11. O.S.I., O'Toole interview, Feb. 11, 1990, AE H-110, pp. 71–72; Walter W. Stewart, Notes of Telephone Conversation with Margot O'Toole, Aug. 25, 1986, Onek Files.

12. O.S.I., Interview with Charles Maplethorpe, Dec. 27, 1989, Transcript, AE H-113, pp. 23–24; N.I.H. Staff, O'Toole interview, May 17, 1988, Transcript, AE H-109, pp. 11–12.

13. Walter W. Stewart, Notes of Telephone Conversation with Margot O'Toole, Aug. 25, 1986, Onek Files; N.I.H. Staff, O'Toole interview, May 17, 1988, Transcript, AE H-109, pp. 12–14; author's interview with Imanishi-Kari, Oct. 29, 1996.

14. Statement by Thereza Imanishi-Kari, *Hearings, Scientific Fraud*, May 4, 1989, pp. 141–142; author's interview with Imanishi-Kari, Oct. 6, 1992.

15. N.I.H. Staff, O'Toole interview, May 17, 1988, Transcript, AE H-109, p. 16. Weaver, who was also at the meeting, recalled that O'Toole was "really into this, and this was going to be her thing and it was going to be exciting." Author's interview with David Weaver, Dec. 11, 1992.

16. Author's interview with Margot O'Toole, Oct. 5, 1992; Statement by Thereza Imanishi-Kari; Henry Wortis, Testimony, *Hearings, Scientific Fraud*, May 4, 1989, pp. 136, 241; copy of O'Toole's *curriculum vitae* in Baltimore Files.

17. Thereza Imanishi-Kari, Testimony, *Hearings, Scientific Fraud*, May 4 1989, pp. 137–138; author's interview with O'Toole, Oct. 5, 1992. The immunologist Leonore Herzenberg of Stanford University says that her group eventually showed that bone-marrow transfers give only conventional B cells. Since conventional B cells only express the transgene, O'Toole's experiment was doomed to failure, but no one could have known that fact at the time. Author's interview with Leonore and Leonard Herzenberg, July 2, 1996.

18. Author's interview with O'Toole, Oct. 5, 1992.

19. Weaver *et al.*, "Altered Repertoire . . . ," *Cell*, 45 (April 25, 1986), 249.

20. *Ibid.*; author's interview with O'Toole, Oct. 5, 1992. O'Toole told N.I.H. staff investigators that Bet-1 "never, ever, ever worked for me . . . as a specific *mu-a* reagent." N.I.H. Staff, O'Toole interview, May 17, 1988, Transcript, AE H-109, p. 23. Reis later testified that what was contaminated was the plate on which Bet-1 was grown. Reis, Testimony, *Appeal Proceedings*, June 29, 1995, pp. 2510–2511.

21. Author's interview with O'Toole, Oct. 5, 1992; N.I.H. Staff, O'Toole interview, May 17, 1988, Transcript, AE H-109, p. 24.

22. N.I.H. Staff, O'Toole interview, May 17, 1988, Transcript, AE H-109, p. 7; O.S.I., Meeting of a Special Scientific Panel in re M.I.T. [with Moema Reis], Transcript, April 23, 1990, AE H-104, pp. 14–15, 19–20. O'Toole recalls having said to her, " 'Moema, I've told you for months that, in my hands the *mu-b* standard [i.e., the standard for the normal B cell antibody] reacts with Bet-1.' She said, 'I know you did, Margot, but I had an experiment where it didn't.' " Author's interview with O'Toole, Oct. 5, 1992.

23. Margot O'Toole, Testimony, *Hearings, Scientific Fraud*, May 9, 1989, p. 206; see also Margot O'Toole, Testimony, *Hearing, Fraud in N.I.H. Grant Programs*, April 12, 1988, pp. 86–87.

24. Author's interview with O'Toole, Oct. 5, 1992.

25. N.I.H. Staff, O'Toole interview, May 17, 1988, Transcript, AE H-109, p. 36; author's interview with O'Toole, Oct. 5, 1992.

26. Author's interviews with Henry Wortis, Oct. 2, 1992; with Melvin Bosma, June 22, 1993; N.I.H. Staff, Notes of Interview with Brigitte Huber, May 20, 1988, Onek Files.

27. Thereza Imanishi-Kari, Statement and Testimony, *Hearings, Scientific Fraud*, May 4, 1989, pp. 132, 135; author's interviews with Imanishi-Kari, Oct. 6, 1992; with Wortis, Oct. 2, 1992. O'Toole says that she was upset that she was not invited to a discussion of Rabin's work. N.I.H. Staff, O'Toole interview, May 17, 1988, Transcript, AE H-109, pp. 17–18.

28. Author's interviews with Imanishi-Kari, March 20, 1996, Oct. 29, 1996. An assay of Bet-1 that O'Toole performed on Jan. 2, 1986, showed that the reagent distinguished between *mu-a* and *mu-b* with a thousand-fold specificity. James B. Wyngarden to David Baltimore, Jan. 31, 1989, and attached [Report of the Scientific Panel], p. 3, copy in Baltimore Files, item 130.

29. Author's interviews with Imanishi-Kari, Oct. 6, 1992, March 20, 1996; Imanishi-Kari, Testimony, *Hearings, Scientific Fraud*, May 4, 1989, pp. 137–138; N.I.H. Staff, O'Toole interview, May 17, 1988, Transcript, AE H-109, p. 26.

30. Imanishi-Kari, Testimony, *Hearings, Scientific Fraud*, May 4, 1989, p. 132; author's interview with Imanishi-Kari, March 20, 1996; O'Toole, Testimony, *Appeal Proceedings*, June 16, 1995, p. 885. O'Toole says that Imanishi-Kari had promised to fund her at Tufts until she could obtain her own grant money and had even signed a remortgage application for O'Toole attesting that she had a job at Tufts. Author's interview with O'Toole, Oct. 5, 1992.

31. Author's interview with Imanishi-Kari, March 20, 1996.

32. Author's interview with O'Toole, Nov. 8, 1992; Sarasohn, *Science on Trial*, p. 3.

33. Author's interview with Imanishi-Kari, Oct. 6, 1992.

34. Author's interview with Vivian Igras, Dec. 11, 1992.

35. Author's interviews with Imanishi-Kari, Oct. 6, 1992, March 20, 1996; with Igras, Dec. 11, 1992. Henry Wortis testified to a congressional hearing that Imanishi-Kari expressed "concern over O'Toole's willingness to put in the necessary and demanding time." At the same hearing, Imanishi-Kari said that O'Toole "indicated to me that she was not at all sure that she didn't want to spend more time with her family, and was not sure that she wanted to continue her career in science." Wortis, Testimony; Imanishi-Kari, Testimony, *Hearings, Scientific Fraud*, May 4, 1989, pp. 230, 137–138.

36. Author's interview with O'Toole, Nov. 8, 1992; Sarasohn, *Science on Trial*, p. 3.

37. Author's interview with O'Toole, Oct. 5, 1992; Imanishi-Kari, Testimony, *Hearings, Scientific Fraud*, May 4, 1989, pp. 132–133.

38. *Ibid.*, pp. 137–138; O'Toole, Testimony, *Hearing, Fraud in N.I.H. Grant Programs*, April 12, 1988, pp. 101–102; author's interview with O'Toole, Oct. 6, 1992; N.I.H. Staff, O'Toole interview, May 17, 1988, Transcript, AE H-109, pp. 21–22.

39. Author's interview with Imanishi-Kari, Oct. 29, 1996.

40. Imanishi-Kari, Testimony, *Hearings, Scientific Fraud*, May 4, 1989, pp. 137–138. Imanishi-Kari says that she had neither the time nor the money to follow up the cell-transfer experiments in 1986 but she added in 1996 that she still hopes to pursue them. Author's interview with Imanishi-Kari, Oct. 6, 1992.

41. Author's interview with O'Toole, Oct. 5, 1992.

42. *Ibid.*

43. N.I.H. Staff, O'Toole interview, May 17, 1988, Transcript, AE H-109, p. 32. A few weeks earlier, in the public forum of a congressional committee, O'Toole testified: "By manipulating the data, she forced the results to conform to her hypothesis. She pointed out that, if my data were manipulated in the same way, they also would support her hypothesis. She said that I was too 'nit-picky' and that 'better reagents' would confirm her results." O'Toole, Chronology of Events, *Hearing, Fraud in N.I.H. Grant Programs*, April 12, 1988, pp. 92–93. For still another version, see O'Toole, Testimony, *Appeal Proceedings*, June 16, 1995, pp. 883–884.

44. Author's interview with Imanishi-Kari, Oct. 6, 1992.

45. O'Toole, Chronology of Events, *Hearing, Fraud in N.I.H. Grant Programs*, April 12, 1988, p. 92; author's interview with O'Toole, Oct. 5, 1992.

46. Author's interviews with Martin Flax, Oct. 7, 1992; with O'Toole, Oct. 5, 1992.

47. Author's interview with O'Toole, Oct. 5, 1992; N.I.H. Staff, O'Toole interview, May 17, 1988, Transcript, AE H-109, pp. 29–30; O'Toole, Testimony, *Hearing, Fraud in N.I.H. Grant Programs*, April 12, 1988, p. 87.

48. N.I.H. Staff, O'Toole interview, May 17, 1988, Transcript, AE H-109, pp. 6–7; author's interview with Charles Maplethorpe, March 23, 1993.

49. Author's interview with Charles Maplethorpe, March 23, 1993; author's interviews, by telephone, with Michael Bevan, Oct. 14, 1996; with Malcolm Gefter, Oct. 16, 1996.

50. N.I.H. Staff, Interview with Charles Maplethorpe, May 20, 1988, Notes, AE H-112, p. 7; O.S.I., Charles Maplethorpe interview, Dec. 27, 1989, Transcript, AE H-113, pp. 6–7, 9; author's interviews with Maplethorpe, March 23, 1993; with O'Toole, Oct. 5, 1992; with Maurice Fox, Oct. 2, 1992; with Bevan, by telephone, Oct. 14, 1996, Maplethorpe, Testimony, *Appeal Proceedings*, Sept. 13, 1995, p. 5742.

51. N.I.H. Staff, Maplethorpe interview, May 20, 1988, Notes, AE H-112, p. 7; O.S.I., Maplethorpe interview, Dec. 27, 1989, AE H-113, pp. 6–7, 9; author's interviews with Maplethorpe, March 23, 1993; with Eisen, Oct. 2, 1992; Bernard Davis notes of telephone conversations with Nancy Hopkins, April 5 and 8, 1991, Davis Files.

52. Pantridge, "In the Eye of the Storm," *Boston Magazine*, July 1991, pp. 41, 104; O.S.I., Maplethorpe interview, Dec. 27, 1989, Transcript, AE H-113, pp. 10–11, 18; author's interview with Maplethorpe, March 23, 1993; Foreman, "Fraud Charge Leaves a Career in Shambles," *Boston Globe*, May 6, 1991, p. 29; Eisen remarks on Maplethorpe, Maggie Hassan to File, Re: Subcommittee Interviews of Dr. Herman Eisen and Dean Gene Brown, May 11, 14, 1988, p. 4, Eisen Files.

53. Imanishi-Kari, Statement, *Hearings, Scientific Fraud*, May 4, 1989, pp. 139–140;

Pantridge, "In the Eye of the Storm," *Boston Magazine,* July 1991, p. 104; author's interviews with Eisen, Oct. 2, 1992; with Imanishi-Kari, March 20, 1996.

54. Author's interview with Imanishi-Kari, March 20, 1996; N.I.H. Staff, Notes of Interview with Brigitte Huber, May 20, 1988, Onek Files; Eisen remarks on Maplethorpe, Maggie Hassan to File, Re: Subcommittee Interviews of Dr. Herman Eisen and Dean Gene Brown, May 11, 1988, May 14, 1988, p. 4, Eisen Files.

55. Hopkins says that Imanishi-Kari once rushed into her office, shutting the door behind her and saying "that Charlie had . . . raised his hand/arm as if to strike her and she had come to my office to hide from him . . . until he was gone." Hopkins adds that she wanted to call the campus police but that Imanishi-Kari wouldn't let her. Imanishi-Kari says that she has no recollection of the incident. She does remember telling Hopkins that something was wrong with Maplethorpe but that she didn't know what to do about it. Davis notes of telephone conversations with Nancy Hopkins, April 5 and 8, 1991, Davis Files; Hopkins to the author, Oct. 15, 1996; author's interview with Imanishi-Kari, Oct. 29, 1996; Imanishi-Kari, Testimony, *Appeal Proceedings,* Aug. 31, 1995, pp. 5062–5069. Maplethorpe's apparent hatred for Imanishi-Kari was noted by Eisen, Igras, and even O'Toole. Author's interviews with Eisen, Oct. 2, 1992; Igras, Dec. 11, 1992; O'Toole, Oct. 5, 1992; Maplethorpe, March 23, 1993.

56. O.S.I., Interview with Chris Albanese, Feb. 2, 1990, Transcript, AE H-118, pp. 24–26.

57. [Charles Maplethorpe], "Notes of My Meeting with Thereza I-K. and Martina B.-S., Nov. 15, 1984," Onek Files.

58. Author's interview with Imanishi-Kari, Oct. 29, 1996; Martina Boersch-Supan, Testimony, *Appeal Proceedings,* June 23, 1995, pp. 1841–1842. The text of the article credits the sequence data in the standard way: "(Maplethorpe and Imanishi-Kari, unpublished results)". Martina Boersch-Supan *et al.,* "Heavy Chain Variable Region: Multiple Gene Segments Encode Anti-4-(hydroxy-3-nitrophenyl) acetyl Idiotypic Antibodies," *Journal of Experimental Medicine,* 161 (June 1985), 1272–1292. See pp. 1280 and 1289 for the credits to Maplethorpe. Luria is dead. Brown does not recall having had a conversation with Maplethorpe over this issue, though he says he could have had one, since he would see students and postdocs on issues like this all the time. Normally he would say, See if you can work it out with the professor, and he would then call the professor to encourage working it out. Maurice Fox, of the M.I.T. biology department, recalls that Imanishi-Kari had to be pushed into crediting Maplethorpe for his sequence. Author's interview with Fox, Oct. 2, 1992; author's telephone conversation with Gene Brown, Oct. 16, 1996.

59. Maplethorpe, Testimony, *Hearing, Fraud in N.I.H. Grant Programs,* April 12, 1988, pp. 106–107; author's interview with Maplethorpe, March 23, 1993.

60. O.S.I., Albanese interview, Feb. 2, 1990, Transcript, AE H-118, pp. 22–24.

61. O.S.I., Maplethorpe interview, March 27, 1989, Transcript, AE H-113, pp. 63–66.

62. *Ibid.,* pp. 63–66; Maplethorpe, Testimony, *Hearing, Fraud in N.I.H. Grant Programs,* April 12, 1988, p. 107.

63. Author's interviews with Imanishi-Kari, Oct. 29, 1996, with Nancy Hopkins, by telephone, Oct. 15, 1996; O.S.I., David Weaver interview, June 14, 1990, Transcript, AE H-116, pp. 51–53; "Partial Transcript of 'Haseltine/Idiots,' prepared by C. Maplethorpe on 5.10.88," copy in Onek Files. O'Toole later recalled that Imanishi-Kari told her that some of the Bet-1 preparations were contaminated and that Reis was purifying them. O'Toole interview, Oct. 5, 1992.

64. O.S.I., Maplethorpe interview, March 27, 1989, Transcript, AE H-113, pp. 85–86.

65. O.S.I., Special Scientific Panel, Charles Maplethorpe interview, Feb. 11, 1990, Transcript, AE H-114, pp. 26, 56, 58. Maplethorpe later told a reporter for the *New York Times* that he discovered some data in a notebook that contradicted the results in the *Cell* paper. Philip M. Boffey, "Nobel Winner Is Caught Up in a Dispute Over Study," *New York Times*, April 12, 1988, p. C10.

66. Author's interview with Igras, Dec. 11, 1992; Igras, Testimony; Maplethorpe, Testimony, *Appeal Proceedings*, Sept. 14, 1995, pp. 6148, 6167–6168.

67. The other professor was Frank Solomon. Bernard Davis notes of telephone conversations with Nancy Hopkins, April 5 and 8, 1991, Davis Files; Maplethorpe, Testimony, House, *Hearing, Fraud in N.I.H. Grant Programs*, April 12, 1988, pp. 107–108; author's interview with Maplethorpe, March 23, 1993.

68. Bernard Davis, notes of telephone conversations with Nancy Hopkins, April 5 and 8, 1991, Davis Files; author's interviews with Imanishi-Kari, March 23, 1993; with Maplethorpe, March 23, 1993.

69. Author's interview with Imanishi-Kari, Oct. 29, 1996.

70. Author's interviews with Maplethorpe, March 23, 1993; with Imanishi-Kari, Oct. 6, 1992; Bernard Davis notes of telephone conversations with Nancy Hopkins, April 5 and 8, 1991, Davis Files; Imanishi-Kari, Statement, House, *Hearings, Scientific Fraud*, May 4, 1989, pp. 139–140.

71. Author's interview with Maplethorpe, March 23, 1993; Pantridge, "In the Eye of the Storm," p. 105.

72. Maplethorpe told an early federal investigation that he spoke with O'Toole about January or February 1986, but he corrected himself in a later investigation, saying that he had talked with her probably in April. N.I.H. Staff, Maplethorpe interview, May 20, 1988, AE H-112, pp. 2–4; O.S.I., Maplethorpe interview, Dec. 27, 1989, Transcript, AE H-113, pp. 61–62; author's interviews with Maplethorpe, March 23, 1993; with O'Toole, Nov. 8, 1992; Maplethorpe, Testimony, *Hearings, Fraud in N.I.H. Grant Programs*, April 12, 1988, p. 108.

73. N.I.H. Staff, Weaver interview, May 18, 1988, Transcript, AE H-115, pp. 6–7; O.S.I., Weaver interview, June 14, 1990, AE H-116, pp. 51, 59–60; author's interview with Weaver, Dec. 11, 1992.

74. Wortis said that he had taken the initiative of suggesting O'Toole to Imanishi-Kari in the first place because O'Toole "wanted to leave Leskowitz, because she was unhappy with the quality of the work being done in the lab." Author's interviews with Wortis, Oct. 2, 1992, Dec. 11, 1992; N.I.H. Staff, O'Toole interview, May 17, 1988, p. 31; author's interviews with Brigitte Huber, Oct. 6, 1992; with O'Toole, Oct. 5, 1992; N.I.H. Staff, Notes of Interview with Wortis, May 20, 1988, Onek Files.

75. Margot O'Toole, [Statement: Rebuttal to Baltimore], *Nature*, 351 (May 16, 1991), 183; O'Toole, Testimony, *Appeal Proceedings*, June 16, 1995, p. 943; O.S.I., Interview with Herman Eisen, May 30, 1991, AE H-117, p. 9; author's interviews with Herman Eisen, Oct. 2, 1992; with O'Toole, Oct. 5, 1992.

76. O'Toole, Testimony, *Hearing, Fraud in N.I.H. Grant Programs*, April 12, 1988, p. 87.

77. Author's interview with O'Toole, Oct. 5, 1992.

78. O'Toole, Chronology of Events, *Hearings, Fraud in N.I.H. Grant Programs*, April 12, 1988, p. 93; author's interviews with O'Toole, Oct. 5, 1992; by telephone, April 8, 1996.

NOTES FOR CHAPTER THREE

1. Author's interview with O'Toole, Oct. 5, 1992; *New York Times,* April 12, 1988, p. C10; Philip Weiss, "Conduct Unbecoming?" *New York Times Magazine,* Oct. 21, 1989, p. 68; N.I.H. Staff, O'Toole interview, May 17, 1988, Transcript, AE H-109, p. 42; Sarasohn, *Science on Trial,* p. 7.

2. Sarasohn, *Science on Trial,* p. 6; Weaver *et al.,* "Altered Repertoire . . . ," *Cell,* 45(April 25, 1986), 250, 252; the seventeen pages comprise pages 18 to 34 of Moema Reis's notebook, which as copied by O'Toole are AE H-15A.

3. N.I.H. Staff, O'Toole interview, May 17, 1988, Transcript, AE H-109, pp. 48–49, 51.

4. Author's interview with O'Toole, Oct. 5, 1992; O'Toole, Testimony and Statement, *Hearing, Fraud in N.I.H. Grant Programs,* April 12, 1988, pp. 87–88, 93–94; O'Toole, Testimony, *Hearings, Scientific Fraud,* May 9, 1989, pp. 1818–1882.

5. Author's interview with O'Toole, Oct. 5, 1992; N.I.H. Staff, O'Toole interview, May 17, 1988, Transcript, AE H-109, p. 62; O'Toole, Testimony, *Appeal Proceedings,* June 16, 1995, pp. 941, 947. In 1990, O'Toole wrote to a federal investigator that Imanishi-Kari told her that "she believed the notebook had been stolen," which departs from all other evidence, including O'Toole's. O'Toole to Suzanne Hadley, Jan. 10, 1990, in AE H-110, pp. 17–18.

6. Author's interviews with O'Toole, Nov. 8, 1992; with Charles Maplethorpe, Washington, D.C., March 23, 1993.

7. Author's interviews with Brigitte Huber, Oct. 6, 1992; with O'Toole, Oct. 5, 1992; with Joan Press, June 16, 1996. O'Toole, Chronology of Events, *Hearing, Fraud in N.I.H. Grant Programs,* April 12, 1988, pp. 94–95; Huber, Testimony, *Appeal Proceedings,* June 23, 1995, pp. 1801–1802. Wortis had known Imanishi-Kari for several years in their common scientific context, had collaborated with her in some research that had not been published, and had been involved in recruiting her to Tufts. N.I.H. Staff, Notes of Interview with Wortis, May 20, 1988, Onek Files.

8. Author's interviews with Huber, Oct. 6, 1992; with O'Toole, Oct. 5, 1992; with Henry Wortis, Dec. 11, 1992; O'Toole, Chronology of Events; Henry Wortis to Dean Banks, June 17, 1987, *Hearing, Fraud in N.I.H. Grant Programs,* April 12, 1988, pp. 94–95, 209–210; Walter W. Stewart, Notes of Telephone Conversation with Margot O'Toole, April 7, 1987; N.I.H. Staff, Notes of Interview with Brigitte Huber, May 20, 1988, Onek Files; O'Toole, Testimony; Wortis, Testimony, *Appeal Proceedings,* June 16, 1995, pp. 953–955; June 30, 1995, p. 2798.

9. O'Toole later told federal investigators that she talked with Flax at the urging of a friend from graduate school who encountered her after the meeting with Wortis, saw that she was upset, and learned from her why. She told me only that she went to Flax immediately after talking with Wortis. N.I.H. Staff, O'Toole interview, May 17, 1988, Transcript, AE H-109, pp. 60–66; author's interview with O'Toole, Oct. 5. 1992; N.I.H. Staff, Notes of Interview with Henry Wortis, May 20, 1988, copy in Onek Files.

10. Author's interviews with O'Toole, Oct. 5, 1992; with Martin Flax, Oct. 7, 1992; N.I.H. Staff, O'Toole interview, May 17, 1988, Transcript, AE H-109, pp. 60–66; N.I.H. Staff, Maplethorpe interview, May 20, 1988, Notes, AE H-112; O.S.I., Maplethorpe interview, Dec. 27, 1989, AE H-113, pp. 92–93; Henry Wortis to Dean Banks, June 17, 1987, *Hearing, Fraud in N.I.H. Grant Program,* April 12,

1988, pp. 209–210; Martin Flax, "Testimony," House, *Hearings, Scientific Fraud,* May 4, 1989, p. 225

11. Author's interview with O'Toole, Oct. 5, 1992; N.I.H. Staff, O'Toole interview, May 17, 1988, Transcript, AE-H109, pp. 60–66.

12. Author's interviews with Wortis, Oct. 2, 1992, Dec. 11, 1992; with Martin Flax, Oct. 7, 1992; with Leonard and Lenore Herzenberg, July 2, 1996.

13. Wortis, Testimony; O'Toole, Testimony, *Hearings, Scientific Fraud,* May 9, 1989, pp. 182, 239–240; O'Toole, Chronology of Events, *Hearing, Fraud in N.I.H. Grant Programs,* April 12, 1988, pp. 94–95; Huber, Testimony, *Appeal Proceedings,* June 23, 1995, p. 1779. Notes by federal investigators of an interview with Huber in 1988 have her saying that when O'Toole first approached her, she suggested fraud, but that piece of evidence is contradicted by all others. N.I.H. Staff, Notes of Interview with Brigitte Huber, May 20, 1988, Onek Files.

14. Author's interviews with Brigitte Huber, Oct. 6, 1992; with Wortis, March 20, 1996; with O'Toole, Oct. 5, 1992; Henry Wortis to Dean Banks, June 17, 1987; O'Toole, Chronology of Events, *Hearing, Fraud in N.I.H. Grant Programs,* April 12, 1988, pp. 94–95, 209–210; Walter Stewart, Notes of Telephone Conversation with Margot O'Toole, Sept. 8, 1996, Onek Files; N.I.H. Staff, O'Toole interview, May 17, 1988, AE H-109, p. 56; O'Toole, Testimony; *Appeal Proceedings,* June 16, 1991, pp. 962–963.

15. Shortly after the 1988 hearing, O'Toole testified to N.I.H. staff that Imanishi-Kari "threatened me with a suit." In a congressional hearing in 1989, she truncated her report of the conversation and gave the impression that she had been fired unconditionally, testifying, "She theatened legal action against me and ended the conversation with saying I was not to return to the laboratory." In 1995, O'Toole provided another slant on the matter, testifying that when Imanishi-Kari told her to return to the lab, she responded, "I said I would come into work, but I wanted it agreed between us in front of a witness, that she wouldn't continue to pressure to drop my pushing for the data review. And I said I wouldn't agree to that. I wanted an agreement that I would work without her demanding that I call off the data review." Whatever transpired between Imanishi-Kari and O'Toole, O'Toole's postdoctoral appointment in Imanishi-Kari's lab of course had only two weeks more to run, since it was scheduled to terminate on May 31, 1986. O'Toole, Chronology of Events, *Hearing, Fraud in N.I.H. Grant Programs,* April 12, 1988, pp. 94–96; O'Toole, Testimony, *Hearings, Scientific Fraud,* May 9, 1989, p. 182; Imanishi-Kari interview, Oct. 6, 1992; Margot O'Toole, Testimony, June 16, 1995, *Appeal Proceedings,* p. 965; author's interviews with Wortis, Dec. 11, 1992, March 20, 1996, and telephone conversation with Wortis, Dec. 18, 1992; N.I.H. Staff, O'Toole interview, May 17, 1988, Transcript, AE H-109, pp. 65, 79; Thereza Imanishi-Kari, "Transcript of Press Conference, May 10, 1990," AE H-102, p. 39; N.I.H. Staff, Notes of Interview with Henry Wortis, May 20, 1988, copy in Onek Files. Lawyers say that, according to the doctrines of common law, one is not vulnerable to suit for taking an action—say, blowing a whistle on suspected scientific fraud—that falls within the category of performance of duties.

16. Author's interview with Robert Woodland, by telephone, Dec. 3, 1992; Wortis, Testimony, *Appeals Proceedings,* June 30, 1995, p. 2796. O'Toole later testified to federal investigators that Wortis "went to her [Imanishi-Kari] with the 17 pages" and that she knew that O'Toole had them, but all other testimony contradicts O'Toole's statement. O.S.I., O'Toole interview, Feb. 11, 1990, Transcript, AE H-110, p. 167.

17. Woodland to Wortis, June 10, 1987, Onek Files. Imanishi-Kari later attested to having shown the Wortis committee all of her own relevant data plus the data in the notebooks of her coworkers, including Reis, Albanese, and O'Toole. "Response to Questions for Dr. Thereza Imanishi-Kari," July 17, 1990, attached to Bruce A. Singal to Suzanne Hadley, July 16, 1990, AE H-286, p. 27.

18. Author's interview with Brigitte Huber, Oct. 6, 1992. On the discovery of the mistyping, see O.S.I., Moema Reis interview, May 4, 1990, Transcript, AE H-105, pp. 19–20.

19. Wortis, Testimony, *Hearings, Scientific Fraud*, May 4, 1989, pp. 230–231; Brigitte Huber, Robert Woodland, and Henry Wortis, ["Reply to Margot O'Toole"], *Nature*, 351 (June 13, 1991), 514; author's interviews with Wortis, Oct. 2, 1992, Dec. 11, 1992; with Huber, Oct. 6, 1992; with Woodland, by telephone, Dec. 3, 1992; with Imanishi-Kari, Oct. 6, 1992; Brigitte Huber, Robert Woodland, Henry Wortis, "Minutes of Ad-Hoc Committee Meeting," June 4, 1986 [prepared June 1987], AE H-230.

20. Brigitte Huber, Testimony, *Hearings, Scientific Fraud*, May 9, 1989, pp. 255–256; author's interviews with Wortis, Dec. 11, 1982; with Woodland, Dec. 3, 1992.

21. Author's interviews with Woodland, by telephone, Dec. 3, 1992; with Wortis, Oct. 2, 1992; with Imanishi-Kari, Oct. 6, 1992; O.S.I., Imanishi-Kari interview, Oct. 13, 1990, Transcript, AE H-103, pp. 41–44, 68; Wortis, Testimony, *Hearings, Scientific Fraud*, May 9, 1989, pp. 255–256. Imanishi-Kari discussed the subcloning in a detailed letter about the experiment to the N.I.H.: Thereza Imanishi-Kari to Katherine L. Bick, March 25, 1988, AE H-245.

22. N.I.H. Staff, Interview with Brigitte Huber, May 20, 1988, Notes, copy in Onek Files; author's interviews with Huber, Oct. 6, 1992; with Wortis, Oct. 2, 1992, Dec. 11, 1992; with Woodland, by telephone, Dec. 3, 1992.

23. Weaver *et al.*, "Altered Repertoire . . . ," *Cell*, 45 (April 25, 1986), 252; author's interviews with Imanishi-Kari, Oct. 6, 1992; with Henry Wortis, March 20, 1996; with Brigitte Humber, Oct. 6, 1992. Albanese later testified that he did indeed have the radiographs and that Imanishi-Kari called, asking him to return them. Albanese, Testimony, *Appeal Proceedings*, June 29, 1995, pp. 2562–2563.

24. O'Toole interview, Oct. 5, 1992; N.I.H. Staff, O'Toole interview, May 17, 1988, Transcript, AE H-109, p. 65; author's interview with Huber, Oct. 6, 1992; O'Toole, Chronology of Events, *Hearing, Fraud in N.I.H. Grant Programs*, April 12, 1988, p. 96; O'Toole, Testimony, House, *Hearings, Scientific Fraud*, May 9, 1989, pp. 182–183. In her "Chronology," O'Toole said she told Huber on the telephone that the Wortis committee "should disqualify themselves." Huber, laughing, told me "that's patently wrong," explaining, "Why would we have gone on then? If we were unqualified, fine. She could find somebody else." Huber interview, Oct. 6, 1992. O'Toole later testified that Huber told her that the Wortis committee had not looked at, for example, idiotype data for the Table 3 hybridomas because Imanishi-Kari said that Albanese had it. However, the only data that Imanishi-Kari thought was in Albanese's possession was the radiographic data. O'Toole continued that it was she who insisted on a second meeting at which she would be present because she had the impression that the Wortis committee had been "snowed." Imanishi-Kari later testified that she had shown the Wortis committee idiotype data for Table 3 at the meeting on May 16. Margot O'Toole, Testimony,; Imanishi-Kari, Testimony, *Appeal Proceedings*, June 16, 1995, pp. 966–999; Sept. 15, 1995, pp. 6474–6476.

25. Copy of Wortis's note—"Margot 5-22-86 phone"—original in possession of Henry

Wortis; author's interviews with Wortis, Oct. 2, 1992, Dec. 11, 1992. In a telephone conversation on May 11, 1993, Wortis told me that he could not remember the nature of the moral dilemma. O'Toole says that she raised all of the issues eventually covered in the Tufts investigation when she first went to see Wortis and that at the meeting on May 16 it was agreed that there would be a second meeting because Imanishi-Kari said that she did not have all the data conveniently available and would have to gather it together. She further says that, when told this by Wortis, she insisted on being present at the second meeting. However, Huber's recollections differ from O'Toole's, and Wortis's contemporary note of his telephone conversation with her indicates that O'Toole's recollection is mistaken. Author's interview with O'Toole, Oct. 5, 1992.

26. Author's interview with Henry Wortis, Oct. 2, 1992; N.I.H. Staff, O'Toole interview, May 17, 1988, p. 63; O'Toole, Chronology of Events, *Hearing, Fraud in N.I.H. Grant Programs*, April 12, 1988, pp. 96–97; author's interviews with Martin Flax, Oct. 7, 1992; with Imanishi-Kari, Oct. 6, 1992; with Robert Woodland, Dec. 3, 1992.

27. Wortis, Testimony, *Hearings, Scientific Fraud*, May 4, 1989, pp. 230–231; Brigitte Huber, Robert Woodland, and Henry Wortis, ["Reply to Margot O'Toole"], *Nature*, 351 (June 13, 1991), 514; author's interviews with Wortis, Oct. 2, 1992, Dec. 11, 1992; with Huber, Oct. 6, 1992; and with Imanishi-Kari, Oct. 6, 1992; Imanishi-Kari, Testimony, *Appeal Proceedings*, Aug. 30, 1995, pp. 4967–4970.

28. Author's interviews with Wortis, Oct. 2, 1992, Dec. 11, 1992; with Huber, Oct. 6, 1992; with Imanishi-Kari, Oct. 6, 1992; Imanishi-Kari, Testimony, *Appeal Proceedings*, Aug. 30, 1995, pp. 4967–4970.

29. Imanishi-Kari, Testimony, *Appeal Proceedings*, Aug. 30, 1995, pp. 4967–4970, 4975; Sept. 15, 1995, pp. 6476–6483; N.I.H. Staff, O'Toole interview, May 17, 1988, Transcript, AE H-109, pp. 48–49, 51–54; author's interviews with O'Toole, Oct. 5, 1992; with Imanishi-Kari, Oct. 6, 1992, March 20, 1996.

30. Author's interviews with O'Toole, Oct. 5, 1992; with Imanishi-Kari, Oct. 6, 1992, March 20, 1996; with Wortis, March 20, 1996; N.I.H. Staff, O'Toole interview, May 17, 1988, Transcript, AE H-109, pp. 48–49, 51–54; O'Toole, Testimony, *Hearings, Scientific Fraud*, May 9, 1989, pp. 182–183. During a later investigation, O'Toole would say that she saw a good many more sheets of Imanishi-Kari's data on May 23, 1986, but when pressed on the point in 1995, she seemed to adhere to her earlier claim that she had seen only two sheets relevant to the idiotyping of the Table 3 hybridomas. See O'Toole, Testimony, *Appeal Proceedings*, June 16, 1995, pp. 1060–1063.

31. Author's interviews with Imanishi-Kari, Oct. 6, 1992; with O'Toole, Oct. 5, 1992, Nov. 8, 1992; N.I.H. Staff, O'Toole interview, May 17, 1988, Transcript, AE H-109, pp. 48–49, 51–54. Federal investigators would designate the two sheets I-1: 41 and I-1:42. See AE H-12.

32. Wortis, Testimony, *Hearings, Scientific Fraud*, May 4, 1989, p. 231; Wortis, Testimony; Huber, Testimony, *Appeal Proceedings*, June 23, 1995, p. 1786, June 30, 1995, pp. 206–210; author's interviews with Huber, Oct. 6, 1992; with Imanishi-Kari, March 20, 1996; with Wortis, by telephone, June 12, 1997; Mark Ptashne, letter to the editor, *Nature*, 352 (July 11, 1991), 101; Margot O'Toole, "More Detailed Comments," [on Eisen's statement in *Nature*], attached to O'Toole to Hadley, fax, June 13, 1991, AE H-295. The report of the Wortis committee, written a year later, declares: "The possibility was raised that a small amount of transgene transcription could allow a small amount of hybrid trans/endogenous

molecules to be made and secreted." The point occurs in the report in the discussion of the Table 3 hybridomas, which strongly implies that the hybrids posed were *mu-gamma*. Huber, Woodland, and Wortis, "Minutes of Ad-Hoc Committee Meeting," June 4, 1986 [prepared June 1987], *Hearings, Fraud in N.I.H. Grant Programs*, April 12, 1988, pp. 211–213.

33. Wortis, Testimony, *Hearings, Scientific Fraud*, May 4, 1989, p. 231; author's interviews with Huber, Oct. 6, 1992; with Imanishi-Kari, March 20, 1996. Woodland says that when he heard about the denouement of the meeting from Huber over the telephone, he assumed that O'Toole was "fairly satisfied" and that Thereza was "holding a grudge about it." Author's interview with Woodland, by telephone, Dec. 3, 1992. Imanishi-Kari testified that she refused to shake O'Toole's hand because she considered O'Toole's having first approached people other than the coauthors about her concerns "completely inappropriate." Imanishi-Kari, Testimony, *Appeal Proceedings*, Aug. 30, 1995, p. 4974.

34. O'Toole, Testimony, *Hearing, Fraud in N.I.H. Grant Programs*, April 12, 1988, p. 88; N.I.H. Staff, O'Toole interview, May 17, 1988, Transcript, AE H-109, pp. 57–58; author's interview with O'Toole, Oct. 5, 1992; O'Toole, "Testimony," *Hearings, Scientific Fraud*, May 9, 1989, pp. 182–183; O'Toole, Testimony, *Appeal Proceedings*, June 16, 1995, pp. 995–997.

35. Author's interviews with Wortis, Dec. 11, 1992; with Huber, Oct. 6, 1992; Huber, Woodland, and Wortis, "Minutes of Ad-Hoc Committee Meeting," June 4, 1986 [prepared June 1987], *Hearings, Fraud in N.I.H. Grant Programs*, April 12, 1988, pp. 211–213; author's interviews with Wortis, Oct. 2, 1992; with Huber, Oct. 6, 1992; Wortis, Testimony; Huber, Testimony, *Hearings, Scientific Fraud*, May 4, 1989, pp. 227–228.

36. Author's interviews with Flax, Oct. 7, 1992; with Wortis, Oct. 2, 1992; Louis Lasagna, "Testimony," House, *Hearings, Scientific Fraud*, May 9, 1989, p. 238; N.I.H. Staff, O'Toole interview, May 17, 1988, Transcript, AE H-109, p. 66; Sarasohn, *Science on Trial*, p. 11; memo, unsigned, "Appointment of Dr. Imanishi-Kari in Pathology Department," May 1986, copy in Onek Files..

37. N.I.H. Staff, O'Toole interview, May 17, 1988, Transcript, AE H-109, pp. 68–70; Sarasohn, *Science on Trial*, p. 12; Nicholas Yannoutsos to Ursula Storb, May 24, 1989, AE R-17.

38. O'Toole, Testimony, *Hearings, Scientific Fraud*, May 9, 1989, p. 183; O'Toole, handwritten notes from conversation with Mary Rowe, [May 29, 1986], together with handwritten notes of directions to Eisen's home in Woods Hole, AE, unnumbered, copy in author's possession.

39. Author's interview with O'Toole, Oct. 5, 1992.

40. N.I.H. Staff, O'Toole interview, May 17, 1988, Transcript, AE H-109, pp. 68–70; O'Toole, Testimony, *Hearing, Fraud in N.I.H. Grant Programs*, April 12, 1988, p. 88; O'Toole, Testimony, *Hearings, Scientific Fraud*, May 9, 1989, p. 183; O'Toole, handwritten notes from conversation with Mary Rowe, [May 29, 1986], together with handwritten notes of directions to Eisen's home in Woods Hole, AE, unnumbered, copy in author's possession.

41. N.I.H. Staff, O'Toole interview, May 17, 1988, Transcript, AE H-109, pp. 68–70.

42. O.S.I., Interview with Gene Brown, May 1991, AE R-46, pp. 1–5, 20–21.

43. Brown recalls that he also called the head of the biology department, the provost, and, he thinks, Mary Rowe, too, to tell them what he had done, and they all pronounced it the right course of action. *Ibid.*, pp. 6–8, 13; author's telephone conversation with Gene Brown, Oct. 16, 1996.

44. O.S.I., Interview with Herman Eisen, May 30, 1991, Transcript, AE H-117, pp. 6, 9; Pantridge, "In the Eye of the Storm," *Boston Magazine*, July 1991, pp. 105–106; author's interview with Herman Eisen, Oct. 2, 1992.

45. Author's interviews with Herman Eisen, July 10, 1996; with O'Toole, Oct. 5, 1992; O.S.I., Eisen interview, May 30, 1991, Transcript, AE H-117, pp. 13, 14, 37–38; N.I.H. Staff, O'Toole interview, May 17, 1988, Transcript, AE H-109, pp. 71–72; Eisen, Testimony, *Hearings, Scientific Fraud*, May 4, 1989, p. 288; Bernard Davis to Walter Stewart, May 3, 1987; Eisen to O'Toole, May 11, 1987, copy in Eisen Files; O'Toole, Testimony, *Appeal Proceedings*, June 16, 1995, p. 1001.

46. Margot O'Toole to Herman Eisen, June 6, 1986, AE H-231.

47. N.I.H. Staff, O'Toole interview, May 17, 1988, Transcript, AE H-109, pp. 71–72; O'Toole, Testimony and Chronology of Events, *Hearing, Fraud in N.I.H. Grant Programs*, April 12, 1988, pp. 103, 97–98; O'Toole to Eisen, June 6, 1986, AE H-231; Eisen interview, Oct. 2, 1992; author's interview with O'Toole, Oct. 5, 1992; Margot O'Toole to Dean Brown, June 9, 1986, AE H-232; Mark Ptashne, letter to the editor, *Nature*, 352(July 11, 1991), 101; O'Toole, Testimony, *Appeal Proceedings*, June 16, 1995, p. 13; Eisen, Testimony, *Appeal Proceedings*, June 26, 1995, pp. 1890–1900; Weaver *et al.*, "Altered Repertoire . . . ," *Cell*, 45(April 25, 1986), 250. In 1984, in one of the studies that led to the disputed *Cell* paper, Grosschedl, Weaver, and Baltimore reported that the transgenic Black/6 mice had normal levels of antibodies such as IgG and IgA circulating in their blood and said that they did not know whether these native antibodies also displayed characteristics of the transgene. Grosschedl *et al.*, "Introduction of a μ Immunoglobulin Gene into the Mouse Germ Line . . . ," *Cell*, 38(Oct. 1984), 655.

48. O'Toole to Eisen, June 6, 1986; Eisen interview, Oct. 2, 1992; O'Toole interview, Oct. 5, 1992; O.S.I., Eisen interview, May 30, 1991, AE H-117, pp. 72, 73; Sarasohn, *Science on Trial*, p. 14; O'Toole, Testimony, *Appeal Proceedings*, June 16, 1995, p. 1005.

49. O'Toole to Eisen, June 6, 1986, AE H-231; Eisen interview, Oct. 2, 1992; author's interviews with O'Toole, Oct. 5, 1992, Nov. 8, 1992; Herman N. Eisen, Letter to the Editor, *Science*, 245(Sept. 15, 1989), 1166–1167; O.S.I., Eisen interview, May 30, 1991, Transcript, AE H-117, pp. 19–30, 37, 87–88; N.I.H. Staff, Baltimore and Weaver interviews, May 18, 1988, AE H-107, p. 6 and AE H-115, p. 9. O'Toole says that Eisen opened the meeting with a report of the outcome at Tufts, but Wortis and Huber as well as Eisen testified that they had no contact with Eisen at the time of the Tufts and M.I.T. reviews. See N.I.H. Staff, O'Toole interview, May 17, 1988, AE H-109, p. 73; N.I.H. Staff, Notes of Interviews with Brigitte Huber and Henry Wortis, May 20, 1988, Onek Files; Maggie Hassan to File, Re Subcommittee Interviews of Dr. Herman Eisen and Dean Gene Brown, May 11, 1988, May 14, 1988, pp. 13, 17, Eisen Files.

50. Author's interviews with David Weaver, Dec. 11, 1992; with Baltimore, Jan. 23, 1992; with Eisen, Dec. 10, 1992; O'Toole, Chronology of Events, *Hearings, Fraud in N.I.H. Grant Programs*, April 12, 1988, pp. 98–99; Eisen, tape of a meeting with Paul Doty *et al.*, Harvard, June 4, 1991, copy in author's possession.

51. Author's interviews with Baltimore, Jan. 23, 1992; with Weaver, Dec. 11, 1992; with Eisen Oct. 2, 1992, Dec. 10, 1992; with Imanishi-Kari, Oct. 6, 1992; Eisen, Testimony, *Hearings, Scientific Fraud*, May 9, 1989, pp. 293–294; O'Toole, Chronology of Events, *Hearing, Fraud in N.I.H. Grant Programs*, April 12, 1988, pp. 98–99; O.S.I., Baltimore interview, April 30, 1990, Transcript, AE H-108, p. 57. According to O'Toole, Imanishi-Kari said that Reis "must have 'got that result

once' " and that Baltimore told her " 'that was not acceptable and he would talk to her in private.' " Baltimore says he does "not remember her saying, 'once,' with the implication being not any other time. She may have said that . . . it's in the paper the way it is because we got that result. But she certainly didn't say . . . only once." O'Toole, Chronology of Events, *Hearing, Fraud in N.I.H. Grant Programs*, April 12, 1988, pp. 98–99; author's interviews with O'Toole, Oct. 5, 1992; with Baltimore, Jan. 23, 1992.

52. O.S.I., Eisen interview, May 30, 1991, Transcript, AE H-117, pp. 51–52; author's interview with Eisen, Oct. 2, 1992; Herman Eisen to Gene Brown *et al.*, June 17, 1986, "Re: Margot O'Toole and the paper by D. Weaver . . . ," AE H-233. O'Toole remained convinced of the merits of her claim. N.I.H. Staff, O'Toole interview, May 17, 1988, Transcript, AE H-109, p. 74; author's interview with O'Toole, Oct. 5, 1992.

53. O'Toole to Eisen, June 6, 1986, AE H-231; O.S.I., Eisen interview, May 30, 1991, Transcript, AE H-117, p. 59; N.I.H. Staff, Baltimore interview, May 18, 1988, Transcript, AE H-107, p. 11.

54. N.I.H. Staff, Weaver interview, May 18, 1988, AE H-115, pp. 11–12; N.I.H. Staff, O'Toole interview, May 17, 1988, Transcript, AE H-109, pp. 51, 76; author's interviews with O'Toole, Oct. 5, 1992; with Baltimore, Jan. 23, 1992; with Imanishi-Kari, Oct. 6, 1992; with Weaver, Dec. 11, 1992; with Eisen, Oct. 2, 1992; the seventeen pages, AE H-15A; Imanishi-Kari, Testimony, *Appeal Proceedings*, Aug. 30, 1995, pp. 4982–4983.

55. Author's interviews with Imanishi-Kari, Oct. 6, 1992; with Weaver, Dec. 11, 1992; with O'Toole, Nov. 8, 1992; O'Toole, Chronology of Events, *Hearing, Fraud in N.I.H. Grant Programs*, April 12, 1988, pp. 98–99; O'Toole, Testimony, *Hearings, Scientific Fraud*, May 9, 1989, pp. 184–185; N.I.H. Staff, O'Toole interview, May 17, 1988, Transcript, AE H-109, p. 75; O'Toole, Testimony, *Appeal Proceedings*, June 16, 1995, pp. 1025–1026.

56. Author's interview with Imanishi-Kari, Oct. 6, 1992.

57. Author's interviews with Baltimore, Jan. 23, 1992; with Weaver, Dec. 11, 1992; with Eisen, Oct. 2, 1992; O'Toole, Chronology of Events, *Hearing, Fraud in N.I.H. Grant Programs*, April 12, 1988, pp. 98–99.

58. Baltimore added, "I would think there are probably a fair number of people whose experiments have been screwed up because they took the numbers for real. Everybody knows, or should know, that when you try to repeat something that doesn't repeat, the best thing to do is to call up the authors and say, 'Give me the details.' Increasingly, the details of how things are done are not in the literature." Author's interviews with Baltimore, Jan. 23, 1992; with O'Toole, Oct. 5, 1992.

59. O'Toole, Testimony, *Hearing, Fraud in N.I.H. Grant Programs*, April 12, 1988, p. 104; N.I.H. Staff, O'Toole interview, May 17, 1988, pp. 44–45; O.S.I., Eisen interview, May 30, 1991, Transcript, AE H-117, pp. 43–44; Imanishi-Kari, Testimony, *Appeal Proceedings*, Aug. 30, 1995, pp. 4984–4989. In 1989, led by Dingell, O'Toole testified that Baltimore indicated that he would oppose the publication of her letter by submitting "something saying I was wrong." In testimony before the Ryan Commission on Research Integrity in 1994, she declared that she "was told that any effort I made to publish a correction would be thwarted." Baltimore says that he "certainly didn't suggest it [her letter and his rebuttal] as a threat, but rather as a procedure that seemed to me an appropriate ventilation of what I saw as the issues that she'd raised." Which is how Weaver remembered the matter, too. O'Toole, Testimony, *Hearings, Scientific Fraud*, May 9, 1989, pp. 218–

220; Margot O'Toole, "Testimony before the Commission on Research Integrity," Dec. 1, 1994, obtained from the Internet; author's interview with Baltimore, Jan. 23, 1992; with Weaver, Dec. 11, 1992. Imanishi-Kari later testified that she consulted Mary Rowe about whether O'Toole's concerns had been resolved and that Rowe said she had spoken with O'Toole right after the meeting with Eisen and, yes, her concerns were resolved. Imanishi-Kari, Testimony, *Appeal Proceedings*, Aug. 30, 1995, pp. 4989–4990.

60. O'Toole, Testimony, *Hearing, Fraud in N.I.H. Grant Programs*, April 12, 1988, pp. 104–105; N.I.H. Staff, O'Toole interview, May 17, 1988, Transcript, AE 109, p. 77; author's interview with Weaver, Dec. 11, 1992.

61. Eisen to Gene Brown *et al.*, June 17, 1986, AE H-233; Maggie Hassan to File, Re: Interviews of Dr. Herman Eisen and Dean Gene Brown by Staff from National Institutes of Health, May 23, 1988, Eisen Files, p. 19.

62. O'Toole interview, Oct. 5, 1992; author's interview with Mary Rowe, by telephone, Sept. 17, 1992; O'Toole, Testimony, *Hearings, Scientific Fraud*, May 9, 1989, p. 185.

63. Eisen to Gene Brown *et al.*, June 17, 1986, AE H-233. Eisen says that he did not send his report to anyone because the inquiry, having not dealt with a charge of fraud, was unofficial. However, since it was addressed to Gene Brown and other M.I.T. officials, he probably wrote it with the intention of sending it out and then had second thoughts, possibly prompted by Mary Rowe. Author's interviews with Eisen, Oct. 5, 1992; with Mary Rowe, by telephone, Sept. 17, 1992.

64. O.S.I., Eisen interview, May 30, 1991, Transcript, AE H-117, pp. 65–66.

65. Author's interview with Eisen, Oct. 2, 1992.

66. Author's interview with O'Toole, Oct. 5, 1992; O'Toole, [Statement: Rebuttal to Baltimore], *Nature*, 351(May 1991), 180; Diana West, "Anatomy of a Scientific Scandal," *Washington Times*, June 16, 1989, pp. E1, E3.

67. Author's interview with Eisen, Oct. 2, 1992; Margot O'Toole to Olga Joly, May 13, 1987, copy in Onek Files.

68. N.I.H. Staff, O'Toole interview, May 17, 1988, Transcript, AE H-109, pp. 80–81; O'Toole, Testimony, *Hearings, Scientific Fraud*, May 9, 1989, p. 185; N.I.H. Staff, Notes of Interview with Henry Wortis, May 20, 1988, Onek Files.

69. Author's interviews with Flax, Oct. 7, 1992; with Wortis, Oct. 2, 1992; with Imanishi-Kari, Oct. 6, 1992; N.I.H. Staff, Notes of Interview with Henry Wortis, May 20, 1988, Onek Files.

70. O'Toole, Testimony, *Hearing, Fraud in N.I.H. Grant Programs*, April 12, 1988, p. 89; O'Toole interview, Oct. 5, 1992.

71. Author's interview with Huber, Oct. 6, 1992; Sarasohn, *Science on Trial*, p. 18.

Notes for Chapter Four

1. Philip M. Boffey, "Major Study Points to Faulty Research at Two Universities," *New York Times*, April 22, 1986, pp. C1, C11; Philip M. Boffey, "Two Critics of Science Revel in Role," *New York Times*, April 19, 1988, p. C1; author's interview with Charles Maplethorpe, March 23, 1993.

2. Author's interview with Maplethorpe, March 23, 1993; Stewart, notes of telephone conversations with Charles Maplethorpe, May 17, 1986, May 18, 1986, Onek Files.

3. Stewart, Testimony, *Hearing Fraud in N.I.H. Grant Programs*, April 12, 1988, pp. 115–116; Stewart, notes of telephone conversation with Maplethorpe, May 20,

1986, Onek Files; author's interview with Charles Maplethorpe, March 23, 1993; Sarasohn, *Science on Trial*, pp. 33–34. A journalist later wrote of Stewart and Feder that when whistle-blowers contact them, they provide anything from "sympathy and understanding to the name and phone number of a journalist who might be willing to publicize the case and embarrass the institution involved." Gary Taubes, "Fraud Busters," *Lingua Franca*, Sept./Oct. 1993, p. 51.

4. Ned Feder, Testimony, *Hearing, Fraud in N.I.H. Grant Programs*, April 12, 1988, p. 6; author's interview with Feder and Stewart, Sept. 16, 1992. A reporter wrote: "Ned doesn't talk and Walter doesn't listen." Taubes, "Fraud Busters," *Lingua Franca*, Sept./Oct. 1993, p. 50. Stewart's wife once described him as "submanic." Edward Dolnick, "Science Police," *Discover*, Feb. 1994, p. 59.

5. Author's interview with Feder and Stewart, Sept. 16, 1992; copy of Feder's *curriculum vitae*, U.S. Congress, House, Task Force on Science Policy of the Committee on Science and Technology, *Hearing, Science Policy Study—Hearings Volume 22: Research and Publications Practices*, 99th Cong., 2nd Sess. (hereafter, *Hearing, Science Policy Study*), May 14, 1986 p. 16; author's conversation with Howard Berg of Harvard University, Oct. 29, 1996.

6. Dolnick, "Science Police," *Discover*, Feb. 1994, p. 59; author's interview with Stewart and Feder, Sept. 16, 1992; Howard Berg to Wassily Leontief, Oct. 12, 1970, copy in possession of Berg; Howard C. Berg, Edward M. Purcell, and Walter W. Stewart, "A Method for Separating According to Mass a Mixture of Macromolecules or Small Particles Suspended in a Fluid, II. Experiments in a Gravitational Field," *Proceedings of the National Academy of Sciences of the United States of America*, 58(Oct. 1967), 1286–1291.

7. Walter Stewart to Howard Berg, n.d. [Aug. 1967]; Berg to Leontieff, Oct. 12, 1970, in the possession of Berg; Walter Stewart, *curriculum vitae*, House, Task Force on Science Policy of the Committee on Science and Technology, *Hearing, Science Policy Study*, May 14, 1986, p. 16; author's conversation with Leroy Hood, Nov. 19, 1996; author's interview with Rollin Hotchkiss and Magda Gabor, by telephone, Nov. 22, 1996. A recruitment brochure, *The Commissioned Officer in the U.S. Public Health Service* (Washington, D.C.: Public Health Service Publication No. 1681, 1967), p. 5, noted that "two years of active duty with the Commissioned Corps fulfills Selective service obligations." Fewer then 4 percent of commissioned officers were scientists, and fewer than 2 percent were assigned to the NIH. *Report of the Secretary's Committee To Study the Public Health Service Commissioned Corps* (Washington, D.C.: U.S. Department of Health, Education, and Welfare, 1971), pp. 21, 24. Joseph E. Rall, who headed the institute that included Feder's lab, recalled that Stewart "was about to be drafted . . . and so got a commission in the Public Health Service, avoided the draft." Author's interview with Joseph E. Rall, March 25, 1996.

8. Author's interview with Stewart and Feder, Sept. 16, 1992; Walter W. Stewart, "Isolation and Proof of Structure of Wildfire Toxin," *Nature*, 229(Jan. 15, 1971), 174–178.

9. Author's interview with Stewart and Feder, Sept. 16, 1992; Stewart to Howard Berg, Dec. 23, 1970; author's interview with Hotchkiss and Gabor, by telephone, Nov. 22, 1996; Sarasohn, *Science on Trial*, p. 32.

10. Walter W. Stewart, "Comments on the Chemistry of Scotophobin," *Nature*, 238(July 28, 1972), 202–209; Barbara J. Culliton, "A Bitter Battle over Error (II)," *Science*, 241(July 1, 1988), 20; author's interview with Stewart and Feder, Sept. 16, 1992; Stewart to Howard Berg, Dec. 23, 1970; clippings on the scotophobin

claims, including from *Time,* the *Boston Globe,* and the *San Francisco Chronicle,* in the possession of Berg; Frank Kuznik, "Fraud Busters," *Washington Post Magazine,* April 14, 1991, p. 24.

11. Berg to Leontieff, Oct. 12, 1970, in possession of Berg; author's interview with Stewart and Feder, Sept. 16, 1972; author's conversations with Berg, Oct. 29, 1996, and with Gerald Holton, Oct. 28, 1996; author's interview with Rall, March 25, 1996. The Harvard Society of Fellows records show that Stewart was a Junior Fellow from 1971 to 1974, which is the normal term, but his *curriculum vitae* shows that he held the fellowship from 1972 to 1974. The *c.v.* may be mistaken, or he may have been offered the fellowship in 1971 and not have accepted it until 1972.

12. Author's interview with Stewart and Feder, Sept. 15, 1992; W. W. Stewart, "Functional Connections Between Cells as Revealed by Dye-Coupling with a Highly Fluorescent Napthalimide Tracer," *Cell,* 14 (1978), 741–759; W. W. Stewart, "Synthesis of 3, 6-Disulfonated 4-Aminophthalimides," *Journal of the American Chemical Society,* 103(1981), 7615–7620; Walter W. Stewart, "Lucifer Dyes—Highly Fluorescent Dyes for Biological Tracing," *Nature,* 292(July 2, 1981), 17–21; W. W. Stewart and N. Feder, "Lucifer Dyes as Biological Tracers: A Review," in H. J. Marthy, ed., *Cellular and Molecular Control of Direct Cell Interactions* (NATO ASI; New York: Plenum Publishing, 1986), pp. 297–312; Walter W. Stewart and Ned Feder, "Attempts To Synthesize a Red-Fluorescing Dye for Intracellular Injection," in P. De Weer and B. M. Salzberg, eds., *Optical Methods in Cell Physiology* (New York: John Wiley, 1986), pp. 65–68. See the cover of *Science,* 229(Aug. 30, 1985) for a dramatic depiction of the use of Lucifer Yellow.

13. Author's interview with Stewart and Feder, Sept. 16, 1992; Dolnick, "Science Police," *Discover,* Feb. 1994, p. 63; Boffey, "Two Critics of Science Revel in Role," *New York Times,* April 19, 1988, pp. C1, C7; Peter Gwynne, "Have the Fraudbusters Gone Too Far," *Scientist,* 2(July 11, 1988), 8–9; Barbara J. Culliton, "A Bitter Battle over Error (II)," *Science,* 241(July 1, 1988), 20; Kuznik, "Fraud Busters," *Washington Post Magazine,* April 14, 1991, p. 25. It was later reported that Stewart and Feder had simply grown bored with their snail research. Taubes, "Fraud Busters," *Lingua Franca,* Sept./Oct 1993, p. 48.

14. Michael Schaller, *Reckoning with Reagan: America and Its President in the 1980s* (New York: Oxford University Press, 1992), pp. 102–103; Suzanne Garment, *Scandal: The Crisis of Mistrust in American Politics* (New York: Times Books, 1991), pp. 7–10 and *passim.*

15. U.S. Congress, House, Subcommittee on Investigations and Oversight of the Committee on Science and Technology, *Hearings, Fraud in Biomedical Research,* 97th Cong., 1st Sess. (hereafter, *Hearings, Fraud in Biomedical Research*) March 31, 1981, April 1, 1981; Philip J. Hilts, "Science Confronted with 'Crime Wave' of Researchers Faking Data in Experiments," *Los Angeles Times,* March 4, 1981, p. 6. A list of fraud cases is given in William Broad and Nicholas Wade, *Betrayers of the Truth* (New York: Simon and Schuster, 1982), pp. 225–232. In addition to Broad and Wade's own articles, see especially, Anne C. Roark, "Scientists Question Profession's Standards amid Accusations of Fraudulent Research," *Chronicle of Higher Education,* Sept. 2, 1980, p. 5, and Morton Hunt, "A Fraud That Shook the World of Science," *New York Times Magazine,* Nov. 1, 1981, pp. 42ff. For a brief overview of the emergence of scientific fraud and misconduct during the 1980s, see Marcel C. LaFollette, *Stealing into Print: Fraud, Plagiarism, and Mis-*

conduct in Scientific Publishing (Berkeley: University of California Press, 1992), pp. 1–31.

16. See Steven Shapin, A *Social History of Truth: Civility and Science in Seventeenth-Century England* (Chicago: University of Chicago Press, 1994). William Raub, a high official at N.I.H., remarked, "We have no way of knowing if there is more [fraud] out there or simply more is being reported. I think it's probable there has been a more aggressive reporting of scientific fraud by the press." Hilts, "Science Confronted with 'Crime Wave . . . ,' " *Los Angeles Times*, March 4, 1981, p. 6.

17. In a somewhat uncommon variation on this theme, the science writers Nicholas Wade and William Broad declared that science holds a high place in contemporary society not because of it practical benefits but because it "seems to represent an idea, a set of values, an ethical example of how human affairs could and should be conducted were reason to be man's guide. In the secular world of the twentieth century, science performs part of the inspirational function that myths and religions play in less developed societies." Broad and Wade, *Betrayers of the Truth*, p. 130.

18. Calculated from tables in *Science Indicators, 1978* (Washington, D.C.: Report of the National Science Board, 1979), p. 238, and *Science Indicators, 1980* (Washington, D.C.: Report of the National Science Board, 1981), p. 272.

19. Michael Rogers, "The Pandora's Box Congress," in James D. Watson and John Tooze, eds., *The DNA Story: A Documentary History of Gene Cloning* (San Francisco: W. H. Freeman, 1981), p. 29.

20. See, for example, Watson and Tooze, eds., *The DNA Story*, pp. 1–62, and Sheldon Krimsky, *Genetic Alchemy: The Social History of the Recombinant DNA Controversy* (Cambridge: MIT Press, 1982).

21. Alfred E. Vellucci to Philip Handler, May 16, 1977; Nicholas Wade, "Recombinant DNA: NIH Group Stirs Storm by Drafting Laxer Rules," *Science*, 190(Nov. 21, 1975), both in Watson and Tooze, eds., *The DNA Story*, pp. 206, 68.

22. Robert E. Pollack to J. D. Watson, June 27, 1977, in Watson and Tooze, eds., *The DNA Story*, p. 192.

23. *Ibid.*; Norton D. Zinder, "The Gene, The Scientists and The Law"; Watson, editorial, "Trying To Bury Asilomar," *Clinical Research*, 26(April 1978), in Watson and Tooze, eds. *The DNA Story*, pp. 199, 346.

24. Norton Zinder to [Paul] Berg *et al.*, Sept. 6, 1977; Paul Berg to Senator Harrison Schmitt, Jan. 5, 1979, in Watson and Tooze, eds., *The DNA Story*, pp. 258, 390–391.

25. *The Star-Ledger*, Newark, N.J., April 2, 1991, files of Walter Stewart and Ned Feder (hereafter Stewart and Feder Files), clippings.

26. Gore, Statements, *Hearings, Fraud in Biomedical Research*, March 31, 1981, April 1, 1981, pp. 1, 39–40. The science journalist Morton Hunt noted the connection between the issue of fraud and "a growing fear of the consequences of scientific knowledge. Nuclear accidents and the potential of nuclear warfare are the most obvious sources of that fear, but perhaps more deep-seated is the alarm engendered in the public by the genetic engineering research that is now finding ways of modifying the forms of life itself." Hunt, "A Fraud That Shook the World of Science," *New York Times Magazine*, Nov. 1, 1981, p. 75.

27. John Long, Ronald Lamont-Havers, and Philip Felig, Testimony; Maddox to Felig, June 24, 1980, *Hearings, Fraud in Biomedical Research*, March 31, 1981, April 1, 1981, pp. 53–54, 59, 74, 80, 83–84, 99. Long had also studied immunofluorescence in collaboration with David Baltimore, who was reported to have concluded that

the data in Long's notebooks seemed to "vary randomly from the published article." *Ibid.*, p. 63.

28. Felig, Testimony, *Hearings, Fraud in Biomedical Research*, March 31, 1981, April 1, 1981, pp. 82, 85, 104.

29. Gore, Statement; Long, Testimony; Felig, Testimony, *Hearings, Fraud in Biomedical Research*, March 31, 1981, April 1, 1981, pp. 1, 41, 53–64, 79–112; Broad and Wade, *Betrayers of the Truth*, p. 179 and *passim*; Katherine Bick, Testimony, U.S. Congress, House, Subcommittee [on Human Resources and Intergovernmental Relations] of the Committee on Government Operations, *Hearing, Scientific Fraud and Misconduct and the Federal Response*, 100th Cong., 2nd Sess. (hereafter, *Hearing, Scientific Fraud and Misconduct and the Federal Response*), April 11, 1988, p. 138. See also Alan Mazur, "Allegations of Dishonesty in Research and Their Treatment by American Universities," *Minerva*, XXVII(Spring 1989), 176–194.

30. Handler and Frederickson, Testimony, *Hearings, Fraud in Biomedical Research*, March 31, 1981, April 1, 1981, pp. 28–29, 40–41, 43, 46. According to information that Frederickson supplied the subcommittee, the Division of Management Survey and Review at the N.I.H. had been involved in only three cases of scientific fraud since 1970. "Response of the Director to Questions . . . ," *ibid.*, pp. 47–48.

31. Frederickson, Testimony, *Hearings, Fraud in Biomedical Research*, March 31, 1981, April 1, 1981, p. 42.

32. Frederickson, Handler, and William Raub, Testimony, *Hearings, Fraud in Biomedical Research*, March 31, 1981, April 1, 1981, pp. 28–29, 43, 167, 175.

33. Walker, Statement, *Hearings, Fraud in Biomedical Research*, March 31, 1981, April 1, 1981, pp. 76–78. Walker noted that if exposure of fraud had to await work by other labs, then "several hundred thousand more Federal dollars would have been spent on an experiment based upon falsified data." *Ibid.*, p. 63.

34. *Ibid.*, pp. 43, 44.

35. Gore, Statement, *Hearings, Fraud in Biomedical Research*, March 31, 1981, April 1, 1981, pp. 2–3.

36. Broad and Wade, *Betrayers of the Truth*, pp. 7, 86, 73. On the issue of replication, see Harriet Zuckerman, "Deviant Behavior and Social Control in Science," in Edward Sagarin, ed., *Deviance and Social Change* (Beverly Hills: Sage, 1977), pp. 94–95. Stewart and Feder later pointed out that the methods of science for detecting error were not designed to expose fraud. Stewart, Testimony, *Hearing, Scientific Fraud and Misconduct and the Federal Response*, April 11, 1988, p. 114.

37. Broad and Wade, *Betrayers of the Truth*, pp. 7–9, 19, 36–37, 86–87, 140, 214–217; Gore, Statement, *Hearings, Fraud in Biomedical Research*, March 31, 1981, April 1, 1981, pp. 2–3.

38. Broad and Wade, *Betrayers of the Truth*, pp. 22–37.

39. *Ibid.*, pp. 85–87.

40. *Ibid.*, p. 96; Frederickson, Testimony, *Hearings, Fraud in Biomedical Research*, March 31, 1981 April 1, 1981, pp. 28–29; Marcia Angell, "Fraud in Science," *Science*, 219(March 25, 1983), 1417–1418; David Joravasky, "Unholy Science," *New York Review of Books*, 30(Oct. 13, 1983), pp. 3–5. On issues connected with pressure to publish, see Patricia Woolf, "Pressure To Publish and Fraud in Science," *Annals of Internal Medicine*, 104(1986), 254–256.

41. Angell, "Fraud in Science," *Science*, 219(March 25, 1983), 1417–1418, Joravasky, "Unholy Science," *New York Review of Books*, 30(Oct. 13, 1983), pp. 3–5.; John Ziman, "Fudging the Facts," *Times Literary Supplement*, Sept. 9, 1983, p. 955;

Robert C. Cowen, "The Not-So-Hallowed Halls of Science," *Technology Review*, 86(May 1983), 8, 85; Peter David, "The System Defends Itself," *Nature*, 303(June 2, 1983), 369; "Fraud and Secrecy: The Twin Perils of Science," *New Scientist*, June 9, 1983, pp. 712–713.

42. Rita Levi-Montalcini, *In Praise of Imperfection* (New York: Basic Books, 1988), p. 158; Frederick Grinnell, "Ambiguity in the Practice of Science," Editorial, *Science*, 272(April 19, 1996), 333; Richard Lewontin, "Billions and Billions of Demons," *New York Review of Books*, Jan. 9, 1997, pp. 30–31.

43. Joravasky, "Unholy Science," *New York Review of Books*, 30(Oct. 13, 1983), p. 3; Thomas Hager, *Force of Nature: The Life of Linus Pauling* (New York: Simon and Schuster, 1995), pp. 374, 430–431.

44. Zuckerman, "Deviant Behavior and Social Control in Science," in Edward Sagarin, ed., *Deviance and Social Change*, pp. 105–106. Millikan's treatment of data in his experiment is incisively analyzed in Gerald Holton, "Subelectrons, Presuppositions, and the Millikan-Ehrenhaft Dispute," in Gerald Holton, *The Scientific Imagination: Case Studies* (Cambridge: Cambridge University Press, 1978), pp. 25–83, especially pp. 61–71. Holton quarreled with Broad and Wade's presentation of Millikan's use of data, which was based on Holton's analysis, but they ignored his attempt at correction. Holton, personal communication, by telephone, May 15, 1996.

45. Ziman, "Fudging the Facts," *Times Literary Supplement*, Sept. 9, 1983; Angell, "Fraud in Science," *Science*, 219(March 25, 1983), 1417–1418.

46. Zuckerman, "Deviant Behavior and Social Control in Science," in Edward Sagarin, ed., *Deviance and Social Change*, p. 98. Handler testified at the Gore hearings: "One can only judge the rare such acts [of fraud] that have come to light as psychopathic behavior originating in minds that made very bad judgments— ethics aside—minds which in at least this one regard may be considered as having been temporarily deranged." Handler, Testimony, *Hearings, Fraud in Biomedical Research*, March 31, 1981, April 1, 1981, p. 12.

47. Walter Stewart, Statement, *Hearing, Scientific Fraud and Misconduct and the Federal Response*, April 11, 1988, pp. 121, 130; Broad and Wade, *Betrayers of the Truth*, pp. 14–15; William J. Broad, "U.S. To Penalize Heart Researcher on Fraudulent Project at Harvard," *New York Times*, Feb. 16, 1983, p. 1; Barbara J. Culliton, "Coping with Fraud: The Darsee Case," *Science*, 220(April 1, 1983), 31–35.

48. Walter W. Stewart and Ned Feder, "Summary," Nov. 23, 1985, copy in Berg Files; Jim Henderson, "When Scientists Fake It," *American Way*, March 1, 1990, p. 62; Walter Stewart and Ned Feder, "Professional Practices among Biomedical Scientists: A Study of a Sample Generated by an Unusual Event," copy in House, Task Force on Science Policy of the Committee on Science and Technology, *Hearing, Science Policy Study*, May 14, 1986, pp. 83, 103, 130.

49. Stewart and Feder, "Professional Practices among Biomedical Scientists," pp. 62–63, 74; Walter W. Stewart and Ned Feder, "The Integrity of the Scientific Literature," *Nature*, 325(Jan. 15, 1987), 207–214.

50. Stewart and Feder, "Professional Practices among Biomedical Scientists," pp. 88, 93–95. Stewart and Feder later told a congressman: "The pressure to publish large numbers of papers has a corrupting effect on scientific practices . . . ; the problem is pervasive." Stewart and Feder to Jim Lightfoot, Sept. 12, 1988, in *Hearing, Scientific Fraud and Misconduct and the Federal Response*, April 11, 1988, p. 224. Author's interview with Stewart and Feder, Sept. 16, 1992.

51. John Maddox, "Fraud, Libel, and the Literature," editorial, Nat
    1987), 181; Boffey, "Major Study Points to Faulty Research at 7
    New York Times, April 22, 1986, pp. C1, C11; Stewart and 1
    "Appendix B: Threats of a Libel Suit," U.S. Congress, House,
    Civil and Constitutional Rights of the Committee on the Judic__
    99th Cong., 2nd Sess., Feb. 26, 1986, pp. 122–126; Patricia K. Woolf, "Ensu___ ,
    Integrity in Biomedical Publication," Journal of the American Medical Association,
    258(Dec. 18, 1987), 3425; Dolnick, "Science Police," Discover, Feb. 1994, p. 58;
    Walter W. Stewart and Ned Feder, "We Must Deal Realistically with Fraud and
    Error," Scientist, Dec. 14, 1987, p. 13; Stewart, Testimony, House, Task Force on
    Science Policy of the Committee on Science and Technology, Hearing, Science
    Policy Study, May 14, 1986, p. 20; Sarasohn, Science on Trial, p. 29.
52. Comments on the Darsee paper, Hearing, Science Policy Study, May 14, 1986,
    pp. 42–55; Boffey, "Major Study Points to Faulty Research at Two Universities,"
    New York Times, April 22, 1986, pp. C1, C11; Kuznik, "Fraud Busters," Washing-
    ton Post Magazine, April 14, 1991, p. 26; Ned Feder, Testimony, Hearing, Libel,
    Feb. 26, 1986, p. 103; author's interview, by telephone, with Hotchkiss and Gabor,
    Nov. 22, 1986.
53. Arnold Relman to Stewart, March 2, 1988, Onek Files.
54. Eugene Braunwald, "On Analysing Scientific Fraud"; John Maddox, "Fraud, Libel,
    and the Literature," editorial, Nature, 325(Jan. 15, 1987), 215, 181; Stewart, Tes-
    timony, Hearing, Science Policy Study, May 14, 1986, p. 44; Boffey, "Major Study
    Points to Faulty Research at Two Universities," New York Times, April 22, 1986,
    pp. C1, C11; Howard Berg to Stewart, Jan. 14, 1986, Berg Files.
55. See, for example, Stewart and Feder to Eisen, Jan. 15, 1987, Stewart and Feder
    Files; author's interview, by telephone, with Hotchkiss and Gabor, Nov. 22, 1996;
    Stewart, Testimony, Hearing, Science Policy Study, May 14, 1986, p. 21; Kuznik,
    "Fraud Busters," Washington Post Magazine, April 14, 1991, p. 25; author's inter-
    view with Rall, March 25, 1996.
56. Stewart, Testimony, Hearing, Science Policy Study, May 14, 1986, p. 21; Kuznik,
    "Fraud Busters," Washington Post Magazine, April 14, 1991, p. 25.
57. Stewart, Testimony, Hearing, Science Policy Study, May 14, 1986, p. 21; author's
    interview with Stewart and Feder, Sept. 16, 1992. Culliton, "A Bitter Battle over
    Error (II)," Science, 241(July 1, 1988), 20.
58. Walter Stewart and Ned Feder to Jesse Roth, Oct. 10, 1985, Oct. 31, 1986; Hear-
    ing, Scientific Fraud and Misconduct and the Federal Response, April 11, 1988,
    pp. 187–195; Jesse Roth, Testimony and accompanying documents concerning the
    removal of the equipment, Hearing, Fraud in N.I.H. Grant Programs, April 12,
    1988, pp. 215–220.
59. Floyd Abrams, "Why We Should Change the Libel Law," New York Times Mag-
    azine, Sept. 29, 1985, pp. 87, 90; Hearing, Libel, Feb. 26, 1986, pp. 102–140; Sar-
    asohn, Science on Trial, pp. 30–31.
60. Author's interview with Stewart and Feder, Sept. 16, 1992.
61. "Tale of the Fraud Study That's Too Hot to Publish"; "Part II: The Fraud Story
    That's Too Hot to Publish," Science and Government Report, XVI (March 15, 1986;
    April 1, 1986).
62. Author's interview with Stewart and Feder, Sept. 16, 1992; Boffey, "Major Study
    Points to Faulty Research at Two Universities," New York Times, April 22, 1986,
    pp. C1, C11.
63. Boffey, "Major Study Points to Faulty Research at Two Universities," New York

420

Times, April 22, 1986, p. C11; Patricia Woolf, "Fraud in Science: How Much, How Serious?" *Hastings Center Report,* 11(1981), 13.

64. Patricia Woolf, Testimony, *Hearing, Science Policy Study,* May 14, 1986, p. 142; June Price Tangney, Testimony; Katherine Bick, Testimony, *Hearing, Scientific Fraud and Misconduct and the Federal Response,* April 11, 1988, pp. 98, 138, 140–141, 157; Mary Miers, Testimony, *Hearing, Fraud in N.I.H. Grant Programs,* April 12, 1988, p. 192.

65. Bick, Testimony, *Hearing, Scientific Fraud and Misconduct and the Federal Response,* April 11, 1988, p. 137; Robert E. Windom, Testimony, *Hearing, Fraud in N.I.H. Grant Programs,* April 12, 1988, pp. 172–173, 192–193; author's interview with Mary Miers, March 24, 1993.

66. Stewart, Testimony, *Hearing, Science Policy Study,* May 14, 1986, pp. 20, 139; Stewart, Testimony, *Hearing, Libel,* Feb. 26, 1986, p. 106; Stewart and Feder to Jim Lightfoot, Sept. 12, 1988, in *Hearing, Scientific Fraud and Misconduct and the Federal Response,* April 11, 1988, p. 219; author's interview with David Baltimore, March 20, 1996; O.S.I., Baltimore interview, April 30, 1990, Transcript, AE H-108, p. 35; Bernard Davis to Stewart and Feder, May 3, 1987, Baltimore Files.

67. Stewart, Testimony, *Hearing, Science Policy Study,* May 14, 1986, pp. 20, 139.

NOTES FOR CHAPTER FIVE

1. Author's interview with Maplethorpe, March 23, 1993; Stewart, notes of telephone conversations with Maplethorpe, July 10, 1986, July 12, 1986, July 9, 1986; with O'Toole, Aug. 8, 1986, Onek Files; Sarasohn, *Science on Trial,* pp. 34–35; Stewart, Testimony, House, *Hearing, Fraud in N.I.H. Grant Programs,* April 12, 1988, pp. 115–116.

2. Walter W. Stewart and Ned Feder, Chronology of Events, *Hearing, Fraud in N.I.H. Grant Programs,* April 12, 1988, p. 62; author's interviews with O'Toole, Oct. 5, 1992; with Walter Stewart and Ned Feder, Sept. 16, 1992.

3. Stewart, notes of telephone conversations with O'Toole, Sept. 8, 9, and 10, 1986, Onek Files; Stewart and Feder, Chronology of Events; O'Toole, Chronology of Events, *Hearing, Fraud in N.I.H. Grant Programs,* April 12, 1988, pp. 62, 99–100; author's interview with O'Toole, Oct. 5, 1992. The immunologist Leonore Herzenberg recalls that when she visited Tufts some years later she had a long talk with O'Toole's husband, Peter Brodeur, while he rode out to the airport with her and kept her company until she boarded her plane. She says, "He told me that [Margot] was really pushed, she was forced to turn over her notebooks [the seventeen pages] by Stewart and Feder. They told her, 'If you don't do that, we're going to subpoena you. It's going to be much worse on you. You're going to need a lawyer. It's going to be expensive. You've got to do this.' And so she did it." Author's interview with Leonore and Leonard Herzenberg, July 2, 1996.

4. Stewart, note, Sept. 16, 1986, Onek Files.

5. Author's interview with O'Toole, Oct. 5, 1992; with Stewart and Feder, Sept. 16, 1992.

6. Walter W. Stewart and Ned Feder, "Original Data Contradict Published Claims: Analysis of a Recent Paper," Sept. 30, 1987, p. 12, copy in Stewart and Feder Files.

7. Author's interviews with O'Toole, Oct. 5, 1992; with Stewart and Feder, Sept. 16, 1992; Terence Speed, Testimony; Imanishi-Kari, Testimony, *Appeal Proceedings,*

Aug. 21, 1995, pp. 2997–2999; Sept. 15, 1995, p. 6458. Stewart and Feder later told a congressman that, while science is self-correcting, it does not follow that it's unnecessary to correct "known errors" in the literature. Not to correct them is "contrary to the scientific ethic and carries with it a high cost." Stewart and Feder to Jim Lightfoot, Sept. 12, 1988; *Hearings, Scientific Fraud and Misconduct and the Federal Response*, April 11, 1988, p. 212.

8. Author's interviews with O'Toole, Oct. 5, 1992; with Stewart and Feder, Sept. 16, 1992; Walter W. Stewart and Ned Feder, "Original Data Contradict Published Claims: Analysis of a Recent Paper," Sept. 30, 1987, in *Hearing, Fraud in N.I.H. Grant Programs*, April 12, 1988, pp. 41–44. Stewart later testified that "it proved possible to understand [the seventeen pages] free of any other assertions of the history of the matter or anything like that, and to relate them quite clearly to the published paper." Stewart, Testimony, *ibid.*, April 12, 1988, p. 116.

9. Copy of request for approval of submission of the paper to *Cell*, Oct. 31, 1986; Stewart and Feder to Joseph E. Rall, Nov. 3, 1986, Stewart and Feder Files.

10. Stewart and Feder to Mary Miers, Institutional Liaison Officer, Office of Extramural Research and Training, Nov. 10, 1986; George J. Galasso to Stewart and Feder, Oct. 17, 1986, Stewart and Feder Files; Stewart and Feder, Chronology of Events, *Hearing, Fraud in N.I.H. Grant Programs*, April 12, 1988, pp. 64–66; author's interviews with Stewart and Feder, Sept. 16, 1992; with Mary Miers, March 24, 1993.

11. Author's interview with Joseph E. Rall, March 25, 1996; Stewart and Feder to Joseph E. Rall, Nov. 3, 1986, Stewart and Feder Files. Rall had previously handled Stewart and Feder's request to publish the Darsee article.

12. Boffey, "Two Critics of Scientific Research Revel in Role," *New York Times*, April 4, 1988, p. C7; author's interviews with Rall, March 25, 1996; with Stewart and Feder, Sept. 16, 1992.

13. Stewart and Feder to Joseph E. Rall, Nov. 3, 1986, Stewart and Feder Files.

14. Rall to Stewart and Feder, Dec. 12, 1986, and attached referees' reports, Stewart and Feder Files.

15. Stewart and Feder to Baltimore, Dec. 18, 1986, Stewart and Feder Files.

16. Author's interview with Stewart and Feder, Sept. 16, 1992.

17. Stewart to Baltimore, March 18, 1987, Baltimore Files, item 25; Stewart and Feder to Baltimore, Dec. 18, 1986, Stewart and Feder Files. "We're not investigators *for* N.I.H.," they characteristically told a reporter, "We look into facts as scientists." Culliton, "A Bitter Battle over Error (II)," *Science*, 241 (July 1, 1988), 18.

18. Author's interview with Rall, March 25, 1996. In conversation with me, Stewart and Feder said they were sure they had sent the seventeen pages and found it difficult to believe they had not. Author's interview with Stewart and Feder, Sept. 16, 1992.

19. Maddox, "Fraud, Libel, and the Literature," editorial; Eugene Braunwald, "On Analysis Scientific Fraud," *Nature*, 325 (Jan. 15, 1987), 181, 215–16.

20. Author's interview with Eisen, Oct. 2, 1992; O.S.I., Eisen interview, May 30, 1991, AE H-117, pp. 67–68; Eisen, [note of conversation with Weaver], Aug. 4, 1986, AE H-234; Eisen, Testimony, *Hearings, Scientific Fraud*, May 4, 1989, pp. 293–294; Eisen, Maggie Hassan to File, Re: Interviews of Dr. Herman Eisen and Dean Gene Brown by Staff from National Institutes of Health, May 23, 1988, Eisen Files, p. 13.

21. Author's interview with Eisen, Oct. 2, 1992; O.S.I., Eisen interview, May 30, 1991, AE H-117, pp. 60–61; Eisen, [note re she "knew it all along"], n.d., AE H-234;

Eisen, Testimony, *Hearings, Scientific Fraud,* May 4, 1989, pp. 293–294; Maggie
Hassan to File, Re: Interviews of Dr. Herman Eisen and Dean Gene Brown by
Staff from National Institutes of Health, May 23, 1988, Eisen Files, p. 16.

22. Baltimore to Eisen, Sept. 9, 1986, AE H-235.

23. Author's interview with Eisen, Oct. 2, 1992; O.S.I., Eisen interview, May 30, 1991,
Transcript, AE H-117, pp. 35–36, 60–61; Baltimore, Testimony, *Hearings, Scientific Fraud,* May 4, 1989, p. 91; O.S.I., Baltimore interview, May 18, 1988, Transcript, AE H-107, p. 17.

24. Eisen, Testimony; Baltimore, Testimony, *Hearings, Scientific Fraud,* May 4, 1989,
pp. 91, 167, 295–296; author's interview with Eisen, Oct. 2, 1992; O.S.I., Eisen
interview, May 30, 1991, Transcript, AE H-117, pp. 83–84; Eisen, "Summary of
Investigation," May 13, 1988, Baltimore Files, item 79.

25. Eisen to Fox, Dec. 30, 1986, AE H-236; author's interviews with Eisen, Oct. 2,
1992; with Maurice Fox, Oct. 2, 1992.

26. Baltimore to Stewart and Feder, Jan. 21, 1987, Stewart and Feder Files.

27. *Ibid.*

28. Taubes, "Fraud Busters," *Lingua Franca,* Sept./Oct. 1993, p. 51.

29. Baltimore, Statement, *Hearings, Scientific Fraud,* May 4, 1989, p. 98; Baltimore to
Stewart and Feder, March 24, 1987, Stewart and Feder Files. In a letter to Stewart
and Feder, a copy of which he sent to Baltimore, Bernard Davis, a prominent
biologist at Harvard and an admirer of their work in the Darsee affair, declared,
"What you call your research is more like a legal proceeding, where reputations
are at stake, and procedures designed to ensure fairness are in order." Davis also
called their attention to "the problems that would be created by a policy that
allowed any self-appointed vigilantes to invade any laboratory" and concluded that
they had "lost their sense of balance over the issue of scientific fraud and error."
Baltimore told Davis that his letter "mirrors closely my own comments to them."
Davis to Stewart and Feder, May 3, 1987; Baltimore to Davis, May 19, 1987,
Baltimore Files, item 29.

30. Stewart to Baltimore, March 18, 1987, Baltimore Files, item 25; author's interview
with Baltimore, March 19, 1996.

31. Author's interview with Eisen, Oct. 2, 1992; Eisen to Stewart and Feder, March
10, 1987; Wortis to Stewart and Feder, March 2, 1987, Stewart and Feder Files.

32. Author's interview with Baltimore, Jan. 23, 1992; O.S.I., Baltimore interview, April
30, 1990, Transcript, AE H-108, pp. 32–34.

33. Stewart to Baltimore, March 18, 1987; Baltimore to Wortis and Eisen, March 24,
1987, Baltimore Files, items 25, 27. Baltimore says that he did not threaten to
sue Stewart and Feder and that Stewart had put words in his mouth, but that he
had captured "the tone of my feelings." Author's interview with Baltimore, Jan.
23, 1992.

34. Author's interview with Rall, March 25, 1996; Baltimore to Rall, March 17, 1987,
Stewart and Feder Files; Stewart to Baltimore, March 18, 1987, Baltimore Files,
item 25.

35. Stewart and Feder, Chronology of Events, *Hearing, Fraud in N.I.H. Grant Programs,* April 12, 1988, p. 70; Rall to Baltimore, April 2, 1987, Stewart and Feder
Files. Stewart and Feder responded to Baltimore's proposal by urging that the
N.I.H. permit them access to the evidence developed in an inquiry so that they
could conduct their own independent and parallel investigation of the *Cell* paper.
They would be free to say about it what they wished, but they would publicly
apologize to the coauthors if they concluded their original suspicions of the paper

were wrong. Author's interview with Stewart and Feder, Sept., 16, 1992; Stewart to Baltimore, March 18, 1987, Stewart and Feder Files.

36. Stewart and Feder to Rall, April 9, 1987, Stewart and Feder Files.

37. Culliton, "A Bitter Battle over Error (II)," *Science*, 241(July 1, 1988), 19–20; Rall to Stewart and Feder, April 20, 1987; Stewart and Feder to Rall, May 1, 1987, with attached referees' report from c. July 1, 1987, Stewart and Feder Files; author's interview with Rall, March 25, 1996.

38. Stewart and Feder to _____, May 20, 1987, and attached chronological summary and new referee's reports, copy in Stewart and Feder Files.

39. Author's interview with Rall; Stewart, Testimony, *Hearing, Scientific Fraud and Misconduct and the Federal Response*, April 11, 1988, p. 133.

40. G. Brian Busey to Robert Lanman, July 14, 1987; Rall to Stewart and Feder, July 17, 1987; C. Brian Busey to Lanman, Sept. 9, 1987, Stewart and Feder Files; author's interview with Miers, March 24, 1993.

41. Leonore Herzenberg et al., "Depletion of the Predominant B-cell Population in Immunoglobulin Mu Heavy-Chain Transgenic Mice," *Nature*, 329(Sept. 3, 1987), 71–73.

42. Author's interview with Leonore and Leonard Herzenberg, July 2, 1996; N.I.H. Staff, Baltimore interview, May 18, 1988, Transcript, AE H-107, pp. 12–13.

43. Author's interview with the Herzenbergs, July 2, 1996; N.I.H. Staff, Baltimore interview, May 18, 1988, Transcript, AE H-107, pp. 12–13.

44. Herzenberg et al., "Depletion of the Predominant B-cell Population in Immunoglobulin Mu Heavy-Chain Transgenic Mice," *Nature*, 329(Sept. 3, 1987), 71, 73; N.I.H. Staff, Baltimore interview, May 18, 1988, Transcript, AE H-107, pp. 12–13; N.I.H. Staff, Interview with David Weaver, May 18, 1988, Transcript, AE H-115.

45. Herzenberg et al., "Depletion of the Predominant B-cell Population in Immunoglobulin Mu Heavy-Chain Transgenic Mice," *Nature*, 329(Sept. 3, 1987), 72; author's interview with the Herzenbergs, July 2, 1996.

46. Author's interview with the Herzenbergs, July 2, 1996; author's telephone conversation with Lee Herzenberg, Feb. 11, 1997; Stewart, notes of telephone conversation with O'Toole, Nov. 11, 1986, Onek Files.

47. Author's interview with the Herzenbergs, July 2, 1996; telephone conversation with Lee Herzenberg, Feb. 11, 1997.

48. Stewart and Feder to Leonore Herzenberg, Sept. 21, 1987, Stewart and Feder Files.

49. Leonard and Leonore Herzenberg to Baltimore, Nov. 6, 1987, AE H-241.

50. *Ibid.*

51. Baltimore to Leonard and Leonore Herzenberg, Nov. 12, 1987, AE H-242.

52. Leonard A. Herzenberg and Leonore A. Herzenberg to Baltimore, Nov. 24, 1987, AE H-243.

53. Alan M. Stall et al., "Rearrangement and Expression of Endogenous Immunoglobulin Genes Occur in Many Murine B Cells Expressing Transgenic Membrane IgM," *Proceedings of the National Academy of Sciences of the United States of America*, 85(May 1988), 3546–3550.

54. Leonore A. Herzenberg and Leonard Herzenberg to Stewart and Feder, Jan. 14, 1988, Stewart and Feder Files.

55. Stewart and Feder, statement, *Hearing, Fraud in N.I.H. Grant Programs*, April 12, 1988, pp. 20–21; Dolnick, "Science Police," *Discover*, Feb. 1994, p. 63; Boffey, "Two Critics of Science Revel in Role," *New York Times*, April 19, 1988, pp. C1, C7; author's interview with Stewart and Feder, Sept. 16, 1992; Kuznik, "Fraud

Busters," *Washington Post Magazine*, April 14, 1991, pp. 25, 32; Stewart, Testimony, *Hearings, Scientific Fraud and Misconduct and the Federal Response*, April 11, 1988, p. 128.

56. Baltimore, Statement, *Hearing, Scientific Fraud*, May 4, 1989, p. 99; author's interviews with Stewart and Feder, Sept. 16, 1992; with Patricia Woolf, by telephone, Nov. 18, 1996; with Al Kildow, June 7, 1996; Walter W. Stewart to Arthur Spitzer, Nov. 25, 1987, Onek Files.

57. Patricia A. Morgan to Stewart, Dec. 16, 1987; Benjamin Lewin to Stewart, Oct. 19, 1987, Stewart and Feder Files.

58. Katherine L. Bick to David Baltimore, May 22, 1987; Baltimore to Bick, June 15, 1987, Baltimore Files, items 33, 43.

59. Author's interview with Wortis, Oct. 2, 1992; [Katherine] Bick to [Dean] Banks, May 22, 1987, listed and summarized in Index of ORI File on "TIK," attached to Marcus H. Christ, Jr., to Joseph Onek, May 1, 1995, Onek Files; Eisen notes, May 14, 1987, June 4, 1987, Eisen Files; Wortis Committee report, n.d., attached to Wortis to Baltimore, June 20, 1987, Baltimore Files, item 46; Deputy Director for Extramural Research to Director, N.I.H., Jan. 18, 1989, "Review of Alleged Inaccuracies in *Cell* Paper—DECISION," attached to [Final Report of the Scientific Panel, Jan. 31, 1989], Baltimore Files, item 130.

60. Banks to Bick, June 22, 1987; Bick to Banks, July 28, 1987; Byrne to Bick, Oct. 2, 1987, listed and summarized in Index of ORI Documents on "TIK," attached to Marcus H. Christ, Jr., to Joseph Onek, May 1, 1995, Onek Files.

61. Mary Miers, Testimony, *Hearing, Fraud in N.I.H. Grant Programs*, April 12, 1988, pp. 197–198, 206, 353–356; author's interview with Mary Miers, March 24, 1993; Katherine Bick to David Baltimore, Jan. 13, 1988, listed in Office of Research Integrity, Index of Documents in the Case of Thereza Imanishi-Kari, copy in Onek Files.

62. N.I.H. Staff, Baltimore interview, May 18, 1988, Transcript, AE H-107, p. 21; Susan Okie, "When Researchers Disagree," *Washington Post*, April 11, 1988, pp. A1, A12. Although Baltimore had not yet received his doctorate when he worked with Darnell at M.I.T., in 1963/64, he was de facto a postdoctoral fellow because he had completed all the requirements for his Ph.D. at Rockefeller.

63. Sarasohn, *Science on Trial*, pp. 57–58; Stewart to Mary Miers, Feb. 17, 1988, Stewart and Feder Files.

64. Author's interview with Mary Miers, March 24, 1993. Miers to Stewart, March 11, 1988, Stewart and Feder Files; Miers and her superiors had known about Alt and Darnell's relationships with Baltimore. In Alt's case at least, one of them, Katherine Bick, who was deputy director for extramural affairs, had concluded to leave it up to Alt to "balance [friendship] with objectivity." Mary [Miers], Note to Dr. Bick, Feb. 12, 1988, and Bick's handwritten response on the note, Office of Research Integrity Files, Case 072, Thereza Imanishi-Kari (hereafter, O.R.I. Files).

NOTES FOR CHAPTER SIX

1. Dingell, "Statement," *Hearings, Scientific Fraud*, May 4, 1989, p. 5; Dingell, "Opening Statement," *Hearing, Fraud in N.I.H. Grant Programs*, April 12, 1988, pp. 1, 3; Sarasohn, *Science on Trial*, pp. 46–49; Joseph Palca, "Bernadine Healy: A New Leadership Style at NIH," *Science*, 253(Sept. 6, 1991), 1988.

2. "J. Edgar Dingell," editorial, *Wall Street Journal*, March 15, 1989, p. A16; Fred Barnes, "Bad Cop," *New Republic*, Oct. 23, 1989, pp. 10–11; Dolnick, "Science Police," *Discover*, Feb. 1994, p. 61; author's conversation with a congressional staff member, Aug. 17, 1992; author's interview with Stockton and Chafin, March 24, 1993; Sarasohn, *Science on Trial*, pp. 49–52; Ronald Kessler, *Inside Congress: The Shocking Scandals, Corruption, and Abuse of Power Behind the Scenes on Capitol Hill* (New York: Pocket Books, 1997), p. 84; Garment, *Scandal*, pp. 143 ff.

3. "J. Edgar Dingell," editorial, *Wall Street Journal*, March 15, 1989, p. A16; Barnes, "Bad Cop," *New Republic*, Oct. 23, 1989, pp. 10–11; Dolnick, "Science Police," *Discover*, Feb. 1994, p. 61; author's conversation with a congressional staff member, Aug. 17, 1992; author's interviews with Bernadine Healy, Aug. 27, 1996; with Jules Hallum, April 24, 1996; with Bruce Maurer, Nov. 1, 1996; Sarasohn, *Science on Trial*, pp. 51–55; Terence Moran, "Specializing in Dingell," *Legal Times*, May 28, 1990, p. 1.

4. Stewart and Feder, Statement, *Hearing, Fraud in N.I.H. Grant Programs*, April 12, 1988, pp. 5–6; Robert Bell, *Impure Science: Fraud, Compromise, and Political Influence in Scientific Research* (New York: John Wiley, 1992), pp. 105–111; Ted Weiss, Statement, *Hearing, Scientific Fraud and Misconduct and the Federal Response*, April 11, 1988, p. 103.

5. "Q&A: With Director Wyngaarden on Political Troubles at N.I.H.," *Science and Government Report*, June 1, 1988, p. 4; Wyden, Statement, *Hearing, Fraud in N.I.H. Grant Programs*, April 12, 1988, p. 4.

6. Sprague, Testimony; Conyers, Statements, *Hearing, Scientific Fraud and Misconduct and the Federal Response*, April 11, 1988, pp. 3, 8, 112–113, 184.

7. Stewart, Testimony; Bick, Testimony, *Hearing, Scientific Fraud and Misconduct and the Federal Response*, April 11, 1988, pp. 133, 157.

8. Eckart, Statement, *Hearings, Scientific Fraud*, May 4, 1989, pp. 35–36.

9. Author's interview with Peter Stockton and Bruce Chafin, March 24, 1993; Sarasohn, *Science on Trial*, pp. 55–56.

10. Author's interview with Stockton and Chafin, March 24, 1993; with Stewart and Feder, Sept. 16, 1992; David Baltimore to Colleague, May 19, 1988, Baltimore Files..

11. Author's interviews with O'Toole, Oct. 5, 1992; with Stewart and Feder, Sept. 5, 1992; Bernard D. Davis, "Dingell's Witness for the Persecution," *Wall Street Journal*, July 22, 1991, p. A8. O'Toole moved to Brookline some time between March 17, 1987, and April 23, 1987. See the return addresses on her letters to Eisen of those dates, AE H-237 and H-238.

12. Author's interviews with O'Toole, Oct. 5, 1992; with Stewart and Feder, Sept. 5, 1992.

13. Stewart and Feder to O'Toole, Jan. 28, 1987, March 10, 1987; O'Toole to Stewart and Feder, March 17, 1987 (cc: Dr. Herman Eisen), Stewart and Feder Files; O'Toole to Herman Eisen, March 17, 1987, April 23, 1987, AE-H237 and H-238. Stewart and Feder reported their conversation with Baltimore in January in a round-robin letter dated May 20, 1987, that detailed their difficulties in getting permission to publish their analysis of the *Cell* paper and that they sent to many scientists. Copy in Baltimore Files.

14. Eisen interview, Dec. 10, 1992; Eisen to O'Toole, May 11, 1987, Eisen Files.

15. Eisen, [notes of telephone conversations with Imanishi-Kari and with Wortis], May 14, 1987, June 4, 1987, Eisen Files; O.S.I., Eisen interview, May 30, 1991, AE H-117, pp. 60–61.

16. O'Toole to Eisen, July 7, 1987, Eisen Files; O'Toole to Eisen, Oct. 11, 1987, AE H-240.

17. O'Toole to Eisen, Oct. 11, 1987, with typed copy of Eisen's marginal notes, AE H-240; N.I.H. Staff, O'Toole interview, May 17, 1988, Transcript, AE H-109.

18. Author's interviews with Stockton and Chafin, March 24, 1993; with Stewart and Feder, Sept. 16, 1992.

19. Author's interviews with Stockton and Chafin, March 24, 1993; with Stewart and Feder, Sept. 16, 1992; Sarasohn, *Science on Trial*, p. 57. Stockton and Chafin told me that Maplethorpe was called to testify so that the subcommittee could obtain a copy of the tape that he had made of Imanishi-Kari, probably of her conversation with Weaver in June 1985. They say that they didn't use it in the hearing because it would have been improper. They were unclear about what they meant, but may have been reluctant to use it because the conversation had been recorded illegally and it contained nothing incriminating.

20. Mary Miers, Testimony; Stewart, Testimony, *Hearing, Scientific Fraud and Misconduct and the Federal Response*, April 11, 1988, pp. 131, 165; Miers to Stewart, March 11, 1988, Stewart and Feder Files; Dingell to Walter W. Stuart [sic], March 23, 1988, Davis Files; author's interview with Stewart and Feder, Sept. 16, 1992; Helen Gavaghan, "American Researchers 'Covered Up' Scientific Fraud," *New Scientist*, 118(April 14, 1988), 22.

21. Author's interviews with David Baltimore, March 20, 1996; with Al Kildow, June 7, 1996; Judy Foreman, "A Noted Researcher's Disputed Work Lands in Congress," *Boston Sunday Globe*, April 10, 1988, p. 3; David Baltimore to Colleague, May 19, 1988, copy in Baltimore Files.

22. O'Toole, Testimony and Chronology of Events, *Hearing, Fraud in N.I.H. Grant Programs*, April 12, 1988, pp. 86, 88–89, 91–100, 120; Sarasohn, *Science on Trial*, p. 64; "Fraud Inquiry: Harsh Treatment for N.I.H. on Capitol Hill," *Science and Government Report*, April 15, 1988, p. 1. O'Toole's account of events at Tufts and M.I.T. was remarkably detailed. She says that after the meeting at M.I.T. in June 1986, she had begun tape-recording her recollections of what had happened. Author's interviews with O'Toole's, Oct. 5, 1992, Nov. 8, 1992.

23. O'Toole, Testimony and Chronology of Events, *Hearing, Fraud in N.I.H. Grant Programs*, April 12, 1988, pp. 89, 100, 119–120.

24. O'Toole, Testimony, *Hearing, Fraud in N.I.H. Grant Programs*, April 12, 1988, pp. 112, 119–120; Sarsasohn, *Science on Trial*, p. 64. O'Toole's omission of Lanman's telephone call evidently bothered her. A month later, during the course of an interview with N.I.H. officials in which Lanman participated, she asked him "whether my response to Congress was in any way inaccurate given that you had spoken to me" and whether she should write a letter of correction. Lanman told her that she need not worry about the matter; his conversation with her had not been a real contact. N.I.H. Panel, O'Toole interview, AE H-109, May 17, 1988, pp. 7–8.

25. Author's interview with Stockton and Chafin, March 24, 1993; O'Toole, Testimony, *Hearing, Fraud in N.I.H. Grant Programs*, April 12, 1988, p. 89.

26. Walter W. Stewart and Ned Feder, "Original Data Contradict Published Claims: Analysis of a Recent Paper," Sept. 30, 1987, *Hearing, Fraud in N.I.H. Grant Programs*, April 12, 1988, pp. 26–61.

27. Stewart, Testimony; Maplethorpe, Testimony, *Hearing, Fraud in N.I.H. Grant Programs*, April 12, 1988, pp. 9–11, 107, 114.

28. Dingell, Wyden, *Hearing, Fraud in N.I.H. Grant Programs*, April 12, 1988, pp. 222–

223, 247–248, 253–256. One reporter later noted that the hearing was "raucous" and that the N.I.H. officials were "visibly shaken by the Chairman's anger." "House Chairman Accuses N.I.H. of Leaking Data to Inquiry Target," *Science and Government Report*, Nov. 15, 1988, p. 3.

29. *Hearing, Fraud in N.I.H. Grant Programs*, April 12, 1988, p. 256; "Fraud Inquiry: N.I.H. on the Capitol Hill Griddle (continued)," *Science and Government Report*, May 1, 1988, p. 5; author's interview with Miers, March 24, 1993.

30. Dingell, Wyden, *Hearing, Fraud in N.I.H. Grant Programs*, April 12, 1988, pp. 253–256; author's interview with Mary Miers, March 24, 1993; Miers, Testimony, *Hearing, Scientific Fraud and Misconduct and the Federal Response*, April 11, 1988, p. 179. Miers says that the congressional browbeating convinced her that she no longer wanted to be in the business of investigating scientific misconduct, so when in July, the opportunity to take another job at N.I.H. came along, she leaped at it. Author's interview with Miers, March 24, 1993.

31. *Hearing, Fraud in N.I.H. Grant Programs*, April 12, 1988, pp. 267–268; "Fraud Inquiry: Harsh Treatment for N.I.H. on Capitol Hill," "Fraud Inquiry: N.I.H. on the Capitol Hill Griddle (continued)," *Science and Government Report*, April 15, 1988, p. 1; May 1, 1988, p. 5.

32. Author's interviews with Stewart and Feder, Sept. 16, 1992; with Stockton and Chafin, March 24, 1993; Boffey, "Two Critics of Science Revel in Role," *New York Times*, April 19, 1988, p. C1; Nicholas McBride, "Pair of Researchers Put Scientific Cheaters under the Miscroscope," *Christian Science Monitor*, May 11, 1988, p. 5; "Congressional Fraud Probers Seek Data at MIT, Tufts," *Science and Government Report*, May 15, 1988, p. 1; Michael F. Barrett, Jr. to Robert B. Lanman, April 25, 1988, Onek Files.

33. "When Scientists Fudge . . . ," editorial, *Washington Post*, April 17, 1988, p. C6. A comprehensive collection of newspaper stories on the Weiss and Dingell hearings is in the Stewart and Feder File.

34. Author's interview with Stockton and Chafin, March 24, 1993.

35. Peter Gwynne, "Have the Fraudbusters Gone Too Far?" *Scientist*, 2(July 11, 1988), 1, 8; Constance Holden, "Whistle-Blowers Air Cases at House Hearings," *Science*, 240(April 22, 1988), 387.

36. Daniel E. Koshland, Jr., "Science, Journalism, and Whistleblowing," *Science*, 240(April 29, 1988), 585; Paul Berg to John D. Dingell, April 26, 1988; Maxine Singer to the *Washington Post*, April 21, 1988, copies in files of Maxine Singer (thereafter Singer Files); John T. Edsall, "The Nature of Whistleblowing," letter, *Science*, 241(July 1, 1988), pp. 11–12; Henry Wortis, Brigitte Huber, and Robert Woodland, "Fraud Allegations," letter, *Science*, 240(May 20, 1988), 968; Henry Wortis, Brigitte Huber, and Robert Woodland, "Unjust Congress," letter, *Nature*, 333(May 26, 1988), 293; David Baltimore to Colleague, May 19, 1988, Baltimore Files; "Summary of Meeting . . . ," attached to Louis Lasagna to Henry Banks, April 22, 1988, Onek Files.

37. Author's interview with Baltimore, March 20, 1996.

38. "Curriculum Vitae, David Baltimore"; Sondra Schlesinger, "David Baltimore, Oral History," Transcript, Feb. 7, 1994, April 13, 1995; Tim Beardsley, "Profile: David Baltimore," *Scientific American*, Jan. 1992, p. 36.

39. Howard M. Temin, "Homology Between RNA from Rous Sarcoma Virus and DNA from Rous Sarcoma Virus-Infected Cells," *Proceedings of the National Academy of Sciences of the United States of America*, 52(1964), 323–339; Howard M. Temin and Satoshi Mizutani, "RNA-Dependent DNA Polymerase in Virions of Rous Sar-

coma Virus," *Nature*, 226(1970), 1211–1213; author's interview with Howard Temin, Nov. 2, 1991.

40. David Baltimore, "Viruses, Polymerases and Cancer," Nobel Lecture, December 12, 1975, in *Les Prix Nobel en 1975* (Stockholm: The Nobel Foundation, 1976), pp. 159–160; Sondra Schlesinger, "David Baltimore, Oral History," Transcript, April 13, 1995.

41. David Baltimore, "Viral RNA-Dependent DNA Polymerase"; Temin and Mizutani, "RNA-Dependent DNA Polymerase in Virions of Rous Sarcoma Virus," *Nature*, 226(June 27, 1970), 1209–1211, 1211–13.

42. Author's interview with Baltimore, March 20, 1996. Baltimore's post-Nobel productivity is probably higher than that of most laureates, whose productivity in research tends to fall after they win the prize but to remain far higher than scientists of comparable age. See Harriet Zuckerman, *Scientific Elite: Nobel Laureates in the United States* (New York: Free Press, 1977), esp. chapter 7.

43. Colin Norman, "MIT Agonizes over Links with Research Unit"; "Whitehead Link Approved"; Michelle Hoffman, "The Whitehead Institute Reaches Toward Adulthood," *Science*, 214(Oct. 23, 1981), 416–417; 1104; 256(April 3, 1992), 25–27.

44. Schlesinger, "David Baltimore, Oral History," Transcript, April 13, 1995; Sarasohn, *Science on Trial*, pp. 79–81; "Baltimore Attacks 'Professional Guardians of the Status Quo,'" *Science*, 239(Feb. 26, 1988), 972; David Baltimore, "Limiting Science: A Biologist's Perspective," *Limits of Scientific Inquiry, Daedalus*, 107(Spring 1978), 37–46.

45. Author's interview with David Baltimore, Aug. 17, 1992; David Baltimore to Colleague, May 19, 1988, Baltimore Files; Judy Foreman, "Science Case: Who Knew What When?" *Boston Globe*, May 3, 1989, p. 14; Paul A. Gigot, "Latest Chapter in the Fine Science of the Smear," *Wall Street Journal*, May 5, 1989, p. A14; Baltimore, Statement, *Hearings, Scientific Fraud*, May 4, 1989, p. 99; David Baltimore, "Baltimore's Travels," *Issues in Science and Technology*, V(No. 4, 1989), 50.

46. Author's interview with O'Toole, Oct. 5, 1992; N.I.H. Staff, Interview with Baltimore, May 18, 1988. In 1992, Imanishi-Kari illustrated the kind of difficulty she encountered in assessing her hybridoma data, pointing to some of the numbers and saying, "You would say maybe these are positive, real, and maybe there is something. But this—we knew all along that this is transgene positive by Northern [blot]. So there are things here that you could say, 'Yeah, could be.' But it doesn't look good, because if you find that the control itself is not working, it's very hard to compare the experiment or say anything about it. Now, six years later, a lot of these things make a lot of sense. But that's six years later, because now I understand exactly what's going on with these cells. When I look today at this data and some other data that I couldn't make too much sense out of [then], it makes much more sense now." Author's interview with Imanishi-Kari, Oct. 6, 1992.

47. In a colloquy with O'Toole about Imanishi-Kari's data, Dingell said that he wanted to ensure that published "scientific data upon which later judgments are made are correct." *Hearing, Fraud in N.I.H. Grant Programs*, Apri. 12, 1988, p. 112.

48. "Congressional Fraud Probers Seek Data at MIT, Tufts," *Science and Government Report*, May 15, 1988, p. 1; Baltimore to Colleague, May 19, 1988, Baltimore Files.

49. David Baltimore, "Dear Colleague," May 17, 1988, copy in Baltimore Files; David Baltimore, "Baltimore's Travels," *Issues in Science and Technology*, V(No. 4, 1989), 50–51.

50. Imanishi-Kari to Katherine L. Bick, March 28, 1988, AE H-245; Baltimore to Col-

league, May 19, 1988, Baltimore Files; Baltimore to James Wyngaarden, May 4, 1988, Baltimore Files, item 77.

51. David Baltimore, "Dear Colleague," May 17, 1988, copy in Baltimore Files; David Baltimore, "Baltimore's Travels," *Issues in Science and Technology*, V(No. 4, 1989), 50–51; Baltimore, Statement, *Hearings, Scientific Fraud*, May 4, 1989, pp. 102–103.

NOTES FOR CHAPTER SEVEN

1. "Congressional Fraud Probers Seek Data at MIT, Tufts"; "Q&A: With Director Wyngaarden on Political Troubles at N.I.H.," *Science and Government Report*, May 15, 1988, p. 1; June 1, 1988, p. 5; author's interview with Mary Miers, March 24, 1993; Deputy Director for Extramural Research to Director, N.I.H., Jan. 18, 1989, "Review of Alleged Inaccuracies in *Cell* Paper—DECISION," attached to [Final Report of the Scientific Panel], Jan. 31, 1989, and James B. Wyngaarden to David Baltimore, Jan. 31, 1989 (hereafter, DECISION, Jan. 18, 1989), Baltimore Files, item 130.

2. N.I.H. Staff, O'Toole interview, May 17, 1988, Transcript, AE-H-109, pp. 36, 65, 74, 83.

3. *Ibid.*, pp. 44–45, 56, 79.

4. *Ibid.*; author's interview with Bruce Maurer, Nov. 1, 1996.

5. Author's interview with Maurer, Nov. 1, 1996; Maplethorpe to Mary Miers, Aug. 1, 1988; N.I.H. Staff, Interview with Charles Maplethorpe, May 20, 1988, Transcript, AE H-251, pp. 5–6; DECISION, Jan. 18, 1989, Baltimore Files, item 130.

6. N.I.H. Staff, Interviews with David Weaver, May 18, 1988, Transcript, AE H-115, pp. 9–13; with Imanishi-Kari, May 19, 1988, Summary, AE H-100; with David Baltimore, May 18, 1988, Transcript, AE H-107; with Huber, Woodland, and Wortis, May 20, 1988, handwritten notes; with Gene Brown, May 18, 1988, handwritten notes, Onek Files; Maggie Hassan to File, May 23, 1988, Re: Interviews of Dr. Gene Brown and Dr. Herman Eisen by Staff from the National Institutes of Health, Eisen Files.

7. N.I.H. Staff, O'Toole interview, May 17, 1988, Transcript, AE H-109, p. 84; author's interviews with Hugh McDevitt, July 2, 1996; with Joseph Davie, Oct. 28, 1996.

8. Author's interviews with Storb, April 8, 1997; with Bruce Maurer, Nov. 1, 1996; with McDevitt, July 2, 1996; with Davie, Oct. 28, 1996.

9. N.I.H. Panel, [Final Draft Report], attached to Janet M. Newburgh to David Baltimore, Nov. 18, 1988, Baltimore Files, item 119; author's interview with McDevitt, July 2, 1996, and, by telephone, March 12, 1997. The documents the panel ultimately used are listed as an appendix to the final draft report. Maurer emphasizes that the panel's charge was "not to do a data audit"—that is, a complete review of every record—but only to examine the data relevant to O'Toole's critique. Author's interview with Maurer, Nov. 1, 1996.

10. N.I.H. Staff, Baltimore interview, May 18, 1988, Transcript, AE H-107, p. 20; Thereza Imanishi-Kari, Statement, *Hearings, Scientific Fraud*, May 4, 1989, pp. 144–145; O.S.I., Imanishi-Kari interview, Oct. 13, 1990, Transcript, AE H-103, pp. 47–48.

11. O.S.I., Baltimore interview, April 30, 1990, Transcript, AE H-108, pp. 70–73; O.S.I., Imanishi-Kari interview, Oct. 13, 1990, Transcript, AE H-103, pp. 44–45,

49–51, 54–58; author's interview with Imanishi-Kari, Oct. 6, 1992. On May 2, 1988, according to a later draft of a federal report on Imanishi-Kari, "NIH received an anonymous telephone call reporting, 'There is tremendous activity at Tufts regarding Dr. Imanishi-Kari's data. The original coverup committee is now helping her get her data ready. . . . The data was xeroxed on Friday, April 29, 1988, and brought to the interested parties at MIT.' " Typescript partial draft of report, n.d., p. 17, attached to Ursula Storb to Barbara Williams, June 21, 1990, O.R.I. Files.

12. Author's interview with Davie, Oct. 28, 1996; DECISION, Jan. 18, 1989, Baltimore Files, item 130. O'Toole recorded her session with the panel, but she has not released the tape. Author's interview with Ursula Storb, April 8, 1997; O'Toole, Testimony, *Appeal Proceedings*, June 16, 1995, p. 1058.

13. [Final Report of the Scientific Panel], Jan. 31, 1989, attached to Wyngaarden to Baltimore, Jan. 31, 1989, Baltimore Files, item 130; author's interview with Maurer, Nov. 1, 1996; O'Toole to Brian Kimes, Nov. 6, 1989, AE H-276.

14. Meeting of N.I.H. Panel and Authors, May 3, 1989, Transcript, AE H-101, pp. 93–95; author's interview with McDevitt, by telephone, June 15, 1997. Davie does not remember O'Toole's proposing the *mu-gamma* idea when she met with the panel but says that he respects McDevitt's memory and has no reason to say she didn't. Storb says she remembers the idea being discussed at the meeting with O'Toole but not who brought it up. In 1995, O'Toole said that she "never had a mu gamma heterodimer theory." O'Toole, Testimony, *Appeal Proceedings*, June 16, 1995, p. 1060; author's interviews with Davie, by telephone, June 23, 1997; with Storb, April 8, 1997.

15. Author's interviews with O'Toole, Oct. 5, 1992; with Davie, Oct. 28, 1996, and, by telephone, March 12, 1997; with McDevitt, July 23, 1996, and, by telephone, March 12, 1997; with Bruce Maurer, Nov. 1, 1996; [Final Report of the Scientific Panel], Jan. 31, 1989, attached to Wyngaarden to Baltimore, Jan. 31, 1989, Baltimore Files, item 130; O'Toole to M. Janet Newburgh, Dec. 2, 1988, AE H-260. O'Toole later complained that she was not told about the subcloning data that Imanishi-Kari presented in support of the paper. It is likely that she was not shown it because the panel did not become aware of it until Imanishi-Kari, who, as O'Toole knew, met with the panel after she did, called it to their attention. See O'Toole to Janet Newburgh, Nov. 28, 1988, AE H-258; O'Toole to Brian Kimes, Nov. 6, 1989, AE H-276.

16. Author's interview with Bruce Maurer, Nov. 1, 1996.

17. Weaver to Miers, June 13, 1988, AE H-247; author's interviews with Davie, Oct. 28, 1996; with Maurer, Nov. 1, 1996; with Storb, April 8, 1997; DECISION, Jan. 18, 1989, Baltimore Files, item 130.

18. Meeting of N.I.H. Panel and Authors, May 3, 1989, Transcript, AE H-1-1, pp. 10–11, 76–77, 102–104; Ursula Storb, Testimony, *Hearings, Scientific Fraud*, May 4, 1989, p. 34; [Final Report of the Scientific Panel], Jan. 31, 1989, attached to Wyngaarden to Baltimore, Jan. 31, 1989, Baltimore Files, item 130.

19. Author's interviews with McDevitt, July 2, 1996; with Davie, Oct. 28, 1996; author's interview, by telephone, with McDevitt, July 9, 1996; O.S.I. Scientific Panel, Interview of Imanishi-Kari, Oct. 13, 1990, AE H-103, pp. 81–88; Weaver *et al.*, "Altered Repertoire . . . ," *Cell*, 45 (April 25, 1986), 250, Table 2.

20. O.S.I. Scientific Panel, Imanishi-Kari interview, Oct. 13, 1990, Transcript, AE H-103, pp. 88–90; author's interview with Storb, April 8, 1997.

21. Author's interviews with McDevitt, July 2, 1996; with Davie, Oct. 28, 1996;

author's telephone conversation with McDevitt, July 9, 1996; O.S.I. Scientific Panel, Interview of Imanishi-Kari, Oct. 13, 1990, AE H-103, pp. 81–88.

22. Author's interview with Shirley Tilghman, March 22, 1996. According to the Poisson formula, about 37 percent of the wells on the plate would contain one hybridoma clone; about 18 percent, two; and about 6 percent, three. About ninety clones would thus be growing on such a plate. The denominator in a truly clonal frequency calculation would thus be 90 instead of 63, which would make the frequency two thirds what the calculation using wells produced, or roughly 33 percent. Imanishi-Kari's data showed that the frequency with which normal mouse produced the idiotypically birthmarked antibodies was no more than 1 percent. The factor of three higher for the transgenic mice is based on allowing a rate for the normal mice of 10 percent.

23. Meeting of N.I.H. Panel and Authors, May 3, 1989, Transcript, AE H-101, pp. 14–16; O.S.I. Scientific Panel, Imanishi-Kari interview, Oct. 13, 1990, Transcript, AE H-103, pp. 7–11, 88–90; O.S.I., Baltimore interview, April 30, 1990, Transcript, AE H-108, pp. 31–32; author's interview with Imanishi-Kari, Oct. 6, 1992; Imanishi-Kari to Bruce Maurer, Oct. 3, 1988, AE H-255.

24. Author's interviews with McDevitt, July 2, 1996, and, by telephone, July 9, 1996; with Davie, Oct. 28, 1996; with Storb, April 8, 1997; O.S.I. Scientific Panel, Interview of Imanishi-Kari, Oct. 13, 1990, AE H-103, pp. 81–88; Imanishi-Kari to Maurer, Oct. 3, 1988, AE H-255; Sarasohn, *Science on Trial*, pp. 109–110.

25. Imanishi-Kari to Bick, March 28, 1988, AE H-245; Imanishi-Kari to Maurer, Oct. 3, 1988, AE H-255; [Final Report of the Scientific Panel], Jan. 31, 1989, attached to Wyngaarden to Baltimore, Jan. 31, 1989, Baltimore Files, item 130; author's interviews with Imanishi-Kari, Nov. 6, 1992, March 20, 1996, Oct. 29, 1996; O.S.I., Imanishi-Kari interview, Oct. 13, 1990, Transcript, AE H-103, pp. 81–88, 94–95, 134.

26. Author's interviews with McDevitt, July 2, 1996, and, by telephone, March 12, 1997; with Davie, Oct. 28, 1996, and, by telephone, March 12, 1997; with Storb, April 8, 1997; O.S.I. Scientific Panel, Imanishi-Kari interview, Oct. 13, 1990, Transcript, AE H-103, pp. 88–90; Davie, Testimony, *Appeal Proceedings*, June 20, 1995, p. 1355.

27. B[arbara] J. C[ulliton], "Panel Completes Interviews in 'Baltimore Case,'" *Science*, 241 (July 15, 1988), 286; Maplethorpe to Miers, Aug. 1, 1988, AE H-251; Imanishi-Kari, "Answers to Questions Posed by Scientific Panel . . . ," July 25, 1988, AE H-250; [Final Report of the Scientific Panel], Jan 31, 1989, attached to Wyngaarden to Baltimore, Jan. 31, 1989, Baltimore Files, item 130.

28. [Final Report of the Scientific Panel], Jan. 31, 1989, attached to Wyngaarden to Baltimore, Jan. 31, 1989, Baltimore Files, item 130; author's interview with Bruce Maurer, Nov. 1, 1996; Joseph Davie, Testimony, *Hearings, Scientific Fraud*, May 4, 1989, pp. 15–16.

29. Imanishi-Kari to Maurer, Oct. 3, 1988, AE H-255; Meeting of N.I.H. Panel and the Authors, May 3, 1989, Transcript, AE H-101, pp. 12, 67–68; Weaver *et al.*, "Altered Repertoire . . . ," *Cell*, 45 (April 25, 1986), 250.

30. Imanishi-Kari to Maurer, Oct. 3, 1988, AE H-255; Meeting of N.I.H. Panel and the Authors, May 3, 1989, Transcript, AE H-101, pp. 12, 67–68; Joseph Davie, Testimony, *Hearings, Scientific Fraud*, May 4, 1989, pp. 14–15; author's interview with McDevitt, July 2, 1996.

31. Author's interview with Baltimore, Jan. 23, 1992, and telephone conversation with Baltimore, April 24, 1996; [Final Report of the Scientific Panel], Jan. 31, 1989,

attached to Wyngaarden to Baltimore, Jan. 31, 1989, Baltimore Files, item 130; "House Chairman Accuses N.I.H of Leaking Data to Inquiry Target," *Science and Government Report*, Nov. 15, 1988, pp. 1, 3; James B. Wyngaarden to John D. Dingell, Nov. 18, 1988, Bernard Davis Files; Thereza Imanishi-Kari, Moema Reis, David Weaver, and David Baltimore, "Altered Repertoire of Endogenous Immunoglobulin Gene Expression in Transgenic Mice Containing a Rearranged Mu Henry Chain Gene," *Cell*, 55 (Nov. 18, 1988), 541.

32. John Butler, "Memo to the Record," Aug. 23, 1988, Onek Files; Bruce Maurer to the author, March 4, 1997.

33. Dingell to Otis R. Bowen, Nov. 10, 1988, Davis Files; "House Chairman Accuses N.I.H. of Leaking Data to Inquiry Target," *Science and Government Report*, Nov. 15, 1988, pp. 1, 3; Alun Anderson and Joseph Palca, "US Congressman Attacks N.I.H. Investigation," *Nature*, 336 (Nov. 24, 1988), 295; Judy Foreman, "Scientists Acknowledge Mistakes in an Article," *Boston Globe*, Nov. 17, 1988, p. 24.

34. "House Chairman Accuses N.I.H. of Leaking Data to Inquiry Target," *Science and Government Report*, Nov. 15, 1988, pp. 1, 3; M. Janet Newburgh to Baltimore, Nov. 18, 1988, Baltimore Files, item 119; Wyngaarden to Dingell, Nov. 18, 1988, Davis Files.

35. Newburgh to Baltimore and attached draft final report, Nov. 18, 1988, Baltimore Files, item 119. McDevitt wrote the initial draft of the report's conclusion and observed there that apart from recommending a formal correction of Table 2, the results of the panel's review were similar to "those of the review carried out by Dr. Eisen." McDevitt, untitled draft, attached to Karen W. Moody to Bruce Maurer, Aug. 22, 1988, O.R.I. Files.

36. Author's interviews with Maurer, Nov. 1, 1996; with Davie, Oct. 28, 1996; Alan M. Stall *et al.*, "Rearrangement and Expression of Endogenous Immunoglobulin Genes . . . ," *Proceedings of the National Academy of Sciences of the United States of America*, 85 (May 1988), 3546–3550; M. Janet Newburgh to David Baltimore and attached draft final report, Nov. 18, 1988, Baltimore Files, item 119; Davie, Testimony; *Appeal Proceedings*, June 20, 1995, pp. 1418–1421.

37. M. Janet Newburgh to David Baltimore and attached draft final report, Nov. 18, 1988, Baltimore Files, item 119.

38. David Baltimore *et al.*, "Memorandum," Nov. 28, 1988, attached to Baltimore *et al.* to Janet Newburgh, Nov. 29, 1988, attached to Normand Smith to Newburgh, Nov. 29, 1988, AE H-259; author's interview with Baltimore, Jan. 23, 1992.

39. Baltimore *et al.*, "Memorandum," Nov. 28, 1988.

40. *Ibid.*; author's interview with Baltimore, Jan. 23, 1992; Meeting of N.I.H. Panel and Coauthors, May 3, 1989, Transcript, AE H-101, pp. 9–10; Imanishi-Kari, Testimony, *Appeal Proceedings*, Aug. 31, 1995, pp. 5272–5273.

41. Margot O'Toole to Janet Newburgh, Nov. 28, 1988, AE H-258. O'Toole sent a two-page supplementary letter four days later. O'Toole to M. Janet Newburgh, Dec. 1988, AE H-260.

42. "Error but No Fraud, N.I.H. Panel Rules on Hotly Disputed Paper," *Science and Government Report*, Dec. 1, 1988, pp. 1, 3; David L. Wheeler, "Panel Investigating Scientists on Possible Misconduct Is Accused of Improperly Sharing Its Conclusions," *Chronicle of Higher Education*, Nov. 30, 1988, p. A4.

43. Warren E. Leary, "Panel Finds 'Serious Errors' in Paper on Gene Experiment," *New York Times*, Dec. 3, 1988, p. 7; "How Better To Police Research," editorial, *Nature*, 336 (Dec. 8, 1988), 503–504.

44. Baltimore to the Editor, *Nature*, Dec. 1988, draft; Marilyn [Smith] to the Gang, Dec. 7, 1988; anon., "Comments on Draft Letter," Baltimore Files.

45. Author's interviews with Davie, Oct. 28, 1996; with McDevitt, July 2, 1996; with Maurer, Nov. 1, 1996; with Mary Miers, March 24, 1993. McDevitt says that the sets of subclones derived from each original well produced antibodies with a variety of characteristics—for example, some were *mu-a*, some were *mu-b*, and still others had isotypes other than *mu*. Such variety, he argues, meant that the original well must therefore have contained many more than one or two different hybridoma cell lines. Such variety, however, does not bother Ursula Storb, who points out that a hybridoma has two pairs of most chromosomes instead of the usual one and says that she "could easily understand that an original cell could make three different heavy chains." Author's interviews with McDevitt, by telephone, March 12, 1997; with Storb, April 8, 1997; Meeting of N.I.H. Panel and Authors, May 3, 1989, Transcript, AE H-101, pp. 17–20, 23–24.

46. Author's interviews with McDevitt, July 2, 1996; with Maurer, Nov. 1, 1996; DECISION, Jan. 18, 1989, Baltimore Files, item 130; B[arbara] J. C[ulliton], "Credit for Whistle-Blower Vanishes," *Science*, 244(May 12, 1989), 643.

47. DECISION, Jan. 31, 1989, Baltimore Files, item 130; Katherine Bick, Testimony, *Hearing, Scientific Fraud and Misconduct and the Federal Response*, April 11, 1988, p. 140.

48. Imanishi-Kari to Robert Lanman, Dec. 30, 1988, AE H-261.

49. Brigitte Huber to Lanman, Jan. 9, 1989; Henry Wortis to Lanman, Jan. 9, 1989; Robert Woodland to Lanman, Jan. 11, 1989, Onek Files.

50. DECISION, Jan. 31, 1989, Baltimore Files, item 130.

51. James B. Wyngaarden to David Baltimore, with attachments, Jan. 31, 1989, Baltimore Files, item 130.

52. Robert Steinbrook, "U.S. Study Clears Noted Scientist of Alleged Research Misconduct," *Los Angeles Times*, Feb. 2, 1989, pp. 3, 25; David L. Wheeler, "Nobelist Found Guilty of Errors in Paper, but Innocent of Fraud," *Chronicle of Higher Education*, Feb. 8, 1989, p. A9; "N.I.H. Clears but Chastises Baltimore and Coauthors," *Science and Government Report*, Feb. 15, 1989, p. 5; Margot O'Toole, "Scientists Must Be Able To Disclose Colleagues' Mistakes Without Risking Their Own Jobs or Financial Support," *Chronicle of Higher Education*, Jan. 25, 1989, p. A44.

53. David Baltimore, Thereza Imanishi-Kari, and David Weaver to James B. Wyngaarden, Feb. 15, 1989, AE H-264.

54. Author's interview with Baltimore, Jan. 23, 1992; O.S.I., Baltimore interview, April 30, 1990, Transcript, AE H-108.

55. Baltimore, Imanishi-Kari, and Weaver to James B. Wyngaarden, Feb. 15, 1989, AE H-264; "Baltimore Sends 'Correction' Demanded by N.I.H.," *Science and Government Report*, March 15, 1989; James B. Wyngaarden to Baltimore, to Imanishi-Kari, and to Weaver, March 10, 1989, Baltimore Files, item 139 and AE H-266.

56. Imanishi-Kari, Weaver, and Baltimore, "On the Specificity of Bet-1 Antibody." *Cell*, 57 (May 19, 1989), 515–516; Benjamin Lewin, "Travels on the Fraud Circuit," editorial, *Cell*, 57 (May 19, 1989), 513. The Harvard biologist Bernard Davis called the decision to force the coauthors to submit the additional letter to *Cell* "regrettable." "It deprives a distinguished scientist of his right to expert judgment about what is a significant error, and to have the outcome of the dispute determined in the traditional way by further work, rather than by legal pressure." Davis,

"Fraud vs. Error: The Dingelling of Science," *Wall Street Journal*, March 8, 1989, p. A14.

57. Author's interview with Imanishi-Kari, March 20, 1996; Imanishi-Kari to the author, March 10, 1997; Sarasohn, *Science on Trial*, p. 89; Philip Boffey, "Nobel Winner Is Caught Up in a Dispute Over Study," *New York Times*, April 12, 1988, p. C10.

NOTES FOR CHAPTER EIGHT

1. Author's interview with Stewart and Feder, Sept. 16, 1992; Patricia K. Woolf, "Science Needs Vigilante Not Vigilantes," *Journal of the American Medical Association*, 260(Oct. 7, 1988), 1940; Normand Smith to the author, March 14, 1997.

2. Maggie Hassan to File, Re: Subcommittee Interviews of Dr. Herman Eisen and Dean Gene Brown on May 11, 1988, May 14, 1988, Eisen Files, pp. 11, 14; author's interview with Stockton and Chafin, March 24, 1993. A subcommittee spokesman told a reporter a few days after the release of the N.I.H. report, "We are not finished with [the matter]." Robert Steinbrook, "U.S. Study Clears Noted Scientist of Alleged Research Misconduct," *Los Angeles Times*, Feb. 2, 1989, p. 25.

3. Office of Scientific Integrity, "Investigation Report Concerning the Weaver et al 1986 Cell Paper," Reference 072, n.d. [March 14, 1991], copy in Baltimore Files, p. 30; author's interviews with Bruce Singal, March 20, 1996; with Imanishi-Kari, March 20, 1996. Singal's firm was Ferriter, Scobbo, Sikora, Caruso, & Rodelphe, at One Milk Street in Boston.

4. Author's interviews with Stewart and Feder, Sept. 16, 1992; with Imanishi-Kari, March 20, 1996; author's telephone conversation with Normand Smith, July 24, 1992; Robert E. Sullivan to Peter Stockton, May 16, 1988, Onek Files; Judy Sarasohn, "Akin, Gump's Science Project," *Legal Times: Law and Lobbying in the Nation's Capital*, July 4, 1988, p. 7; Robert B. Lauman to Bruce A. Singal, May 24, 1988; Bick to Dingell, July 21, 1988, O.R.I. Files; John D. Dingell, [subpoena of Imanishi-Kari's materials], June 30, 1988, AE H-248; Bruce Singal to Robert Lanman, July 19, 1988, AE H-249; author's interview with Bruce Singal, March 20, 1996. In late April, the subcommittee was content to receive copies of the data on the understanding that the N.I.H. would obtain the originals from Imanishi-Kari. It probably decided to ask for the originals itself once it learned that Imanishi-Kari had kept them. Michael F. Barrett, Jr. to Robert Lanman, April 25, 1988, Onek Files.

5. Sarasohn, "Akin, Gump's Science Project," *Legal Times: Law and Lobbying in the Nation's Capital*, July 4, 1988, pp. 1, 6; Sarasohn, *Science on Trial*, pp. 82–83.

6. John D. Dingell to David Weaver, Aug. 8, 1988; Weaver to Dingell, Aug. 22, 1988, AE H-252, H-254; Baltimore to Dingell, Aug. 25, 1988, attached to Baltimore to Michael Barrett, Jr., Baltimore Files, items 108, 109; David Baltimore To Those Who Have Sent Me Letters of Support, April 27, 1989, Singer Files. Bruce Chafin says that Dingell was approached by people under investigation constantly; that his staff had to protect his time; and that they did not treat Baltimore any differently from the way they dealt with the CEO of a major corporation. Author's interview with Bruce Chafin and Peter Stockton, March 24, 1993.

7. Author's interview with Stewart and Feder, Sept. 16, 1992; Kuznik, "Fraud Busters," *Washington Post Magazine*, April 14, 1991, p. 32; Philip Weiss, "Conduct Unbecoming?" *New York Times Magazines*, Oct. 21, 1989, p. 68.

8. Stewart contends that Imanishi-Kari threw out all the pages in the notebooks that proved the truth of what O'Toole claimed and that she even edited O'Toole's notebook to the same end. Author's interviews with Stewart and Feder, Sept. 16, 1992; with Stockton and Chafin, March 24, 1993; Sarasohn, *Science on Trial*, pp. 98–99.

9. Sarasohn, *Science on Trial*, pp. 97–100; Hargett, Testimony; Larry Stewart, Testimony, *Hearings, Scientific Fraud*, May 4, 1989, pp. 50–51, 59; John Hargett, Testimony; Larry F. Stewart, Testimony, Aug. 23, 1995, *Appeal Proceedings*, pp. 3559, 3984; Jock Friedly, "How Congressional Pressure Shaped the 'Baltimore Case,'" *Science*, 273(Aug. 16, 1996), 874. Larry Stewart is no relation to Walter Stewart.

10. Sarasohn, *Science on Trial*, pp. 98–99; Special Meeting of N.I.H. Office of Scientific Integrity with U.S. Secret Service Representatives, Transcript, July 14, 1988, pp. 8–10, Onek Files.

11. Special Meeting of N.I.H. Office of Scientific Integrity with U.S. Secret Service Representatives, Transcript, July 14, 1988, pp. 47–54.

12. Feder thought that they all lied, but doubted that there was a conspiracy. Author's interviews with Stewart and Feder, Sept. 16, 1992; with Stockton and Chafin, March 24, 1993. At the hearing of the Dingell subcommittee in 1989, Congressman Axel McMillan, a subcommittee member from North Carolina, queried Baltimore: "So it'd be very easy for someone who does not have the scientific experience and discipline to come in and pick up those notes, and pick and choose, and draw erroneous conclusions." To which Baltimore responded: "Absolutely. I doubt that there's a scientific study of this sort in existence that doesn't have within the laboratory notebooks contradictory data." Baltimore, Testimony, *Hearings, Scientific Fraud*, May 4, 1989, p. 157.

13. William Booth, "A Clash of Cultures at Meeting on Misconduct," *Science*, 243 (Feb. 3, 1989), 598; author's interviews with Stewart and Feder, Sept. 16, 1992; with Norton Zinder, Nov. 19, 1996, by telephone; Sarasohn, *Science on Trial*, pp. 115–116. Stewart suggested to me that the comment was off the cuff, made in connection with a previous speaker's mentioning the Holocaust in another context. However, Stewart and Feder had previously made the same analogy in a letter to Elizabeth Neufeld, a professor of biological chemistry at UCLA medical school. Neufeld told a *New York Times* reporter that the "clear implication of their letter [was] that if you don't agree with them, you must be like one of those German scientists. I was deeply offended." Boffey, "Two Critics of Science Revel in Role," *New York Times*, April 19, 1988, p. C7.

14. O'Toole to M. Janet Newburgh, Feb. 5, 1989, March 30, 1989, AE H-246, H-263; O'Toole to Bick, March 6, 1989, O.R.I. Files; DECISION, Jan 18, 1989, Baltimore Files, item 130.

15. Moema H. Reis F. Soares to Imanishi-Kari, April 19, 1988, AE H-268; Wyngaarden, Testimony, *Hearings, Scientific Fraud*, May 4, 1989, p. 10; Meeting of N.I.H. Panel and Authors, May 3, 1989, transcript, AE H-101, pp. 66–67.

16. Dingell to James B. Wyngaarden, Jan. 31, 1989; Dingell to Lanman, April 3, 1989, Davis Files; James B. Wyngaarden to David Baltimore, April 28, 1989, Baltimore Files, item 160; Wyngaarden, Testimony; Davie, Testimony, *Hearings, Scientific Fraud*, May 4, 1989, pp. 10, 22, 26–27, 33, 84.

17. Special Meeting of N.I.H. Office of Scientific Integrity with the U.S. Secret Service Representatives, Transcript, July 14, 1989, pp. 8–10, 17, Onek Files; Larry F. Stewart, Testimony, John Hargett, Testimony, *Hearings, Scientific Fraud*, May 4, 1989, pp. 51, 59–60, 65. It is interesting to note that an FBI scientist advised Janet

Newburgh at the N.I.H. that it would be very difficult to determine when within a period of two years various documents originated. Janet Newburgh, Memorandum of Telephone Call, between Newburgh and Jim Lile, Dec. 21, 1988, Onek Files.

18. See Weaver's Testimony on Figure 4, *Hearings, Scientific Fraud,* May 4, 1989, p. 120.

19. Sarasohn, *Science on Trial,* p. 103; author's interview with David Weaver, Dec. 11, 1992; Steven Herzog, Testimony, *Hearings, Scientific Fraud,* May 4, 1989, pp. 51–59, 76–78.

20. James B. Wyngaarden, Testimony, *Hearings, Scientific Fraud,* May 4, 1989, p. 10; Brian W. Kimes to Baltimore, May 31, 1989, Baltimore Files, item 171; Brian W. Kimes to Imanishi-Kari, May 31, 1989, AE H-269.

21. Bob Sullivan to Professor Herman Eisen, April 25, 1989, Eisen Files; author's interview with Robert Woodland, by telephone, Dec. 3, 1992. Stockton said, groping for an explanation of why Woodland wasn't included, that he hadn't been "as much of a player as Wortis and Huber," adding, "He was kind of a wimp anyway." Author's interview with Stockton and Chafin, March 24, 1993.

22. Dingell to Baltimore, May 2, 1989, Baltimore Files, item 162; Dingell, Statement, *Hearings, Scientific Fraud,* May 4, 1989, p. 4; author's interview with Baltimore, Jan. 23, 1992; Baltimore, "Baltimore's Travels," *Issues in Science and Technology,* V (No. 4, 1989), 51; Sarasohn, *Science on Trial,* pp. 118–119.

23. Statement from David Baltimore, press release, May 2, 1989, Singer Files. Weaver later explained the technical basis of his exasperation: "The endogenous [i.e., native] gene expression is incredible relative to the transgene. And not only is the endogenous gene a different isotype, but the variable region of the gene is very similar to the transgene variable [region. There's] almost no difference on the sequence level which, from my point of view, means that the idiotype is going to be essentially the same. The idiotype reagent is not unique; it recognizes a kind of general structure of the variable regions. What Imanishi-Kari's preparation of this antibody meant was that it would recognize a family of antibodies that were related, but not of absolute identity. Here we had a clone that overproduced an endogenous gene of virtually identical idiotype and a different isotype. What we said in the paper was that the endogenous gene gave that cell line the idiotype property of the cell. The controversy or what they exaggerated—I heard, mainly by Stewart—was that there might be a [band showing a] little bit of expression of the transgene. It was the transgene size, but I still think it was a blur. . . . So I don't necessarily believe the difference; I think they're exaggerating [the significance of the blur]. And, secondly, it doesn't even make any sense on a scientific level because, with that clone, there's no question where the idiotype comes from. You know, it's from this overexpressed gene that's virtually identical at the sequence level to the transgene. That's just ridiculous." Author's interview with David Weaver, Dec. 11, 1992.

24. Meeting of N.I.H. Panel and Authors, May 3, 1989, Transcript, AE H-101, pp. 3–4; Special Meeting of N.I.H. Office of Scientific Integrity with U.S. Secret Service Representatives, Transcript, July 14, 1988, pp. 2–3, Memo of Telephone Call, Bruce Singal to Janet Numburgh, May 2, 1989, O.R.I. Files.

25. Meeting of N.I.H. Panel and Authors, May 3, 1989, Transcript, AE H-101, pp. 3–4.

26. Dingell, Statement; Baltimore, Testimony, *Hearings, Scientific Fraud,* May 4, 1989, pp. 4, 172–174; Baltimore, "Baltimore's Travels," *Issues In Science and Technology,* V (No. 4, 1989), 51; Dingell to Baltimore, May 2, 1989, Baltimore Files, item 162;

Larry Thompson, "Science under Fire," *Washington Post, Health, Science, and Society,* May 9, 1989, p. 12; author's interviews with Stewart and Feder, Sept. 16, 1992; with Baltimore, Jan. 23, 1992; Bruce Singal to Michael F. Barrett, Jr., April 28, 1989, *Hearings on Scientific Fraud (Part II),* 101st Cong., 2nd, Sess., (hereafter Hearings on Scientific Fraud (Part II)), May 14, 1990, pp. 158–160; Peter G. Gosselin, "Probe by Dingell Raises Issue: Who Polices Scientists?" *Boston Sunday Globe,* April 30, 1989, pp. 1, 18. The complexity of the forensic information presented to the coauthors is evident from the extensive notes on the briefing that were taken by Normand Smith, Baltimore's lawyer. Copy of the notes in author's possession.

27. Author's interviews with Davie, Oct. 28, 1996; with McDevitt, July 2, 1996; with Storb, April 27, 1997. Jock Friedly, "How Congressional Pressure Shaped the Baltimore Case," *Science,* 273 (Aug. 16, 1996), 874. Davie's recollection of a late-night meeting is confirmed by Chafin's remark at the hearing that he and Davie talked "until the wee hours the other night." Davie remembers the panel's being briefed a day or two before the hearing, but at the hearing, which took place on Thursday, McDevitt referred to a briefing the previous Friday. See McDevitt, Testimony, *Hearings, Scientific Fraud,* May 4, 1989, pp. 17–18, 71–72. McDevitt says that, when told about the cropping of the radiograph, the panel remarked, "You know, everybody does that." McDevitt interview, July 2, 1996. He later told a reporter, "I kept saying, 'If you want to make a presumption of guilt, yeah, this fits with it. . . . But it fits equally well with other explanations.'" Friedly, "How Congressional Pressure Shaped the Baltimore Case," *Science,* 273 (Aug. 16, 1996), p. 874.

28. Author's interviews with McDevitt, July 2, 1996; with Davie, Oct. 28, 1996; Dingell, Statement, *Hearings, Scientific Fraud,* May 4, 1989, p. 4. McDevitt remembers being told about the limited briefing of the coauthors in a second meeting with the subcommittee staff prior to the hearing. However, Storb does not recall a second meeting. McDevitt probably confused the meeting with the subcommittee staff with a meeting at the N.I.H. the day before the hearing, which he mentioned in Feb. 1990 and which Storb also recalls. Author's interview with Ursula Storb, April 8, 1997; O.S.I., "Special Meeting of Scientific Panel [with Margot O'Toole]," Feb. 11, 1990, Transcript, AE H-110, pp. 160–162.

29. Author's interviews with McDevitt, July 2, 1996; with Davie, Oct. 28, 1996; with Storb, April 8, 1997; Memorandum of Telephone Call, Singal to Newburgh, May 2, 1989, O.R.I. Files. See Sarasohn, *Science on Trial,* p. 122.

30. Meeting of N.I.H. Panel and Authors, May 3, 1989, Transcript, AE H-101, pp. 3–4, 33–36, 55–56. The conversation also covered the issue of wells v. clones. "I think we disagree," Davie said, "but its an honest disagreement," to which Baltimore responded, "That's all I want to make sure, that it's an honest disagreement." *Ibid.,* pp. 29–32.

31. *Ibid.,* pp. 35–36; McDevitt, Testimony, *Hearings, Scientific Fraud,* May 4, 1989, pp. 71–72.

32. Meeting of N.I.H. Panel and Authors, May 3, 1989, Transcript, AE H-101, pp. 74–76; Bruce Singal to Michael F. Barrett, Jr., April 28, 1989, Onek Files.

33. Meeting of N.I.H. Panel and Authors, May 3, 1989, Transcript, AE H-101, pp. 5–6, 52, 100–102.

34. Sharp to Dear Colleague, April 18, 1989, copy in author's possession; author's telephone conversation with Phillip Sharp, Oct. 5, 1992; Berg to Gerry Sikorski, April 26, 1989, Singer Files; Warren E. Leary, "Congress' Vigilance May Stifle

Research, Science Leaders Fear," *Phoenix Arizona Gazette*, May 19, 1989, Stewart and Feder Files, clippings.

35. Peter G. Gosselin, "Panel Will Focus on Scientist's Notes,"*Boston Globe*, May 3, 1989; Robert E. Pollack, "In Science, Error Isn't Fraud," *New York Times*, May 2 1989, p. A25; Leonore Herzenberg to Baltimore, May 1, 1989, Baltimore Files, item 161.

36. Mike Berman to Participants . . . , May 3, 1989, attached to Bob Sullivan to John M. Deutch, Herman N. Eisen, Gene M. Brown, Mary P. Rowe, May 3, 1989, Eisen Files.

37. Author's interview with Baltimore, March 20, 1996; Thompson, "Science under Fire," *Washington Post, Health, Science, and Society*, May 9, 1989, p. 12. Baltimore later said, "I wasn't planning anything dramatic. In fact, I wasn't planning any-thing specific. I simply wasn't going there to eat crow." Jock Friedly, "Trial and Error," *Boston Magazine*, Jan. 1997, p. 50.

38. Sarasohn, *Science on Trial*, p. 114; Dingell, Statement, *Hearings, Scientific Fraud*, May 4, 1989, p. 1, 3–4.

39. Hargett, Testimony, *Hearings, Scientific Fraud*, May 4, 1989, pp. 65–67; Joseph Palca, "Disputed Paper Takes Centre Stage in Congress," *Nature*, 339 (May 11, 1989), 83.

40. Author's interview with McDevitt, July 2, 1996; Wyngaarden, Testimony, *Hear-ings, Scientific Fraud*, May 4, 1989, pp. 11–12, 82–83.

41. Davie, Testimony, *Hearings, Scientific Fraud*, May 4, 1989, pp. 13–14, 18–20, 67.

42. *Ibid.*, pp. 15–16, 22–23, 76–80.

43. Imanishi-Kari, Testimony and Statement, *Hearings, Scientific Fraud*, May 4, 1989, pp. 133, 145–149.

44. *Ibid.*, p. 133.

45. Baltimore, Statement and Testimony; Selsing to Baltimore, May 1, 1989, *Hearings, Scientific Fraud*, May 4, 1989, pp. 89, 92, 105, 109, 209–210, 212. Baltimore also entered a letter from Philip Leder, a professor at Harvard Medical School, which informed him that experiments with transgenic mice similar to his own had not yielded the detection of double producers. Leder to Baltimore, May 1, 1989, *ibid.*, p. 211. The *Cell* paper had received a considerable degree of scientific attention, having been cited thirty times in three years in other scientific papers, including only four by any of the coauthors, a rate ten times above average. Thompson, "Science under Fire," *Washington Post, Health, Science, and Society*, May 9, 1989, p. 12.

46. Baltimore, Statement and Testimony, *Hearings, Scientific Fraud*, May 4, 1989, pp. 88–92, 101–102, 154.

47. *Ibid.*, pp. 88–92, 102, 165–166.

48. *Ibid.*, pp. 171–172.

49. *Ibid.*, pp. 172–174; Thompson, "Science under Fire," *Washington Post, Health, Science, and Society*, May 9, 1989, p. 12; author's telephone interview with Daniel Joseph, March 25, 1989.

50. Author's telephone interview with Joseph, March 25, 1989; author's interview with Albert Kildow, June 7, 1996; Barbara J. Culliton, "The Dingell Probe Finally Goes Public," *Science*, 244(May 12, 1989), 646.

51. Author's interview with O'Toole, Oct. 5, 1992; O'Toole, Testimony, *Hearings, Scientific Fraud*, May 9, 1989, p. 181. Bruce Chafin says that O'Toole had been scheduled to testify before the May 4 hearing, but O'Toole's recollection is sharper. Author's interview with Stockton and Chafin, March 24, 1993.

52. O'Toole, Testimony, *Hearings, Scientific Fraud*, May 9, 1989, pp. 181, 185–187, 205, 207, 208. An unsigned, internal N.I.H. memorandum written just before the hearings noted that O'Toole's previous statements "were much more qualified" than those she was then making. Anon., "The Evolution of Dr. O'Toole's Allegations," April 24, 1989, Onek Files.

53. O'Toole, Testimony, *Hearings, Scientific Fraud*, May 9, 1989, pp. 203–204, 298–299. Dingell had observed to Wyngaarden on May 4 that O'Toole has "been destroyed" as a scientist, that she "has been driven out of research." *Ibid.*, May 4, 1989, p. 39.

54. John Deutch, Testimony; Louis Lasagna, Testimony, *Hearings, Scientific Fraud*, May 9, 1989, pp. 256, 262, 299–300.

55. Huber, Testimony; Wortis, Testimony, *Hearings, Scientific Fraud*, May 9, 1989, pp. 228–229, 242, 244. Wortis says that one of the unpublished experiments to which he alluded had been performed by Imanishi-Kari and the other by one of his postdocs. Author's interview with Wortis, Oct. 2, 1992.

56. Huber, Testimony; Wortis, Testimony, *Hearings, Scientific Fraud*, May 9, 1989, pp. 254–255, 245.

57. *Ibid.*, pp. 247–248.

58. *Ibid.*, pp. 250–251.

59. *Ibid.*, pp. 177–178, 301. See also the comments of Alex McMillan, pp. 43–44, 74–75.

60. Philip Weiss, "Conduct Unbecoming?" *New York Times Magazine*, Oct. 21, 1989, p. 71; Diana West, "Anatomy of a Scientific Scandal: O'Toole's Whistle," *Washington Times*, June 15, 1989, pp. E1, E3; "Baltimore Wins PR Battle, but Key Issues Remain," *Science and Government Report*, May 15, 1989, p. 5; B[arbara] J. C[ulliton], "Whose Notes Are They?" *Science*, 244 (May 19, 1989), 765; Herman N. Eisen, "O'Toole's Charges," letter, *Science*, 245 (Sept. 15, 1989), 1166–1167.

61. Stephen Jay Gould, "Judging the Perils of Official Hostility to Scientific Error," *New York Times*, July 30, 1989, Sec. 4, p. 6; "The Science Police," editorial, *Wall Street Journal*, May 15, 1989, p. A12; Sarasohn, *Science on Trial*, pp. 161–163; "Baltimore Wins PR Battle, but Key Issues Remain," *Science and Government Report*, May 15, 1989, p. 5; "Policing Scientific Research," *Washington Post*, May 9, 1989, p. A22; Judy Foreman, "The Bruises of a Battle over Science," *Boston Globe*, May 30, 1989; "Dingell's New Galileo Trial," *Detroit News*, May 17, 1989, Stewart and Feder Files, clippings.

62. Warren E. Leary, "Disputed M.I.T. Gene Research Is Facing Investigation," *New York Times*, April 30, 1989, p. 19; Barbara J. Culliton, "Dingell v. Baltimore," *Science*, 244 (April 28, 1989), 413; West, "Anatomy of a Scientific Scandal: O'Toole's Whistle," *Washington Times*, June 16, 1989, pp. E1, E3; Sarasohn, *Science on Trial*, pp. 84–85.

63. B[arbara] J. C[ulliton], "Whose Notes Are They?" *Science*, 244(May 19, 1989), 765; Margot O'Toole, "The Dingell Investigation," letter, *Science*, 244(June 16, 1989), 1243.

64. Gregory Gordon, "Cover-up Charge Puts Scientists under Microscope," *Detroit News*, Oct. 29, 1989, p. 22A; Bruce Singal to John D. Dingell, March 27, 1989, AE H-267; Dingell, *Hearings, Scientific Fraud*, May 4, 1989, p. 174; Dingell, Opening Statement; Dingell to Imanishi-Kari, July 14, 1989, *Hearings, Scientific Fraud (Part II)*, May 14, 1990, pp. 92, 229; Notes on Stewart and Feder and the Subcommittee, attached to William Raub to John Dingell, Dec. 15, 1989, Davis Files; O'Toole, Testimony, *Appeal Proceedings*, Sept. 14, 1995, pp. 6125–6126.

65. Greenberg, *Baltimore Sun*, May 22, 1989; "Baltimore Wins PR Battle, but Key Issues Remain"; "The Hearings According to Chairman John Dingell," *Science and Government Report*, May 15, 1989, p. 5; June 15, 1989, pp. 2, 4.

66. Author's telephone conversation with Jan Witkowski, Dec. 2, 1992; Booth, "A Clash of Cultures at Meeting on Misconduct," *Science*, 24(Feb. 3, 1989), 598; Palca, "Research, Misconduct and Congress," *Nature*, 337(Feb. 9, 1989), 503; author's interview with Stockton and Chafin, March 24, 1993; Benjamin Lewin, "Travels on the Fraud Circuit," editorial, *Cell*, 57(May 19, 1989), 513–514; "Rockefeller U. Faculty Cool to President Baltimore," *Science and Government Report*, Oct. 15, 1989, p. 6.

67. Author's interviews with Walter Gilbert, Oct. 2, 1992; with Brigitte Huber, Oct. 7, 1992; Weiss, "Conduct Unbecoming," *New York Times Magazine*, Oct. 29, 1989, p. 71; Booth, "A Clash of Cultures at Meeting on Misconduct," *Science*, 24(Feb. 3, 1989), 598; Gordon, "Cover-up Charge Puts Scientists under Microscope," *Detroit News*, Oct. 29, 1989, p. 22A; Palca, "Research, Misconduct and Congress," *Nature*, 337(Feb. 9, 1989), 503; Sarasohn, *Science on Trial*, pp. 167–168.

68. Barbara J. Culliton, "Baltimore to Succeed Lederberg?" *Science*, 245(Sept. 29, 1989), 1441; William K. Stevens, "Fraud Inquiry Figure Asked To Lead Rockefeller U."; "Dispute on New President Shatters Tranquil Study at Rockefeller U.," *New York Times*, Oct. 4, 1989, p. 18; Oct. 10, 1989, pp. 1, C6; Sarasohn, *Science on Trial*, pp. 169–171.

69. Author's interview with David Baltimore, by telephone, June 17, 1996; Sarasohn, *Science on Trial*, pp. 169–171; Barbara J. Culliton, "Rockefeller Braces for Baltimore," *Science*, 247(Jan. 12, 1990), 148.

70. William K. Stevens, "Baltimore Accepts Post Despite Faculty Outcry," *New York Times*, Oct. 18, 1989, p. B9; Sarasohn, *Science on Trial*, pp. 169–171.

71. Wyngaarden to Dingell, June 30, 1989, Davis Files; author's interviews with Brian Kimes, March 25, 1996; with Alex Rich, by telephone, May 31, 1996; with Frank Press, by telephone, Oct. 15, 1997. Rich recalls Press's saying that Dingell told him, Press, directly that he was going to get Baltimore. Press says that he doesn't remember whether Dingell made the remark directly to him or to another scientist who then recounted it to him. Albert Kildow, Baltimore's press aide, recalls that shortly after the hearing, in a meeting with several subcommittee staff, he remarked that he assumed it was all over now and that he was told, "No way. We've only begun." Author's interview with Kildow, June 7, 1996.

NOTES FOR CHAPTER NINE

1. Department of Health and Human Services, Public Health Service, "Announcement of Development of Regulations Protecting Against Scientific Fraud or Misconduct; Request for Comments," *Federal Register*, 53(Sept. 19, 1988), 36344–36347; Robert Steinbrook, "U.S. Proposes Rules to Deter Scientific Fraud," *Los Angeles Times*, Sept. 19, 1988, p. 1. The N.I.H. had wanted simply to issue a set of interim guidelines on scientific misconduct but was told by the Office of Management and Budget that it had first to publish the proposed regulations for comment. David L. Wheeler, "Proposed N.I.H. Rules on Science Fraud Are Sent Back to Drawing Board," *Chronicle of Higher Education*, June 1, 1988, p. A4.

2. Lewin, "Travels on the Fraud Circuit," editorial, *Cell*, 57(May 19, 1989), 514; "Baltimore Sends Correction Demanded by N.I.H.," *Science and Government Report*, March 15, 1989, p. 6; Barbara J. Culliton, "Fraud Review May Be Taken from N.I.H.," *Science*, 243(March 24, 1989), 1545; "Statement of Organization, Functions and Delegation of Authority," *Federal Register*, 54(March 16, 1989), 11060–11061; Sarasohn, *Science on Trial*, pp. 105–106; "The Science Police," editorial, *Wall Street Journal*, May 15, 1989.

3. Author's interview with Brian Kimes, March 25, 1996; Wheeler, "Proposed N.I.H. Rules on Science Fraud Are Sent Back to the Drawing Board," *Chronicle of Higher Education*, June 1, 1988, p. A4; Department of Health and Human Services, Public Health Service, "Responsibilities of Awardee and Applicant Institutions for Dealing with and Reporting Possible Misconduct in Science," *Federal Register*, 54(Aug. 8, 1989), 32446–32451.

4. Author's interviews with Suzanne Hadley, Sept. 15, 1996, and, by telephone, April 9, 1996. The third staffer was Martin Blumsack.

5. Author's interviews with Kimes, March 25, 1996; with Suzanne Hadley, Sept. 15, 1992; *Science and Government Report*, May 1, 1989, p. 1; Alun Anderson and Joseph Palca, "Disputed Paper Still Causing Problems," *Nature*, 339(May 4, 1989), 3.

6. Wyngaarden to McDevitt, May 18, 1989; Hadley to Moema Reis, April 17, 1990, Onek Files; Wyngaarden to Dingell, May 23, 1989; Kimes to Baltimore, May 31, 1989, AE H-269; William Raub to Dingell, June 30, 1989, Davis Files; Kimes to Imanishi-Kari, May 31, 1989, *Hearings, Scientific Fraud (Part II)*, May 14, 1990, pp. 161–162.

7. O'Toole to Bernard Davis, May 14, 1989, Baltimore Files, item 166; O'Toole to Kimes, June 9, 1989, AE H-270; O'Toole to Janet Newburgh, May 15, 1989; Bruce Singal to Kimes, July 27, 1989, Onek Files.

8. Kimes to Baltimore, Aug. 10, 1989, Baltimore Files, item 182; Kimes to Imanishi-Kari, Aug. 10, 1989, *Hearings, Scientific Fraud (Part II)*, May 14, 1990, pp. 165–166; Baltimore to Kimes, Sept. 26, 1989, AE H-273.

9. Wyngaarden to McDevitt, May 18, 1989; Hadley to Moema Reis, April 17, 1990, Onek Files; Wyngaarden to Dingell, May 23, 1989; William Raub to Dingell, June 30, 1989, Davis Files; author's interview with Hadley, Sept. 15, 1992; Jeffrey Mervis, "N.I.H. Disarray Stymies *Cell* Paper Investigation," *Scientist*, Nov. 27, 1989, p. 4; Bruce Singal to Kimes, July 27, 1989, Onek Files; Office of Scientific Integrity, "Investigation Report Concerning the Weaver Et. Al. 1986 Cell Paper," Reference 072, Draft, n.d. [March 14, 1991], p. 12. The panelist who made the remark about the Gestapo was most likely McDevitt. He says, "The quotation sounds like me—but I don't really remember." McDevitt to the author, July 29, 1997.

10. Singal to Dingell, July 27, 1989; Singal to Kimes, July 27, 1989; Aug. 23, 1989, *Hearings, Scientific Fraud (Part II)*, May 14, 1990, pp. 163–164, 170, 230–233; author's interview with Imanishi-Kari, March 20, 1996.

11. Robert Charrow, "Scientific Misconduct: Sanctions in Search of Procedures," *Journal of NIH Research*, 2(June 1990), 21. It is interesting to note that Congressman Norman F. Lent, a New York Republican, faulted the previous investigation with "a three-man panel, no transcript being taken, no attorney on behalf of the subject being present." *Hearings, Scientific Fraud*, May 4, 1989, pp. 32–33. In September, Normand Smith expressed worry to Robert Lanman, the N.I.H. legal adviser, that unless the O.S.I. installed regular procedures, there might never "be any end to the number and variety of charges which Dr. O'Toole will [be] allowed to bring

or the number of forums in which she will be allowed to bring them." Smith to Lanman, Sept. 26, 1989, O.R.I. Files.

12. Kimes to Baltimore, Aug. 10, 1989, Baltimore Files, item 182; Kimes to Imanishi-Kari, Aug. 10, 1989; Kimes to Singal, Aug. 10, 1989; Lanman to Singal, Aug. 14, 1989, *Hearings, Scientific Fraud (Part II)*, May 14, 1990, pp. 165–169; author's interview with Kimes, March 25, 1996; O'Toole to Kimes, Sept. 12, 1989, AE H-272. See Kimes's similar assurances in Kimes to Singal, Sept. 14, 1989, Onek Files; Kimes to Baltimore, Oct. 23, 1989, Baltimore Files, item 197.

13. Author's interviews with Kimes, March 25, 1996; with O'Toole, Oct. 5, 1992; Jeffrey Mervis, "N.I.H. Disarray Stymies *Cell* Paper Investigation," *Scientist*, Nov. 27, 1989, p. 4; O'Toole to Kimes, June 9, 1989, AE H-270.

14. O'Toole to Kimes, Sept. 12, 1989, AE H-270, H-272; Kimes to O'Toole, Oct. 13, 1989, Onek Files; Record of Telephone Conversations, Oct. 6, 1989, Dec. 14, 1989, copy obtained from the O.S.I. files, Jock Friedly Files; (hereafter, Friedly Files); author's interview with O'Toole, Oct. 5, 1992, O'Toole, Testimony, *Appeal Proceedings*, Sept. 14, 1995, pp. 6115–6116.

15. Mervis, "N.I.H. Disarray Stymies *Cell* Paper Investigation," *Scientist*, Nov. 27, 1989, p. 4; O'Toole to Kimes, Oct. 23, 1989, Onek Files.

16. Mervis, "N.I.H. Disarray Stymies *Cell* Paper Investigation," *Scientist*, Nov. 27, 1989, p. 4; author's interviews with Kimes, March 25, 1996; with Stewart and Feder, Sept. 16, 1992; Kimes to Stewart, Sept. 19, 1989, Onek Files.

17. Author's interviews with Kimes, March 25, 1996; with William Raub, by telephone, Oct. 22, 1997.

18. Mervis, "N.I.H. Disarray Stymies *Cell* Paper Investigation," *Scientist*, Nov. 27, 1989, p. 4; O'Toole to Kimes, Nov. 6, 1989, AE H-276.

19. Record of Telephone Conversation, Kimes, Barbara Williams, and O'Toole, Oct. 31, 1989, Onek Files; Margot O'Toole, "Allegations Concerning the Cell Paper and Laboratory Records Submitted in Support of the Claims of the Cell Paper," attached to O'Toole to Kimes, Nov. 6, 1989, AE H-276.

20. O'Toole, Testimony, *Hearing, Fraud in N.I.H. Grant Programs*, April 12, 1988, p. 86; author's interview with Wortis, March 20, 1996.

21. Author's interview with O'Toole, Oct. 5, 1992; O'Toole to Eisen, October 11, 1987, with Eisen's marginalia, AE H-240; O'Toole to Kimes, Nov. 6, 1989, AE H-276.

22. The claim of damaged reputation appears in, among other places, O'Toole to Kimes, Sept. 12, 1989, AE H-272; O'Toole to Hadley, March 20, 1990, AE H-284; O'Toole, Testimony, *Hearings, Scientific Fraud*, May 9, 1989, pp. 185–186.

23. O'Toole, Testimony and Statement, *Hearings, Scientific Fraud*, May 9, 1989, pp. 185–187, 198; Baltimore, "Dear Colleague," Baltimore Files; O'Toole to Kimes, June 6, 1989, AE H-270. Baltimore's press representative, Albert Kildow, says he had strict instructions from Baltimore never to put out anything disparaging about O'Toole. Author's interview with Kildow, June 7, 1996.

24. William P. Homans, Jr., to Robert E. Sullivan, June 6, 1989, copy in Provost's Office, M.I.T.; John Deutch to the author, Aug. 9, 1996.

25. Homans to Sullivan, June 6, 1989, copy in Provost's Office, M.I.T.; John Deutch to the author, Aug. 9, 1996; Eisen, [Reply to O'Toole], *Nature*, 351 (May 30, 1991), 344; O'Toole to Kimes, June 9, 1989, AE H-270.

26. O'Toole to Newburgh, Feb. 5, 1989, March 30, 1989; O'Toole to Kimes, June 6,

1989, Nov. 6, 1989; O'Toole to Hadley, March 20, 1990, AE H-263, H-246, H-270, H-276, H-284.

27. O'Toole to Newburgh, Feb, 5, 1989, March 30, 1989; O'Toole to Kimes, June 6, 1989, Nov. 6, 1989; O'Toole to Hadley, March 20, 1990, AE H-246, H-263, H-270, H-276, H-284.

28. O'Toole to Hadley, March 20, 1990, AE H-284; O'Toole to Kimes, Oct. 6, 1989, Onek Files; O'Toole, "Allegations Concerning the Cell Paper and Laboratory Records Submitted in Support of the Claims of the Cell Paper," attached to O'Toole to Kimes, Nov. 6, 1989, AE H-276; author's interviews with Kimes, March 25, 1996; with O'Toole, Oct. 5, 1992; Reis to Kimes, Oct. 16, 1989, Baltimore Files, item 196; O.S.I., Record of Telephone Conversation with O'Toole, Dec 11, 1989, Onek Files.

29. O'Toole, "Allegations Concerning the Cell Paper and Laboratory Records Submitted in Support of the Claims of the Cell Paper," attached to O'Toole to Kimes, Nov. 6, 1989, AE H-276, Kimes to O'Toole, Oct. 26, 1989, O.R.I. Files.

30. Author's interview with O'Toole, Oct. 5, 1992; O'Toole to Suzanne Hadley, March 20, 1990, and attached memoranda to N.I.H. Panel and the O.S.I. Staff, March 18, 1990, AE H-280, H-284, H-285.

31. O.S.I., Interview with Charles Maplethorpe, Dec. 27, 1989, Transcript, AE H-113, pp. 3–4; Office of Scientific Integrity, "Investigation Report Concerning the Weaver Et. Al. 1986 Cell Paper," Reference 072, Draft, n.d. [March 14, 1991], p. 11.

32. M. Janet Newburgh, Memo of Telephone Call, O'Toole to Newburgh, May 15, 1989, Onek Files; O.S.I., Record of Telephone Conversation with Charles Maplethorpe, Dec. 11, 1989, Onek Files; O.S.I., Interview with Charles Maplethorpe, Dec. 27, 1989, Transcript, AE H-113, pp. 27, 33, 41, 63–66; Maplethorpe to Hadley, Feb. 15, 1990, AE H-279.

33. Author's interview with O'Toole, Oct. 5, 1992; Office of Scientific Integrity, "Investigation Report Concerning the Weaver Et. Al. 1986 Cell Paper," Reference 072, Draft, n.d. [March 14, 1991], p. 12.

34. O'Toole to Hadley, Jan. 10, 1990, AE H-277.

35. O.S.I., Record of Telephone Conversation with O'Toole, Dec 11, 1989, Onek Files; O'Toole to Hadley, Jan. 10, 1990, AE H-110, p. 1.

36. O'Toole to Hadley, Jan. 10, 1990, AE H-110, pp. 13–15.

37. *Ibid.*, pp. 15–16.

38. *Ibid.*, pp. 16–17.

39. *Ibid.*

40. *Ibid.*, p. 17.

41. Yannoutsos to Ursula Storb, May 24, 1989, AE R-17, attached to Storb to Hadley, Feb. 12, 1990, Onek Files.

42. *Ibid.*, pp. 8–9, 11–13, 22–25, 199–201, 203–205.

43. O.S.I., Interview with Albanese, Feb. 2, 1990, Transcript, AE H-118, pp. 198, 218–219.

44. Maplethorpe, Statement of Receipt of Albanese's Notebooks, Jan. 9, 1990, Onek Files; O.S.I., "Meeting of Special Scientific Panel [with Charles Maplethorpe]," Feb. 11, 1990, Transcript, AE H-110, pp. 58–59; Maplethorpe to Hadley, Feb. 15, 1990, AE H-279. The panel members presumably knew about the overheard conversation because they said they had read the transcript of Maplethorpe's interview with the O.S.I. at the end of December.

45. O.S.I., "Meeting of Special Scientific Panel [with Margot O'Toole]," Feb. 11, 1990, Transcript, AE H-110, pp. 83–84, 183–184, 160–162.

46. *Ibid.*, pp. 165–167.

47. O.S.I., "Meeting of Special Scientific Panel [with Margot O'Toole]," Feb. 11, 1990, Transcript, AE H-110, p. 102; author's interview with O'Toole, Oct. 5, 1992; O'Toole, "Allegations . . . ," attached to O'Toole to Kimes, Nov. 6, 1989, AE H-276. On this issue, see O'Toole's analysis of the grant application, O'Toole to Hadley, Jan. 10, 1990, AE H-277, especially pp. 7–9.

48. O.S.I., "Meeting of Special Scientific Panel [with Margot O'Toole]," Feb. 11, 1990, Transcript, AE H-110, pp. 102, 103–104; author's interview with O'Toole, Oct. 5, 1992.

49. O.S.I., "Meeting of Special Scientific Panel [with Margot O'Toole]," Feb. 11, 1990, Transcript, AE H-110, p. 190.

50. Maplethorpe to Hadley, Feb. 15, 1990, AE H-279; O'Toole to N.I.H. Panel Members and O.S.I. Staff Investigating the 1986 *Cell* Paper, March 18, 1990, Memorandum 4, AE H-283.

51. Author's interview with Brian Kimes, March 25, 1996; John T. Kung to Hadley, March 6, 1990; William E. Paul to Hadley, March 14, 1990, AE R-15, R-16.

52. O.S.I., Meeting of Special Scientific Review Panel in re: M.I.T. [telephone interview with Moema Reis], April 23, 1990, Transcript, AE H-104, p. 21; May 4, 1990, Transcript, AE H-105, pp. 16–18.

53. O.S.I., Meeting of Special Scientific Review Panel in re: M.I.T. [telephone interview with Moema Reis], May 14, 1990, Transcript, AE H-106, pp. 11–17.

54. Author's interviews with Kimes, March 25, 1996; with Hadley, Sept. 15, 1992.

55. O'Toole to Kimes, June 9, 1989, AE H-270.

56. *Ibid.*

57. Margot O'Toole, "Allegations . . . ," attached to O'Toole to Kimes, Nov. 6, 1989, AE H-276; O.S.I., "Meeting of Special Scientific Panel [with Margot O'Toole]," Feb. 11, 1990, Transcript, AE H-110, pp. 190, 121, 167–168; Kimes to Deutch, Sept. 12, 1989.

58. O.S.I., "Meeting of Special Scientific Panel [with Margot O'Toole]," Feb. 11, 1990, Transcript, AE H-110, pp. 168–171; O'Toole to O.S.I. Staff and Scientific Panel, "Re: My Bet-1 Data; the Recent Secret Service Report," June 21, 1990, AE H-285; author's interview with Hadley, Sept. 15, 1992. In a hearing on the case in 1995, O'Toole was asked whether she had received reports of the Secret Service findings either directly or via the O.S.I., the presumable implication being that such information might have drawn her attention to the tapes. At one point she said that she did not, except that she was once asked by Peter Stockton for the meaning of certain notations on the notebook pages that the Secret Service was investigating. However, at another point she said she was told about the developing tape evidence by Walter Stewart. O'Toole, Testimony, *Appeal Proceedings*, June 16, 1995, pp. 1064–1067, Sept. 14, 1995, pp. 6116–6119, 6121–6129.

59. Kimes to Deutch, Sept. 12, 1989; Deutch to Kimes, Oct. 4, 1989, AE H-278, H-274; Hadley to Deutch, Feb. 28, 1990, Onek Files; Hadley to Maplethorpe, Feb. 22, 1990, AE R-19, p. 2; O.S.I., Interview with Maplethorpe, Dec. 27, 1989, Transcript, AE H-113, pp. 107–108.

60. Deutch to Hadley, March 5, 1990; O.S.I., Record of Telephone Conversation, Maplethorpe and Hadley, March 26, 1990, Onek Files; Richard P. Kusserow to William F. Raub, April 13, 1990; Hadley, Note to Members of the O.S.I. Expert Advisory Panel . . . , April 20, 1990, Onek Files.

61. Author's interview with Kimes, March 25, 1996; Kimes to Walter Stewart, Sept. 19, 1989, Onek Files; Kimes to Baltimore, Aug. 10, 1989, Baltimore Files, item 182; James E. Mosimann, "Personal Information and Education," AE H-81.

62. Maplethorpe to Hadley, Feb. 15, 1990, AE H-279; William McClure to O.S.I. Staff and Scientific Panel Members of the "Tufts/MIT" Investigation, April 9, 1990, attached to McClure to Hadley, April 9, 1990. The O.S.I. later held that the use of the background counts avoided "the significant component of the behavior of high counts that reflects actual experimental variation," adding, "Thus, the central aim of the statistical examinations was to determine if the behavior of the handwritten counts . . . is consistent with that expected from gamma counters, in the absence of major experimental effects." Office of Scientific Integrity, "Investigation Report Concerning the Weaver Et. Al. 1986 Cell Paper," Reference 072, Draft, n.d. [March 14, 1991], p. 45.

63. William McClure to O.S.I. Staff and Scientific Panel Members of the "Tufts/ MIT" Investigation, April 9, 1990, attached to McClure to Hadley, April 9, 1990; McClure to Hadley, April 12, 1990; McClure to Hadley, April 17, 1990; James E. Mosimann, [Preliminary Draft Report on Statistical Analysis], April 19, 1990; Hadley, Note to Members of the O.S.I. Expert Advisory Panel . . . , April 20, 1990, Onek Files.

64. Dingell, *Hearings, Scientific Fraud (Part II)*, May 14, 1990, p. 136.

65. John Hargett and Larry Stewart, Testimony, *Hearings, Scientific Fraud (Part II)*, May 14, 1990, pp. 138–141, 147–148.

66. *Ibid.*

67. *Ibid.*, p. 150.

68. *Hearings on Scientific Fraud (Part II)*, May 14, 1990, pp. 204–206; Peter G. Gosselin, "U.S. Panel Orders Criminal Probe of Ex-MIT Researcher," *Boston Globe*, May 15, 1990, p. 16; "US Attorney Following the Case," *Science and Government Report*, May 15, 1990, p. 3.

NOTES FOR CHAPTER TEN

1. Hadley to Baltimore, Feb. 1, 1990, Baltimore Files; Robert B. Lanman and Hadley to Singal, Jan. 10, 1990; Hadley to Imanishi-Kari, with attached list of issues, Feb. 1, 1990, *Hearings on Scientific Fraud (Part II)*, May 14, 1990, pp. 177–178, 180–183.

2. Author's interviews with Imanishi-Kari, Oct. 6, 1992, March 20, 1996; Hadley to Imanishi-Kari, with attached list of issues, Feb. 1, 1990; Lanman and Hadley to Singal, Feb. 21, 1990; Singal to Hadley, April 20, 1990, *Hearings, Scientific Fraud (Part II)*, May 14, 1990, pp. 181–184, 202–203.

3. Hadley to Imanishi-Kari, Jan. 12, 1990; John W. Diggs to Joseph J. Byrne, April 11, 1990; Singal to Hadley, April 20, 1990, *Hearings on Scientific Fraud (Part II)*, May 14, 1990, pp. 179, 197–198, 202–203; author's interview with Imanishi-Kari, Oct. 6, 1992. During fiscal 1989, not a single N.I.H. grant had been terminated for cause. Al Mason, for Sonny Kreitman, "Note to Dr. Suzanne Hadley," May 23, 1990, *Hearings on Scientific Fraud (Part II)*, May 14, 1990, p. 204.

4. Hadley to Singal, April 19, 1990; Singal to Hadley, April 20, 1990, May 1, 1990, *Hearings on Scientific Fraud (Part II)*, May 14, 1990, pp. 186, 188–189, 202–203.

5. Dingell to Singal, Dec. 11, 1989; Michael Barrett, Jr., to Singal, Jan. 11, 1990;

Dingell to Singal, Jan. 26, 1990, *Hearings on Scientific Fraud (Part II)*, May 14, 1990, pp. 236–267, p. 241, 243–244.

6. Singal to Dingell, March 6, 1990, *Hearings on Scientific Fraud (Part II)*, May 14, 1990, pp. 247–249.

7. Garry M. Jenkins to Dingell, April 16, 1990, and May 4, 1990, each with attached report; Singal to Dingell, May 10, 1990, *Hearings on Scientific Fraud (Part II)*, pp. 252–258, 260–264.

8. Singal to Dingell, May 10, 1990, *Hearings on Scientific Fraud (Part II)*, May 14, 1990, pp. 260–264.

9. Thereza Imanishi-Kari, Press Conference, May 10, 1990, Transcript, AE H-102, pp. 3–4.

10. "Secret Service Says Data Faked in Baltimore Case," *Science and Government Report*, May 15, 1990, p. 1.

11. O.S.I., Interview with Weaver, June 14, 1990, Transcript, AE H-116, pp. 91–95.

12. *Ibid.*

13. *Ibid.*

14. Author's interview with McDevitt, July 2, 1996.

15. Jules Hallum, "Foreword" to an unpublished memoir, copy in author's possession; Barbara J. Culliton, "Scientists Confront Misconduct," *Science*, 241(Sept. 30, 1988), 1748; author's interviews with Hadley, Sept. 15, 1992; with Jules Hallum, by telephone, April 24, 1996.

16. Author's interviews with Kimes, March 25, 1996; with Hallum, April 24, 1996; with Hadley, Sept. 15, 1992; Hallum, "Cloaks and Daggers: the X, Y, Z Affair," Chapter IX of an unpublished memoir, p. 19, copy in author's possession.

17. Author's interviews with Kimes, March 25, 1996; with Hadley, Sept. 15, 1992; Suzanne W. Hadley, "Can Science Survive Integrity?" lecture presented at the University of California, San Diego, Oct. 17, 1991, copy in author's possession.

18. Author's interviews with Kimes, March 25, 1996; with McDevitt, July 2, 1996; with Storb, April 8, 1997. Storb told Hadley in February 1990 that she was "very impressed with the way you are handling the investigation," adding that "the clarity of presentation by the O.S.I. is far superior to Walter Stewart's somewhat muddled accounts we heard last year." Storb to Hadley, Feb. 12, 1990, Onek Files.

19. O.S.I., Interview with Reis, May 14, 1990, Transcript, AE H-106, pp. 11–17; O.S.I., Interview with Maplethorpe, Dec. 27, 1989, AE H-113, pp. 103–104.

20. Normand Smith to the author, Sept. 2, 1996.

21. O.S.I., Interview with David Baltimore, April 30, 1990, Transcript, AE H-108, pp. 45–46.

22. *Ibid.*, pp. 53–55, 59–60.

23. *Ibid.*, pp. 64–68.

24. *Ibid.*, pp. 69–70; Baltimore to the author, April 15, 1997.

25. O.S.I., Interview with Baltimore, April 30, 1990, Transcript, AE H-108, pp. 75–76.

26. *Ibid.*, pp. 76–79; Normand Smith to the author, Sept. 2, 1996.

27. Author's interviews with Kimes, March 25, 1996; with Hadley, by telephone, May 8, 1996; with Hallum, April 24, 1996. At a hearing of the Dingell subcommittee in August 1991, William Raub explained that Hadley had not gotten too close to O'Toole but had only worked successfully to win her trust and her active involvement in the case. Raub, Testimony, U.S. Congress, House, Subcommittee on Oversight and Investigations of the Committee on Energy and Commerce, *Hearings, Scientific Fraud*, 102nd Cong., 1st Sess., March 6 and Aug. 1, 1991, p. 149 (hereafter, *Hearings, Scientific Fraud*, Aug. 1, 1991).

28. Kimes to O'Toole, Oct. 13, 1989, handwritten postscript, Onek Files; O'Toole to O.S.I. Staff and Scientific Panel, March 18, 1990, "Memorandum 2," AE H-281; author's interviews with Suzanne Hadley, Sept. 15, 1992, and, by telephone, April 9, 1996.

29. Author's interview with Suzanne Hadley, Sept. 15, 1992; Record of Telephone Conversation with O'Toole, March 8, 1990, Onek Files; O'Toole to Kimes, Nov. 6, 1989, AE H-276. The solicitousness with which the O.S.I. dealt with O'Toole is evident in Hadley to O'Toole, May 3, 1990, O.R.I. Files.

30. O'Toole to O.S.I. Staff and Scientific Panel, "Memorandum 1," "Memorandum 2," "Memorandum 3," "Memorandum 4," March 18, 1980; O'Toole, "Re: My Bet-1 Data; the Recent Secret Service Report," June 21, 1990; O'Toole to Hadley, March 20, 1990, AE H-280, H-281, H-282, H-283, H-284, H-285. The burden of O'Toole's "powderkeg" claim, which was developed in Memorandum 1, was to bolster her contention that the Table 3 hybridomas had really been characterized with a reagent that tested for the presence of mouse antibodies rather than of idiotype. Imanishi-Kari later provided extensive evidence to the contrary. She also said she doubted that her laboratory had even possessed the reagent for mouse antibodies that O'Toole claimed had been used. "Response to Questions for Thereza Imanishi-Kari," July 17, 1990, pp. 21–25; O.S.I., Interview of Thereza Imanishi-Kari, Oct. 13, 1990, Transcript, AE H-286, pp. 156–158.

31. O'Toole to Kimes, Oct. 13, 1989, Onek Files; Raub to Dingell, Dec. 15, 1989; Dingell to Raub, Jan. 9, 1990, Davis Files; Kimes, Report of Telephone Conversation with O'Toole, Oct. 26, 1989, O.R.I. Files.

32. Author's interviews with Kimes, March 25, 1996; with Hadley, Sept. 15, 1992, and, by telephone, May 8, 1996; Office of Scientific Integrity, "Investigation Report Concerning the Weaver Et. Al. 1986 Cell Paper," Reference 072, Draft, n.d. [March 14, 1991], copy in Baltimore Files, p. 13. Kimes told Singal that the O.S.I. was getting "as much information as possible from the subcommittee staff." Kimes to Singal, Aug. 14, 1989, *Hearings on Scientific Fraud (Part II)*, May 14, 1990, pp. 168–169.

33. Dingell to Raub, Jan. 4, 1990; Michael F. Barrett, Jr., to Raub, Jan. 4, 1990, *Hearings, Scientific Fraud (Part II)*, May 14, 1990, pp. 239–240; Barrett to Hadley, Sept. 6, 1990; Raub to Dingell, Sept. 24, 1990, Onek Files; Raub to Dingell, Jan. 17, 1990, O.R.I. Files; author's interviews with Kimes, March 25, 1996; by telephone, with Hallum, April 24, 1996. During the N.I.H. staff's interview with Herman Eisen and Gene Brown in Boston, in the spring of 1988, Lanman remarked: "I should note that we are really at the Committee's mercy here. We have to give them any documents that we collect. We also haven't gotten lots of support within N.I.H. for resisting the Committee. . . ." Maggie Hassan to File, Re: Interviews of Dr. Herman Eisen and Dean Gene Brown by Staff from the National Institutes of Health, May 23, 1988, Eisen Files, p. 2.

34. Author's interviews with Kimes, March 25, 1996; with Hadley, Sept. 15, 1992; Friedly, "How Congressional Pressure Shaped the 'Baltimore Case,'" *Science*, 273 (Aug. 16, 1996), 874–875; Barbara J. Culliton, "Fraudbusters Back at N.I.H," *Science*, 248 (June 29, 1990), 1599. Apparent evidence of such apparent leaks to the press include various articles in the *Detroit News* such as Gordon, "Cover-up Charge Puts Scientists under Microscope," Oct. 29, 1989, pp. 17A, 22A, and "N.I.H.'s Baltimore Inquiry near Completion of Report," *Science and Government Report*, July 1, 1990, pp. 2–3. Kimes noted, following a telephone conversation with Gordon about the Imanishi-Kari case, that Gordon's questions revealed he

had "information which could *only have been obtained from Dingell's staff.*" A biologist named Howard Schachman says that sometime in the winter or early spring of 1989, after giving a lecture on scientific misconduct at Berkeley, the *Chicago Tribune* reporter John Crewdson came up to talk with him. Schachman recalls that Crewdson, who was known to be close to the Dingell subcommittee, told him about detailed evidence that had not yet been made public concerning the Secret Service's forensic analysis of Imanishi-Kari's notebooks. Kimes, Report of Telephone Conversation, Oct. 25, 1989, O.R.I. Files; author's telephone conversation with Schachman, Oct. 23, 1997.

35. Author's interview with Bruce Singal, March 20, 1996.
36. Author's interviews with Jules Hallum, by telephone, April 24, 1996; with Bruce Maurer, Nov. 1, 1996.
37. Hadley, Testimony, *Hearings, Scientific Fraud (Part II)*, May 14, 1990, pp. 163–189, 204–206; pp. 163–189 (correspondence); author's interview with Hadley, Sept. 15, 1992.
38. Gosselin, "US Panel Orders Criminal Probe of Ex-MIT Researcher," *Boston Globe*, May 15, 1990, pp. 1, 16; Barbara J. Culliton, "Rockefeller Braces for Baltimore," *Science*, 247(Jan. 12, 1990), 148–149; Sarasohn, *Science on Trial*, pp. 190, 216; David Hamilton, "White Coats, Black Deeds," *Washington Monthly*, April 1990, pp. 27, 29; author's conversation with Maxine Singer, March 23, 1996.
39. Hadley to Singal, May 10, 1990, *Hearings, Scientific Fraud (Part II)*, May 14, 1990, p. 259. Hadley sent nine questions on June 6 and another five on June 27. Hadley to Singal, June 6, 1990, Onek Files; "Response to Questions for Thereza Imanishi-Kari," July 17, 1990, attached to Singal to Hadley, July 16, 1990, AE H-286, "Table of Contents," p. 1.
40. "Response to Questions for Thereza Imanishi-Kari," July 17, 1990, attached to Singal to Hadley, July 16, 1990, AE H-286, "Table of Contents," pp. 1–3.
41. "Response to Questions for Thereza Imanishi-Kari," July 17, 1990; O.S.I., Interview of Thereza Imanishi-Kari, Oct. 13, 1990, Transcript, AE H-286; author's interview with Singal, March 20, 1996.
42. "Response to Questions for Thereza Imanishi-Kari," July 17, 1990, pp. 16–19, 21–25; O.S.I., Interview of Thereza Imanishi-Kari, Oct. 13, 1990, Transcript, AE H-286, pp. 156–158.
43. "Response to Questions for Thereza Imanishi-Kari," July 17, 1990, p. 13. See also, O.S.I., Interview of Thereza Imanishi-Kari, Oct. 13, 1990, Transcript, AE H-286, pp. 97–99, 101–102.
44. "Response to Questions for Thereza Imanishi-Kari," July 17, 1990, pp. 3–4. See also Bruce Singal's comments, O.S.I., Interview with Imanishi-Kari, Oct. 13, 1990, Transcript, AE H-103, pp. 8–10, 189–192.
45. "Response to Questions for Thereza Imanishi-Kari," July 17, 1990, pp. 7–9.
46. *Ibid.*
47. O.S.I., Interview of Thereza Imanishi-Kari, Oct. 13, 1990, Transcript, AE H-286, pp. 90–92, 152–153.
48. The two Secret Service reports were dated, respectively, Sept. 10, 1990, and Sept. 28, 1990. John W. Hargett and Larry F. Stewart to Charles C. Maddox, Sept. 10, 1990, Onek Files; O.S.I., Interview of Thereza Imanishi-Kari, Oct. 13, 1990, Transcript, AE H-286, pp. 63–67.
49. "Response to Questions for Thereza Imanishi-Kari," July 17, 1990, pp. 10–11; O.S.I., Interview of Thereza Imanishi-Kari, Oct. 13, 1990, Transcript, AE H-286, pp. 63–67, 79; author's interview with Imanishi-Kari, Oct. 6, 1992.

50. O.S.I., Interview of Thereza Imanishi-Kari, Oct. 13, 1990, Transcript, AE H-286, pp. 18–19, 23–24; author's interview with Imanishi-Kari, Oct. 6, 1992.
51. O.S.I., Interview of Thereza Imanishi-Kari, Oct. 13, 1990, Transcript, AE H-286, pp. 63–67, 70–71, 194–196.
52. *Ibid.*, pp. 63–67; "Response to Questions for Thereza Imanishi-Kari," July 17, 1990, pp. 12–13, 26–27.
53. Singal to Hadley, Nov. 26, 1990, AE H-103; Hadley to Singal, Nov. 6, 1990, Onek Files.
54. *Ibid.*
55. O.S.I., Interview of Thereza Imanishi-Kari, Oct. 13, 1990, Transcript, AE H-286, pp. 187–188; author's interview with Hadley, Sept. 15, 1992; Hadley to the O.S.I. Expert Advisory Panel, Aug. 31, 1990, O.R.I. Files; Record of Telephone Conversation, Hadley and O'Toole, Nov. 28, 1990; "Summary of Secret Service Findings Reported 1/9/91," attached to John Hargett and Larry Stewart to Dr. Hadley, Jan. 10, 1991, Onek Files.
56. Hadley to Weaver, Jan. 14, 1991; Weaver to Hadley, Feb. 1, 1991, Onek Files.
57. Office of Scientific Integrity, "Investigation Report Concerning the Weaver Et. Al. 1986 Cell Paper," Reference 072, n.d. [March 14, 1991], pp. 44–57; author's interview with Davie, Oct. 28, 1996. The draft report initially said that the statistical analyses on the January fusion data had been done on the green tapes. It turned out that they had been done on the yellow tapes, and the O.S.I. issued a correction to the draft report on March 20, 1991. Hadley to Baltimore, March 20, 1991, and attached correction for Finding "B," Baltimore Files.
58. Office of Scientific Integrity, "Investigation Report . . . ," pp. 57–58, 74–75.
59. *Ibid.*, pp. 75–76, 80–81, 93–95.
60. *Ibid.*, pp. 98–99.
61. *Ibid.*, pp. 101, 103.
62. *Ibid.*, pp. 114–116; O.S.I., Interview with Imanishi-Kari, Oct. 13, 1990, Transcript, AE H-286, p. 94.
63. Office of Scientific Integrity, "Investigation Report . . . ," pp. 112–113.
64. *Ibid.*, pp. 9, 10, 37, 41, 44; Davie, Testimony, *Hearings, Scientific Fraud*, May 4, 1989, pp. 22–23.
65. Office of Scientific Integrity, "Investigation Report . . . ," pp. 1, 2, 37; author's interviews with McDevitt, July 2, 1996; with Davie, Oct. 28, 1996; with Storb, April 8, 1997.
66. Office of Scientific Integrity, "Investigation Report . . . ," pp. 116–118.
67. *Ibid.*, pp. 119–121.
68. *Ibid.*, pp. 120–121.
69. Author's interviews with Davie, Oct. 29, 1996; with McDevitt, July 2, 1996, and, by telephone, March 12, 1997; with Hadley, Sept. 15, 1992.
70. Author's interview with McDevitt, July 2, 1996.
71. Author's interview with Storb, April 8, 1977.
72. Author's interview with McDevitt, July 2, 1996.
73. *Ibid.*; author's interview, by telephone, with McDevitt, June 15, 1997; author's interview with Storb, April 8, 1997.
74. Author's interview with McDevitt, July 2, 1996.
75. Author's interviews with Davie, Oct. 28, 1996; with McDevitt, July 2, 1996.
76. "Minority Opinion Submitted by Drs. Hugh O. McDevitt and Ursula Storb . . . ," in Office of Scientific Integrity, "Investigation Report. . . . "
77. Author's interview with McDevitt, July 2, 1996; Office of Scientific Integrity, "Investigation Report. . . . " The O.R.I. index of its document files on Imanishi-

Kari lists a record of a telephone conversation among Raub, Hadley, and Stockton on Friday, March 8, 1991, informing Stockton that the draft report would be out the following Wednesday or Thursday. It would not be surprising if the subcommittee asked for a copy of the draft report and expected to receive one, since it had asked for and received a copy of the N.I.H. draft report in November 1988, despite the restrictions of confidentiality that were supposed to apply to such documents, and since it held that it had a right to obtain any documents it wanted from the O.S.I. In any case, it is evident that the subcommittee had received a copy of the draft report by March 20, 1991, the day before its contents were reported in the press, but perhaps earlier. In a letter to Peter Stockton written on March 20, 1991, Raub noted that "it has come to my attention that the Subcommittee on Oversight and Investigation is in possession of the confidential draft report . . . concerning the OSI investigation of possible scientific misconduct in connection with the 1986 Weaver et al. *Cell* paper." Raub said that "the NIH did not provide the draft report to the Subcommittee"; he was nevertheless sending a correction to it that was "transmitted today to the recipients of the draft report." Raub says that he does not remember how he knew that Dingell had a copy of the draft report but thinks the information must have come from the staff of the O.S.I. He notes, "In those days, there was no question that if a congressman insisted on obtaining a document that didn't jeopardize national security, the congressman would get it." The correction was sent to the recipients on March 20, 1991. Raub's letter was apparently sent on March 21, 1991, but the phrase "transmitted today" as well as the dated sign-offs on the file copy of the letter place the writing of it at the day before. Hadley to Baltimore, March 20, 1991; Raub to Stockton, March 21, 1991, O.R.I. Files; author's interview with Raub, by telephone, Oct. 22, 1997.

NOTES FOR CHAPTER ELEVEN

1. Malcolm Gladwell, "Scientist Retracts Paper amid Allegations of Fraud," *Washington Post*, March 21, 1991, pp. A1, A6; Malcolm Gladwell, remarks, Symposium at M.I.T. on the Baltimore Case in American Political Culture, Oct. 28, 1996. Gladwell, in his own story, mentioned that the draft report had been provided to news organizations together with the twenty-four photocopied articles. He says that about 9 A.M. on the morning his story appeared, Stewart telephoned one of the editors at the *Post* in a rage, declaring that Gladwell had betrayed the confidence usually awarded leakers. He had exposed the source of the draft document by reporting that it had arrived with the articles because Stewart and Feder were well known to provide photocopies of background materials. Stewart says that Malcolm Gladwell was one of the worst reporters he had ever encountered. Gladwell, remarks, Oct. 28, 1996; author's interview with Stewart and Feder, Sept. 16, 1992.
2. Christopher Anderson, "New Evidence Emerges in Tufts Misconduct Case," *Nature*, 347(Sept. 27, 1990), 317; David P. Hamilton, "Verdict in Sight in the 'Baltimore Case,'" *Science*, 251(March 8, 1991), 1168–1172. The gist of the article in *Nature* was reported in Richard Saltus, "Journal Cites New Evidence ex-MIT Scientist Faked Data," *Boston Globe*, Sept. 28, 1990, p. 32.
3. Author's interview with Stewart and Feder, Sept. 16, 1992; Dingell to Louis Sullivan, Aug. 30, 1991; Wilford J. Forbush to Director, Office of Management,

Sept. 17, 1992, Friedly Files. The day that the findings of the draft report appeared in the press, Hadley told the expert advisory panel that "the leak *did not* come from this office." That assurance did not, however, preclude the possibility that someone at the O.S.I. had given the report to the subcommittee. Transmitting documents to the subcommittee was not regarded as leaking them. O'Toole says she did not want the draft report leaked, fearing that its unauthorized release would delay the conclusion of the process, but she expressed no objection to the leak in principle. Hadley, Note to Members of the Expert Advisory Panel, March 21, 1991, O.R.I. Files; author's interview with O'Toole, Oct. 5, 1992.

4. "All Things Considered," March 22, 1991, Transcript, Stewart and Feder Files; "The Ten O'Clock News," with Hope Kelly, Boston, March 27, 1991, tape of the broadcast in author's possession.

5. "Full Press on Dr. Baltimore," editorial, *Wall Street Journal*, April 5, 1991, p. A14; David L. Wheeler, "President of Rockefeller University Retracts Scientific Paper That N.I.H. Office Says Contains Fabricated Data," *Chronicle of Higher Education*, March 27, 1991, p. A11; Malcolm Gladwell, remarks, Symposium at M.I.T. on the Baltimore Case in American Political Culture, Oct. 28, 1996. Bruce Singal said that for years he kept running into scientists, who, assuming the draft report was definitive, would commiserate that his client had been found guilty of fraud. Author's interview with Singal, March 20, 1996.

6. "American Heroine," *Ladies Home Journal*, Nov. 1991, p. 273; Philip Hilts, "Hero in Exposing Science Hoax Paid Dearly," *New York Times*, March 21, 1991, p. 1; "Bad Science, Bad Response," editorial, *Baltimore Sun*, March 28, 1991; "Hero," *People*, April 15, 1991, pp. 51–52; Daniel S. Greenberg, "Squalid Science," *Baltimore Sun*, March 27, 1991, p. 15A; "Squalor in Science: A Review of the Baltimore Case"; "Public Trust Creates a Heavy Burden for Science," *Science and Government Report*, April 1, 1991, p. 2; May 1, 1991, pp. 6–7; Abigail Trafford, "Ivory Tower Whistle-Blowers," *Washington Post Health Magazine*, March 26, 1991, p. 6; Ted Weiss, "Point of View: Too Many Scientists Who 'Blow the Whistle' End Up Losing Their Job and Careers," *Chronicle of Higher Education*, June 26, 1991, p. A36.

7. Author's interview with O'Toole, Oct. 5, 1992; N.I.H. Staff, Interview with O'Toole, May 17, 1988, Transcript, AE H-109, pp. 1, 4; "The Antibodies That Weren't," *Newsweek*, April 1, 1991, p. 60; O'Toole, Interview with Hope Kelly, "The Ten O'Clock News," Boston, March 27, 1991; Bernard D. Davis, "Dingell's Witness for the Persecution," *Wall Street Journal*, July 22, 1991, p. A8.

8. "All Things Considered," March 22, 1991, Transcript, Stewart and Feder Files; Barbara Carton, "The Mind of a Whistleblower," *Boston Globe*, April 1, 1991, pp. 42–43; "Hero," *People*, April 15, 1991, p. 51; Pamela S. Zurer, "Scientific Whistleblower Vindicated," *Chemical and Engineering News*, April 8, 1991, pp. 35–36, 40; O'Toole, Interview with Hope Kelly, "The Ten O'Clock News," Boston, March 27, 1991.

9. Anthony Gottlieb, on the Chris Lydon Show, n.d. [March 1991].

10. David Warsh, "Rep. John Dingell: Who Will Watch This Watchman?" *Boston Globe*, March 31, 1991, pp. 1, 3; reprinted in the *Washington Post*, April 3, 1991; Pantridge, "In the Eye of the Storm," *Boston Magazine*, July 1991, p. 108; "The Whistleblower on Fraud at MIT," editorial, *Boston Globe*, March 25, 1991, p. 14; "A Sad Case of Scientific Hubris," *Plain Dealer*, March 27, 1991, p. 8C; "Scientific Fraud," editorial, *Washington Post*, March 28, 1991, p. A22; "A Scientific Water-

gate," editorial, *New York Times*, March 26, 1991, p. 22; Ehrenreich, "Science, Lies, and Ultimate Truth," *Time*, May 20, 1991, p. 66.

11. *Economist*, March 30, 1991, p. 83; Gregory Gordon, " 'Whistleblower' Backed in Science Inquiry," *Detroit News*, March 21, 1991, p. 18A; David P. Hamilton, "N.I.H. Finds Fraud in *Cell* Paper," 251(March 29, 1991), 1552–1554; Pantridge, "In the Eye of the Storm," *Boston Magazine*, July 1991, p. 108; "Dealing with Fraud in Science," editorial, *Chicago Tribune*, April 6, 1991, p. 16; *Time*, April 1, 1991, p. 65; John Maddox, "Dr. Baltimore's Experiment in Hubris," *New York Times*, March 31, 1991, p. E13; John Maddox, "Greek Tragedy Moves on One Act," editorial, *Nature*, 350(March 28, 1991), 269.

12. "Misconduct Update: Slow Progress on Big Cases," *Science and Government Report*, Oct. 1, 1990, p. 3; "A Scientific Watergate," editorial, *New York Times*, March 26, 1991, p. 22; author's interviews with Kildow, June 7, 1996; with Boffey, by telephone, Oct. 27, 1997.

13. Frank Kuznik, "Fraud Busters," *Washington Post Magazine*, April 14, 1991, p. 32; Walter W. Stewart and Ned Feder, "Analysis of a Whistle-blowing," *Nature*, 351(June 27, 1991), 687–91.

14. Editorial, *Detroit News*, March 25, 1991, p. 8A; "Dingell Sounds Off on Indirect Costs, Scientific Fraud," *Science and Government Report*, May 15, 1991, pp. 1, 4; Richard L. Berke, "A Crusader Tilts at the Ivory Towers Looking for Old-Fashioned Corruption," *New York Times*, April 21, 1991, Section 4, p. 4; "All Things Considered," March 22, 1991, Transcript, Stewart and Feder Files.

15. John Walsh, "John Dingell: Demanding Humility and Fair Play from Scientists," *Journal of NIH Research*, 2(Oct. 1990), 43; David P. Hamilton, "Verdict in Sight in the 'Baltimore Case,' " *Science*, 251(March 8, 1991), 1172; Peter G. Gosselin, "Rep. Dingell Asks Scrutiny of MIT, Tufts . . . ," *Boston Globe*, March 22, 1991, p. 12; Christopher Anderson, "N.I.H.: Imanishi-Kari Guilty," *Nature*, 350(March 28, 1991), 262; David P. Hamilton, "Baltimore Throws in the Towel," *Science*, 252(May 10, 1952), 770.

16. Judy Foreman, "MIT Institute Used Funds Wrongly," *Boston Globe*, April 17, 1991, pp. 1, 8; Alice Huang, memo to files, April 17, 1991; Huang, Re: Dingell's Tightening Noose, memo to files, April 29, 1991, files of Alice Huang (hereafter Huang Files); Pantridge, "In the Eye of the Storm," *Boston Magazine*, July 1991, p. 109.

17. Huang, Re: Dingell's Tightening Noose, memo to files, April 29, 1991, Huang Files; Pantridge, "In the Eye of the Storm," *Boston Magazine*, July 1991, p. 109.

18. Jock Friedly, "Trial and Error," *Boston Magazine*, Jan. 1997, p. 68; Judy Foreman, "Probes Go Deeper in Baltimore Case," *Boston Globe*, April 1, 1991, p. 37; Walter Gilbert to the author, Oct. 14, 1992.

19. Natalie Angier, "University Hurt as Leader Endures Scientific Dispute," *New York Times*, April 1, 1991, p. 7.

20. Leon Jaroff, "Crisis in The Labs," *Newsweek*, Aug. 21, 1991, p. 45; Daniel J. Kevles and Leroy Hood, "Reflections," in Daniel J. Kevles and Leroy Hood, eds., *The Code of Codes: Scientific and Social Issues in the Human Genome Project* (Cambridge: Harvard University Press, 1992), pp. 301–303.

21. Michael Spector, "The Case of Dr. Gallo," *New York Review of Books*, 38(Aug. 15, 1991), 51; Barbara J. Culliton, "Gallo Inquiry Takes Puzzling New Turn," *Science*, 250(Oct. 12, 1990), 202; David P. Hamilton, "What Next in the Gallo Case," *Science*, 254(Nov. 15, 1991), 944; Randy Shilts, *And the Band Played On:*

Politics, People, and the AIDS Epidemic (New York: St. Martin's Press, 1987), pp. 445, 460, 529.

22. U.S. Congress, House, Subcommittee on Oversight and Investigation of the Committee on Energy and Commerce, *Hearings, Financial Responsibility at Universities*, 102nd Cong., 2nd Sess., March 13 and May 9, 1991, pp. 1, 3–4, 141, 213; Marcia Barinaga, "The Rise and Fall of Donald Kennedy," *Science*, 253(Aug. 9, 1991), 617; Colleen Cordes, "Angry Lawmakers Grill Stanford's Kennedy on Research Costs," *Chronicle of Higher Education*, 37(March 20, 1991), p. A27; "A Resignation at Stanford," editorial, *Washington Post*, July 31, 1991, p. A20. Ultimately, Stanford was obligated to repay less than half a percent—but still some $3 million—of the total it had charged the federal government for overhead from 1981 to 1992. William Cellis, III, "Navy Settles a Fraud Case on Stanford Research Costs," *New York Times*, Oct. 19, 1994, p. 16; Jock Friedly, "University Research in Danger," *The Palo Alto Weekly*, Jan. 8, 1992, pp. 19ff; Jock Friedly, "Judge Dismisses Suit Against Stanford," *Science*, 273(Sept. 13, 1996), 1488. Stanford was much less at fault than either Dingell or the media claimed, but Kennedy has ruefully conceded that the university had left itself "open to a painful trial by media." Donald Kennedy, *Academic Duty* (Cambridge: Harvard University Press, 1997), pp. 167–175.

23. "A Sad Case of Scientific Hubris," *Plain Dealer*, March 27, 1991, p. 8C; Edwin Yoder, "In Scientific Research We Trust . . . Usually," *Star Ledger*, April 2, 1991; "Thin Skins and Fraud at M.I.T.," *Time*, April 1, 1991, p. 65; "The Antibodies That Weren't," *Newsweek*, April 1, 1991, p. 60; Daniel S. Greenberg, "Soiled Lab Coats and Science's Alibis," *Baltimore Sun*, April 9, 1991; "Fraud in Science," *The Journal of NIH Research*, 3(Aug. 1991), 30; Philip J. Hilts, "Can the Scales of Justice Be Calibrated for Scientific Fraud?" *New York Times*, March 31, 1991, p. E7.

24. Author's interview with Eisen, July 10, 1996; Philip J. Hilts, *Scientific Temperaments: Three Lives in Contemporary Science* (New York: Simon and Schuster, 1982), pp. 132, 139; Friedly, "Trial and Error," *Boston Magazine*, Jan. 1997, pp. 46–47.

25. Hilts, *Scientific Temperaments*, pp. 114, 117, 131–132, 139, 141, 156, 167.

26. Harvard first attempted to capitalize on Ptashne's work itself by forming a company to exploit it, but the university quickly dropped the proposal in the face of a hurricane of criticism from its faculty and the press that the administration was trying to turn the university into a profit-making enterprise. Martin Kenney, *Biotechnology: The University-Industrial Complex* (New Haven, Conn: Yale University Press, 1986), pp. 78–80; *Wall Street Journal*, Nov. 18, 1980, p. 21; Nicholas Wade, "Gene Goldrush Splits Harvard, Worries Brokers," *Science*, 210(Nov. 21, 1980), 878; "Harvard Backs off Recombinant DNA," *Nature*, Dec. 4, 1980, p. 423; *New York Times*, Oct. 27, 1980, p. 1; Nov. 18, 1980, p. 1; Hilts, *Scientific Temperaments*, pp. 167–168, 178–182.

27. Hilts, *Scientific Temperaments*, pp. 104–105, 109, 117, 141.

28. Sarasohn, *Science on Trial*, pp. 114–115; author's interview, by telephone, with Henry Wortis, June 12, 1997; tape recording of a discussion among Paul Doty, John Edsall, Walter Gilbert, Herman Eisen, Stephen Harrison, and Mark Ptashne, June 4, 1991, copy in author's possession (hereafter, Doty *et al.* tape).

29. Sarasohn, *Science on Trial*, pp. 165–166; author's interviews with Stewart and Feder, Sept. 16, 1992; with Stockton and Chafin, March 24, 1993.

30. Sarasohn, *Science on Trial*, pp. 165–166; author's interview with Stewart and Feder, Sept. 16, 1992; Carton, "The Mind of a Whistleblower," *Boston Globe*, April 1, 1991, pp. 42, 43; Hilts, "Hero in Exposing Science Hoax Paid Dearly," *New York Times*, March 21, 1991, p. 1; Jeffrey Mervis, "N.I.H. Cuts Research Funding of Scientist under investigation for *Cell* Paper Data," *Scientist*, June 11, 1990, p. 9.

31. John T. Edsall, "Government and the Freedom of Science," *Science*, 121(April 29, 1955), 615–618; John T. Edsall, "Two Aspects of Scientific Responsibility," *Science*, 212(April 3, 1981), 11–14; author's interview with John T. Edsall, Oct. 1, 1992; John Edsall, "Some Thoughts on Scientific Fraud and Misconduct," *Hearing, Fraud in N.I.H. Grant Programs*, April 12, 1988, p. 149; Bernard Davis, draft manuscript of a book on the Baltimore case, Chap. 9, pp. 4–5, Davis Files; Sarasohn, *Science on Trial*, pp. 39–40.

32. Edsall mistakenly believed that the N.I.H. panel had objected to Table 2 in the *Cell* paper for the same reason O'Toole had criticized it; and he incorrectly believed that Dingell had called in the Secret Service in response to O'Toole's specific dissents. He was also under the misimpression that most of the O.S.I. draft report had been written by the scientific panel rather than by the O.S.I. staff. Author's interview with Edsall, Oct. 1, 1992; John T. Edsall, "On Margot O'Toole and the Baltimore Case: A Personal Note on the Evolution of My Involvement," *Ethics and Behavior*, 4 (No. 3, 1994), 239–247.

33. John T. Edsall, "Some Thoughts on Science and Ethics," *Journal of NIH Research*, 3(Aug. 1991), 31; author's interview with Edsall, Oct. 1, 1992; Judy Foreman, "Baltimore Seems to Be Moving to Clear His Name," *Boston Globe*, April 24, 1991, p. 13.

34. Gilbert, Interview with Chris Lydon, n.d. [March 1991], tape of the show in author's possession; author's interview with Paul Doty, June 3, 1997.

35. Author's interview with James D. Watson, by telephone, May 28, 1997; Alice Huang to the author, Sept. 2, 1996; Horace Judson, transcript of interview with Howard Temin, March 15 and 16, 1993. Temin had publicly speculated that RNA might generate DNA, but he told Bernard Davis in a telephone conversation that he had not publicly speculated whether a polymerase existed in the viral particle or was somehow made. Davis, notes of telephone conversation with Temin, Dec. 29, 1992, Davis Files. At a scientific meeting in Budapest during the summer of 1992, Judson, a historian and writer about science, asked Baltimore if he knew that Watson had been accusing him of having stolen the Nobel Prize by doing the experiment that detected reverse transcriptase after Baltimore had heard Temin report his results at the Houston meeting. Alice Huang, who was present at the conversation, shortly thereafter wrote a memo to files about it: "What a shocker the last question was. We knew that Watson had been making these comments but never what he based them on. Judson said that he had done some investigating on dates of submission to Nature, etc. and told Watson that he was wrong and Watson had shrugged and said, 'OK, so I am wrong this time.' ... Judson was surprised that DB wasn't even at the Houston meeting." Huang, memo to file, Aug. 27, 1992: DB's Meeting with Horace Freeman [Freeland] Judson in Budapest, Huang Files.

36. Temin recalled that in their telephone conversation he suggested to Baltimore that they publish their papers together, since they confirmed each other, but that Baltimore "refused because he said he had been stung by doing that previously." Baltimore does not recall refusing for that reason or any other. He believes that

he had already sent his paper to *Nature*. He likely had, since it arrived at the journal the next day, although he conceivably could have sent it the same day to the recently opened office of the journal in Washington, D.C. He remembers subsequently asking the editors to delay publication of it until Temin finished, so that the two papers could be published simultaneously. Temin further recalled that he "submitted the paper to *Nature* with a cover letter from a colleague, Waclaw Szybalski, telling the editors that these two papers were coming and it would be best to publish them together." But by the time Temin's paper with the covering letter arrived, Baltimore's paper had been at the journal almost two weeks, which suggests at the least he had agreed to a delay in publication. The editorial records of *Nature* might clear up the matter, which in any case has no bearing on the mutual independence of the two discoveries. See the interviews with Howard Temin and David Baltimore, *Journal of NIH Research*, 5(July 1993), 85–88; author's interviews with Baltimore, June 16, 1997, and, by telephone, May 31, 1997.

37. Author's interview with James D. Watson, by telephone, May 28, 1997; Bernard Davis, draft manuscript of a book on the Baltimore case, Chap. 9, pp. 21–22, Davis Files; Bernadine Healy, "The Dangers of Trial by Dingell," *New York Times*, July 3, 1996, p. 11; Elizabeth Davis to the author, June 11, 1997.

38. The contemporary record comprises a summary of the conversation that Alice Huang wrote down in a memo to files, April 29, 1991, Huang Files; Maxine Singer to Amy, June 16, 1991, Singer Files; author's interview, by telephone, with Mark Ptashne, May 8, 1997; B. Davis, notes on telephone conversation with Roy Vagelos, April 28, 1991, Davis Files. Vagelos says that he has no recollection of the conversation or of much else that went on concerning Baltimore after the leak of the draft report but that he does remember feeling that what was happening to Baltimore was "horrible." Author's conversation with Roy Vagelos, June 12, 1997. Watson says that he recalls Vagelos telephoning him one day but that he doesn't recall the conversation. Author's interview with Watson, by telephone, May 28, 1997. Ptashne told the Harvard biologist Bernard Davis that most people initially connected with the case knew the paper was fraudulent and had lied under oath to cover up. Bernard Davis to Horace Judson, April 19, 1991, Davis Files.

39. Friedly, "Trial and Error," *Boston Magazine*, Jan. 1997, pp. 47–48; author's interviews with Maurice Fox, Oct. 2, 1992; with Alexander Rich, by telephone, May 31, 1996; Eisen tape, June 1991; author's interviews with Baltimore, June 16, 1997, Oct. 27, 1997, and by telephone, June 17, 1996; with Kildow, June 7, 1996; author's conversation with Alice Huang, Oct. 26, 1992.

40. The meeting with Eisen, held on June 4, 1991, included Edsall, Gilbert, Ptashne, and Paul Doty, another of Baltimore's Harvard critics. Their conversation was taped. Excerpts from it are referenced as Doty *et al.* tape, June 4, 1991. Bernard Davis advanced an interesting take on the possibility of a Harvard-M.I.T. rivalry in a book that he was writing on the Baltimore case that death prevented him from completing. He speculated: "The location of Harvard and MIT in the same city might also play a role, though hardly a major one. It would be natural to be particularly aware of, and interested in, the activities of neighboring colleagues. But a deeper factor could be the invidious relations that sometimes develop between neighboring institutions. This case provided more than the usual basis for such comparisons. The Department at Harvard College had been a leader in the early development of what one might call pure molecular biology, with studies of [Paul] Doty on the physical chemistry of DNA strands, Gilbert and Ptashne on

the interactions of DNA with regulatory proteins, and members of Watson's laboratory on the ribosome; and Edsall had earlier been a pioneer in studying the physical chemistry of amino acids and their polypeptide chains. In contrast, MIT, beginning somewhat later, emphasized a different approach, applying the tools of molecular biology to complex cellular functions at higher levels of organization—studies of cancer, the immune system, embryonic development, or viral infection. For some years those at Harvard who had been so successful in 'pure' molecular biology discouraged the department from moving into these 'messy' areas. However, as this expansion of the scope of molecular and then cellular biology became increasingly fruitful biochemistry at Harvard College was no longer the center of the most exciting developments, which lay in these newer areas and grew rapidly in the Medical School. Today the college department has expanded into the more complex areas; but meanwhile Baltimore's flourishing research was a prime example of the competition. . . . It is striking that the severe critics at Harvard are all from the earlier, 'conservative' school, while many of the younger members disagree. Indeed, within the Department the intense advocacy of the critics so marred the informal lunches that the chairperson eventually had to rule the subject off limits." Davis, draft manuscript of a book on the Baltimore case, Chap. 9, pp. 3–4, Davis Files.

41. Author's interviews with Gilbert, Oct. 2, 1992; with Watson, by telephone, May 28, 1997; Doty *et al.* tape, June 4, 1991.
42. Author's interviews with Gilbert, Oct. 2, 1992; with Watson, by telephone, May 28, 1997; with Berg, by telephone, April 8, 1996; Doty *et al.* tape, June 4, 1991.
43. Author's interviews with Gilbert, Oct. 2, 1992; with Ptashne, by telephone, May 28, 1997; Doty *et al.* tape, June 4, 1991; Foreman, "Probes Go Deeper in Baltimore Case"; Foreman, "Baltimore Seems to Be Moving to Clear His Name", *Boston Globe*, April 1, 1991, p. 37; April 24, 1991, p. 13; "The Antibodies That Weren't," *Newsweek*, April 1, 1991; Friedly, "Trial and Error," *Boston Magazine*, Jan. 1997, p. 50; Davis, note on telephone conversation with Ptashne, April 12, 1991, Davis Files.
44. Doty *et al.* tape, June 4, 1991; Mark Ptashne, letter, *Nature*, 353(Oct. 10, 1991), 495; author's interviews with Ptashne, by telephone, May 28, 1997; with Wortis, by telephone, June 12, 1997.
45. Author's interview with Imanishi-Kari, Oct. 29, 1996; David Weaver, Testimony, *Appeal Proceedings*, June 28, 1995, pp. 2415–2416.
46. Author's interview with Wortis, by telephone, June 12, 1997.
47. Author's interview with Baltimore, by telephone, May 31, 1997; Baltimore, "Dear Colleague," May 19, 1988, Baltimore Files; Baltimore, Statement, *Hearings, Scientific Fraud*, May 4, 1989, p. 105.
48. Author's interview with Ptashne, by telephone, May 28, 1997; Doty *et al.* tape, June 4, 1991; Bernard Davis, note of a telephone conversation with Ptashne, April 12, 1991, Davis Files.
49. Richard P. Feynman, *"Surely You're Joking Mr. Feynman!" Adventures of a Curious Character* (New York: W. W. Norton, 1985), p. 341; Feynman, "The Uncreative Scientist," talk delivered at the Y Dinner Forum, California Institute of Technology, Feb. 17, 1967, taped by Greg Evans, partial transcription provided by Judith R. Goodstein, California Institute of Technology Archives; James Gleick, *Genius: The Life and Science of Richard Feynman* (New York: Pantheon, 1992), pp. 349–350.
50. James Woodward and David Goodstein, "Conduct, Misconduct, and the Struc-

ture of Science," *American Scientist*, 84(Sept.–Oct. 1996), 483–485; author's interview with Baltimore, by telephone, May 31, 1997.

51. Doty *et al.* tape, June 4, 1991.
52. Gilbert to the author, Oct. 14, 1992; Sarasohn, *Science on Trial*, pp. 167–168; Friedly, "Trial and Error," *Boston Magazine*, Jan. 1997, p. 50; author's interview with Watson, by telephone May 28, 1997.
53. John Lear, review of *The Double Helix*, *Saturday Review*, 51(March 16, 1968), 36, 86.
54. Maxine Singer to Amy, June 16, 1991, Singer Files; J. D. Watson, Cold Spring Harbor Laboratory, *Annual Report, 1990*, April 1991, p. 19; John Lear, "Heredity Transactions," *Saturday Review* 51(March 1968), 36, 86. In a telephone conversation, Howard Temin remarked to Bernard Davis, according to Davis's notes: "Jim long concerned with pub[lic] image of sci[ence], took job of guarding purity." B. Davis, "Watson and Nobel Prize, Temin 12/29/92," Davis Files. Watson may have been specially sensitive to maintaining congenial relations with the Congress in 1991 because many scientists had recently attacked the genome project on grounds that it was taking too large a portion of federal funds for biomedical research. Watson says that the dissidents represented no real threat, a judgment supported by the healthy state of the genome budget in fiscal 1991, and he did not have their attacks in mind when he worried about the impact of Baltimore's behavior on congressional attitudes towards science. Robert Cook-Deegan, *The Gene Wars: Science, Politics, and the Human Genome* (New York: Norton, 1994), pp. 170–178; author's interview with Watson, by telephone May 28, 1997.
55. Huang, memo to files, April 29, 1991, Huang Files; author's interview with Ptashne, by telephone, May 28, 1997; Bernard Davis, notes on telephone conversation with Roy Vagelos, April 28, 1991, Davis Files; Maxine Singer to Amy, June 16, 1991, Singer Files.
56. Huang, memo to files, including the capitalized phrase, April 29, 1991, Huang Files. In a meeting in June, Ptashne asked rhetorically how science would look to the lay world when Imanishi-Kari went to jail. Doty *et al.* tape, June 4, 1991.
57. Huang, memo to files, April 29, 1991, Huang Files.

NOTES FOR CHAPTER TWELVE

1. "An Address by David Baltimore on the Occasion of His Installation as President of The Rockefeller University," Sept. 13, 1990, Rockefeller Archive Center, Pocantico Hills, North Tarrytown, NY; Culliton, "Rockefeller Braces for Baltimore," *Science*, 247(Jan. 12, 1990), 149; author's interviews with James Darnell, by telephone, June 19, 1996; with Baltimore, Oct. 31, 1997.
2. Stephen S. Hall, "David Baltimore's Final Days," *Science*, 254(Dec. 13, 1991), 1577; Natalie Angier, "University Hurt as Leader Endures Scientific Dispute," *New York Times*, April 1, 1991, p. 7; author's interview with Baltimore, by telephone, June 17, 1996.
3. Author's interviews with Baltimore, by telephone, June 17, 1996; with Lauren Baltimore, by telephone; Oct. 26, 1997; Huang, Re: Dingell's Tightening Noose, April 17, 1991, Huang Files; Huang to the author, July 14 and July 20, 1997. Huang recalls that she and her family were able to return to Kweiyang after a month or so because fear of further Japanese advance had dissipated by then.
4. Gladwell, "Scientists Retracts Paper amid Allegations of Fraud," *Washington Post*,

March 21, 1991, pp. A1, A6; Anderson, "N.I.H.: Imanishi-Kari Guilty," *Nature*, 350(March 28, 1991), 263; Angier, "Rockefeller U. Is Hurt as President Feels Pain of Science-Fraud Case"; Philip Hilts, "How Charges of Lab Fraud Grew into a Cause Celebre," *New York Times*, April 1, 1991, pp. 1, 12; March 26, 1991, p. B6; Hamilton, "Baltimore Throws in the Towel," *Science*, 252(May 10, 1991), 768; Friedly, "Trial and Error," *Boston Magazine*, Jan. 1997, p. 68; author's interviews with Baltimore, Jan. 23, 1992, and, by telephone, June 17, 1996; with Kildow, June 7, 1996.

5. Reis to Baltimore, April 1, 1991, Baltimore Files; "Scientist Defends Position on Data," *New York Times*, May 17, 1991, p. 7; Weaver, Albanese, Constantini, and Baltimore, Letter to the Editor, "Retraction: Altered Repertoire of Endogenous Immunoglobulin Gene Expression in Transgenic Mice Containing a Rearranged Mu Heavy Chain Gene," *Cell*, May 17, 1991, p. 536.

6. David Baltimore, "Specific Responses to N.I.H. Draft Report," attached to Gerard F. Treanor, Jr., to Suzanne Hadley, May 2, 1991, AE H-289.

7. *Ibid.*; David Baltimore, [Statement], *Nature*, 351(May 9, 1991), 94–95. The part of the statement where Baltimore said he had "come to appreciate" the government's obligation to exercise oversight of public funds expended in science is striking. Baltimore had said as much in the Dingell subcommittee hearings in 1989, and it is difficult to see how he could not have previously appreciated the government's duty to ensure proper stewardship of public funds, including in science. It is likely that this part of the statement, at least, was simply poorly drawn.

8. Baltimore, [Statement], *Nature*, 351(May 9, 1991), 94–95; Baltimore, "Specific Responses to N.I.H. Draft Report," attached to Gerard F. Treanor, Jr., to Suzanne Hadley, May 2, 1991, AE H-289.

9. "The End of the Baltimore Saga," editorial, *Nature*, 351(May 9, 1991), 85. The statement was covered in the *New York Times*, the *Boston Globe*, and the *Washington Post*. Philip J. Hilts, "Nobelist Apologizes for Defending Research Paper with Fraudulent Data," *New York Times*, May 4, 1991, p. 1; Judy Foreman, "Baltimore Admits an 'Excess of Trust,'" *Boston Globe*, May 4, 1991, pp. 1, 28. The Associated Press carried the story of Baltimore's apology and the *Washington Post*, for one, published it. See "Nobel Scientist Apologizes to Whistleblower," *Washington Post*, May 4, 1991, p. A6.

10. Hilts, "Nobelist Apologizes for Defending Research Paper with Fraudulent Data," *New York Times*, May 4, 1991, p. 1; Margot O'Toole, "Margot O'Toole's Responses to Draft Investigative Report, Reference 072, from the Office of Scientific Integrity," May 13, 1991, AE H-290, p. 6; Margot O'Toole, "Margot O'Toole's Record of Events," *Nature*, 351(May 16, 1991), 180–183; Judy Foreman, "Whistleblower Contends Baltimore Was Warned of Fraud," *Boston Globe*, May 16, 1991, p. 17.

11. O'Toole, "Margot O'Toole's Responses to Draft Investigative Report, Reference 072, from the Office of Scientific Integrity," May 13, 1991, AE H-290, p. 6; O'Toole, "Margot O'Toole's Record of Events," *Nature*, 351(May 16, 1991), 180–183.

12. O'Toole claimed that her account in the spring of 1991 was the same as what she had provided in June 1986 and since. See her Statement, *Nature*, 351 (June 27, 1991), 693.

13. O'Toole, "Margot O'Toole's Record of Events," *Nature*, 351(May 16, 1991), 180–183; O'Toole, [Reply to Eisen], *Nature*, 351(June 27, 1991), 693.

14. O'Toole, "Margot O'Toole's Record of Events," *Nature*, 351(May 16, 1991), 180–

183. O'Toole enlisted in support of her charge that the Wortis committee had not seen the June subcloning data the fact that it was not mentioned in the committee's report to the N.I.H. in June 1987. She also pointed to a discussion of the Imanishi-Kari case held at Tufts in May 1988, of which she had a tape recording. She said that neither Wortis, Huber, nor Woodland had mentioned the subcloning data at the Tufts discussion and she inferred that they therefore had not seen it. O'Toole had made this argument during the course of the O.S.I. investigation, but the O.S.I. had evidently not accepted it not only because the Wortis committee members had testified that they had seen subcloning data but also because Moema Reis said she had done the subcloning with Imanishi-Kari. The O.S.I. allowed that the subcloning had been done but that Imanishi-Kari had later altered the numbers. O'Toole continued to insist that it had not ever been done. See Margot O'Toole, [Reply to Wortis et al], *Nature*, 353(Aug. 15, 1991), 560; O'Toole to N.I.H. Panel Members and O.S.I. Staff Investigating the 1986 *Cell* Paper, "Memorandum 3," March 18, 1990, AE H-282; O'Toole to Hadley, Response to the Wortis Committee's Statement in *Nature*, fax, June 28, 1991, AE H-296; O'Toole to Hadley, July 3, 1991, Onek Files. In 1991, O'Toole privately declared to Hadley that she had told the N.I.H. panel in June 1988 that the subcloning had not been done, a claim that she did not mention in her response to the draft report of the N.I.H. panel in November 1988. She also said that the objection the panel raised against Table 2 in 1988—wells versus clones—she had raised in 1986. There is no evidence that she had faulted Table 2 for that reason in 1986. She made the claim now in the service of trying to argue that Imanishi-Kari had known about the wells-versus-clones issue for two years but did not try to answer it with subcloning data until 1988. O'Toole to Hadley, Response to the Wortis Committee's Statement in *Nature*, fax, June 28, 1991, AE H-296. O'Toole mentions having taped the meeting with the N.I.H. panel in her Testimony, *Appeal Proceedings*, June 16, 1995, p. 1045.

15. O'Toole, "Margot O'Toole's Record of Events," *Nature*, 351(May 16, 1991), 183; O'Toole, [Reply to Eisen], *Nature*, 351(June 27, 1991), 693; Margot O'Toole, "More Detailed Coments [Eisen's Statement in *Nature*]," attached to O'Toole to Hadley, fax, June 13, 1991, AE H-295.

16. O'Toole, "Margot O'Toole's Record of Events," *Nature*, 351(May 16, 1991), 183; [O'Toole, reply to Eisen], *Nature*, 351(June 27, 1991), 693. O'Toole provided no evidence that she had been professionally slandered either in her statements in *Nature* or in the full version of the response that she sent to the O.S.I., See "Margot O'Toole's Response to Draft Investigative Report . . . ," May 13, 1991, AE H-290, pp. 21–22. Eisen says that he would have written a recommendation for O'Toole based on the cogency of her memorandum to him of June 6, 1986, but that she never asked him for a letter of support. Author's conversation with Eisen, July 10, 1996.

17. Huang, memo to files, Nov. 6, 1991, Huang Files.

18. Statements by David Baltimore and Herman Eisen; Thereza Imanishi-Kari, [Abridged Response to the O.S.I. Report]; Statements by Henry Wortis, Brigitte Huber, and Robert Woodland, *Nature*, 351(May 30, 1991), 341–345; 351(June 13, 1991), 514; author's interview with Baltimore, Jan. 23, 1992.

19. Statements by David Baltimore and Herman Eisen; Thereza Imanishi-Kari, [Abridged Response to the O.S.I. Report]; Statements by Henry Wortis, Brigitte Huber, and Robert Woodland, *Nature*, 351(May 30, 1991), 341–345; 351(June 13, 1991), 514.

20. Statements by David Baltimore and Herman Eisen; Thereza Imanishi-Kari, [Abridged Response to the O.S.I. Report]; Statements by Henry Wortis, Brigitte Huber, and Robert Woodland, *Nature*, 351(May 30, 1991), 341–345, 351(June 13, 1991), 514.

21. Imanishi-Kari, "O.S.I's Conclusions Wrong," *Nature*, 351(May 30, 1991), 344–345.

22. *Ibid.*, pp. 344–345.

23. Davis to Philip Weiss, Sept. 25, 1989, Davis Files; Bernard Davis, Lecture Delivered at the Marine Biology Laboratory, Woods Hole, Mass., July 20, 1992, tape recording and transcript in author's possession; Bernard D. Davis, *Storm over Biology: Essays on Science, Sentiment, and Public Policy* (Buffalo: Prometheus Books, 1986); Bernard Davis, "Science and Politics: Tensions Between the Head and the Heart," *Annual Reviews of Microbiology*, 46(1992), 1–33.

24. Davis, on "The Ten O'Clock News," Boston, [May] 1989, tape of the broadcast in author's possession.

25. Davis to Philip Weiss, Sept. 25, 1989; Bernard Davis, draft manuscript of a book on the Baltimore case, Speech re His Book, Jan. 13, 1993, Davis Files; Davis, "Is the Office of Scientific Integrity Too Zealous?" *Scientist*, May 13, 1991, p. 12; Philip J. Hilts, "Can the Scales of Justice Be Calibrated for Scientific Fraud," *New York Times*, March 31, 1991, p. E7; Foreman, "Baltimore Seems to Be Moving to Clear His Name"; Davis, letter, *Boston Globe*, April 24, 1991, p. 13; May 8, 1991, p. 22; Jules Hallum, "Foreword," to an unpublished manuscript of a book on scientific fraud, copy in my possession.

26. Foreman, "Baltimore Seems to Be Moving to Clear His Name," *Boston Globe*, April 24, 1991, p. 13; Davis, notes on telephone conversation with Roy Vagelos, April 28, 1991; Davis to Richard Furlaud, May 1, 1991, Davis Files.

27. Davis to Furlaud, May 1, 1991, and attached copy of statement, April 29, 1991; Davis to Baltimore, May 1, 1991, Davis Files; "Baltimore's Backers Rally to Retain Rockefeller Post," *Science and Government Report*, June 1, 1991, pp. 4–5.

28. Davis to Furlaud, May 1, 1991; Ptashne to Davis, April 10, 1991; Davis, Memorandum, "Meeting with Margot O'Toole, April 16, 1991, at Genetics Institute"; Davis, Lecture Delivered at the Marine Biology Laboratory, Woods Hole, Mass., July 20, 1992, tape recording and transcript in author's possession, Davis Files; Elizabeth Davis to the author, June 9, 1997. Davis's impression of O'Toole's rigidity of mind was based initially on her having testified in April 1988 that "it is extremely counterproductive to science that known errors remain published as truths." As Davis well knew, the reality is that scientists permit countless errors to remain uncorrected in the scientific literature on the understanding that the course of research will simply pass the errors by.

29. Davis to Furlaud, May 1, 1991, Davis Files; Davis, "On the Injustice to Margot O'Toole," draft, June 19, 1991, Baltimore Files; Davis, "Dingell's Witness for the Persecution," *Wall Street Journal*, July 22, 1991, p. A8; author's conversation with Bernard Davis, Aug. 26, 1991.

30. Furlaud to Davis, May 22, 1991; Alice Huang to Derek Bok, May 22, 1991, Davis Files; Maxine Singer to Amy, June 16, 1991; note of phone call from Paul Berg, June 20, 1991, Singer Files; author's interview with Rall, March 25, 1996.

31. Maxine Singer to Amy, June 16, 1991, Singer Files.

32. Davis, "On the Injustice to Margot O'Toole," draft, June 19, 1991, Baltimore Files; Davis, Lecture Delivered at the Marine Biology Laboratory, Woods Hole, Mass., July 20, 1992, p. 11, Davis Files.

33. Author's interview with McDevitt, July 2, 1996; Hilts, "Crucial Research Data in Report Biologist Signed Are Faked"; Hilts, "Hero in Exposing Science Hoax Paid Dearly," *New York Times*, May 4, 1991, p. 1; March 21, 1991, p. 1. Donald Kennedy later wrote that the tone of the *Times*'s coverage was "relentlessly negative," a characteristic that he attributed explicitly to Hilts's reporting. Donald Kennedy, "The Baltimore Case: Let's Not Forget What Went Wrong," *Nature Medicine*, 2(Aug. 1996), 843–844.

34. Hilts, " 'I Am Innocent,' Embattled Biologist Says," *New York Times*, June 4, 1991, pp. C7, C10.

35. *Ibid.*

36. *Ibid.*, p. C10; Imanishi-Kari to Editor in Chief of Science Section, draft, June 1991, Baltimore Files; author's interview with Imanishi-Kari, Oct. 6, 1992.

37. Hilts, "Nobelist Apologizes for Defending Research Paper with Faulty Data," *New York Times*, May 4, 1991, p. 1; Ptashne to Baltimore, July 3, 1991, Huang Files.

38. Foreman, "Whistleblower Contends Baltimore Was Warned of Fraud," *Boston Globe*, May 16, 1991, p. 17; Hamilton, "Baltimore Throws in the Towel," *Science*, 252(May 10, 1991), 768, 770; Walter Gilbert, "Faulty Behavior at Issue in Imanishi-Kari/Baltimore Case," *Genetic Engineering News*, July 8, 1991, p. 4.

39. Davis, draft manuscript of a book on the Baltimore case, Chap. 9, p. 26; Doty *et al.* tape, June 4, 1991.

40. Doty *et al.* tape, June 4, 1991; author's interview with Gilbert, Oct. 2, 1992. Several years later, Gilbert said that he thought Eisen had "deliberately misunderstood" the kind of hybrid molecules O'Toole had hypothesized. Gilbert, Testimony, *Appeal Proceedings*, June 21, 1995, p. 1529.

41. "Baltimore Case—In Brief," *Science*, 253(July 5, 1991), 25; O'Toole, [Reply to Eisen]; John Cairns, "To an Officer of the National Academy of Sciences, June 28, 1991," *Nature*, 351(June 27, 1991), 692; 352(July 11, 1991), 101.

42. O'Toole, "More Detailed Comments," [on Eisen's Statement in *Nature*], attached to O'Toole to Hadley, fax, June 13, 1991, AE H-295; Mark Ptashne, letter; Herman Eisen, letter, *Nature*, 352(July 11, 1991), 101–102; "Baltimore Case—In Brief," *Science*, 253(July 5, 1991), 25.

43. Author's conversation with Herman Eisen, July 10, 1996; Davis, draft manuscript of a book on the Baltimore case, Chap. 9, pp. 15–16; Eisen to the author, July 3, 1997. Eisen concedes that O'Toole may have had *mu-mu* hybrids in mind, but he thought then that her terminology was misleading: "She was very familiar with the immunological literature, and any reasonably informed immunologist would probably refer to such a *mu-mu* molecule as a 'mixed IgM' or a 'mixed *polymer*' and surely not as a 'heterodimer.' " Eisen adds that he is "perfectly willing to be charitable . . . and accept that the situation in Fig. 1 was (is)ambiguous." But he says that he further realizes now that Figure 2 poses "less ambiguity," pointing out that the reactivity to the idiotype in a mixed *mu* molecule would have been at least ten times greater than what the data in Figure 2 showed the reactivity actually was. Eisen to the author, June 11, 1997.

44. "Responsibility and Weaver *et al*," [commentary by Paul Doty], *Nature*, 352(July 18, 1991), 183–184; author's interview with Paul Doty, June 3, 1997.

45. Doty *et al.* tape, June 4, 1991; author's interview with Paul Doty, June 3, 1997.

46. Author's interview with Paul Doty, June 3, 1997; "Responsibility and Weaver *et al*," [commentary by Paul Doty], *Nature*, 352(July 18, 1991), 183–184. Doty also talked with William McClure, the member of the O.S.I. scientific panel who had done part of the statistical analysis of the handwritten digits and whom Doty

knew from McClure's days as a junior faculty member at Harvard. Doty says, "I knew what a careful and honest chap he was, so I respected his view on it pretty strongly, even though there were mistakes." Author's interview with Paul Doty, June 3, 1997. Bernard Davis had the impression that Doty was a guest in the home of John Maddox, the editor of *Nature*, when he wrote the letter. Doty, however, says that he had been traveling in England for two weeks and that, although he stayed in Maddox's home one night, he had already written the letter by then. Davis, draft manuscript of a book on the Baltimore case, Chap. 9, n. 6; author's telephone conversation with Doty, June 16, 1997.

47. Author's interview with Doty, June 3, 1997; "Responsibility and Weaver *et al*," [commentary by Paul Doty], *Nature*, 352(July 18, 1991), 183–184.

48. Author's interviews with Doty, June 3, 1997; with Baltimore, June 16, 1997; "Responsibility and Weaver *et al*," [commentary by Paul Doty], *Nature*, 352(July 18, 1991), 183–184; Nicholas Yannoutsos, letter, *Nature*, 352(July 11, 1991), 102.

49. Author's interview with Doty, June 3, 1997; "Responsibility and Weaver *et al*," [commentary by Paul Doty], *Nature*, 352(July 18, 1991), 183–184.

50. Baltimore, "Open Letter to Paul Doty," *Nature*, 353(Sept. 5, 1991), 9; author's interview with David Baltimore, June 16, 1997.

51. Doty and Ptashne, letters, *Nature*, 353(Oct. 10, 1991), 495. Baltimore says that he had retracted the paper "out of a sense of recognition of the [investigative] process" and that his retraction was "purely contingent" upon the outcome of the process. The letter reflected his continuing disbelief in the charges against Imanishi-Kari and his belief in the substance of the paper. Author's interview with Baltimore, June 16, 1997.

52. Stephen S. Hall, "David Baltimore's Final Days," *Science*, 254(Dec. 13, 1991), 1578–1579; Richard Saltus, "Baltimore, Citing Furor, Quits as Head of University," *Boston Globe*, Dec. 3, 1991, pp. 1, 17; Sarasohn, *Science on Trial*, pp. 246–247; author's conversation with Norton Zinder, Oct. 30, 1993; interviews with Baltimore, June 17, 1996, Oct. 31, 1997; with James Darnell, by telephone, June 19, 1996; with Alexander Bearn; by telephone, Oct. 28, 1997; Debra E. Blum, "Rockefeller U., Still Affected by Charges Involving Its President . . . ," *Chronicle of Higher Education*, Sept. 25, 1991, p. A28; Philip J. Hilts, "Science and the Stain of Scandal," *New York Times*, Dec. 4, 1991, pp. B1, B11.

53. Author's interviews with Baltimore, by telephone, June 17, 1996; with Paul Berg, by telephone, June 27, 1996; with James Darnell, by telephone, June 19, 1996; Huang, memo to files, Nov. 6, 1991, Huang Files; Blum, "Rockefeller U., Still Affected by Charges Involving Its President . . . ," *Chronicle of Higher Education*, Sept. 25, 1991, p. A28; Tim Beardsley, "Profile: David Baltimore," *Scientific American*, Jan. 1992, p. 36; Angier, "Rockefeller U. Is Hurt as President Feels Pain of Science-Fraud Case," *New York Times*, April 1, 1991, pp. 1, 7.

54. Hall, "David Baltimore's Final Days," *Science*, 254(Dec. 13, 1991), 1577; Angier, "University Hurt as Leader Endures Scientific Dispute," *New York Times*, April 1, 1991, pp. 1, 7; author's interviews with Baltimore, by telephone, June 17, 1996; with Roy Vagelos, by telephone, June 12, 1997; with Richard Furlaud, by telephone, June 11, 1997; Blum, "Rockefeller U., Still Affected by Charges Involving Its President . . . ," *Chronicle of Higher Education*, Sept. 25, 1991, p. A28; Baltimore, "President's Letter to the Board of Trustees," Aug. 28, 1991, Huang Files; Hilts, "Science and the Stain of Scandal," *New York Times*, Dec. 4, 1991, p. B1.

55. Author's interviews with Paul Berg, by telephone, June 27, 1996; with two other senior members of the Rockefeller community, who prefer to remain anonymous;

Hall, "David Baltimore's Final Days," *Science*, 254(Dec. 13, 1991), 1578; John Noble Wilford, "Second Research Group in a Year Is Leaving Rockefeller University," *New York Times*, Oct. 9, 1991, p. B8.

56. Author's interviews with Vagelos, June 12, 1997; with Berg, by telephone, April 8, 1996; Hall, "David Baltimore's Final Days," *Science*, 254(Dec. 13, 1991), 1578.

57. Hall, "David Baltimore's Final Days," *Science*, 254(Dec. 13, 1991), 1578–1579; Saltus, "Baltimore, Citing Furor, Quits as Head of University," *Boston Globe*, Dec. 3, 1991, pp. 1, 17; Sarasohn, *Science on Trial*, pp. 246–247; author's conversation with Norton Zinder, Oct. 30, 1993; interviews with Baltimore, June 17, 1996; with James Darnell, by telephone, June 19, 1996; Blum, "Rockefeller U., Still Affected by Charges Involving Its President . . . ," *Chronicle of Higher Education*, Sept. 25, 1991, p. A28; Hilts, "Science and the Stain of Scandal," *New York Times*, Dec. 4, 1991, pp. B1, B11.

58. Author's interview with Furlaud, by telephone, June 11, 1997; Huang, notes, Sept./Oct. 1991, Huang Files; Blum, "Rockefeller U., Still Affected by Charges Involving Its President . . . ," *Chronicle of Higher Education*, Sept. 25, 1991, p. A28; Hall, "David Baltimore's Final Days," *Science*, 254(Dec. 13, 1991), 1578. According to Hall, Wiesel's poll showed that Baltimore had lost the support of 70 percent of the senior faculty. Apart from the fact that the figure does not square with what Wiesel evidently reported to Baltimore, it was obtained after Baltimore's resignation and may have been a post hoc misrepresentation by Hall's source. It seems unlikely that, faced with such overwhelming opposition, David Rockefeller would have announced his huge gift to the university at the meeting of the trustees on October 17 and coupled it with his strong endorsement of Baltimore.

59. Kathleen Teltsch, "The Founding Grandson Gives to Rockefeller U.," *New York Times*, Oct. 18, 1991, p. B3; Huang, memo to files, Nov. 6, 1991, Huang Files; author's interview with Furlaud, by telephone, June 11, 1997; C[olin] N[orman], "Rockefeller's $20 Million Gift," *Science*, 254 (Oct. 25, 1991), 512.

60. Huang, memo to files, Nov. 6, 1991, Huang Files; author's interview with Furlaud, by telephone, June 11, 1997.

61. Author's interviews with Furlaud, by telephone, June 11, 1997; with Vagelos, by telephone, June 12, 1997; Hall, "David Baltimore's Final Days," *Science*, 254 (Dec. 13, 1991), 1576.

62. Huang, memo to files, Nov. 11, 1991, Huang Files. Furlaud recalls that he and Rockefeller essentially reported the faculty's views to Baltimore, asking him, What should we do? Author's interview with Furlaud, by telephone, June 11, 1997.

63. Huang, memos to files, Nov. 11, 1991, Nov. 30, 1991, Huang Files; author's interview with Kildow, June 7, 1996.

64. Huang, memo to files, Nov. 30, 1991; Darnell to Furlaud, Nov. 27, 1991, Huang Files.

65. Author's interview with Baltimore, June 17, 1996; Huang, memo to files, Nov. 30, 1991, Huang Files; Baltimore to Furlaud and Rockefeller, Dec. 2, 1991, copy in Stewart and Feder Files. The same day that Darnell sent Furlaud his own resignation letter, he sent Baltimore a copy of it with a handwritten covering note telling Baltimore that he didn't deserve such a fate. Since he mentioned in the letter that he had told Furlaud Baltimore should resign, he probably said the same thing in his telephone call to Baltimore at Wood's Hole. Darnell to Baltimore, Nov. 27, 1991, Huang Files. Furlaud says that the decision to resign was Baltimore's and that the trustees endorsed it. Author's interview with Furlaud, by telephone, June 11, 1997.

66. Hilts, "Nobelist Caught Up in Fraud Case Resigns as Head of Rockefeller U.",
    Hilts, "Science and the Stain of Scandal," *New York Times*, Dec. 3, 1991, p. A10;
    Dec. 4, 1991, p. B1; Eisen to Rockefeller, Oct. 25, 1991; Rockefeller to Eisen, Dec.
    5, 1991, Eisen Files.

67. Editorial, *Wall Street Journal*, Dec. 4, 1991, p. A16; Richard Saltus, "Baltimore's
    Legacy: Concern about Oversight of Scientists," *Boston Globe*, Dec. 5, 1991.
    Nature noted in an editorial on the resignation, "Complaints abound that Dingell
    is a bully." "Baltimore Defeat a Defeat for Research," editorial, *Nature*, 354 (Dec.
    12, 1991), 419–420.

68. "Baltimore Steps Down from Rockefeller Presidency," *Science and Government
    Report*, Dec. 15, 1991, p. 4; "Baltimore Defeat a Defeat for Research," editorial,
    Nature, 354 (Dec. 2, 1991), 419; "Rough Justice for Dr. Baltimore," editorial, *New
    York Times*, Dec. 5, 1991, p. 20.

Notes for Chapter Thirteen

1. Joan L. Press to the editor, *Boston Globe*, March 21, 1997, files of David Parker
   and Joan Press (hereafter Parker and Press Files); Judy Foreman, "Fraud Charge
   Leaves a Career in Shambles," *Boston Globe*, May 6, 1991, p. 27. Press says, "The
   notion that any and all of the three people on the Tufts committee would have
   any rationale, reason, or whatever to lie for Thereza about anything in that paper
   is absurd. The notion that in addition to those three people, the M.I.T. people
   were also doing the same is ridiculous. Good God, what did they have to gain?"
   Author's interview with Joan Press, June 16, 1996.

2. Barbara Mishkin, "Responding to Scientific Misconduct, Due Process and Pre-
   vention," *Journal of the American Medical Association*, 260 (Oct. 7, 1988), 1934–
   1935.

3. Sarasohn, *Science on Trial*, pp. 105–106; Robert Charrow, "Scientific Misconduct:
   Sanctions in Search of Procedures," *Journal of NIH Research*, 2 (June 1990), 91–
   92; David P. Hamilton, "A Question of 'Fitness,'" *Science*, 248 (June 29, 1990),
   1598. The sociologist Patricia K. Woolf had earlier called attention to the violations of due process inherent in the kind of inquiry into scientific fraud that the
   Dingell subcommittee had conducted in April 1988. Woolf, "Science Needs Vigilance Not Vigilantes," *Journal of the American Medical Association*, 260 (Oct. 7,
   1988), 1939.

4. *Abbs v. Sullivan*, Civil Case 90-C-0470-C, transcript of Motion Hearing (W. D.
   Wisc., Aug. 2, 1990), quoted in Barbara Mishkin, "Ethics, Law, and Public Policy,"
   *Professional Ethics Report, Newsletter of the American Association for the Advance-
   ment of Science Committee on Scientific Freedom & Responsibility Professional Society Ethics Group*, IV (Spring 1991), 4–5; Robb London, "Judge Strikes Rules on
   Scientific Fraud Inquiries," *New York Times*, May 3, 1991, p. B13; Robert P. Charrow, "Scientific Misconduct Revisited: O.S.I. on Trial," *Journal of NIH Research*,
   2(Oct. 1990), 83–84.

5. *Abbs v. Sullivan*, Civil Case 90-C-0470-C, transcript of Motion Hearing (W. D.
   Wisc., Aug. 2, 1990), quoted in Mishkin, "Ethics, Law, and Public Policy," *Pro-
   fessional Ethics Report, Newsletter of the American Association for the Advancement
   of Science Committee on Scientific Freedom & Responsibility Professional Society
   Ethics Group*, IV (Spring 1991), 4–5; *Abbs v. Sullivan*, 756 F. Supp. 1172 (W. D.
   Wisc. 1990); London, "Judge Strikes Rules on Scientific Fraud Inquiries," *New*

York Times, May 3, 1991, p. B13; Charrow, "Scientific Misconduct Revisited: O.S.I. on Trial," *Journal of NIH Research*, 2(Oct. 1990), 83–84.

6. Public Health Service, "Policies and Procedures for Dealing with Possible Misconduct in Extramural Research," Aug. 29, 1990, copy dated Aug. 30, 1990, copy in author's possession, pp. 1, 9.

7. London, "Judge Strikes Rules on Scientific Fraud Inquiries," *New York Times*, May 3, 1991, p. B13; Christopher Anderson, "Back to the Drawing Board," *Nature*, 350 (March 14, 1991), 100; Hallum and Hadley, letter, *Science*, 251 (March 15, 1991), 1296; Warren E. Leary, "On The Trail of Misconduct in Science Where U.S. Billions Can Be at Stake," *New York Times*, March 25, 1991, p. A10. The O.S.I.'s parent office, the Office of Scientific Integrity Review, published the O.S.I.'s procedures on June 13, 1991, in the *Federal Register*. At the same time, the Department of Justice appealed Judge Crabb's ruling, claiming that the O.S.I.'s procedures are not federal rules requiring public notice and comment. The rules the O.S.I. published were, except for minor changes, those that the N.I.H. had issued for investigations of misconduct in 1986. The O.S.I. won on appeal in May 1992, on the technical grounds that Abbs had no standing to protest its rules because he had not suffered a demonstrable injury from them. United States Court of Appeals for the Seventh Circuit, *Abbs v. Sullivan*, Nos. 91-1923, 91-1924, 91-2149, May 1, 1992. In 1996, the case approached closure. See Jock Friedly, "After 9 Years, a Tangled Case Lurches Toward a Close," *Science*, 272 (May 17, 1996), 947–948.

8. Leary, "On the Trail of Misconduct in Science Where U.S. Billions Can Be at Stake," *New York Times*, March 25, 1991, p. A10; Charrow, "Scientific Misconduct Revisited," *Journal of NIH Research*, 2(Oct 1990), 84–85; Jules V. Hallum and Suzanne W. Hadley, "OSI: Why, What, and How," *ASM [American Society of Microbiology] News*, 56(No. 12, 1990), 647–651.

9. Leary, "On the Trail of Misconduct in Science Where U.S. Billions Can Be at Stake," *New York Times*, March 25, 1991, p. A10; Charrow, "Scientific Misconduct Revisited," *Journal of NIH Research*, 2 (Oct 1990), 84–85; Hallum and Hadley, "OSI: Why, What, and How," *ASM [American Society of Microbiology] News*, 56 (No. 12, 1990), 647–651; David L. Wheeler, "N.I.H. Office That Investigates Scientists' Misconduct Is Target of Widespread Charges of Incompetence," *Chronicle of Higher Education*, May 15, 1991, p. A5.

10. Hamilton, "Can O.S.I. Withstand a Scientific Backlash?" *Science*, 253 (Sept. 6, 1991), 1084–1085; David L. Wheeler, "President of Rockefeller University Retracts Scientific Paper That N.I.H. Office Says Contains Fabricated Data," *Chronicle of Higher Education*, March 27, 1991, pp. A8, A11; Jules V. Hallum and Suzanne Hadley, "Scientific Misconduct," *Journal of NIH Research*, 2 (Sept. 1990), 12.

11. London, "Judge Strikes Rules on Scientific Fraud Inquiries," *New York Times*, May 3, 1991, p. B13; interview with Hallum, Clyde Watkins, and Alan Price, of the O.S.I., *Science and Government Report*, Oct. 1, 1991, pp. 3–6. Hallum, with the Abbs ruling in mind, disparaged the due-process issue because a grant is not given to benefit the scientist but to benefit society. Since there's no liberty or property issue, introducing due process into the matter is "a bit strange." *Ibid.*, pp. 3–4.

12. Wheeler, "N.I.H. Office . . . ," *Chronicle of Higher Education*, May 15, 1991, pp. A5, A8; "The Antibodies That Weren't," *Newsweek*, April 1, 1991, p. 60; London, "Judge Strikes Rules on Scientific Fraud Inquiries," *New York Times*, May 3, 1991, p. B13; Hamilton, "Can OSI Withstand a Scientific Backlash?" *Science*, 253

(Sept. 6, 1991), 1085. *Nature* editorialized, "OSI's powers are so close to those of a prosecutor and jury combined that it may be time to face the implications of that reality." "Even Misconduct Trials Should Be Fair," *Nature*, 350 (March 28, 1991), 259–260.

13. "Response Submitted by Thereza Imanishi-Kari, Ph.D. to Draft O.S.I. Report on Alleged Scientific Misconduct," May 23, 1991, including Bruce A. Singal to Suzanne W. Hadley, May 23, 1991, copy in AE H-292; author's interview with Bruce A. Singal, March 20, 1996.

14. Author's interview with Press, June 16, 1996.

15. *Ibid.*; "An Open Letter on O.S.I.'s Methods," *Nature*, 357 (June 27, 1991), 693; David P. Hamilton, "Did Imanishi-Kari Get a Fair Trial," *Science*, 252 (June 21, 1991), 1607.

16. Hallum to Parker and Press, June 28, 1991, Parker and Press Files; author's interview with Press, June 16, 1996; "Dingell Gets Baltimore," editorial, *Wall Street Journal*, Dec. 4, 1991, p. A16; Department of Health and Human Services, Public Health Service, "Policies and Procedures for Dealing with Possible Scientific Misconduct in Extramural Research," *Federal Register*, 56 (June 13, 1991), 27384–27394; Hamilton, "Did Imanishi-Kari Get a Fair Trial?" *Science*, 252 (June 21, 1991), 1607. Singal recalls that he would regularly encounter scientists who would tell him, Too bad that Imanishi-Kari had been found guilty of fraud, revealing that they were unaware that the draft report was just a tentative finding, let alone that its conclusions rested on profoundly faulty procedures. Author's interview with Singal, March 20, 1996.

17. Department of Health and Human Services, Public Health Service, "Policies and Procedures for Dealing with Possible Scientific Misconduct in Extramural Research," *Federal Register*, 56(June 13, 1991), 27384–27394; Robert J. Cousins to Members of the FASEB Societies, July 24, 1991, Parker and Press Files.

18. Cousins to Members of the FASEB Societies, July 24, 1991, Parker and Press Files.

19. Hamilton, "Can O.S.I. Withstand a Scientific Backlash?" *Science*, 253 (Sept. 6, 1991), 1085; "FASEB Society Members Respond"; "Federal Scientific Misconduct Policy . . . ," *FASEB Newsletter*, Sept. 1991, p. 4; April/May 1992, p. 4; Davis and Guenin to Lyle Bivens, Aug. 9, 1991, Press and Parker Files.

20. Erik Eckholm, "A Tough Case for Dr. Healy," *New York Times Magazine*, Dec. 1, 1991, pp. 67, 118, 122.

21. *Ibid.*; Marlene Cimmons, "A Straight Shooter Energizes Troubled N.I.H.," *Los Angeles Times*, April 14, 1992, p. E1; Stephen Burd and David L. Wheeler, "In Her First Years, N.I.H. Director Moves Swiftly on Planning and Women's Health . . . ," *Chronicle of Higher Education*, April 8, 1992, pp. A28–A29; author's interview with Bernadine Healy, by telephone, Aug. 27, 1996.

22. Bernadine Healy, Testimony and Statement, *Hearing, Scientific Fraud*, Aug. 1, 1991, pp. 177–181, 226.

23. *Ibid.*; Robert Lanman, Testimony, *Hearings, Scientific Fraud*, Aug. 1, 1991, pp. 134–135.

24. Author's interview with Healy, by telephone, Aug. 27, 1996; [anon.], "Chronology"; Healy, Statement, *Hearings, Scientific Fraud*, Aug. 1, 1991, pp. 163–164, 197–198.

25. Healy, Testimony and Statement, *Hearings, Scientific Fraud*, Aug. 1, 1991, pp. 170–171, 185–186; Joseph Palca, "Scientists Get Mad at OSI," *Science*, 252(June 21, 1991), 1606.

26. Healy, Testimony, *Hearings, Scientific Fraud*, Aug. 1, 1991, pp. 170–171.

Endnotes (pages 299-302) 467

27. Author's interview with Healy, by telephone, Aug. 27, 1996; Tribe to the author, Oct. 27, 1997. Tribe's text is *American Constitutional Law* (Mineola, N.Y.: Foundation Press, 1978; 2nd ed., 1988). The Privacy Act is P.L. 93-579, Dec. 31, 1974.

28. Author's interview with Healy, by telephone, Aug. 27, 1996; Healy, Statement and Testimony; [anon.], "Chronology," *Hearings, Scientific Fraud*, Aug. 1, 1991, pp. 152, 188, 205–207, 163–165. Malcolm Gladwell, then a reporter at the *Washington Post*, recalls that Hadley was absorbed in Popovic's personal history as a Czech Jewish refugee. "She said to me—and I'll never forget this—'It sounds like something out of a novel. It happens to be true.' What's interesting in that statement is that in this long period she talked to me, she never once talked about the scientific issues that were involved. She was living out some kind of grandiose literary fantasy about her role in unlocking this fabulous story that could just as easily have been written as a novel." Gladwell, remarks, M.I.T. Symposium on the Baltimore Case in American Political Culture, Oct. 28, 1996.

29. Author's interview with Hadley, Sept. 15, 1992; Hallum, draft of a book on scientific fraud, Chapter IX, p. 20; Hamilton, "Healy Returns Fire at Dingell Hearing"; Hamilton, "Can OSI Withstand a Scientific Backlash," *Science*, 253 (Aug. 9, 1991), 618; (Sept. 6, 1991), 1085–1086; Healy, Statement, *Hearings, Scientific Fraud*, Aug. 1, 1991, p. 199.

30. Storb to Martin Flax, May 12, 1986, Onek Files; Sarasohn, *Science on Trial*, pp. 253–255; author's interview with Storb, April 8, 1997.

31. Sarasohn, *Science on Trial*, pp. 253–255; author's interview with Storb, April 8, 1997; Healy, Testimony; [anon.], "Chronology," *Hearings, Scientific Fraud*, Aug. 1, 1991, pp. 162–163, 171.

32. Healy, Testimony; [anon.], "Chronology," *Hearings, Scientific Fraud*, Aug. 1, 1991, pp. 162–165, 171, 194–195; Hilts, "Conflict of Interest Is Charged in Inquiry on Research Paper," *New York Times*, June 14, 1991, p. D18; David P. Hamilton, "O.S.I. Investigator Reined In"; David P. Hamilton "Post-Mortem on Storb Resignation," *Science*, 253 (July 26, 1991), 372; (Aug. 22, 1991), 850; Bernard Davis, notes on telephone conversation with Storb, Oct. 21, 1993, Davis Files; author's interview with Storb, April 8, 1997; William McClure to Hadley, June 17, 1991, O.R.I. Files.

33. Joseph Palca, "Bernadine Healy: A New Leadership Style at N.I.H.," *Science*, 253(Sept. 6, 1991), 1088; author's interview with Healy, by telephone, Aug. 27, 1996.

34. Author's interview with Healy, by telephone, Aug. 27, 1996; Sarasohn, *Science on Trial*, pp. 255–256; [anon.], "Chronology"; Healy, Statement, *Hearings, Scientific Fraud*, Aug. 1, 1991, pp. 165–166, 172, 200–201. Raub later said that he was one of the staff who told Healy that Hadley had gotten close to O'Toole but that he had meant the report as a compliment to Hadley for having won O'Toole's trust. Raub, Testimony, *Hearings, Scientific Fraud*, Aug 1, 1991, p. 149.

35. Lanman, Hallum, and Hadley, Testimony; [anon.], "Chronology"; Healy, Statement, *Hearings, Scientific Fraud*, Aug. 1, 1991, pp. 150, 151, 164, 200–201; Hallum to Lanman, June 10, 1991, O.R.I. Files.

36. [anon.], "Chronology"; Healy, Statement, *Hearing, Scientific Fraud*, Aug. 1, 1991, pp. 166, 208–210.

37. Sarasohn, *Science on Trial*, p. 256; [anon.], "Chronology," *Hearings, Scientific Fraud*, Aug. 1, 1991, p. 166; author's interviews with Hadley, Sept. 15, 1992; with Hallum, by telephone, April 24, 1996; Hamilton, "OSI Investigator Reined In";

Hamilton, "Can OSI Withstand a Scientific Backlash?" *Science*, 253(July 26, 1991), 372; (Sept. 6, 1991), 1085.

38. Hamilton, "Healy Returns Fire at Dingell Hearing," *Science*, 253(Aug. 9, 1991), 618; Sarasohn, *Science on Trial*, pp. 257–258; author's interview with Healy, by telephone, Aug. 27, 1996; Healy, Statement; Bernadine P. Healy, "Handling Scientific Misconduct Challenges for N.I.H.," July 15, 1991; *Hearings, Scientific Fraud*, Aug. 1, 1991, pp. 203, 211–213, 218–220.

39. Healy, "The Dangers of Trial by Dingell," *New York Times*, July 3, 1996, p. 11; Palca, "Bernadine Healy: A New Leadership Style at N.I.H.," *Science*, 253(Sept. 6, 1991), 1088; author's interview, by telephone, with Healy, Aug. 27, 1996. Jay Moskowitz, who was an associate director of science policy in N.I.H. at the time, confirms Healy's account of the questions that Stockton put to her secretary about Healy's living arrangements. He recalls that Stockton asked him, too, what kind of car he drove and where he lived. He says, "Stockton spared no vulgarity when talking with me. Always threats, constant threats—that we would be dragged down to a hearing with Mr. Dingell. He wanted to make it clear that he had the power to put us before the oversight committee. You know how that works. Dingell sits two stories above you and goes after you relentlessly." Author's interview, by telephone, with Jay Moskowitz, June 30, 1997.

40. Sarasohn, *Science on Trial*, pp. 257–258; Hamilton, "OSI Investigator 'Reined In,'" *Science*, 253(July 26, 1991), 372; Lanman and Healy, Testimony; Healy to Assistant Secretary of Health James O. Mason, July 19, 1991, *Hearings, Scientific Fraud*, Aug. 1, 1991, pp. 142, 170, 255–256.

41. Dingell, Opening Statement; Lent, Statement; [anon.], "Chronology," *Hearings, Scientific Fraud*, Aug. 1, 1991, pp. 125–131, 162–166.

42. Author's interview with Healy, by telephone, Aug. 27, 1996.

43. Healy, Testimony; Mason, Testimony, *Hearings, Scientific Fraud*, Aug. 1, 1991, pp. 169, 228, 251–252; author's interview with Moskowitz, by telephone, June 30, 1997.

44. Palca, "Bernadine Healy: A New Leadership Style at N.I.H.," *Science*, 253(Sept. 6, 1991), 1087; A Friend of Science and NIH to Hallum, Aug. 7, 1991, Friedly Files.

45. "N.I.H.'s Bungling Goes On in the Baltimore Case," *Science and Government Report*, Sept. 15, 1991, pp. 2–3; Willard J. Forbush to Assistant Inspector General for Criminal Investigations, Sept. 17, 1991; Dingell to Louis Sullivan, Aug. 30, 1991, Friedly Files; author's interview with Healy, by telephone, Aug. 27, 1996.

46. "Dr. Healy, Dr. Kennedy, and Mr. Dingell," editorial, *Nature*, 352(Aug. 8, 1991), 457–458. During the fall, even *Science and Government Report* published what amounted to a forum on the matter, giving space to interviews in one issue with Hallum and two colleagues from the N.I.H. misconduct offices, who insisted that the existing system was essentially fair, and in another to Healy, who contended it was not. *Science and Government Report*, Oct. 1, 1991, pp. 1, 3–6; Nov. 1, 1991, pp. 4–5.

47. *Science and Government Report*, Oct. 15, 1991, p. 1.

48. Author's interviews with Healy, by telephone, Aug. 27, 1996; with Moskowitz, June 30, 1997.

49. Author's interview with Healy, by telephone, Aug. 27, 1996; "OSI Reorganization Plan Goes Public," *Science*, 255(March 6, 1992), 1199; Healy, Statement, *Hearings, Scientific Fraud*, Aug. 1, 1991, p. 173.

50. "OSI Reorganization Plan Goes Public"; David P. Hamilton, "OSI: Better the Devil You Know?" *Science*, 255(March 6, 1992), 1199; (March 13, 1992), 1344–1347.

51. Hamilton, "OSI: Better the Devil You Know?" *Science*, 255(March 13, 1992), 1344–1347; David L. Wheeler, "U.S. Agency Proposes Trial-Like Hearings to Judge Cases of Scientific Misconduct," *Chronicle of Higher Education*, March 18, 1992, pp. A8–A9; "A New Beat for the Science Cops," editorial, the *Washington Post*, March 20, 1992, p. A24.

52. "Department of Health and Human Services, Public Health Service; Statement of Organization, Functions and Delegations of Authority," *Federal Register*, 57(June 8, 1992), 24262–24263.

53. "Hallum Resigns, Blasts New Misconduct Office," *Science*, 257(Sept. 4, 1992), 1335; Healy, Testimony, *Hearings, Scientific Fraud*, Aug. 1, 1991, p. 221; author's interview with Chris Pascal, by telephone, May 3, 1996.

54. "Changes in Misconduct Policies on the Way," *Science*, 258(Oct. 23, 1992), 535; Notice, "Opportunity for a Hearing on Office of Research Integrity Misconduct Findings," *Federal Register*, 57(Nov. 6, 1992), 53125.

55. Notice, "Opportunity for a Hearing on Office of Research Integrity Misconduct Findings," *Federal Register*, 57(Nov. 6, 1992), 53125.

56. Author's interview with Pascal, by telephone, May 3, 1996; Stephen Burd, "Public Health Service Plans Hearings for Scientists Accused of Fraud," the *Chronicle of Higher Education*, June 17, 1992, p. A23.

NOTES FOR CHAPTER FOURTEEN

1. Singal to Hadley, March 29, 1991, Bruce Singal's Imanishi-Kari: Case File (hereafter Singal Files).; author's interview with Singal, March 20, 1996.

2. Hadley to Singal, April 9, 1991, Singal Files.

3. Singal to Hadley, May 23, 1991, transmittal letter with Thereza Imanishi-Kari, "Response Submitted by Thereza Imanishi-Kari, Ph.D., to Draft O.S.I. Report . . . ," May 23, 1991, AE H-292.

4. Thereza Imanishi-Kari, "Response Submitted by Thereza Imanishi-Kari, Ph.D., to Draft O.S.I. Report . . . ," May 23, 1991, AE H-292, pp. 1–4, 24, 38–39. The six pages in question were I-1:83–88. In the formal allegations that O'Toole filed with the O.S.I. in November 1989, she said that she had seen those pages during her meeting with the Wortis committee on May 23, 1986. The pages reported experiments that first selected antibodies carrying the idiotypic birthmark and then identified their isotypes. O'Toole said that the experiments actually selected the much broader category of mouse antibodies in general, and then isotyped those, and that Imanishi-Kari had admitted the fact, calling it a mistake. She contended that neither these data nor the other data she saw that day showed that the hybridomas reported in Table 3 produced antibodies from native genes that carried an idiotype related to the transgene's, and she went on that "all agreed that this was a fatal flaw in the study." However, in her memo to Eisen on June 6, O'Toole said nothing about the experiments reported on pages I-1:83–88, nor did she mention a fatal flaw in the serology of the *Cell* paper. Her discussion of Table 3 dwelled entirely on the analytical methods used to characterize the hybridomas reported there and was informed by what she had seen of page I-1:41. In an interview with N.I.H. in 1988, she said, referring to a loose-leaf sheet of data

concerning *mu* and another concerning *gamma,* that "the only data that I saw [at Tufts on May 23] were these two sheets." O'Toole, "Allegations Concerning the Cell Paper . . . ," attached to O'Toole to Kimes, Nov. 6, 1989, AE H-276; O'Toole to Eisen, June 6, 1986, AE H-231; N.I.H. Staff, O'Toole interview, May 17, 1988, AE H-109, pp. 51–53.

5. Thereza Imanishi-Kari, "Response Submitted by Thereza Imanishi-Kari, Ph.D., to Draft O.S.I. Report . . . ," May 23, 1991, AE H-292, pp. 7–8, 21–25, 27, 28–29, 35.

6. *Ibid.,* pp. 9–16.

7. *Ibid.,* p. 45; Singal to Hadley, May 23, 1991.

8. Healy, Statement, *Hearings, Scientific Fraud,* Aug. 1, 1991, pp. 196–197; Thereza Imanishi-Kari, by her attorney, Bruce A. Singal, "Respondent's Memorandum in Support of Her Motion . . . ," Dec. 17, 1993, Singal Files, pp. 14–20; author's interview with Jules Hallum, by telephone, April 24, 1996.

9. Diggs to Parker, Oct. 10, 1991, Parker and Press Files.

10. "N.I.H.'s Bungling Goes On in the Baltimore Case," *Science and Government Report,* Sept. 15, 1991, pp. 3–4; "Daniel S. Greenberg Responds," *Journal of NIH Research,* 3(Nov. 1991), 14; Sarasohn, *Science on Trial,* pp. 259–261; Director, Office of Scientific Integrity to Deputy Director for Extramural Research, Nov. 12, 1991, O.R.I. Files.

11. David P. Hamilton, "FBI Investigates Leaks at O.S.I.," *Science,* 255 (March 20, 1992), 1503; author's interviews with Healy, by telephone, Aug. 27, 1996; with Hallum, by telephone, April 24, 1996; Healy to Alan B. Carroll, March 10, 1992, copy in author's possession.

12. Hamilton, "FBI Investigates Leaks at O.S.I.," *Science,* 255(March 20, 1992), 1503; Phillip Hilts, "F.B.I. Pursues Leak of Documents in Science Misconduct Inquiries," *New York Times,* March 14, 1992, p. 11.

13. Healy, "The Dingell Hearings on Scientific Misconduct: Blunt Instruments Indeed," *New England Journal of Medicine,* 329(Sept. 2, 1993), 726–727. It is obvious from the articles by Hamilton and Hilts that Hadley was the "chief investigator" to whom Healy referred, and Hadley tacitly acknowledges it in denying the truth of the allegation. Author's interviews with Healy, by telephone, Aug. 27, 1996; with Hallum, by telephone, April 24, 1996; with an N.I.H. official, by telephone, June 6, 1996; with Hadley, by telephone, April 9, 1996.

14. Hilts, "F.B.I. Pursues Leak of Documents in Science Misconduct Inquiries," *New York Times,* March 14, 1992, p. 11; John D. Dingell, "Shattuck Lecture—Misconduct in Medical Research," *New England Journal of Medicine,* 328(June 3, 1993), 1613. The printed text of the lecture appears to be close to what Dingell actually said about Healy, Hadley, and Baltimore, if not Gallo. See "Medical Editor Axes Dingell's Remarks about Gallo," *Science and Government Report,* Oct. 1, 1992, pp. 6–7.

15. Philip J. Hilts, "The Science Mob," *New Republic,* May 18, 1992, pp. 24–31.

16. "Science Friday," June 19, 1992, tape of the program in author's possession; "Whistleblower Prize for O'Toole," *Science and Government Report,* July 3, 1992, 27; Sarasohn, *Science on Trial,* pp. 259–261.

17. Friedly, "Trial and Error," *Boston Magazine,* Jan. 1997, pp. 51, 68; author's interview with Baltimore, Aug. 17, 1992.

18. Angier, "Embattled Biologist Will Return to M.I.T.," *New York Times,* May 19, 1992, p. C5; "Baltimore To Move Across the Street?" *Science,* 256(May 1, 1992), 603; Huang, memo to file, Aug. 18, 1993, Huang Files; author's interview with Alexander Rich, by telephone, May 31, 1996; Huang to the author, July 14, 1997.

19. Author's interviews with Imanishi-Kari, Oct. 6, 1992, March 20, 1996, Oct. 29, 1996; with Henry Wortis, March 20, 1996; with Flax, Oct. 7, 1992.

20. Author's interviews with Imanishi-Kari, Oct. 6, 1992, March 20, 1996, Oct. 29, 1996; with Henry Wortis, March 20, 1996; with Flax, Oct. 7, 1992; with Sol Gittleman, by telephone, July 15, 1997; with Mary Lee Jacobs, by telephone, July 17, 1997; Imanishi-Kari, Testimony, *Appeal Proceedings*, Aug. 29, 1995, pp. 4652–4653; Press and Parker to Ada Kruisbeck, July 22, 1991; Press to ___ ___, June 3, 1994, Parker and Press Files. The two papers were: Thereza Imanishi-Kari *et al.*, "Endogenous Ig Production in μ Transgenic Mice," Parts I and II, *Journal of Immunology*, 150(April 15, 1993), 3311–3326, 3327–3346. Both were submitted in May 1992 and acknowledge support of the work in them by American Cancer Society Grant IM 477.

21. Thereza Imanishi-Kari, by her attorney, Bruce A. Singal, "Respondent's Memorandum in Support of Her Motion . . . ," Dec. 17, 1993, Singal Files, pp. 14–20; "Affidavit of Albert H. Lyter, III," June 17, 1992; "Supplementary Affidavit of Albert H. Lyter, III," August 13, 1992, both attached to "Supplemental Response of Thereza Imanishi-Kari, Ph.D. to O.S.I. Draft Report," Aug. 18, 1992.

22. U.S. Department of Justice, United States Attorney District of Maryland, "Press Release," July 13, 1992; David P. Hamilton, "U.S. Attorney Decides Not to Prosecute Imanishi-Kari," *Science*, 257(July 17, 1992), 318; Christopher Anderson and Traci Watson, "US Drops Imanishi-Kari Investigation . . . ," *Nature*, 358(July 16, 1992), 177; Bennett and Garinther to Charles Maddox, July 13, 1992, O.R.I. Files.

23. Author's interview with Singal, March 20, 1996. Baltimore's lawyer, Normand Smith, recalls Singal's saying that a member of the U.S. attorney's office told Singal that at the determining meeting on the case the week before, half the time was given to the decision not to indict, the other half to how to handle the announcement of the decision. Author's conversation with Smith, by telephone, July 15, 1992.

24. "Prosecutors Halt Scientific Fraud Probe," *Washington Post*, July 14, 1992, p. 3; Anderson and Watson, "US Drops Imanishi-Kari Investigation . . . ," *Nature*, 358(July 16, 1992), 177; Anthony Fauci to Jeanette Thorbecke, Oct. 26, 1992, Parker and Press Files.

25. "Supplemental Response of Thereza Imanishi-Kari, Ph.D. to O.S.I. Draft Report," Aug. 18, 1992, attached to Singal to Chris Pascal, Aug. 18, 1992, AE H-298, pp. 14–16.

26. *Ibid.*, pp. 5–7 and n. 1; Dingell to J. Michael McGinnis, Sept. 16, 1992, O.R.I. Files.

27. *Ibid.*, pp. 14–20; Lanman to David Parker, Dec. 2, 1992; Parker and Press to Colleagues, Oct. 3, 1992, Parker and Press Files; Clyde A. Watkins to Imanishi-Kari, Aug. 19, 1992, Onek Files.

28. Parker and Press to Colleagues, Oct. 3, 1992; Fauci to Thorbecke, Oct. 26, 1992, Parker and Press Files.

29. Parker to Fauci, Nov. 14, 1992; John J. McGowan to Parker, Jan. 11, 1993, Parker and Press Files.

30. Kong-Pen Lam, L. A Herzenberg, and A. M. Stall, "A High Frequency of Hybridomas from M54 μ Heavy Chain Transgenic Mice Initially Co-express Transgenic and Rearranged Endogenous μ Genes," *International Immunology*, 5(1993), 1011–1022; Robert Cooke, " 'False' Now Judged True," *New York Newsday*, [1993]; Thereza Imanishi-Kari *et al.*, "Endogenous Ig Production in μ Transgenic Mice," Parts I and II, *Journal of Immunology*, 150(April 15, 1993), 3311–3326, 3327–3346;

John Travis, "Imanishi-Kari Says Her New Data Shows She Was Right," *Science*, 260(May 21, 1993), 1073–1074.

31. Dingell, "Shattuck Lecture—Misconduct in Medical Research," *New England Journal of Medicine*, 328(June 3, 1993), 1614–1615; Philip J. Hilts, "Federal Inquiry Finds Misconduct by a Discoverer of the AIDS Virus," *New York Times*, Dec. 31, 1992, p. 1.

32. Healy, "The Dingell Hearings on Scientific Misconduct: Blunt Instruments Indeed"; Baltimore, Letter to the Editor, *New England Journal of Medicine*, 329(Sept. 2, 1993), 725, 727, 732–733.

33. Department of Health and Human Services, Departmental Appeals Board, Research Integrity Adjudications Panel, "Decision" on Mikulas Popovic, Docket No. A-93–100, Decision No. 1446, Nov. 3, 1993, pp. 1,4.

34. *Los Angeles Times*, Nov. 13, 1993, p. 1.

35. Singal to Lyle Bivens, Dec. 17, 1993, Singal Files; Fred Karush to Singal, Aug. 21, 1992; Dagmar Ringe to Singal, Oct. 14, 1992, attached to Thereza Imanishi-Kari, by her attorneys, Bruce A. Singal, "Respondent's Memorandum in Support of Her Motion . . . ," Dec. 17, 1993, Singal Files.

36. Thereza Imanishi-Kari, by her attorneys, Bruce A. Singal, "Respondent's Memorandum in Support of Her Motion . . . ," Dec. 17, 1993, Singal Files.

37. *Ibid.*; Public Health Service, "Policies and Procedures for Dealing with Possible Misconduct in Extramural Research," Aug. 29, 1990, copy dated Aug. 30, 1990, copy in author's possession, p. 16.

38. *Ibid.*

39. Author's interview with Chris Pascal, by telephone, June 30, 1997; Office of Research Integrity, "Investigation Report: The Massachusetts Institute of Technology, ORI 072," enclosed with Terence J. Tychans and Lyle W. Bivens to Theresa Imanishi-Kari, Ph.D., via Counsel, Oct. 26, 1994, Onek Files, p. 27. The circulation of the draft final report in the summer of 1993 is evident from the index of documents in the Imanishi-Kari case, Onek Files. The heavy load of old and new cases was a point made in an interim response to Singal's motion by the director of the O.R.I., Lyle W. Bivens, in a letter to Singal, March 31, 1994, Singal Files. Bivens advised Singal that the points he had raised were being "carefully considered."

40. Author's interview with Pascal, by telephone, June 30, 1997; Marcus H. Christ, Jr., Stephen M. Godek, and Carol C. Conrad, *Post-Hearing Brief of the Office of Research Integrity*, Dec. 22, 1995, Before the United States Department of Health and Human Services, Departmental Appeals Board, Research Integrity Adjudications Panel, In the Matter of Thereza Imanishi-Kari, Board Docket No. A-95-33, p. 18.

41. Office of Research Integrity, "Investigation Report: Massachusetts Institute of Technology, ORI 072," Oct. 26, 1994, Onek Files.

42. *Ibid.*, pp. 8, 10–12, 13–15, 106, 141.

43. *Ibid.*, pp. 15–20, 205, 209.

44. *Ibid.*, pp. 6, 20, 29–42.

45. *Ibid.*, pp. 183 and appendices; Ursula Storb to Lyle Bivens, June 26, 1994, AE H-299. The Secret Service response to Lyter is attached to Harold J. Grasman to Barbara Williams, Feb. 23, 1992, AE H-504.

46. Singal to Bivens, Oct. 14, 1994, Singal Files. Hugh McDevitt remembers that he was asked to give his opinion of the report within two weeks and that he was "infuriated" that he was allowed so little time to respond to "this great big god-

damn report." He adds that he protested but that he never heard from the O.R.I. Author's interview with McDevitt, July 2, 1996.

47. Terence J. Tychans and Lyle W. Bivens to Theresa Imanishi-Kari, Ph.D., via Counsel, Oct. 26, 1994, Onek Files; *ORI Newsletter*, 3(Dec. 1994), 7. The formal request for an appeal hearing was filed in Joseph Onek and Thomas Watson to Research Integrity Adjudications Panel, Departmental Appeals Board, Department of Health and Human Services, Nov. 23, 1994, Onek Files.

48. Richard Stone and Eliot Marshall, "Imanishi-Kari Case: ORI Finds Fraud," *Science*, 266(Dec. 2, 1994), 1467–1469; author's interview, by telephone, with Sol Gittelman, provost at Tufts, July 15, 1997.

49. Stone and Marshall, "Imanishi-Kari Case: ORI Finds Fraud," *Science*, 266(Dec. 2, 1994), 1467–1469; Lyle W. Birens to Dingell, Oct. 28, 1994, O.R.I. Files.

NOTES FOR CHAPTER FIFTEEN

1. Author's interview with Julius S. Youngner, by telephone, Oct. 23, 1997.
2. Author's interviews with Jeffrey Sacks, by telephone, Sept. 12, 1997; with Leslie Sussan, by telephone, Oct. 6, 1997; with John Settle, Oct. 10, 1997; with Youngner, by telephone, Oct. 23, 1997.
3. Terency J. Tychan and Lyle W. Bivens to Bruce A. Singal, for transmission to Imanishi-Kari, via Counsel, Oct. 26, 1994, copy in author's possession; author's interviews with Jeffrey Sacks, by telephone, Sept. 12, 1997; with Settle, Oct. 10, 1997; Ford, *Appeal Proceedings*, Aug. 22, 1995, pp. 3196–3197; "Departmental Appeals Board," fact sheet from the Board. The panel ruled against Onek's obtaining certain documents concerning the internal deliberations of the O.R.I. or documents relating to the Dingell subcommittee because they were held to be privileged by the U.S. House of Representatives. Ford to Onek and Watson, Christ *et al.*, n.d., Onek Files; "The Panel provides a de novo review. What this means is that a Panel decision is not a review of what ORI did during its investigation or whether what ORI found was reasonable based on the evidence ORI considered." Department of Health and Human Services, Departmental Appeals Board, Research Integrity Adjudications Panel, Subject: Thereza Imanishi-Kari, Ph.D.; Docket No. A-95–3, Decision No. 1582 (hereafter, *Appeal Decision*) June 21, 1996, pp. 2–3
4. Ford, *Appeal Proceedings*, June 12, 1995, pp. 6–7.
5. Christ, opening remarks, *Appeal Proceedings*, June 12, 1995, p. 10; author's interview with Marcus H. Christ, Jr., by telephone, Aug. 29, 1997.
6. Dahlberg testified on June 12, 13, and 14; Mosimann, on June 15 and 19; O'Toole, on June 16; Davic, on June 20; and McClure, on June 22.
7. Gilbert, Testimony, *Appeal Proceedings*, pp. 1465, 1487–1488, 1504–1505, 1534–1536.
8. Author's interview with Joseph N. Onek, July 3, 1995.
9. *Ibid.*
10. *Ibid.*; Onek, *Appeal Proceedings*, June 23, 1995, pp. 1686–1688. On the first day of the hearing, Onek remarked in response to Christ's opening statement, "O.R.I. Counsel has used the same kind of somewhat overheated rhetoric in several other misconduct cases, and in those cases, no misconduct was found. And we are confident that there will be no misconduct found here." *Ibid.*, June 12, 1995, p. 54.

11. **Witnesses for the O.R.I.**: Austin M. Barron, John E., Dahlberg, Leiko Dahlgren, Joseph M. Davie, Walter Gilbert, John W. Hargett, Charles Maplethorpe, William R. McClure, James E. Mosimann, Margot O'Toole, Larry F. Stewart. **Witnesses for Imanishi-Kari**: Christopher Albanese, David Baltimore, Martina E. Boersch-Supan, J. Donald Capra, Elliott W. DeHaro, Martin E. Dorf, Herman N. Eisen, William P. Fitzgerald, Brigitte T. Huber, Vivien Igras, Thereza Imanishi-Kari, John F. Kearney, Norman W. Klinman, Robert L. Kuranz, Phillipa Charlotte Marrack, Edward B. Reilly, Moema H. Reis, Gerald B. Richards, Terence Paul Speed, Susan L. Swain, Reynold Verret, David T. Weaver, Robert T. Woodland, Henry H. Wortis.

12. Martin E. Dorf, Susan L. Swain, Norman W. Klinman, Testimony, *Appeal Proceedings*, June 28, 1995, pp. 2426–2428, 2433–2438, 2483–2487.

13. *Appeal Proceedings*, Aug. 29, 1995, pp. 4641–4642; Sept. 1, 1995, p. 5410; Sept. 14, p. 6015. Officials at Kyoto University had written that they had no record of Imanishi-Kari's having been enrolled there. However, she had a letter from the professors with whom she had worked that she had done enough to complete the "two year Master course," which was the document on which her belief that she had earned a masters degree at Kyoto was based. Mikita Kato and Atsuyosa Hagiwara, To Whom It May Concern, Aug. 26, 1970, AE R-41; Kazuhiro Sakai to John W. Krueger, Aug. 26, 1991, Sept. 16, 1994, O.R.I. Files.

14. Maplethorpe, Testimony, *Appeal Proceedings*, Sept. 13, 1995, pp. 5717–5718, 5767–5771; Sept. 14, 1995, pp. 6167–6178.

15. *Ibid.*, Sept. 13, 1995, pp. 5742–5747; Imanishi-Kari, Testimony *Appeal Proceedings*, Sept. 15, 1995, p. 6456.

16. Reis, Testimony; Imanishi-Kari, Testimony, *Appeal Proceedings*, June 29, 1995, pp. 2511–2512, 2604–2605; Aug. 30, 1995, p. 4934.

17. O'Toole, "Allegations Concerning the Cell Paper . . . ," attached to O'Toole to Kimes, Nov. 6, 1989, AE H-276; 'Toole to Eisen, June 6, 1986, AE H-231; N.I.H. Staff, O'Toole interview, May 17, 1988, AE H-109, pp. 51–53. Panel Member Ballard later asked O'Toole how she knew that the ELISA run on the Table 3 hybridomas had been run for isotype but not idiotype. O'Toole responded, "Well, [Imanishi-Kari] said it to me . . . We had the conversation in detail at the May '86 meeting, but she had said it to me before." O'Toole, Testimony, *Appeal Proceedings*, Sept. 14, 1995, pp. 6104–6105.

18. O'Toole, Testimony, *Appeal Proceedings*, June 16, 1995, pp. 1039–1041, 1046–1049, 1060–63.

19. *Ibid.*, pp. 1057–1060.

20. Huber, Testimony; Eisen, Testimony; Wortis, Testimony; Imanishi-Kari, Testimony, *Appeal Proceedings*, June 23, 1995, pp. 1789–1791; June 26, 1995, pp. 1963–1966; June 30, 1995, pp. 2808–2810; Sept. 15, 1995, pp. 6474–6476.

21. Imanishi-Kari, Testimony, *Appeal Proceedings*, Sept. 15, 1995, pp. 6476–6481.

22. Dahlberg, Testimony, *Appeal Proceedings*, Sept. 15, 1995, pp. 6301, 6304, 6306. Joseph Davie testified that he had come to think Figure 1 was unrepresentative of the behavior of Bet-1, a departure from his view at the time the N.I.H. panel first reported in November 1988 that he justified on the vague grounds that since then "more in-depth analysis has taken place." Davie, Testimony, *ibid.*, June 20, 1995, pp. 1294–1297.

23. Kearney, Testimony; Capra, Testimony, *Appeal Proceedings*, June 28, 1995, pp. 2446–2448; June 30, 1995, pp. 2822–2823.

24. Eisen, Testimony; Marrack, Testimony, *Appeal Proceedings*, June 26, 1995,

pp. 1902–1903, 1975–1976; Aug. 22, 1995, pp. 3134–3135. Henry Wortis acknowledged to Onek that he did not publish the difficulties he encountered when using Bet-1 as a reagent, whereupon Onek, in one of the rare moments of humor in the hearing, suggested, "Perhaps you should have taken the Fifth Amendment before answering that, Dr. Wortis." Wortis, Testimony, *ibid.*, June 30, 1995, p. 2785.

25. Capra, Testimony, *Appeal Proceedings*, June 30, 1995, pp. 2825–2826.

26. Marrack, Testimony, *Appeal Proceedings*, Aug. 22, 1995, pp. 3109–3116. Imanishi-Kari defended her use of the terminology in the same vein when Onek asked her if most practicing immunologists would have known that the hybridomas in Table 2 were uncloned: "Yes, absolutely. . . . You must be crazy if you believe somebody actually took so many . . . original wells and did cloning experiments with each one of them . . . because it's too many, simply too many." Imanishi-Kari, Testimony, *ibid.*, Aug. 30, 1995, pp. 4841–4842.

27. Marrack, Testimony; Imanishi-Kari, Testimony, *Appeal Proceedings*, Aug. 22, 1995, pp. 3113–3116; Aug. 31, 1995, pp. 5310–5312.

28. Gilbert, Testimony, *Appeal Proceedings*, pp. June 21, 1995, pp. 1498–1502.

29. *Ibid.*, pp. 1490–1491, 1494–1498, 1507–1508.

30. Dahlberg, Testimony, *Appeal Proceedings*, June 14, 1995, pp. 457–458. The O.R.I.'s claim that the Table 3 hybridomas had not been tested for idiotype formed the foundation of another charge against Imanishi-Kari. The table reported that the isotype of four of the hybridomas had not been determined. The reason was that the reagent used to capture antibodies grabbed those that were idiotypically birthmarked, then tested them for isotype. The antibodies generated in the four hybridomas, however, did not carry the idiotype; hence, they were not captured and their isotypes could thus not be determined. The O.R.I. insisted that the reagent used in the ELISA captured any mouse antibodies in the supernatant. All, therefore, would have been tested for isotype, which made the report of the nondetermination of the isotypes of the four hybridomas a falsification. Dahlberg conceded that if they had been tested for idiotype, as Imanishi-Kari said they had been, the charge would disappear. Dahlberg, Testimony; Imanishi-Kari, Testimony, *Appeal Proceedings*, June 14, 1995, pp. 519–520; Aug. 29, 1995, 4753–4754.

31. Author's interview with Imanishi-Kari, Oct. 29, 1996.

32. Dahlberg, Testimony; Imanishi-Kari, Testimony, *Appeal Proceedings*, Sept. 15, 1995, p. 6436; Aug. 29, 1995, p. 4748.

33. Dahlberg, Testimony, *Appeal Proceedings*, June 14, 1995, pp. 510–512, 528–529. Gilbert insisted that even if Imanishi-Kari had detected *gamma* antibodies that were idiotypically birthmarked, it would still have been "wrong" not to report evidence for double producers. A scientist reading the paper "would still want to know whether these cells were producing several different antibodies." Gilbert, Testimony, *Appeal Proceedings*, June 21, 1995, pp. 1507–1508.

34. Dahlberg, Testimony; Imanishi-Kari, Testimony, *Appeal Proceedings*, June 14, 1995, pp. 526–528; Aug. 30, 1995, pp. 4805–4806, 4830–4831.

35. Albanese, Testimony; Imanishi-Kari, Testimony, *Appeal Proceedings*, June 29, 1995, pp. 2557, 2572–2573; Aug. 30, 1995, pp. 4830–4831; Sept. 15, 1995, pp. 6464–6465. Onek later pointed out that Albanese had voluntarily testified even though he risked an embarrassment that was, in fact, realized at the hearing. After leaving M.I.T., he had eventually taken a job at Genetics Institute. Marcus Christ cross-examined him: "And you were fired from Genetics Institute for falsifying data on a quality control test, isnt' that right?" Albanese responded, "I was fired because I brought before the people who were in charge of that department some

information that I was subsequently pushed into. . . . In fact, I blew the whistle . . . and I paid the price. . . . I was forced . . . to do something that I thought was not correct. I subsequently went to the people who I thought would be most informed on the situation, and it cost me my job." Albanese, Testimony, *Appeal Proceedings*, June 29, 1995, pp. 2580–2582.

36. Dahlberg, Testimony, *Appeal Proceedings*, Sept. 15, 1995, pp. 6426–6431.
37. Woodland, Testimony; Huber, Testimony, *Appeal Proceedings*, June 23, 1995, pp. 1723, 1738, 1806–1808. O'Toole also thought it significant that the Wortis committee's report, of June 1987, did not mention subcloning. Godek raised that point, too. O'Toole to N.I.H. Panel Members and O.S.I. Staff Investigating the 1986 *Cell* Paper, Memorandum 3, March 18, 1990, AE H-282; O'Toole to Hadley, Response to the Wortis Committee's Statement in *Nature*, fax, June 28, 1991, AE H-296; O'Toole to Hadley, July 3, 1991, Onek Files.
38. Dahlberg, Testimony, *Appeal Proceedings*, June 13, 1995, pp. 343–347, 379–398.
39. Dahlberg, Testimony; Imanishi-Kari, Testimony, *Appeal Proceedings*, June 13, 1995, pp. 379–398; Aug. 31, 1995, pp. 5247–5248.
40. Dahlberg, Testimony, McClure, Testimony, *Appeal Proceedings*, June 13, 1995, p. 392; June 22, 1995, p. 1596.
41. Imanishi-Kari, Testimony, *Appeal Proceedings*, Aug. 30, 1995, pp. 4873–4877, 4887–4891; Aug. 31, 1995, pp. 5247–5248.
42. See, for example, Dahlberg, Testimony, *Appeal Proceedings*, June 13, 1995, pp. 382–398; Sept. 15, 1995, pp. 6304, 6375–6376, 6402–6404.

Notes for Chapter Sixteen

1. Mosimann, Testimony, *Appeal Proceedings*, June 15, 1995, pp. 720–722.
2. Mosimann, Testimony, *Appeal Proceedings*, Sept. 13, 1995, p. 5883; author's interview with Tom Watson, by telephone, March 27, 1995. The O.R.I.'s statistical analyses are presented in Appendix B of its final report on the case: Office of Research Integrity, "Investigation Report . . . , ORI 072," Oct. 26, 1994, Onek Files.
3. Barron, Testimony, *Appeal Proceedings*, June 16, 1995, pp. 830, 843–844.
4. Author's interview with Tom Watson, by telephone, Aug. 29, 1997.
5. Speed, Testimony; Godek, cross examination, *Appeal Proceedings*, June 27, 1995, pp. 2220–2221; Sept. 14, 1995, pp. 5989–5994, 6236; author's interview with Tom Watson, by telephone, Aug. 27, 1997.
6. Speed, Testimony, *Appeal Proceedings*, June 27, 1995, pp. 2123–2127, 2143–2144; author's interview with Terence Speed, by telephone, Aug. 26, 1997.
7. Speed, Testimony, *Appeal Proceedings*, June 27, 1995, pp. 2175–2176, 2180–2181, 2195–2196, 2211–2212; Sept. 14, 1995, pp. 6087–6088. Speed found some of Mosimann's adjustments "amazing." *Ibid.*, Sept. 14, 1995, pp. 2195–2196.
8. Barron, Testimony; Mosimann, Testimony, *Appeal Proceedings*, June 16, 1995, pp. 843–844, 1181; Sept. 13, 1995, pp. 5794–5797.
9. Speed, Testimony; Imanishi-Kari, Testimony, *Appeal Proceedings*, June 27, 1995, pp. 2247–2249; Sept. 14, 1995, pp. 6002–6004; Aug. 31, 1995, p. 5129.
10. Speed, Testimony, *Appeal Proceedings*, Sept. 14, pp. 6010–6013, 6075–6076; author's interview with Speed, by telephone, Aug. 26, 1997. Speed referred the panel to a standard treatment of data snooping, in David Freedman, Robert Pisani,

and Roger Purves, *Statistics* (New York: W. W. Norton, 1978), Chapter 29. See, especially, pp. 496 *ff.*

11. *Ibid.*, June 27, 1995, p. 2185; Aug. 21, 1995, pp. 2974–2978.

12. Stewart, Testimony; Hargett, Testimony; Ford, *Appeal Proceedings*, Aug. 22, 1995, pp. 3229–3230; Aug. 25, 1995, pp. 3851–3852, 3870–3872; Aug. 25, 1995, p. 4131.

13. Stewart, Testimony; Hargett, Testimony, *Appeal Proceedings*, Aug. 22, 1995, p. 3216; Aug. 23, 1995, pp. 3499–3501, 3613–3614, 3617–3619; Aug. 24, 1995, pp. 3646–3647, 3657–3658, 3854–3857; Aug. 25, 1995, pp. 3858–3859.

14. *Ibid.*, Aug. 23, 1995, pp. 3499–3501; Aug. 24, 1995, pp. 3646–3647, 3829–3830; Aug. 25, 1995, pp. 3854–3857, 3983–3984; Singal to Dingell, March 27, 1989, AE H-267.

15. Hargett, Testimony; Stewart, Testimony, *Appeal Proceedings*, Aug. 23, 1995, pp. 3501–3554; Aug. 25, 1995, pp. 3863–3864, 3867.

16. Watson, *Appeal Proceedings*, Aug. 21, 1995, 3084–3088.

17. Stewart, Testimony, *Appeal Proceedings*, Aug. 23, 1995, pp. 3616–3618.

18. Hargett, Testimony; Stewart, Testimony, *Appeal Proceedings*, Aug. 23, 1995, pp. 3511–3516, 3558–3559, 3571–3572; Aug. 25, 1995, pp. 3851–3852, 4131. The breaks in the chain of custody loomed unexpectedly large when O'Toole mentioned in testimony Charles Maplethorpe's delivering some of his notebooks to Walter Stewart in her presence on an occasion in the summer of 1988. Maplethorpe's notebooks were "these looseleaf things, and they had little pieces of gamma counter tapes taped to them," she remarked, adding that she heard Stewart say, " 'Oh, there's some of that vomit green stuff that's in Thereza's notebook.' " Maplethorpe had given the O.S.I. a second batch of tapes in November 1990, but they were all from 1981 to 1983 and provided several of the matches with Imanishi-Kari's green tapes that undergirded the finding of fraud. Onek interjected at the hearing that Maplethorpe had said that this first batch of tapes had been produced in 1984 and 1985—years for which the Secret Service had been unable to find any green tapes like those in Imanishi-Kari's notebooks. O'Toole's revelation pointed to why Imanishi-Kari's lawyers were "so concerned about the chain of custody and the subcommittee," Onek said, adding that it was essential to know what happened to Maplethorpe's first batch of green tapes. Marcus Christ countered that, whatever O'Toole might have heard in 1988, all that Maplethorpe turned in was in the possession of the O.R.I. and every tape was dated. O'Toole, Testimony; Onek, *ibid.*, Sept. 14, 1995, pp. 6126–6131, 6133–6139.

19. Hargett, Testimony; Stewart, Testimony, *Appeal Proceedings*, Aug. 23, 1995, pp. 3501–3507, 3563–3564; Aug. 25, 1995, pp. 3874–3877, 3880.

20. *Ibid.*, Aug. 23, 1995, pp. 3501–3507, 3562, 3613–3614; Aug. 25, 1995, pp. 3863, 4141–4142. The rejection of Reis's notebook and the subcommittee's role in it had come out in a meeting of O.S.I. officials with the Secret Service in mid-1989. Imanishi-Kari's lawyers had gained access to a transcript of that meeting as a result of her appeal. See Special Meeting of N.I.H. Office of Scientific Integrity with U.S. Secret Service Representatives, July 14, 1989, pp. 8–10, Onek Files.

21. Stewart, Testimony, *Appeal Proceedings*, Aug. 25, 1995, pp. 3895, 4137–4139.

22. *Ibid.*, Aug. 25, 1995, pp. 3861–3862, 4028–4035, 4148–4149.

23. Richards, Testimony, *Appeal Proceedings*, Aug. 28, 1995, pp. 4195–4199.

24. Richards, Testimony; Reis, Testimony; Verret, Testimony, *Appeal Proceedings*, June 29, 1995, pp. 2608–2609; Aug. 25, 1995, pp. 4200–4203; Aug. 28, 1995, pp. 4273–4274; Aug. 29, 1995, pp. 4551–4556.

25. Stewart, Testimony; Richards, Testimony, *Appeal Proceedings*, Aug. 24, 1995, pp. 3717–3719; Aug. 25, 1995, pp. 3997–3998; Aug. 28, 1995, p. 4271.
26. Richards, Testimony; Imanishi-Kari, Testimony, *Appeal Proceedings*, Aug. 28, 1995, pp. 4254–4255; Aug. 31, 1995, pp. 5100–5104, 5108, 5113.
27. Hargett, Testimony; Richards, Testimony; Imanishi-Kari, Testimony, *Appeal Proceedings*, Aug. 23, 1995, pp. 3473–3475, 3496–3499; Aug. 28, 1995, pp. 4227–4232; Aug. 31, 1995, pp. 5145–5146, 5149, 5318.
28. Stewart, Testimony; Verret, Testimony; Imanishi-Kari, Testimony, *Appeal Proceedings*, Aug. 25, 1995, pp. 4108–4109; Aug. 29, 1995, pp. 4551–4556; Aug. 31, 1995, pp. 5100–5104.
29. Watson, examination; Kuranz, Testimony, *Appeal Proceedings*, Aug. 22, 1995, pp. 3168–3171, 3180–3183; Aug. 29, 1995, pp. 4480–4482, 4492–4493 4499–4500, 4502, 4517–4518. As Kuranz explained the technical reasons, ribbon inks generally contained pigments, which could not be dissolved and thus could not be transferred to the thin-layer chromatography plates; and they used color elements only in small quantities, which made them light and indistinct on the plates. *Ibid.*, Aug. 29, 1995, pp. 4492–4493. Stewart acknowledged that to a point it would be a good idea to have as a control a blank piece of paper run for each sample, so you could tell if you have stuff in the blank sheet itself. But he added that once he had analyzed one piece of blank paper, he didn't always run others because he knew the results. Stewart, Testimony, *ibid.*, Aug. 25, 1995, p. 4055.
30. Richards, Testimony, *Appeal Proceedings*, Aug. 25, 1995, pp. 4200–4203.
31. *Ibid.*, Aug. 25, 1995, pp. 4200–4203.
32. Stewart, Testimony, *Appeal Proceedings*, Aug. 24, 1995, pp. 3726–3727, 3730–3732, 3736–3738, 3832–3833; Aug. 25, 1995, pp. 4131–4136.
33. *Ibid.*, Aug. 24, 1995, pp. 3726–3727, 3730–3732, 3736–3738, 3832–3833; Aug. 25, 1995, pp. 3902–3904; Sept. 1, 1995, pp. 5532–5535.
34. *Ibid.*, Aug. 23, 1995, pp. 3621–3622; Aug. 24, 1995, pp. 3736–3738, 3740–3742; Aug. 25, 1995, pp. 4129–4132; Sept. 1, 1995, pp. 5532–5535.
35. *Ibid.*, Aug. 25, 1995, pp. 4123–4125, 4128–4129, 4131–4136.
36. Richards, Testimony; Igras, Testimony; DeHaro, Testimony *Appeal Proceedings*, Aug. 28, 1995, pp. 4241–4244, 4458–4466, 4475–4478; Aug. 29, 1995, pp. 4561–4562, 4572–4574, 4607–468, 4612–4613; Sept. 14, 1995, pp. 6179–6180, 6183–6184; Jock Friedly, "Feds Stumble in a Final 'Baltimore Case' Showdown at Appeals Panel," *Probe*, Dec. 1, 1995, p. 6.
37. Richards, Testimony, *Appeal Proceedings*, Aug. 28, 1995, pp. 4257–4259, 4434.
38. Fitzgerald, Testimony, *Appeal Proceedings*, Sept. 14, 1995, pp. 6163–6167, 6173–6174.
39. Dahlgren, *Appeal Proceedings*, Sept. 13, 1995, pp. 5938–5944.
40. Richards, Testimony, *Appeal Proceedings*, Aug. 28, 1995, pp. 4259–62.
41. Igras, Testimony; Richards, Testimony; Fitzgerald, Testimony, *Appeal Proceedings*, Aug. 28, 1995, pp. 4265, 4391–4392; Aug. 29, 1995, pp. 4575–4576; Sept. 14, 1995, pp. 6163–6167.
42. Richards, Testimony, Aug. 28, 1995, pp. 4259–4262, 4266–4267, 4369, 4379–4381, 4389–4391.
43. Richards, Testimony, Aug. 28, 1995, p. 4265.
44. Stewart, Testimony, *Appeal Proceedings*, Sept. 1, 1995, pp. 5540–5543.
45. Richards, Testimony, *Appeal Proceedings*, Sept. 14, 1995, pp. 6185–6196.
46. *Ibid.*, Sept. 14, 1995, pp. 6191, 6195–6199.
47. Dahlberg, Testimony, *Appeal Proceedings*, Sept. 15, 1995, pp. 6414–6417.

48. Jeffrey A. Sacks and Leslie A. Sussan to Joseph Onek and Thomas Watson, Marcus Christ, *et al.*, Oct. 12, 1995, Onek Files.

49. Marcus H. Christ, Jr., Stephen M. Godek, and Carol C. Conrad, *Post-Hearing Brief of the Office of Research Integrity,* Dec. 22, 1995, Before the United States Department of Health and Human Services, Departmental Appeals Board, Research Integrity Adjudications Panel, In the Matter of Thereza Imanishi-Kari, Board Docket No. A-95-33, pp. 19–23. It was apparently significant to the O.R.I. that there had been "no testimony that this [Wortis and/or Huber's informing Imanishi-Kari about the specific issues] did not occur during that week." The O.R.I. created its scenario reluctantly, submitting in a note to its brief that the case should be decided on "the evidence adduced, not on the parties' respective speculations as to how and why things may have occurred." *Ibid.*, pp. 19, n. 6, 23.

50. *Ibid.,* pp. 23–32. According to the O.R.I., Imanishi-Kari must have figured out that the question would cover Bet-1, because O'Toole had frequently complained about the reagent; the idiotype data for Table 3, because she had just asked about it; and the data for Table 2, because she had been working with Reis's mouse-husbandry notebooks. *Ibid.*

51. *Ibid.*, pp. 19–23.

52. *Ibid.*, pp. 10, 12, 23, 55, 60–61, 64–65, 108, 127.

53. *Ibid.*, pp. 49–53, 63.

54. *Ibid.*, pp. 69–70. The O.R.I.'s brief said that agent Stewart, according to his testimony, "saw a Packard and Beckman sharing one printer when he visited M.I.T. in 1989." All that his testimony actually said, however, was that he saw a printer on a cart between a Packard counter and a Beckman counter. Stewart, Testimony, *Appeal Proceedings*, Sept. 1, 1995, p. 5555.

55. Joseph Onek and Tom Watson, *Post-Hearing Brief of Respondent,* Feb. 9, 1996, Before the United States Department of Health and Human Services Departmental Appeals Board, Research Integrity Adjudications Panel, In the Matter of Thereza Imanishi-Kari, Board Docket No. A-95-33.

56. Barron, Testimony, *Appeal Proceedings,* June 16, 1995, p. 5019; author's interview with Tom Watson, by telephone, March 27, 1995; Onek and Watson, *Post-Hearing Brief of Respondent,* Feb. 9, 1996, pp. 8, 12–13.

57. Onek and Watson, *Post-Hearing Brief of Respondent,* Feb. 9, 1996, pp. 24–37.

58. *Ibid.*, pp. 17–24; author's interview with Onek, Oct. 2, 1995.

59. Onek and Watson, *Post-Hearing Brief of Respondent,* Feb. 9, 1996, pp. 37–39.

60. *Ibid.*, pp. 2, n. 2, 41.

61. *Ibid.*, pp. 6–7, 12–13, 85–86, 97–98. The Secret Service found reason to question the authenticity of forty-four pages in the I-1 notebook. Only twenty-five pages had been used for the *Cell* paper and the corrections. Stewart, Testimony; Imanishi-Kari, Testimony, *Appeal Proceedings,* Aug. 24, 1995, pp. 3829–3830, Aug. 31, 1995, p. 5276.

62. Onek and Watson, *Post-Hearing Brief of Respondent,* Feb. 9, 1996, pp. 2–3, 49, 65.

63. Before the United States Department of Health and Human Services, Departmental Appeals Board, Research Integrity Adjudications Panel, In the Matter of Thereza Imanishi-Kari, Ph.D., Board Docket No. A-95-33, *Reply of the Office of Research Integrity to the Post-Hearing Brief of Respondent,* March 1, 1996, p. 4; Christ, Oral Argument; Onek, Oral Argument, *Appeal Proceedings,* March 19, 1996, p. 6537.

64. *Reply of the Office of Research Integrity,* March 1, 1996, pp. 7, 97–98.

NOTES FOR CHAPTER SEVENTEEN

1. Jeffrey Toobin, notes of interviews with David Baltimore and Thereza Imanishi-Kari, n.d.; author's conversations with Baltimore and Imanishi-Kari, June 23, 1996; *Appeal Decision*, June 21, 1996, pp. 1–2. The summary decision conformed to the standard of such proceedings: The O.R.I. "did not prove its charges by a preponderance of the evidence. The Panel recommends that no debarment be imposed and determines that no other administrative actions should be taken." *Ibid.*, p. 2.

2. *Appeal Decision*, June 21, 1996, pp. 11–17, 16, n. 25; Toobin, notes of interviews with David Baltimore and Thereza Imanishi-Kari, n.d.; author's conversations with Baltimore and Imanishi-Kari, June 23, 1996.

3. *Appeal Decision*, June 21, 1996, pp. 5, 14–15, 61, 134–135, 149–151.

4. *Ibid.*, pp. 17, 19, n. 30, 52, n. 76, 75. The panel noted that the two thirds of the I-1 notebook that was unchallenged contained much of the data that had appeared in print. *Ibid.*, p. 13.

5. *Ibid.*, pp. 13, n. 22, 76. Recall that Dahlberg had found scientifically incredible that a hybridoma would produce antibodies with isotypes that increased in diversity over time. He understood that to do so it would have to gain chromosomes. Imanishi-Kari argued the opposite—that they lost chromosomes over time, giving other chromosomes, with genes for different isotypes, a chance to express themselves. *Ibid.*, pp. 94, n. 142, 95.

6. *Ibid.*, pp. 55, 59–60, 67, 79–85, 128, 133. Youngner says that he and Celia Ford "spent days going through those notebooks trying to find some norm to which Imanishi-Kari's notebooks could be compared, and we couldn't find any. Every notebook was a norm in and of itself in many ways. It was hard to find a common normative base." Author's interview with Youngner, by telephone, Oct. 23, 1997.

7. *Appeal Decision*, pp. 12–13, 140, 142–146, 144, n. 210, 149–151, 153, n. 220, 162–163.

8. *Ibid.*, p. 163–164, 168.

9. *Ibid.*, pp. 21, 105, 99–100, n. 149, 107.

10. *Ibid.*, pp. 28, n. 42, 32, 35–36, 40, n. 59, 41–48, 52, 72, 134. Youngner says that he was impressed by the capabilities and honesty of the Secret Service agents, but he thinks that they were given an "impossible" task—that is, the dating of the materials in the notebooks. Author's interview with Youngner, by telephone, Oct. 23, 1997.

11. *Appeal Decision*, p. 11.

12. Author's interview with Marcus Christ, by telephone, Aug. 29, 1997. Stockton is quoted in Kaiser and Marshall, "Imanishi-Kari Ruling Slams ORI," *Science*, 272(June 28, 1996), 1864–1865. Stewart is quoted in "Scientist Cleared of Fraud Charge," *New York Times*, June 22, 1996, p. 1; Dingell expressed his views in John Dingell, Letter to the Editor, *New York Times*, July 2, 1996, p. 10 and John Dingell, "The Elusive Truths of the Baltimore Case," *Washington Post*, July 18, 1996, p. A27; Hadley and O'Toole are quoted in Rick Weiss, "Proposed Shifts in Misconduct Reviews Unsettle Many Scientists," *Washington Post*, June 30, 1996, p. A6.

13. Kaiser and Marshall, "Imanishi-Kari Ruling Slams ORI,"*Science*, 272(June 28, 1996), 1864–1865; Paulette V. Walker, "After a 1-Year Battle, Appeals Panel Clears Tufts U. Biologist," *Chronicle of Higher Education*, July 5, 1996, p. A22;

author's interview with Ursula Storb, April 8, 1997; "Scientist Cleared of Fraud Charge"; "The Fraud Case that Evaporated," editorial, *New York Times*, June 22, 1996, p. 1; June 25, 1996, p. A10.

14. Anthony Lewis, "Tale of a Bully"; Kolata, "Decision in Scientific Misconduct Raises New Questions," *New York Times*, June 24, 1996, p. 15; June 25, 1996, p. B10; Maxine Singer, "Assault on Science," *Washington Post*, June 26, 1991, p. A21; Donald Kennedy, "The Baltimore Affair: Let's Not Forget What Went Wrong," *Nature Medicine*, 2(Aug. 8, 1996), 843–844.

15. Philip Weiss, "The Whistleblower," *Mirabella*, Aug. 1990, p. 152.

16. Author's interviews with Shirley Tilghman, March 22, 1996; with Huber, Oct. 6, 1992 and Oct. 7, 1992; with Eisen, Oct. 2, 1992; O'Toole, Statement, *Hearing, Fraud in N.I.H. Grant Programs*, April 12, 1998, p. 87; O'Toole to Newburgh, Dec. 2, 1988, AE H-260; O'Toole to Kimes, Sept. 12, 1989, AE H-272; *Appeal Decision*, June 21, 1996, p. 168. The appeals panel reasoned that if Imanishi-Kari had been telling O'Toole to misrepresent her data, she would surely have raised the point with Wortis and Huber, which she did not. "Instead, she portrayed herself from the beginning as shocked by the apparent inconsistency of the 17 pages with reported data, which indicates she expected until then that the data were consistent with the paper as she read it." *Appeal Decision*, June 21, 1996, p. 89, n. 135. Wortis says that the issue of wells versus clones never came up at the Tufts inquiries. Wortis, Testimony, *Appeal Proceedings*, June 30, 1995, p. 2800.

17. O'Toole to N.I.H. Panel Members and O.S.I. Staff Investigating the 1986 *Cell* Paper, Memorandum 3, March 18, 1990, AE H-282; *Appeal Decision*, June 21, 1996, p. 93, n. 141.

18. Author's interviews with Imanishi-Kari, Oct. 29, 1996; with Huber, Oct. 6, 1992. Thomas Wegmann, O'Toole's employer at Harvard in 1973–74, remarked in 1993 that she could be accused of "philosophical naivete and fanaticism." Wegmann to Bernard Davis, May 18, 1993, Davis Files.

19. Author's interviews with Woodland, by telephone, Dec. 3, 1992; with Leonore and Leonard Herzenberg, July 2, 1996.

20. Author's interviews with the Herzenbergs, July 2, 1996; with Bruce Maurer, Nov. 1, 1996; O'Toole, interview with Hope Kelly, "The Ten O'Clock News," WGBH, Boston, March 27, 1991.

21. Lewis, "Tale of a Bully," *New York Times*, June 24, 1996, p. 15; Dolnick, "Science Police," *Discover*, Feb. 1994, p. 60; Maggie Hassan to File, Re: Interviews of Dr. Herman Eisen and Dean Gene Brown by Staff from National Institutes of Health, May 23, 1988, Eisen Files, p. 22; Kuznik, "Fraud Busters," *Washington Post Magazine*, April 14, 1991, pp. 24–25.

22. Christopher Anderson, "NIH Fraudbusters Get Busted," *Science*, 260(April 16 1993), 288; author's interviews with Stewart and Feder, Sept. 16, 1992; with Ned Feder, by telephone, March 25, 1996; Paul Gray, "The Purloined Letters," *Time*, April 26, 1983, pp. 59–60. Philip J. Hilts, "Fraud Sleuth Protests Work Halt with Fast,"*New York Times*, May 31, 1993, p. B5; Dolnick, "Science Police," *Discover*, Feb. 1994, p. 63. Stewart and Feder's analysis of Oates's works led the American Historical Association to reopen its investigation of him, this time addressing his books on Faulkner and King. It once again concluded that he had not "committed plagiarism as it is conventionally understood," but that the two books showed evidence of "too great and too continuous dependence, even with attribution, on the structure, distinctive language, and rhetorical strategies of other scholars and sources." Denise K. Magner, "History Association To Prove Accusations of Pla-

giarism Against Stephen Oates"; "Verdict in a Plagiarism Case," *Chronicle of Higher Education*, June 2, 1993, pp. A12–14; Jan. 5, 1994, pp. A17, A20.

23. Charrow, reply to Hadley and Hallum, *Journal of NIH Research*, 2(Sept. 1990), 14; Sarah Glazer, "Combating Scientific Misconduct," *CQ Researcher*, Jan. 10, 1997. A high O.S.I. official acknowledged that there was "something" to the view that the agency was a creature of Congress, explaining, "N.I.H. was responding to a belief that Congress thought we had really screwed up, and they were right. We really had." Author's interview with the official, Sept. 15, 1992.

24. Author's interviews with Hallum, by telephone, April 24, 1996; and with the O.S.I. official, Sept. 15, 1992.

25. *Appeal Decision*, June 21, 1996, p. 168, n. 227; Onek, *Appeal Proceedings*, June 26, 1995, pp. 1982–1983; author's interview with Youngner, by telephone, Oct. 23, 1997.

26. Author's interview with Onek, by telephone, Dec. 12, 1995.

27. Author's interview with the legal officer, by telephone, June 6, 1996.

28. Author's interview with Kimes, March 25, 1996; Barbara J. Culliton, "Dingell Disavows 'Dingell' Report on Gallo," *Nature Medicine*, 1(March 3, 1995), 188. Hadley is quoted in documents that Stewart made available on his Web page. In May 1995, Gallo left the N.I.H. to head a new institute for the study of viral diseases, including AIDS, that would be part of the University of Maryland and funded by the state. Late in the year, Hadley joined in a campaign against the venture, urging Maryland legislators to deny Gallo state money because he had displayed a "disregard of accepted standards for scientific conduct and ethics." Amy Goldstein, "Md. Lawmakers Ponder Virology Institute," *Washington Post*, Dec. 15, 1995, pp. A10–11.

29. Lewis, "Tale of a Bully"; Richard L. Berke, "A Crusader Tilts at the Ivory Towers Looking for Old-Fashioned Corruption," *New York Times*, June 24, 1996, p. 15; April 21, 1991, Section 4, p. 4; Culliton, "Fraudbusters Back at NIH," *Science*, 248(June 29, 1990), 1599; Dingell, "Shattuck Lecture—Misconduct in Medical Research," *New England Journal of Medicine*, 328(June 3, 1993), 1614; "The Hearings According to Chairman John Dingell," *Science and Government Report*, June 15, 1989, pp. 3–4.

30. Author's interview with Stockton and Chafin, March 24, 1993; John D. Dingell, "The Elusive Truths of the Baltimore Case," *Washington Post*, July 18, 1996.

31. John Walsh, "John Dingell: Demanding Humility and Fair Play from Scientists," *Journal of NIH Research*, 2(Oct. 1990), 43; author's interview with Stockton and Chafin, March 24, 1993; Bernadine Healy, "The Dangers of Trial by Dingell," *New York Times*, July 3, 1996, p. 11; author's interview with Kimes, March 25, 1996. The rules are laid out in a small red booklet, *Subcommittee on Oversight and Investigations of the Committee on Energy and Commerce . . .*, 102d Cong., 1991–92, "Selected Provisions of the Rules . . . ," a copy of which Dingell would hand every witness before taking testimony. Edward Richards, a professor of law at the University of Missouri, followed a number of misconduct probes closely enough to conclude that Dingell interfered with the operations of the O.R.I. "It's every bit as bad as McCarthy in a smaller universe," he said. Jock Friedly, "How Congressional Pressure Shaped the Baltimore Case," *Science*, 273(Aug. 16, 1996), 875.

32. K. C. Cole, "Fraud Charge Shakes Faith in Ground Rules of Science," *Los Angeles Times*, July 15, 1996, p. 16; author's interview with Baltimore, March 20, 1996; with Onek, by telephone, Oct. 16, 1997; Darcy Wilson to Colleague, March 10,

1992, Parker and Press Files. Press and Parker were joined by Wilson, one of Imanishi-Kari's friends, in raising money for her defense fund. Wilson, to whom the contributions were sent, says that the effort brought in about $30,000. Author's conversation with Wilson, August 7, 1997. The *Wall Street Journal* columnist Paul A. Gigot noted in 1989 that when he first met Dingell, he was impressed by "his intelligence and rough-hewn candor, but," he had to add, "the more I've watched him and his investigators at work, the more it's clear that they are manning a runaway train. The arrogance bred of unchallenged power has stripped them of self-restraint and distorted their understanding of the public good." Gigot, "Latest Chapter in the Fine Science of the Smear," *Wall Street Journal*, May 5, 1989, p. A14.

33. Kennedy, "The Baltimore Affair: Let's Not Forget What Went Wrong," *Nature Medicine*, 2(Aug. 8, 1996), 843–844.

34. *Ibid.*; Donald Kennedy, Letter to the Editor, *New York Times*, July 2, 1996, p. 10; author's discussions with several reporters who wrote about the case. In one of the rare articles on Imanishi-Kari, Margaret Pantridge observed, "The media have given the scandal so much attention that it has become a modern morality play. . . . But, oddly, in that media produced play, Imanishi-Kari has a bit part. Her story . . . is eclipsed by the giant figures of . . . David Baltimore . . . and John Dingell." Pantridge, "In the Eye of the Storm," *Boston Magazine*, July 1991, p. 40. On the role of the press in the post-Watergate "Culture of Mistrust," see Garment, *Scandal*, pp. 77 ff.

35. Kennedy, "The Baltimore Affair," *Nature Medicine*, 2(Aug. 8, 1996), 843–844; author's interviews with Healy, Aug. 27, 1996, with members of the Whitehead Institute, Sept. 30, 1992 and June 3, 1997; Freeman J. Dyson, "Science in Trouble," *American Scholar*, 62(1993), 515; "The Baltimore Vindication," *Wall Street Journal*, July 2, 1996, p. A14.

36. Author's interview with Kimes, March 25, 1996; David Goodstein, "Conduct and Misconduct in Science," World Wide Web Site http://www.caltech.edu/~goodstein/conduct.html, April 6, 1996, pp. 1, 4; Office of Research Integrity, *ORI Newsletter*, 5(March 1997), 1; Bernard D. Davis, "How Far Should Big Brother's Hand Reach?" *ASM News*, 56(No. 12, 1990), 643; O.R.I., *1996 Annual Report on Possible Research Misconduct*, World Wide Web Site http://phs.os.dhhs.gov/phs/ori/other/assure96.htm. A number of observers fault the O.R.I. for exaggerating the rate at which it finds misconduct. See Jock Friedly, "ORI's Self-Assessment: A Batting Average of .920?" *Science*, 275(Feb. 28, 1997), 1255.

37. Author's interview with Kimes, March 25, 1996; Gary Taubes, *Bad Science: The Short Life and Weird Times of Cold Fusion* (New York: Random House, 1993); David Warsh, "The Fortune That Never Was," *Boston Globe*, June 30, 1996, p. 73. Baltimore, Statement, May 2, 1989, copy in Singer Files. In 1994, misconduct convictions returned about $1.25 million to the N.I.H. Office of Research Integrity, *Annual Report, 1994* (Washington, D.C.: Department of Health and Human Services, April 1995), p. 3.

38. Daniel Koshland, "Zero Fraud—Only with Zero Science,"*New York Times*, Aug. 19, 1989, p. 13; Rick Weiss, "Proposed Shifts in Misconduct Reviews Unsettle Many Scientists," *Washington Post*, June 30, 1996, p. A6.

39. Berke, "A Crusader Tilts at the Ivory Towers Looking for Old-Fashioned Corruption," *New York Times*, April 21, 1991, Sect. 4, p. 4; Christopher Anderson, "NIH: Imanishi-Kari Guilty," *Nature*, 350(March 28, 1991), 263; "Caltech Deals with Fraud Allegations," *Science*, March 1, 1991, p. 1014; Richard Hynes to Mark

Wrighton, May 9, 1991, Eisen Files; author's interview with Eisen, Dec. 10, 1992. As early as 1989, Dingell noted of the accelerating pace of reform on misconduct matters in American universities, "I think that this is a process that is going on inside the scientific community at this minute. And that's really what I want." "The Hearings According to Chairman John Dingell," *Science and Government Report,* June 15, 1989, p. 4.

40. Author's interview with O'Toole, Oct. 5, 1992.

41. "The Fraud Case That Evaporated," editorial; Kolata, "Decision in Scientific Misconduct Raises New Questions"; Dingell, Letter to the Editor, *New York Times,* June 25, 1996, pp. A10, B10; July 2, 1996, p. 10; Kaiser and Marshall, "Imanishi-Kari Ruling Slams ORI," *Science,* 272 (June 28, 1996), 1865; "A Judgment Fit for Prime Time," *Nature,* 381 (June 27, 1996), 717; Rick Weiss, "Proposed Shifts in Misconduct Reviews Unsettle Many Scientists," *Washington Post,* June 30, 1996, p. A6. The National Science Foundation's record in handling cases of scientific misconduct has been unblemished by the kind of controversies that surrounded the high-profile O.R.I. cases. Analysts attributed the absence of upheavals at least in part to the fact that the Foundation dealt with misconduct cases by sending in a team comprising an investigator, a scientist, and a lawyer. "NIH's Bungling Goes On in the Baltimore Case," *Science and Government Report,* Sept. 15, 1991, p. 4.

42. Kuznik, "Fraud Busters," *Washington Post Magazine,* April 14, 1991, p. 31; Glazer, "Combating Scientific Misconduct," *CQ Researcher,* Jan. 10, 1997, p. 6; author's interview with the Boston area scientist; *Integrity and Misconduct in Research: Report of the Commission on Research Integrity* (Washington, D.C.: U.S. Department of Health and Human Services, 1995), p. 21; Panel on Scientific Responsibility and the Conduct of Research, Committee on Science, Engineering, and Public Policy, National Academy of Sciences, National Academy of Engineering, Institute of Medicine, *Responsible Science: Ensuring the Integrity of the Research Process* (vol. I; Washington, D.C.: National Academy Press, 1992), pp. 120–121. Typical of the reportage on the difficulties that were said to have befallen O'Toole as a result of her whistle-blowing are Kathy A. Fackelman, "Trouble in the Laboratory," *Science News,* 137 (March 31, 1990), 205; Philip Weiss, "The Whistleblower," *Mirabella,* Aug. 1990, p. 150.

43. *Integrity and Misconduct in Research,* pp. iii, 10–12, 23, 27–30; Glazer, "Combating Scientific Misconduct," *CQ Researcher,* Jan. 10, 1997, p. 9.

44. Ralph Bradshaw to Donna Shalala, Jan. 4, 1996, July 2, 1996; Bradshaw to William Raub, May 13, 1996; FASEB NEWS, May 13, 1996, all at the Web site of the Federation of American Societies for Experimental Biology (http://www.faseb.org); Glazer, "Combating Scientific Misconduct," *CQ Researcher,* Jan. 10, 1997, pp. 6, 11; Weiss, "Proposed Shifts in Misconduct Reviews Unsettle Many Scientists," *Washington Post,* June 30, 1996, p. A6. John Suttie, the new head of FASEB, remarked, "A charge of misconduct is so disastrous, whether it's true or not. It's like a sexual misconduct case that's eventually proven not guilty. It's never believed." Glazer, "Combating Scientific Misconduct," p. 9. A study released by the O.R.I. in the summer of 1996 found that scientists exonerated of fraud allegations tended to suffer, among other consequences, ostracism, loss of rank, and reductions in support staff. Rick Weiss, "After Misconduct Probes, Some Scientists Are Fighting Back in Court," *Washington Post,* Nov. 29, 1996, p. A25.

45. Implementation Group on Research Integrity and Misconduct, "Implementation Proposals on Recommendations by the Commission on Research Integrity," June

14, 1996, FASEB Web site; author's interview with Raub, by telephone, Oct. 22, 1997. According to a high N.I.H. official, the OSTP effort was converging on a definition of misconduct limited to fabrication, falsification, and plagiarism. Author's conversation with the official, March 8, 1998.

46. Bradshaw to Raub, May 13, 1996, FASEB Web site; Glazer, "Combating Scientific Misconduct," *CQ Researcher*, Jan. 10, 1997, pp. 4–5.

47. *Integrity and Misconduct in Research: Report of the Commission on Research Integrity*, p. 17; Implementation Group on Research Integrity and Misconduct, "Implementation Proposals on Recommendations by the Commission on Research Integrity," June 14, 1996. Bold type is in the original. Author's interview with Raub, by telephone, Oct. 22, 1997. By 1997, scientific fraud and misconduct were capturing some attention in Britain and Germany, but a panel in Germany declined to recommend the establishment of a government agency to investigate particular cases. Robert Koenig, "Panel Proposes Ways to Combat Fraud," *Science*, 278(Dec. 19, 1997), 2049–2050; *ORI Newsletter*, Sept. 1997.

48. "An Address by David Baltimore on the Occasion of His Installation as President of the Rockefeller University," Sept. 13, 1990, Rockefeller Archive Center, Pocantico Hills, North Tarrytown, New York.

49. Author's interviews with Baltimore, Aug. 17, 1992, March 20, 1996; Singer, "Assault on Science," *Washington Post*, June 26, 1996, p. A21; Kolata, "Decision in Scientific Misconduct Raises New Questions," *New York Times*, June 25, 1996, p. B10; Alice Huang, memo to files, Aug. 8, 1992, Huang Files.

50. Kolata, "Decision in Scientific Misconduct Raises New Questions," *New York Times*, June 25, 1996, p. B10; Jon Cohen, "Baltimore to Head New Vaccine Panel," *Science*, 274(Dec. 20, 1996), 2005; Richard Saltus, "MIT Laureate To Lead Caltech," *Boston Globe*, May 14, 1997, p. A3. At the time it was announced that David Baltimore would become the new president of the California Institute of Technology, where I am a professor of humanities, I was completing a two-year term as chairman of the Caltech faculty. In that capacity, I had been involved many months earlier in setting up the search for a new president, but I had been uninvolved in the search itself. The selection of Baltimore was as much news to me as it was to the rest of the Caltech community.

51. Author's interviews with Ptashne, by telephone, May 28, 1997; with Doty, June 3, 1997; Jock Friedly, "Trial and Error," *Boston Magazine*, Jan. 1997, p. 74. John Edsall said the decision was "a shock" to him but that age and infirmity had prevented him from evaluating it. He did remark in 1992, however, that he was "thankful that the [US Attorney] decided not to prosecute [Imanishi-Kari], even though she may be guilty. I have a feeling that she had been through such an ordeal that she should not be made to suffer any more." Edsall to the author, June 9, 1997, June 23, 1997; author's interview with Edsall, Oct. 1, 1992.

52. Author's interview with Watson, by telephone, May 28, 1997. Baltimore says he has no recollection of having been invited to the Banbury meeting and that, in any case, the stated purpose of the gathering was to discuss the issue of ethics in science, not the *Cell* paper. Author's interview with Baltimore, Oct. 31, 1997.

53. Robin Lloyd, "Prize-Winning Virologist To Take Reins This Fall," Pasadena *Star News*, May 14, 1997, p. 1; Robert Lee Hotz, "Biomedicine's Bionic Man," *Los Angeles Times Magazine*, Sept. 28, 1997, p. 13.

54. Baltimore, Statement, *Hearings, Scientific Fraud*, May 4, 1989, pp. 107–109. Baltimore added that the "majority of the criticisms of the *Cell* paper fall within the envelope of this subjectivity." Using the same data, Margot O'Toole might have

"written a very different paper, or none at all . . . and she would have been reflecting her own judgment. In this case, where she was not an author, it seems appropriate that the paper corresponded to the judgment of the authors and not of Dr. O'Toole." *Ibid.* The historian of physics Gerald Holton has noted: "All readings are not data. Sometimes you have to have the feeling in the tips of your fingers to understand what the difference is." He adds that Einstein had a word for it: *Fingerspitzengefuehl.* David Eisenberg, a physical chemist at UCLA, declares, "The facts never speak for themselves. They're always interpreted. Creativity in science consists in being able to see a pattern in those data." K. C. Cole, "Fraud Charge Shakes Faith in Ground Rules of Science," *Los Angeles Times,* July 15, 1996, p. 16. On issues arising in collaborations, see Mario Biagioli, "The Instability of Authorship: Credit and Responsibility in Contemporary Biomedicine," *The FASEB Journal,* 12(Jan. 1998), 3–16.

55. Eliot Marshall, "Disputed Results Now Just a Footnote," *Science,* 273(July 12, 1996), 174–175; Capra, Testimony, *Appeal Proceedings,* June 30, 1995, pp. 2834–2835.

56. Davis, "Fraud vs. Error: The Dingelling of Science," *Wall Street Journal,* March 8, 1989, p. A14; Meeting of N.I.H. Panel and Authors, May 3, 1989, Transcript, AE H-101, p. 32. Nicholas Wade reflected that in a sense the scientific community "was under attack," as Baltimore claimed. Glazer, "Combating Scientific Misconduct," *CQ Researcher,* Jan. 10, 1997, p. 3.

57. Davis to Philip Weiss, Sept. 25, 1989, Davis Files; Dyson, "Science in Trouble," *American Scholar,* 62 (1993), 515–516.

58. Author's interview with Imanishi-Kari, Oct. 29, 1996; Weiss, "After Misconduct Probes, Some Scientists Are Fighting Back in Court," *Washington Post,* Nov. 29, 1996, p. A25.

59. Imanishi-Kari to the author, June 18, 1997.

Essay on Sources

DOCUMENTS AND records accumulated by the federal government constitute a mine of information for the multiple subjects covered in this book. The emergence of scientific misconduct as an issue may be followed in the Gore hearings, U.S. Congress, Subcommittee on Investigations and Oversight of the Committee on Science and Technology, *Hearings, Fraud in Biomedical Research*, 97th Cong., 1st Sess., March 31 and April 1, 1981, and the policy-oriented inquiries of Congressman Donald Fuqua, U.S. Congress, House, Task Force on Science Policy of the Committee on Science and Technology, *Hearing, Science Policy Study—Hearings Volume 22: Research and Publication Practices*, 99th Cong., 2nd Sess., May 14, 1986; and of Congressman Ted Weiss, U.S. Congress, House, Subcommittee of the Committee on Government Operations, *Hearing, Scientific Fraud and Misconduct and the Federal Response*, 100th Cong. 2nd Sess., April 11, 1988. Essential sources of testimony and documents in the Baltimore Case are the four hearings conducted by Congressman John D. Dingell's House Subcommittee on Oversight and Investigations of the Committee on Energy and Commerce: *Hearing, Fraud in NIH Grant Programs*, 100th Cong., 2nd Sess., April 12, 1988; *Hearings, Scientific Fraud*, 101st Cong., 1st Sess., May 4 and 9, 1989; *Hearings, Scientific Fraud (Part II)*, 101st Cong., 2nd Sess., May 14, 1990; and *Hearings, Scientific Fraud*, 102nd Cong., 1st Sess., March 6 and Aug. 1, 1991.

Indispensable to understanding the case are the documents gathered or developed by the Office of Scientific Integrity (O.S.I.) and the Office of Research Integrity (O.R.I.) that became publicly available at the Departmental Appeals Board of the Department of Health and Human Services, in Washington, D.C., as a result of Imanishi-Kari's appeal hearing in 1995. These files include an extensive body of memoranda and correspondence dating from 1986 onward; transcripts or notes of interviews that were conducted by the N.I.H. staff as part of the investigation in 1988; and transcripts or notes of interviews conducted by the O.S.I. as part of its investigation from 1989 through 1990. The appeal files also contain the notebooks of Imanishi-Kari, Moema

Reis, and Margot O'Toole as well as the notebooks of other scientists, such as Charles Maplethorpe, that the Secret Service used as dated comparisons for Imanishi-Kari's radiation-counter tapes. The complete record of the appeal hearing, totaling some 6,500 pages, is the "Transcript of Proceedings before the United States Department of Health and Human Services," In the Matter of: Thereza Imanishi-Kari, Ph.D., Board Docket No.: A-95-33, Case 072, 1995. It provides an adversarial evaluation of the testimony and issues that developed during the many years of the case that is as advantageous to the historian as it was to the judges. The briefs that the lawyers for the O.R.I. and Imanishi-Kari filed on behalf of their respective clients are: Before the United States Department of Health and Human Services Departmental Appeals Board, Research Integrity Adjudications Panel, In the Matter of Thereza Imanishi-Kari, Board Docket No. A-95-33: Marcus H. Christ, Jr., et al., *Post-Hearing Brief of the Office of Research Integrity*, Dec. 22, 1995; Tom Watson and Joseph Onek et al., *Post-Hearing Brief of Respondent*, Feb. 9, 1996; and Marcus H. Christ et al., *Reply of the Office of Research Integrity to the Post-Hearing Brief of Respondent*, March 1, 1996. How the Appeals Board weighed the arguments and reasoned to its decision on each of the counts brought against Imanishi-Kari may be followed in Department of Health and Human Services, Departmental Appeals Board, Research Integrity Adjudications Panel, Thereza Imanishi-Kari, Board Docket No. A-95-33, "Decision," June 21, 1996.

I obtained a good deal of essential information about events in or connected to the case from letters, notes, memoranda, and tape recordings made available to me from several additional sources. David Baltimore and Herman Eisen allowed unrestricted access to their files on the affair, both of which included documents that had not become part of the public record. Eisen also provided a copy of the audiotape record- ing of his discussion at Harvard on June 4, 1991, with Paul Doty *et al.* Alice Huang supplied copies of many of her memoranda to files, which provide a revealing record of developments that led to her husband's resignation from the Rockefeller presidency. I also found generally useful but particularly important for the events of 1991 the notes, documents, and tapes of television and radio programs that the late Bernard Davis accumulated in the course of working on a book about the case and to which, along with chapters he had drafted, I was given carte-blanche access by his widow, Elizabeth Davis. Walter Stewart and Ned Feder sent me a copy of the file of docu- ments that they circulated during the early history of the case and that discloses their views of it. Joseph Onek opened a window onto the internal workings of the O.S.I. by making freely available the documents that he obtained on discovery during Imanishi- Kari's appeal. I also obtained a number of documents from the O.R.I. file of the case through a request under the Freedom of Information Act. The files of David Parker and Joan Press document the efforts they mounted on Imanishi-Kari's behalf. Maxine Singer supplied several documents from her personal files and Jock Friedly shared some important items that he obtained in the course of his own research on the case. Bruce Singal provided his briefs on behalf of Imanishi-Kari as well as several items of his correspondence with the O.S.I., and Normand Smith sent me a copy of the O.S.I. draft report.

I gained considerable knowledge and understanding of the case from my own inter- views and correspondence with the following list of people, which includes all the principals, most of whom I spoke with more than once; many of the investigators of the *Cell* paper; and numerous others whom the case touched in one way or another during its ten-year life and who answered specific questions about it: Judith A. Ballard, Lauren Baltimore, David Baltimore, Alexander Bearn, Paul Berg, Michael Bevan, Philip M. Boffey, Mel Bosma, William Broad, Bruce Chafin, Marcus H. Christ, Jr., Joseph

Davie, Paul Doty, John T. Edsall, Herman Eisen, Ned Feder, Gerald Fink, Martin Flax,
Cecilia Sparks Ford, Maurice Fox, Richard Furlaud, Magda Gabor, Walter Gilbert, Suz-
anne W. Hadley, Jules V. Hallum, Patricia Harsche, Bernadine Healy, Leonard Herzen-
berg, Leonore Herzenberg, Nancy Hopkins, Rollin Hotchkiss, Brigitte Huber, Vivien
Igras, Thereza Imanishi-Kari, Albert Kildow, Brian Kimes, Eric Lander, Charles Maple-
thorpe, Bruce A. Maurer, Hugh McDevitt, Jay McKay, Matthew S. Meselson, Mary
Miers, Donald Mosier, Jay Moskowitz, Joseph N. Onek, Margot O'Toole, Chris Pascal,
Frank Press, Joan Press, Mark P. Ptashne, Joseph E. Rall, William Raub, Jeffrey A. Sacks,
Howard K. Schachman, Norval D. (John) Settle, Bruce Singal, Normand F. Smith, III,
Terence Speed, Alan Stall, Walter Stewart, Peter Stockton, Ursula Storb, Leslie A. Sussan, Lawrence H. Tribe, Roy Vagelos, Nicholas Wade, James D. Watson, Thomas Watson, David Weaver, Robert Woodland, Patricia Woolf, Henry H. Wortis, Julius S.
Youngner, and Norton Zinder. A number of the interviews I conducted were recorded
and transcribed. It is my intention eventually to deposit copies of the tapes and transcriptions in the archives of the California Institute of Technology along with most of
the other materials that I have collected in my research on the history of the case.

The case stimulated extensive coverage and comment in newspapers and magazines.
The articles, columns, and editorials published about it are important sources of information if used with care and, of course, for evidence essential to assessing the media's
treatment of both issues and participants. Much of the print media coverage can be
tracked through the *Reader's Guide to Periodical Literature*, but my own collection of
such materials was enriched by the extensive clipping file that Stewart and Feder
included in the documents they sent me. *Science on Trial: The Whistle Blower, the
Accused, and the Nobel Laureate* (New York: St. Martin's Press, 1993) is a responsible,
informative treatment of the case from an earlier perspective by Judy Sarasohn, an
editor then at *Legal Times*, in Washington, D.C. Suzanne Garment locates the case
in the larger context of her subject in *Scandal: The Crisis of Mistrust in American
Politics* (New York: Times Books, 1991).

A variety of books and reports illuminate the background of the case and the context
in which it developed. For the history of immunology, see Pauline M. H. Mazumdar,
Species and Specificity: An Interpretation of the History of Immunology (New York:
Cambridge University Press, 1995); and Mazumdar's edited volume, *Immunology,
1930–1980* (Toronto: Wall & Thompson, 1989); and Anne Marie Moulin, *Le dernier
langage de la médecine: Histoire de l'immunologie de Pasteur au Sida* (Paris: Presses
Universitaire de France, 1991). For technical issues in molecular immunology in the
period of the *Cell* paper, I found it useful to consult Bruce Alberts *et al.*, *The Molecular
Biology of the Cell* (2nd ed.; New York: Garland, 1989) and Leroy E. Hood *et al.*,
Immunology (2nd ed.; Menlo Park, CA: The Benjamin/Cummings Publishing Co.,
1984).

In their *Betrayers of the Truth* (New York: Simon and Schuster, 1982), William
Broad and Nicholas Wade reliably treat the several cases of scientific fraud that came
to light beginning in the later 1970s, but their use of historical cases and their gen-
eralizations about fraud and misconduct in science are disputable. Several other case-
study explorations of the subject followed theirs, including Alexander Kohn, *False
Prophets* (New York: Basil Blackwell, 1986) and Robert Bell, *Impure Science: Fraud,
Compromise, and Political Influence in Scientific Research* (New York: John Wiley,
1992), which deals superficially with the Baltimore case. Studies that bear on scientific
misconduct and are informed by the realities of scientific practice include Gerald
Holton, "Subelectrons, Presuppositions, and the Millikan-Ehrenhaft Dispute," in Gerald Holton, *The Scientific Imagination: Case Studies* (Cambridge: Cambridge Univer-

sity Press, 1978), pp. 25–83, which is an authoritative counter-treatment of Broad and Wade's take on the same historical case; Harriet Zuckerman, "Deviant Behavior and Social Control in Science," in Edward Sagarin, ed., *Deviance and Social Change* (Beverly Hills: Sage, 1977), pp. 87–138; Patricia Woolf, "Pressure To Publish and Fraud in Science," *Annals of Internal Medicine*, 104(1986), 254–256; James Woodward and David Goodstein, "Conduct, Misconduct and the Structure of Science," *American Scientist*, 84(Sept.-Oct. 1996), 479–490; and Mario Biagioli, "The Instability of Authorship: Credit and Responsibility in Contemporary Biomedicine," *The FASEB Journal*, 12(Jan. 1998), 3–16.

Since the Gore hearings at the opening of the 1980s, the issue of scientific fraud has captured considerable attention and led to a number of publications. An early example is Judith P. Swazey and Stephen R Scher, eds., *Whistleblowing in Biomedical Research: Policies and Procedures for Responding to Reports of Misconduct, Proceedings of a Workshop* (President's Commission for the Study of Ethical Problems in Medicine and Biomedical and Behavioral Research; the American Association for the Advancement of Science, Committee on Scientific Freedom and Responsibility; and Medicine in the Public Interest; Washington, D.C.: Government Printing Office, 1982). A later one is *Responsible Science: Ensuring the Integrity of the Research Process* (Vol. I; Panel on Scientific Responsibility and the Conduct of Research, Committee on Science, Engineering, and Public Policy, National Academy of Sciences, National Academy of Engineering, Institute of Medicine; Washington, D.C.: National Academy Press, 1992). Alan Mazur, "Allegations of Dishonesty in Research and Their Treatment by American Universities," *Minerva*, XXVII(Spring 1989), 176–194, emphasizes that, whatever the incidence of scientific fraud, in the 1980s academic research institutions were responding poorly to accusations of it within their walls. A similar point is made in several of the essays in "Perspectives on Research Misconduct," a special issue of the *Journal of Higher Education*, 65(May/June 1994). Donald Kennedy, *Academic Duty* (Cambridge: Harvard University Press, 1997) reflects on the matter in the larger frame of his subject. A judicious treatment of the problem of safeguarding the integrity of scientific publication is Marcel C. Lafollette, *Stealing into Print: Fraud, Plagiarism, and Misconduct in Scientific Publishing* (Berkeley: University of California Press, 1992).

Developments on the legal and policy fronts of scientific integrity since the late 1980s may be followed in the quarterly *Professional Ethics Report: Newsletter of the American Association for the Advancement of Science, Committee on Scientific Freedom & Responsibility, Professional Society Ethics Group*; in the *Annual Reports* of the Office of Research Integrity; and in the quarterly *ORI Newsletter*, which commenced publication in January 1993. The seminal document for recent attempts to reform both the definitions of scientific misconduct and the procedures for handling allegations of it is the report of the Ryan Commission, *Integrity and Misconduct in Research: Report of the Commission on Research Integrity* (Washington, D.C.: U.S. Department of Health and Human Services, 1995). The sharp criticisms of the commission's proposals were taken into account by the Implementation Group, an intradepartmental group under the chairmanship of William Raub that was appointed by Secretary of Health and Human Services Donna Shalala in 1995 and submitted a report to her in 1996 under the title, "Implementation Proposals on Recommendations by the Commission on Research Integrity," which was made available on its Web site by FASEB, the Federation of American Societies for Experimental Biology.

The Einstein quotation in the epigraph is from John Stachel, "'A Man of My Type'—Editing the Einstein Papers," *British Journal for the History of Science*, 20(1987), 59.

Acknowledgments

THIS BOOK would have been impossible to write without the assistance of a number of individuals and institutions. I am grateful to the many people listed in the Essay on Sources who took time from their busy lives to talk with me, often at length, about their experience in the case or their knowledge of particular aspects of it. A number of people provided notes or documents or helped me gain access to records. For such help I wish to thank David Baltimore, Howard Berg, Robert Cooke, John Deutch, Herman Eisen, Ned Feder, Jock Friedly, Cecilia Sparks Ford, Sol Gittleman, Judith R. Goodstein, Gerald Holton, Leroy Hood, Alice Huang, Horace Freeland Judson, Thereza Imanishi-Kari, Mary Lee Jacobs, Bruce Maurer, John Parascandola, David Parker, Chris Pascal, Joan Press, the Rockefeller Archive Center staff, Jeffrey A. Sacks, Maxine Singer, Bruce Singal, Normand F. Smith, III, Walter Stewart, Jeffrey Toobin, and Jan Witkowski. I owe special thanks to Joseph Onek and Adria Hicks, of Crowell & Moring, for providing copies of documents in the case obtained on discovery during Imanishi-Kari's appeal; to Andrea Selzer, of the Departmental Appeals Board, for arranging to have sent to me copies of many of the documents that came within public reach as a result of the appeal; to Darlene Christian for digging out and processing the documents I asked for under the Freedom of Information Act; to Marilyn Smith, then at Rockefeller University, for sending the numerous documents I requested from David Baltimore's files; to Elizabeth Davis for generously making available the materials that her late husband, Bernard Davis, accumulated in connection with his work on a study of the case that he did not get to complete; and to the editors of *Cell* for permission to reprint tables from the disputed paper.

My work on this book was greatly facilitated by the resources of the California Institute of Technology. I am grateful to the staff of the Millikan Library, particularly those in the Humanities and Social Science Libraries and the Interlibrary Loan department; and to the Division of the Humanities and Social Sciences for its many infrastructural encouragements and for the good offices of John Ledyard and Susan Davis. I benefited greatly from the secretarial support of Ingeborg E. Sepp during most of the course of my work on this study and during the rest of it from the secretarial services of Sheryl Cobb, Marion Lawrence, Gina Morea, Christine Silva, and Margaret York. I am also indebted for assistance in research to Kathy Cooke and Peter Neushul and for preparation of a number of the figures to Robert Turring. I take pleasure in expressing my gratitude to the Andrew W. Mellon Foundation, which has been generous in its support of my work, including the efforts on this project.

Much of the armature of this book was formed by the article on the case that I published in *The New Yorker* in May 1996. My approach to the story profited significantly from the editorial judgment and advice of Sharon Delano as well as Tina Brown, John Bennet, and Henry Finder. My confidence in the soundness and accuracy of the story's treatment was boosted by the fact checking of Susan Choi and Amy Tuebke-Davidson, which was carried out with an engagement in the substance of the matter that was encouraging for the degree it extended beyond the call of duty. Edwin Barber, my editor at W. W. Norton, first suggested that I expand the article into a book. I am grateful for his ongoing appreciation of the story as it grew in scope, detail, and, inevitably, length and for his having been forgiving about the unexpected amount of time it took me to complete the book. I am indebted to Robert Silvers for early encouragement; and I want to thank Ronald Goldfarb, my agent, who provided unfailing support and wise advice.

I profited in developing my understanding of various features of the case from discussions with Jock Friedly and with my Caltech colleagues David L. Goodstein and James F. Woodward, partly in connection with their having invited me several times to lecture about the case in their jointly taught course on scientific ethics. Pamela Bjorkman and Ellen Rothenberg, both members of the biology faculty at Caltech, kindly reviewed my descriptions of various technical matters. Three people each gave the manuscript of the book a critical reading. I am immensely indebted to John L. Heilbron and Alison Winter for their comments and suggestions—and to Bettyann Holtzmann Kevles for the same, but for much more as well.

Index

Page numbers in *italics* refer to figures and illustrations.

Esperanza